Dynamical Properties of Solids

Volume 4

Dynamical Properties of Solids

Volume 4

Disordered Solids, Optical Properties

edited by

G. K. Horton
Rutgers University
New Brunswick, USA

A. A. Maradudin
University of California
Irvine, USA

1980

North-Holland Publishing Company
Amsterdam · New York · Oxford

ISBN 0 444 85315 4 (Vol. 4)
Library of Congress Catalog Card Number: 75-501105

Publishers: North-Holland Publishing Company
Amsterdam · New York · Oxford

Sole distributors for the U.S.A. and Canada:
Elsevier North-Holland, Inc.
52 Vanderbilt Avenue, New York, N.Y. 10017

Printed in The Netherlands

Preface

In this, the fourth volume of our series, two general areas of lattice dynamics are covered. Whereas the theoretical and experimental topics considered in the preceding volumes were discussed on the basis of crystalline solids, the first two contributions to the present volume, by Weaire and Taylor and by Visscher and Gubernatis, are concerned with lattice dynamical properties of noncrystalline and disordered solids, respectively. The emphasis in these chapters is on the fundamentals of this subject, as well as a survey of its present state, with the result that we expect that these two chapters will continue to be standard references to the subject matter even after it is developed further in subsequent work. Given the current interest in amorphous materials such development is inevitable.

The remaining two chapters are devoted to optical properties of crystalline solids. In a long review, Anastassakis presents the theory and existing experimental work on morphic effects on mode properties of solids, i.e. effects induced in crystals by the application of external forces such as electric and magnetic fields and uniaxial stress and hydrostatic pressure, that are not allowed by symmetry in the absence of these forces. Although not all morphic effects are directly optical effects, the overwhelming majority of the many different kinds of morphic effects that have been studied experimentally have been examined either by the experimental techniques of infrared absorption and Raman scattering, or both. This is a subject that will undoubtedly be developed further in the future for its ability to provide information about microscopic properties of crystals unobtainable by other methods.

A topic of considerable technological importance, that at the same time provides an opportunity for fundamental research activity, is infrared absorption by multiphonon processes in highly transparent solids. This subject is reviewed by Mills, Sparks and Duthler in this volume.

We wish to thank our friends from the North-Holland Publishing Company, Dr. W. H. Wimmers, Dr. P. S. H. Bolman, Dr. William Montgomery, and Mrs. J. Kuurman for their cooperation and supervision of the publishing process.

It is also a pleasure to thank Mrs. Kathryn L. Roberts for her help in the preparation of the Subject Index for this volume.

January 1980

G. K. Horton
New Brunswick, N. J.
USA

A. A. Maradudin
Irvine, California
USA

Contents

Volume 4

List of Contributors

E. M. Anastassakis, Physics Department, Polytechnion, Athens 147, Greece

C. J. Duthler, Xonics, Inc., Santa Monica, California 90401, USA

J. E. Gubernatis, Los Alamos Scientific Laboratory, Los Alamos, New Mexico 87545, USA

D. L. Mills, Dept. of Physics, University of California, Irvine, California 92717, USA

M. Sparks, Xonics, Inc., Santa Monica, California 90401, USA

P. C. Taylor, Naval Research Laboratory, Washington, D.C. 20375, USA

W. M. Visscher, Los Alamos Scientific Laboratory, Los Alamos, New Mexico 87545, USA

D. Weaire, Dept. of Physics, Heriot-Watt University, Edinburgh, UK

CHAPTER 1

Vibrational Properties of Amorphous Solids

D. WEAIRE*

Department of Physics
Heriot-Watt University
Edinburgh, UK

and

P. C. TAYLOR

Naval Research Laboratory
Washington, D.C. 20375, USA

Present address: Department of Experimental Physics, University College, Dublin, Ireland.

Dynamical Properties of Solids, edited by
G. K. Horton and A. A. Maradudin

Contents

1. Introduction

It is the nature of this subject that the definitive experiment usually proves elusive. Likewise most theories are of the type which is often entitled "An approach to..." A review of inconclusive data and speculative theory would be unenlightening. For this reason, no attempt will be made to produce an exhaustive compendium of data and hypothetical models on ever more complicated systems. We shall instead focus our attention on those few simple systems for which significant understanding has been obtained.

For details not covered in this chapter and for more complete bibliographies, the reader is referred to a number of recent reviews. The vibrational properties of non-crystalline solids have been discussed by Böttger (1974). Lucovsky (1974) and Lucovsky and Galeener (1976) have treated the vibrational properties of semiconducting glasses, and both Bell (1972) and Dean (1972) have considered the oxide glasses. Raman scattering in amorphous semiconductors is the subject of reviews by Mort (1973), Brodsky (1975) and Solin (1977). Extensive collections of both Raman and infrared results on more "traditional" oxide glasses are contained in two works by Wong and Angell (1974, 1976).

The *ingenu* reader of the current literature of this subject (for whom we are presumably writing) may find himself confronted with obstacles to understanding at a disconcertingly elementary level. The very word "amorphous" means different things to different people. Even familiar words raise questions in the mind of a critical reader. Does the term "phonon" imply the existence of a $\boldsymbol{k}$-vector? What do "acoustic" and "optical" mean when applied to vibrations in an amorphous system? Confusion abounds on such matters and, if nothing else, we hope in this chapter to achieve sufficient clarity of thought and expression to dispel some of it. Hence the short glossary presented in the next section.

1.1. A glossary of dangerous words

Acoustic

Applies to the branch of the dispersion relation for which $\omega \to 0$ as $\boldsymbol{k} \to 0$. Should not be applied to a non-periodic system without suitable apologies.

Amorphous

Amorphous solids are not crystalline on any significant scale, i.e. we exclude polycrystalline solids from this category. See §2.

Band

Leaving aside its looser experimental usage, this means a set of lattice vibrations associated with a single branch of the dispersion relation $\omega(\boldsymbol{k})$. In the absence of periodicity it cannot be used in this or any other precise sense, except in those cases where there are ranges of frequency where the density of vibrational states is non-zero, separating discrete bands.

Disordered

Amorphous solids are a subset of disordered solids. Crystalline solids in which there is substitutional or compositional randomness are termed disordered, but not amorphous.

Glassy / Vitreous

Those amorphous solids which can be prepared by cooling from the liquid state are commonly called glassy or vitreous. Unfortunately this distinction is not always adhered to in the literature.

Lattice

Strictly speaking, a lattice ought to be periodic but the loose application of this word to any infinite three-dimensional structure associated with condensed matter is a relatively harmless indulgence.

Lattice vibration

In keeping with the above, we take this to mean a vibrational normal mode of an infinite three-dimensional structure.

Longitudinal, transverse

Refers to the relative orientation of polarisation vector and $\boldsymbol{k}$-vector. In the absence of either, these terms are doubly devoid of meaning in a non-periodic system, except in the extreme low frequency regime.

Optic

Not *acoustic*. The same caveat applies in general although the existence of a discrete range of vibrational frequencies might provide some justification for the use of the term in a few cases.

Phonon

Quantised lattice vibration. We take the view, denied at least implicitly by some, that periodicity of the structure is not essential to the use of this term.

2. *Theoretical background*

2.1. Introduction

We first consider in § §2.2 and 2.3 the two ingredients of the dynamical matrix, *structure* and *force constants*. At the present stage, only the crudest of force constant schemes are commonly applied to amorphous solids, and it seems better to review these quickly than to refer to detailed accounts elsewhere in this series.

Given both structure and forces, how is one to predict vibrational properties? The remaining sections, § §2.4–2.10, are addressed to this question and the variety of answers so far advanced.

2.2. Structure

Vibrational spectroscopy has often been presented as a probe of structure, capable of revealing the form of local atomic arrangements in cases, such as that of an amorphous solid, in which X-ray diffraction gives only limited information. It is a point of view which has probably been overemphasised and somewhat fanciful interpretations of data have sometimes been made in attempts to justify it! A major question of interest has been – *do amorphous solids have microcrystalline or continuous random structures*?

Some indication of the meaning of this question may be provided by fig. 1. The weight of evidence has in most cases come to rest on the side of

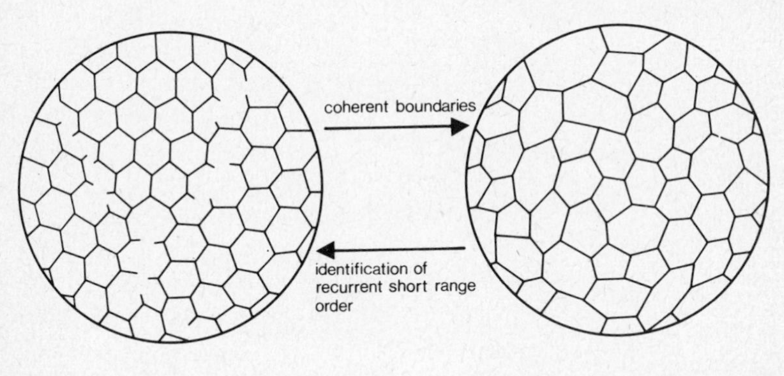

Fig. 1. Schematic diagrams of the local topology in the microcrystallite model (left) and the continuous random network model (right).

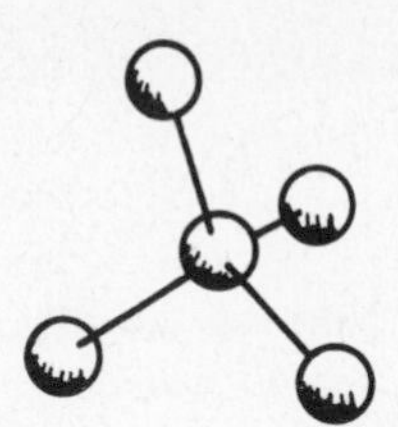

Fig. 2. Local bonding arrangement in Group IV amorphous solids where each atom has four nearest neighbours in a tetrahedral configuration.

the continous random picture, which is generally attributed to Zachariasen (1932).

In considering the details of the continuous random model, it is necessary to distinguish covalently and metallically bonded solids. In the former case it takes the form of a *random network* of nearest-neighbour bonds. For example, the random network model of amorphous Si each atom has four nearest neighbours in a tetrahedral configuration (fig. 2). It is difficult to

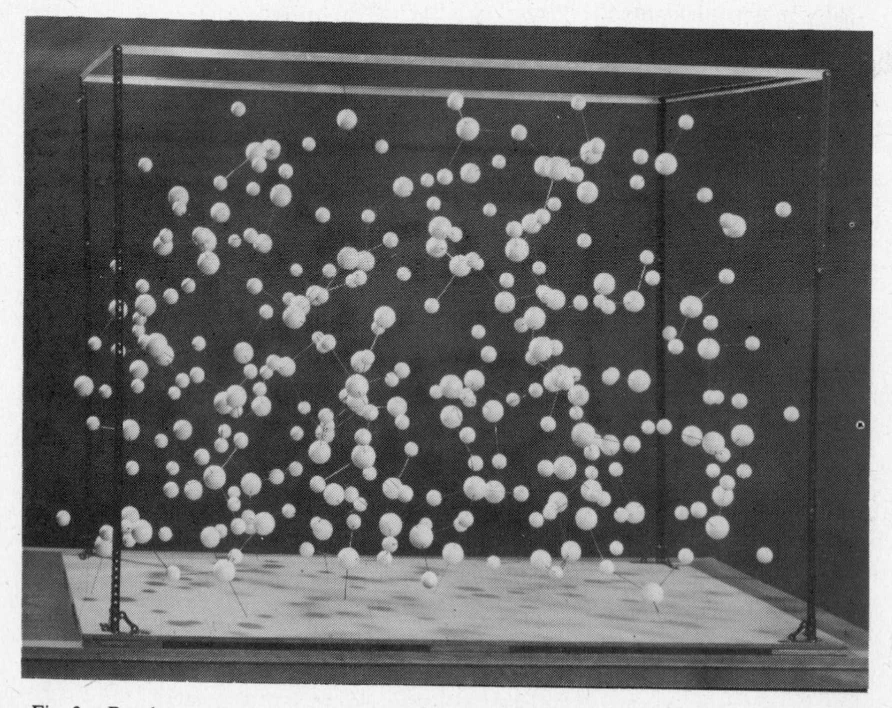

Fig. 3. Random network model of glassy SiO_2 (large spheres represent Si atoms) where the silicon atoms are tetrahedrally coordinated and the oxygen atoms are two-fold coordinated. (Photograph courtesy of Bell and Dean 1972; Crown Copyright.)

convey the nature of this model descriptively. Interested readers are strongly recommended to build one themselves! In doing so, they will be struck by the ease with which such networks can be built for sufficiently low coordination numbers. One does not need to contrive to meet the local bonding requirements by foresight or repeated building. A celebrated handbuilt model is shown in fig. 3.

Ultimately, any particular random network model should be specified by a table of coordinates. Statistical measures of structure, such as the radial distribution function, cannot uniquely specify the model. They are, of course, useful in any attempt to analyse its properties. Ring statistics, in particular, have a peculiar fascination for many people and much emphasis has been laid on them in characterising particular models.

As for amorphous metals, these are now considered to conform rather well to the Bernal model (Bernal and Mason 1960), which is the dense random packing of hard spheres. Here we cannot uniquely define nearest-neighbour shells.

2.3. Force constants

In the course of internecine disputes on the finest details of the dispersion relations of crystalline solids, the wood is often obscured by the trees. Let us therefore state emphatically that *the broad features of the vibrational spectrum of most solids are dominated by nearest-neighbour central forces*. For qualitative purposes, it is therefore forgivable to contemplate the very simple *central force* expression for the potential energy associated with atom displacements u_i,

$$V = \frac{1}{2} \sum_{\substack{ij \\ \text{neighbours}}} \alpha_{ij}\left[(\boldsymbol{u}_i - \boldsymbol{u}_j)\cdot\boldsymbol{r}_{ij}\right]^2. \tag{2.1}$$

Here α_{ij} is the force constant associated with the atoms i and j, often taken (in an elemental amorphous solid) to be equal for all nearest-neighbour pairs ij. The vector $\boldsymbol{r}_{ij}$ connects the equilibrium positions of atoms i and j. Different classes of solid part company when we attempt to improve upon (2.1). For *metals*, longer-range central forces with an oscillatory dependence on range may be incorporated and can in a few instances be calculated from first principles (Heine and Weaire 1970). For covalently bonded solids, semi-empirical schemes are used to incorporate short-range *non-central forces*. Thus, in the simplest (axially symmetric) *Born model*, we add to (2.1) a term of the form

$$\sum_{\substack{ij \\ \text{neighbours}}} \beta_{ij}(\boldsymbol{u}_i - \boldsymbol{u}_j)^2, \tag{2.2}$$

where again β_{ij} might be taken to be equal for all nearest-neighbour pairs ij. This may be called a *bond bending* term, as contrasted with the *bond stretching* term (2.1). The above form is quite satisfactory for many purposes but entails difficulties associated with rotational invariance. This complication may be avoided entirely by using forces which are derived from an *ad hoc* expression for the potential energy (as a function of atom positions) which is *manifestly* rotationally invariant. Following Keating (1966), we may choose to write an expression quadratic in the change of the scalar products of nearest-neighbour bond vectors from their equilibrium values. This, when expanded in terms of displacements, gives for each pair of nearest neighbours a term proportional to

$$\sum_{i\Delta\Delta'} [(\boldsymbol{u}_l - \boldsymbol{u}_{l\Delta})\cdot \boldsymbol{r}_{\Delta'}(l) + (\boldsymbol{u}_l - \boldsymbol{u}_{l\Delta'})\cdot \boldsymbol{r}_{\Delta}(l)]^2. \tag{2.3}$$

Finally, the valence force field model deals with lengths and angles rather than scalar products. Martin (1970) argues persuasively in support of the superiority of the Keating model, at least for Group IV semiconductors, but it is for our purposes a marginal consideration.

Semi-empirical models of such simplicity are rather *passé* in the context of crystalline systems. Much more elaborate schemes have been developed, involving a multiplicity of disposable parameters, although these have in turn been replaced by *shell models* which yield similar results with less extravagant use of adjusted force constants. This is achieved by the use of extra degrees of freedom representing, in a crude classical sense, valence electrons. For Group IV semiconductors Weber's *bond charge model* (1974), which is in the same spirit but uses bond charges rather than atom-centred shells, is the latest and perhaps ultimate refinement of this empirical tradition. For crystalline Si and Ge, it leaves few mysteries unveiled.

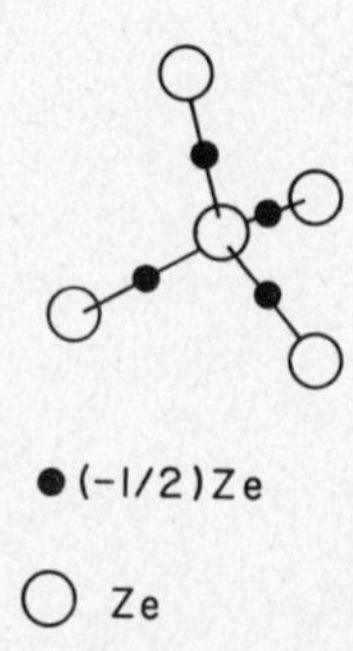

Fig. 4. Schematic representation of the bond charge model as applied to Group IV amorphous semiconductors.

Fig. 4 illustrates the general nature of this model. Recently it has been applied to amorphous Group IV semiconductors by Meek (1977), as described in §4.2.

Finally, highly ionic solids can only be reasonably described by a model which includes long-range coulomb forces between ions. This is elementary in principle but in practice it has been difficult to analyse the effects of such forces in an amorphous solid.

2.4. The problem

How can we analyse the vibrational properties of a given random network model? In discussing this question we shall concentrate for the moment on the phonon density of states $n(\omega)$, which we define to be the density (as a function of frequency) of vibrational normal modes.

In the study of *crystals* this may be obtained by integration of $\boldsymbol{k}$-space, once the dispersion relations $\omega(\boldsymbol{k})$ have been determined. The latter may be calculated by the diagonalisation of a secular matrix of dimensions $3n \times 3n$, where n is the number of atoms per unit cell.

In a disordered system, the absence of periodicity implies that Bloch's theorem is not applicable, and we are confronted with the necessity of using a much larger secular matrix. Strictly (and somewhat absurdly), it may be said that the dynamical matrix must have dimensions dictated by the number of atoms in an experimental sample, but clearly a much more reasonable number (10^2?, 10^3?) should give results characteristic of an infinite sample, for most purposes. Just how many atoms constitute an effectively infinite sample, in this case, depends in part on the *boundary conditions* used. In some early work in this field (Dean 1972, Bell 1972) atoms were either fixed or left free on the surface of a chosen random network cluster. The surface/volume ratio is rarely small enough for this to be done without the introduction of spurious features arising from the surface. Apart from some special methods in which boundary conditions are not relevant, we must therefore ensure that the boundary conditions are such as to minimise surface features and/or make a projection of the *local* density of states at the centre of the cluster. Suitable boundary conditions include the *periodic* case (which entails a somewhat demanding model building exercise, rarely executed satisfactorily!) or the use of effective fields (§2.8).

We thus envisage a calculation of the eigenvalues of a dynamical matrix of dimension 10^2 or more. This is feasible (up to about 10^3 or so), by any one of a number of methods, direct or indirect, to be reviewed in the sections which follow.

What are the limitations of such an approach?

Firstly, calculations for any finite sample of practical size cannot tell us anything about the low-frequency (Debye) regime.

Brute force cannot, of itself, give much qualitative insight, and must therefore be supplemented by more simplistic *ad hoc* models which are analytically tractable and can be tested against it.

Numerical methods can also be expensive, although objections on such grounds are often hypocritical, since it is not exactly cheap to employ theorists chewing pencils in a search for more elegant approaches!

Attempts to devise simpler models fall into various categories. Firstly there are those that are based on an analysis of the normal modes of molecules representative of the local groupings of atoms in the solid. Secondly there are calculations for *pseudo*-lattices, which have tree-like branching structures, and have the advantage of providing analytically soluble models. Thirdly, one may attempt to use calculations for *crystals* as a basis for predicting the properties of an amorphous solid, usually by considering the effects of random local distortions. This last approach is often given a formal basis by means of the "Gubanov transformation" by which atoms of a crystal are placed in 1 : 1 correspondence with those of an amorphous solid. No amount of formal labelling and mathematical window-dressing can make this more than an empty gesture, in our opinion.

There have also been studies (both experimental and theoretical) of crystalline polymorphs with large unit cells with a view to providing a "half-way house" between the simplicity of such crystal structures as diamond cubic and the complexity of the random network.

All of these methods take liberties with the *structure*. Sometimes one can simplify the form of the vibrational Hamiltonian instead, discarding, for example, all but the nearest-neighbour central forces. One may thus arrive at a model for which analytic proofs of exact results are available, providing a skeletal framework for understanding the main features of more realistic calculations.

Finally, the low-frequency regime (§5) requires special arguments based on the continuum approximation and the consideration of the possibility of "tunnelling modes". This remains the least satisfactory area of theory.

2.5. Diagonalisation

It would hardly be appropriate to devote much space here to a discussion of methods of finding eigenvalues, but it does deserve some attention, if only to indicate that considerable progress is still being made in this field.

Two methods offer particularly efficient techniques for finding eigenvalues of large matrices of the type encountered in the study of amorphous solids. The economy of both methods is due in part to the fact that eigen*vectors* are not calculated explicitly or implicitly.

The Sturm sequence algorithm of Dean and Bacon (1964) is closely analogous to the node counting method which is applicable in one dimension. The latter is based on the observation that as the number of nodes of the eigenfunction changes by one from each eigenvalue to the next, so the *integrated* density of states (the number of states *below* a particular eigenvalue) can be obtained by counting nodes at a chosen eigenvalue. Indeed the Sturm sequence method is most efficient if the finite sample to which it is applied is forced into what might be called a one-dimensional form by the choice of "rigid wall" boundary conditions on all dimensions except one (for which a periodic boundary condition may be used). For short range interactions, an appropriate labelling of coordinates will then give a secular matrix M of "banded" form, i.e., non-zero elements occur only close to the diagonal. The essential computational steps consist of the evaluation of determinants of sub-matrices of $M-EI$ along the diagonal and the number of changes of sign of these determinants (analogous to eigenfunction nodes in one dimension) gives the integrated density of states $N(E)$. Individual eigenvalues can be located by a searching process in which E is varied.

The Lanczos method (Wilkinson 1965, Edwards and Thouless 1972) is a relatively recent innovation, yet to be widely applied. It appears to be subject to less restrictive limitations and should be applicable to secular matrices of the order of $10^4 \times 10^4$.

2.6. Efficient methods of determination of $n(\omega)$

Since individual eigenvalues of the secular matrix have no particular significance, methods such as those described in the last section can hardly be the most efficient means of finding $n(\omega)$ for a model amorphous solid.

The *equation-of-motion method*, pioneered by Alben and collaborators (Beeman and Alben 1977) answers the need for a better method. Some of its favourable features are

(a) It generates a smoothed version of $n(\omega)$ directly, with a resolution which improves steadily as the computation proceeds.

(b) It uses relatively little core storage (a table of interactions – usually a nearest-neighbour table – is the main requirement).

(c) It uses relatively little computation time and this increases only linearly with the size of the sample.

(d) It is readily adaptable to calculate properties of eigenvectors in addition to $n(\omega)$ itself.

(e) It is easily understood!

The essential idea is that if the system is set in motion by a set of random displacements, all normal modes will contribute equally (apart from statistical fluctuations) to the ensuing motion. It follows that, if this is calculated by integration of the equation of motion and the time dependence of a chosen coordinate is *Fourier transformed*, the resulting function is closely related to the *local density of states* associated with the coordinate. An average over all coordinates gives the total density of states.

A somewhat similar method is the *recursion method*, developed by Haydock et al. (1972) (see also Meek 1977). Indeed, one may catalogue the same advantages as those of the equation-of-motion method. The essence of this technique is that it forces the problem into the framework of that of a *linear chain*, which provides a very convenient form for the calculation of the density of states. Having chosen some particular coordinate, we imagine it to be at the end of a semi-infinite chain of interactions. Further basis functions, which are combinations of the coordinates of the system under study, are selected, so that the dynamical matrix takes the form of a linear chain. The new basis functions are defined recursively – hence the name of the method.

The density of states may be derived from this new version of the Hamiltonian by expressing the Green's function as a continued fraction. The formula is an *infinite* continued fraction. In practice, special tricks are used in the truncation which is necessarily involved, in order to minimise its spurious effects.

It is a matter of experience that these two methods have proved roughly equally effective. Indeed one suspects that they could be proved to be essentially equivalent, but that has not yet been done. Both methods lose much of their utility when forces are extended much beyond nearest neighbours.

2.7. Molecular modes

The excision, from a solid, of a single molecule to be treated separately is usually a very drastic approximation. Except in the case where the solid is made up of molecules weakly bonded (e.g. by Van der Waals forces) to one another, the shifts of mode frequencies due to the cutting of bonds at the surface are likely to be so large as to obscure any interpretations based on this picture.

Nevertheless Lucovsky (1974) has noted some striking correspondences between molecular normal mode frequencies and peaks of $n(\omega)$. Sen and

Thorpe (1977) have attributed these to the peculiar geometry of the bonding in the solids concerned, leading to an effective decoupling of some molecular modes.

In general, this approach can only be realistic if special boundary conditions are used, that is, "effective fields" are placed on the surface atoms. Such matters are dealt with in the next section.

2.8. Bethe lattices and related methods

Consider the case of a tetrahedrally bonded solid with the force constants associated with the Born model (§2.3). One may consider a given atom to have a shell of tetrahedrally coordinated neighbours each of which in turn has such a shell, and so on, but with *no closed rings of bonds*. We have thus defined a Cayley tree or Bethe lattice (fig. 5). Of course, such a structure cannot exist in the ordinary geometrical sense since different atoms will overlap if we try to construct it. However it is, in a more abstract sense, a reasonable approximation to a real tetrahedrally bonded solid.

Both literally and figuratively, the study of Bethe lattices has involved a lot of loose ends. The formal theory of density of states calculations for tree-like structures has never been very clearly expounded. The difficulty lies in the fact that the surface/volume ratio does not go to zero as the size of the system goes to infinity. Thus, when one speaks of the density of states of an *infinite* Bethe lattice, the meaning is not as clear as in real lattices. However, this has not deterred theorists from writing down equations and solving them.

It might seem that our definition of the Bethe lattice is ambiguous in that we have failed to define the dihedral angles (between the bonds of

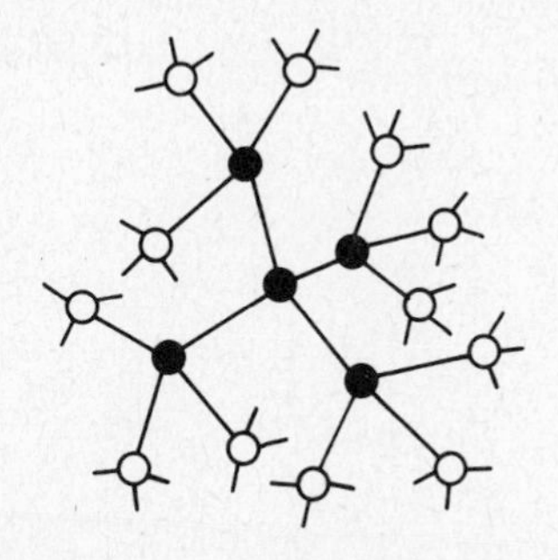

Fig. 5. Schematic representation of a small segment of a Bethe lattice as applied to a tetrahedrally bonded amorphous solid. Every atom is surrounded by a shell of tetrahedrally coordinated neighbours, but there are no closed rings of bonds.

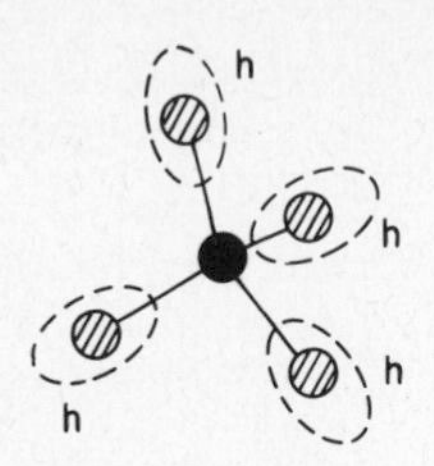

Fig. 6. Effective "molecule" obtained by placing effective fields on the atoms at the surface of a single tetrahedral unit with one interior and four surface atoms. The unit moves in the external potential indicated by the dashed lines (Thorpe 1973).

neighbouring atoms) but it so happens that, for the Born model, the density of states is independent of these!

In a Bethe lattice, as in a Bravais lattice, each site is equivalent. This fact, together with the uncomplicated topology of the structure, makes the possible calculation of many of the properties of this lattice by a simple *self-consistency* argument due to Thorpe (1973). Effective fields (potential wells) are placed on the atoms at the surface of a tetrahedral cluster. These fields, which are functions of frequency, represent missing "branches" of the tree-like structure (fig. 6). Leaving these parameters unspecified the normal modes of the resulting effective "molecule" can be found.

In order to determine these parameters it is sufficient to demand that the mean-square displacements of the central atom and surface atoms are the same, since they are, after all, completely equivalent when viewed as components of the original Bethe lattice. This is only one of a variety of simple arguments which may be concocted to give a solution of this problem.

The results of such calculations can be very helpful but they suffer from an awkward feature – namely that *band edges* fall short of the points which they would achieve in a real lattice. In particular the low-frequency edge of the spectrum is not at $\omega = 0$!

Effective fields may be used for a variety of other purposes. For example, effective fields representing Bethe lattice branches may be grafted onto a random network cluster. This is the "cluster – Bethe" model (Joannopoulos and Yndurain 1974).

The use of effective fields, whether self-consistent or not, is one obvious way to make the molecular models of the last section more realistic.

2.9. Exact results

Interesting exact results can be found for sufficiently elementary vibrational Hamiltonians. For example, in the case of tetrahedrally bonded solids, the neglect of any distortions from exact local tetrahedral symme-

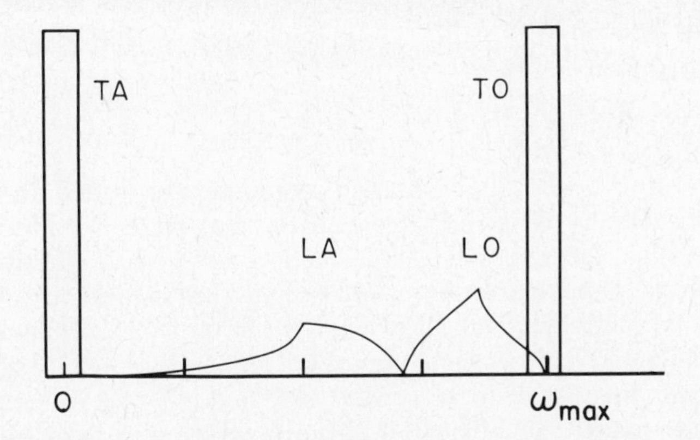

Fig. 7. Phonon density of states of diamond cubic Si as given by the Keating model with only a bond stretching force constant ($\alpha=0.495$ dyn/cm). The contributions of the two delta functions have been spread over a finite frequency interval so that their total weight is apparent (Weaire and Alben 1972).

try, and forces other than central nearest-neighbour forces, enable us to make the following statements (Weaire and Alben 1972). They apply to *any* structure, whether topologically disordered or not. (See fig. 7.)

(a) The vibrational spectrum extends from $\omega=0$ to $\omega=\sqrt{8}\,\alpha$, where α is the central force constant.

(b) It consists of three equal parts, two of which are delta functions at 0 and $\sqrt{8}\,\alpha$, respectively.

The latter is a common occurrence for structures of low coordination. Delta functions (which would correspond to flat dispersion relations in the case of a crystal) appear whenever simplified forces are used, and they are often independent of structure.

Of course we may also calculate *moments* of the spectrum (as a function of ω^2) exactly, by path counting methods.

2.10. Matrix elements

A major stumbling block in the analysis of infrared and Raman spectra of amorphous solids has been the general uncertainty regarding the effects of transition probabilities on the observed spectra. It is not that these are very mysterious, but rather that the historical background of work on crystals has not been helpful.

For infrared absorption, the relevant matrix element (whose squared modulus enters the transition probabilities) may be regarded as arising from the *dipole moment* associated with the set of displacements $x_{i\gamma}$ associated with a given normal mode.

This may be written

$$p^{\gamma} = \sum_{i\alpha} e^{\alpha}_{i\gamma} x_{i\gamma}. \tag{2.4}$$

Here atoms are labelled i and Cartesian coordinate axes with γ. In the case of an ionic solid, the effective charge tensor $e^{\alpha}_{i\gamma}$ may be replaced by a scalar e^*_i, representing the effective charge on the ion i, as an obvious first approximation. However, for homopolar solids such as a-Si, it is far from clear what form $e^{\alpha}_{i\gamma}$ should take. Indeed, it is still the subject of controversy (see, e.g. Klug 1976, and ensuing discussion). Alben et al. (1975) suggested an (intuitively motivated) form for a-Si and Ge (see §4.1). Weber's (1974) bond charge model should provide a more satisfactory basis for future work.

For Raman scattering the important quantity is the displacement dependent polarizability (a second-rank tensor). Again, we may write this in terms of the atomic displacements:

$$p^{\gamma\beta} = \sum_{i\alpha} D^{\alpha\beta}_{i\gamma} x_{i\gamma}. \tag{2.5}$$

Given an approximate form for $e^{\alpha}_{i\gamma}$ or $D^{\alpha\beta}_{i\gamma}$ the contribution of a given (normalised) lattice vibration to the spectrum is given by the squared modulus of (2.4) or (2.5), contracted with unit vectors which represent incident, or incident and scattered, electric fields. Apart from various trivial factors (see e.g. Beeman and Alben 1977), this procedure yields the *absorption coefficient* and *reduced Raman spectrum* (see §3.5), respectively.

The neglect of variation of matrix elements within each vibrational band is an approximation generally attributed to Shuker and Gammon (1970). One might well ask, as in §1.2, what is a "band" in an amorphous solid! Even if one can agree on this question, this is not usually a very good approximation.

Neutron scattering matrix elements are less mysterious, since the scattering centres (nuclei) appear as delta functions to a thermal neutron. Again discarding relatively trivial factors, the contribution of a given vibrational mode of frequency ω to the loss spectrum at energy $\hbar\omega$ is given as a function of scattering vector Q by

$$\left| \sum_{i\alpha} r_{i\alpha} Q_{\alpha} \exp(\mathrm{i} Q_{\alpha} r_{i\alpha}) \right|^2 . \tag{2.6}$$

Here $\boldsymbol{r}_i$ is the position of atom i. We have assumed coherent scattering and used the one-phonon approximation, but (2.6) is otherwise exact and is straightforwardly evaluated. That is not to say, however, that results are easy to interpret on this basis.

The expression simplifies considerably in the *incoherent* case, when the complicated interference effects due to the phase factor in expression (2.6) disappear. The energy loss spectrum then becomes proportional to the *density of states*, apart from the known prefactors. Even for predominantly coherent neutron scatterers, scattering is therefore effectively incoherent in the limit $Q \rightarrow \infty$. Experimentally, the momentum transfer vector Q must be chosen to have a value such that

$$QR \gg 1, \tag{2.7}$$

where R is the shortest interatomic separation. This makes severe demands on most experimental facilities (see §3.4). However, there have recently been a number of examples of the application of high-Q neutron scattering to the determination of densities of states of amorphous solids.

While on the subject of the nature of the eigenstates and their influence on experimental spectra, it is worthy of note that there is, in theory, an important distinction between *localised* and *extended* states. There should be critical frequencies, analogous to "mobility edges" in electronic transport, at which states become localised. Alas, no one has yet devised a means of probing this aspect of lattice vibrations in disordered systems, so it remains merely an occasional preoccupation of theorists. The Shuker–Gammon approximation, mentioned above, is sometimes justified by statements to the effect that vibrational states are localised. Whatever limited validity it has is not (in our opinion) at all dependent on localisation.

3. *Experimental background*

The simplest amorphous solids are of course those which are elemental, such as the semiconductors a-Ge, a-Si, a-As, a-Se and a-S, but there are significant experimental difficulties with all of these. Both a-Ge and a-Si can only be made in thin film form so that both infrared and neutron scattering measurements become difficult. In addition, there is some evidence that amorphous films of these semiconductors have slightly different structures depending upon the method of preparation (evaporation, sputtering, glow discharge, thermal decomposition, etc.) and even on the parameters within a given preparation procedure (substrate temperature, pressure, deposition rate, etc.). The vibrational properties in the high-frequency regime do not depend as dramatically on these variables as do, for example, the transport properties, but one must nonetheless be careful in interpreting details of the experimental spectra. For example, amorphous As can be made in bulk form, but only by hydrogen transport of evaporated material so that there is some residual hydrogen present in the bulk material. As we shall see in §4.2, the properties of this material are significantly different from those of purer thin film samples.

Two-component glasses (such as SiO_2, B_2O_3, GeS_2, As_2S_3 and so forth) can generally be prepared in a more reproducible manner. However, this advantage is obtained at the expense of introducing additional structural and vibrational complexities which can be extremely difficult to handle theoretically. Thus the first, and often the most important, experimental difficulty concerns the choice of suitable materials – structurally simple materials which can be prepared reproducibly in various sizes and shapes.

3.1. Experimental techniques

In the high-frequency region (frequencies $\geqslant 3\times10^{12}$ Hz or reduced frequencies $\geqslant 100\ \mathrm{cm}^{-1}$) the most important experimental techniques for investigating vibrational properties of amorphous solids include Raman, infrared and neutron spectroscopy. For amorphous *metals* both Raman and infrared measurements are impractical because the incident light (infrared or visible) does not penetrate the solid appreciably. For infrared measurements this means that the reflectivity is very close to 100% (owing to the electronic contribution) and the observation of effects due to phonons requires an accurate examination of very small changes in the reflectivity. Even when the appropriate sensitivity is obtained one cannot always be certain that data are not affected by bulk or surface contamination. Raman measurements suffer from the same limitations because of the shallow penetration depth of the incident light. Signals are weak because the active sample volume is very small and surface effects become very important.

Neutron scattering does not yet play as dominant a role in the investigation of amorphous solids as in crystalline solids, since attention has so far been confined mainly to the vibrational density of states. Raman and infrared measurements give at least a rough indication of this.

In what we shall call the low-frequency region (frequencies $\leqslant 3\times10^{12}$ Hz) the three techniques which are most important at high frequencies are supplemented by several additional ones. These include Brillouin and Rayleigh scattering, microwave dielectric loss, ultrasonic attenuation and various phonon echo measurements.

3.1.1. Raman scattering

A modern Raman scattering experimental system usually consists of a laser source, a double grating monochromator, and a detector such as a photomultiplier tube, together with some equipment for processing the scattered light signal, such as photon counting electronics and signal averaging capabilities (see, for example, Anderson 1971). The laser is usually an argon or krypton ion gas laser or a dye laser pumped by an argon laser. There are two standard experimental arrangements for the incident laser light (transmission or reflection geometries) depending upon whether the light is transmitted or absorbed by the sample. In addition there are two possible polarization configurations for amorphous solids where the incident and scattered photons are polarized either parallel or perpendicular to one another.

3.1.2. *Infrared spectroscopy*

Modern instruments for infrared spectroscopy are not dramatically different from their counterparts of thirty or more years ago except that modern electronics are often used to automate data processing.

Infrared measurements are commonly performed using a grating, a prism or an interferometric spectrometer. The grating or prism instruments usually consist of a source (for example, a tungsten or mercury arc lamp or a glow bar) a grating or prism monochromator, a detector (thermocouple, Golay cell, photoconductor, bolometer and so forth), and various electronic components for data accumulation and reduction. In an interferometric spectrometer the monochromator is replaced by an interferometer (usually of the Michelson type) and more sophisticated data processing equipment is necessary to transform the accumulated data into a frequency spectrum. In the very far infrared spectral region ($\leqslant 100\ \mathrm{cm}^{-1}$) a Lamallar interferometer and a polarizing interferometer are sometimes used. For further details see Hadni (1967) and Robinson (1973).

3.1.3. *Neutron scattering*

Neutron scattering experiments on amorphous solids have generally been performed using either a triple-axis crystal spectrometer or a time-of-flight (pulsed beam) spectrometer (see Brockhouse (1961) and other authors in the same work). In the triple-axis spectrometer the collimated neutrons are passed through the sample under investigation and finally through an analyzing crystal to a suitable detector. The angle between the incident beam and the sample can be varied to change the incident monochromatic neutron energy. Similarly, the angle between the sample and the analysing crystal and the angle between the analysing crystal and the detector can also be varied. These adjustments provide great versatility in the measurements which can be made, including the provision for scans over constant wave vector transfer (constant Q scans).

In a time-of-flight spectrometer a pulsed monoenergetic beam of neutrons is scattered by the sample and the energy distributions of the neutrons scattered through various angles are obtained by measuring their time of flight. The pulsed, monoenergetic neutron beam is obtained using a series of mechanical choppers or rotating crystal. Time-of-flight spectrometers are capable of higher resolution for a given intensity than are triple-axis spectrometers, but at least as far as crystalline solids are concerned, they are not as versatile.

3.1.4. Special techniques at low frequencies

At reduced frequencies below approximately 100 cm^{-1} the techniques described above are either supplemented or replaced by several additional techniques. Molecular gas lasers are useful as a source in this range, particularly for investigations of the dielectric properties of amorphous solids where high resolution is not required. By choosing appropriate gases one can obtain close to 2 cm^{-1} resolution in this range without the necessity for either a monochromator or an interferometer (Martin 1967, Robinson 1973, Yamanaka 1976). In the microwave region dielectric loss experiments are performed by either filling a cavity with the sample to be studied or by a cavity perturbation technique (see, for example, Lance 1964).

Low-frequency modes in amorphous solids can be studied using ultrasonic or Brillouin scattering techniques. In the various ultrasonic measurements, which include not only attenuation but also phonon echo experiments, an ultrasonic pulse is applied to the sample through an appropriate transducer and the sound propagation through the sample is measured. This technique requires pulsed radio frequency and microwave sources and detectors (see, for example, Truell et al. 1968). Brillouin scattering is usually accomplished with a Fabry–Perot interferometer (Anderson 1971).

3.2. Data reduction in infrared spectroscopy

3.2.1. Reflection spectra

The interpretation of reflection spectra is usually based on the use of Kramers–Krönig (KK) relations. These relations are quite general and are applicable to any homogeneous and linear physical system. They express the interdependence of the two electrical constants ε' and ε'' which make up the complex dielectric constant $\varepsilon = \varepsilon' - i\varepsilon''$ (or the two optical constants n and k, where the complex index of refraction $N = \varepsilon^{1/2} = n - ik$). In terms of the dielectric constant the KK relations are

$$\begin{aligned}\varepsilon'(\omega)-\varepsilon_\infty&=\frac{2}{\pi}\int_0^\infty \frac{\omega'\varepsilon''(\omega')}{\omega'^2-\omega^2}\,\mathrm{d}\omega',\\ \varepsilon''(\omega)&=-\frac{2\omega}{\pi}\int_0^\infty \frac{\varepsilon'(\omega')}{\omega'^2-\omega^2}\,\mathrm{d}\omega',\end{aligned}\tag{3.1}$$

where $\omega=2\pi\nu$ is the angular frequency and ε_∞ is the real part of the dielectric constant at high frequency, where it is assumed there is no absorption. In practice most amorphous solids have a transparent region of frequencies (above the vibrational absorption and below the interband electronic absorption) within which ε_∞ can be defined. Eq. (3.1) can be equally well written in terms of the real and imaginary parts of the conductivity ($\sigma=\sigma_\mathrm{r}+\mathrm{i}\sigma_\mathrm{i}$) or the optical constants through the relations $\varepsilon'=n^2-k^2=4\pi\sigma_\mathrm{i}/\omega$, $\varepsilon''=2nk=4\pi\sigma_\mathrm{r}/\omega$. At optical frequencies results are often expressed in terms of the absorption coefficient $\alpha=2\omega k/c$, where c is the velocity of light in vacuum. In practical units (resistance expressed in ohms) the absorption coefficient can be converted to a conductivity through the relation $n\alpha(\mathrm{cm}^{-1})=120\,\pi\sigma_\mathrm{r}(\Omega^{-1}\,\mathrm{cm}^{-1})$.

The KK relations can be easily adapted to extract the optical or electrical constants from an observed frequency dependence of the reflectivity *provided that the vibrational absorption is isolated in frequency from competing absorption processes at higher and lower frequencies*. If this situation does not hold then substantial error will be introduced in using the method. We denote the normalized single surface reflectance at normal incidence by R, where in terms of the optical constants,

$$R=\frac{(n-1)^2+k^2}{(n+1)^2+k^2}.\tag{3.2}$$

It can be shown from eq. (3.1) that the phase shift on reflection is given by

$$\theta(\omega)=\frac{2\omega}{\pi}\int_0^\infty \frac{\ln R(\omega')^{1/2}}{\omega^2-\omega'^2}\,\mathrm{d}\omega'.\tag{3.3}$$

One may now use eq. (3.3) to evaluate numerically θ, given the experimental values of R. The optical constants are then determined from the Fresnel formulae

$$\begin{aligned}n&=\frac{1-R}{1+R-2\sqrt{R}\cos\theta},\\ k&=\frac{2\sqrt{R}\sin\theta}{1+R-2\sqrt{R}\cos\theta}.\end{aligned}\tag{3.4}$$

In practice the reflectivity must be well determined over the entire range of vibrational frequencies so that both a low-frequency (ε_0) and a high-frequency (ε_∞) dielectric constant can be defined from the data. Often the determination of ε_0 for amorphous solids requires dielectric response measurements in the microwave frequency region.

The reduction of reflectivity data using KK analysis is in some respects easier for amorphous than for crystalline solids because the spectral features are broader in the former case. Hence one requires fewer numerical points to produce a satisfactory transform. On the other hand, because the normal crystalline selection rules are relaxed in amorphous solids they tend to absorb over a wider frequency range than their crystalline counterparts.

3.2.2. *Transmission spectra*

In spectral regions where the vibrational absorption is weak the optical constants are commonly determined from transmission measurements. The normalized transmittance T through a thin disc or free-standing film of thickness d with plane parallel faces can be expressed in terms of the optical constants n, k as defined above and the wavelength of the incident radiation λ by the usual multiple reflection formula. At normal incidence the expression is

$$T = \frac{(1-R)^2(1+k^2/n^2)^2}{(e^{\beta} - Re^{-\beta})^2 + 4R\sin^2(\delta+\chi)}, \tag{3.5}$$

where $\tan\chi = 2k/(n^2+k^2-1)$, $\beta = 2\pi kd/\lambda$, $\delta = 2\pi nd/\lambda$ and R is the normalised single surface reflectance at normal incidence defined in eq. (3.2). The absorption coefficient α defined above is related to the quantities of eq. (3.5) through the equation $\alpha = 2\beta/d$. When there is essentially no absorption in the sample ($k \approx 0$), T oscillates between unity and some minimum value, which depends on the reflectivity R, with a period in λ^{-1} of $(2nd)^{-1}$. Thus the thickness d can be determined from the period of these interference oscillations provided that n is known. Similar oscillations occur in reflectivity measurements on thin samples where the single surface reflectance formulae of the preceding section break down. For thin amorphous films deposited on substrates both the reflectance and the transmittance data are more difficult to interpret (i.e. KK analysis cannot be applied directly) and one must usually resort to KK analysis which is coupled to some sort of iterative procedure to determine the optical constants self-consistently from the data. (See, for example, Brodsky and Lurio (1974) or Stimets et al. (1973) for specific examples.)

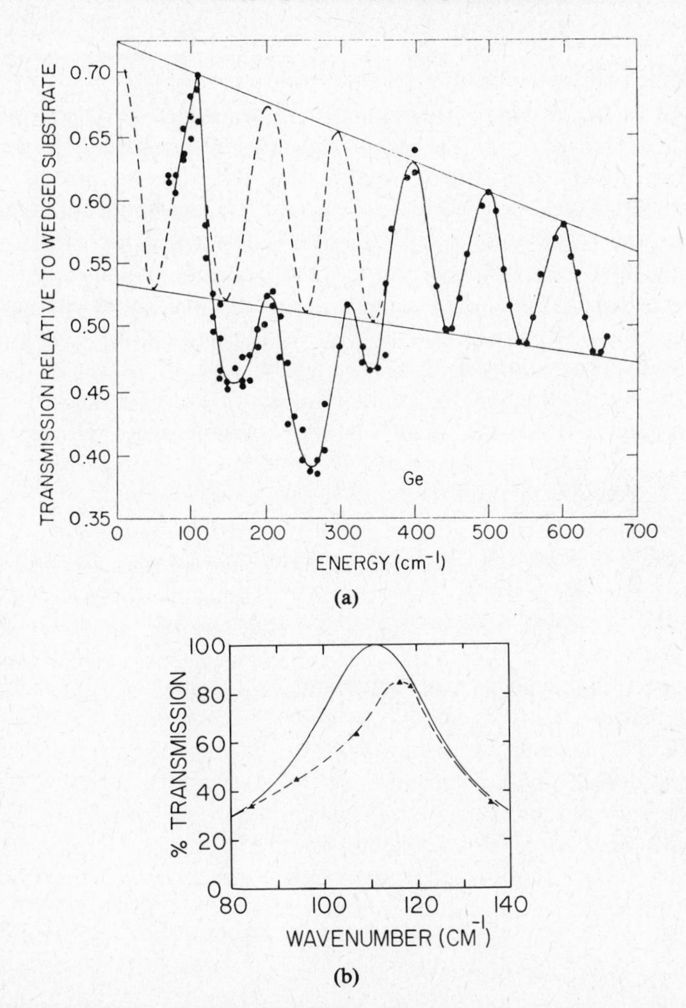

Fig. 8. (a) The experimentally determined transmittance (data points) and calculated transmittance (solid line) of 25 μm of amorphous Ge on both sides of a wedged, high-resistivity crystalline Si substrate relative to a matched, uncoated substrate. The dashed curve is an estimate of the interference fringes in the transmittance in the absence of any absorption in the amorphous Ge film (Brodsky and Lurio 1974). (b) Experimentally determined transmittance (data points) and calculated transmittance (dashed curve) of a 10.4 μm free-standing film of amorphous Ge. The solid curve is the interference fringe in the transmittance in the absence of absorption.

As an example of the use of these procedures we briefly consider the extraction of the optical constants from transmission data on a-Ge films as illustrated in fig. 8. There we show the transmittance of 25 μm of a-Ge, sputtered on *both* sides of a wedged transmitting substrate (after Brodsky and Lurio 1974). These data, which were taken on a normal grating spectrometer, have been normalised to the transmittance of a wedged substrate, and they show quite clearly the interference oscillations in the a-Ge. The solid lines delineate the maxima and minima estimated for the interference oscillations. The dashed lines denote the calculated a-Ge spectrum in the absence of absorption. Also shown in fig. 8 is a similar transmission spectrum for a free-standing film of a-Ge in the low frequency region (after Taylor et al. 1976). For these data eq. (3.5) can be applied directly and the solid curve represents the absorption-free spectrum for the appropriate values of n and d. The triangles denote data taken at discrete frequencies using an optically pumped molecular gas laser as the source.

We should caution that eq. (3.5) provides an accurate measure of the absorption in a sample only when αd (or 2β)$\gtrsim 1$. For example, in figs. 8a and 8b this condition holds only over the approximate reduced frequency ($\bar{\nu}=\lambda^{-1}$) ranges 125–350 cm^{-1} and 85–115 cm^{-1}, respectively. In amorphous solids such as Ge, which only occur in thin film form ($d \leqslant d_{max}$), one is sometimes tempted to extract absorption data in regions where αd is considerably less than 1. However, anyone who recollects the protracted debate on the sharpness of the electronic absorption edge in a-Ge (see, for example, Donovan 1974) will advise restraint!

The form of eq. (3.5) is greatly simplified if either $\alpha d \gg 1$ or the thickness of the sample is wedged so that interference effects are averaged out. In this case the average transmittance is given by

$$T=\frac{(1-R)^2 e^{-\alpha d}}{1-R^2 e^{-2\alpha d}}. \tag{3.6}$$

In the literature the term containing $\exp(-2\alpha d)$ is often neglected, sometimes inappropriately.

3.3. Data reduction in Raman spectroscopy

Despite theoretical misgivings mentioned in §2.10, the scheme of Shuker and Gammon (1970) for reducing the Raman spectra of amorphous solids is often employed by experimentalists. The Shuker–Gammon reduction scheme assumes a constant coupling constant for a given "band" of modes. While this procedure might be close to the truth for glasses which exhibit well-separated Raman peaks, it is crude when applied to the tetrahedrally coordinated amorphous solids.

It is common in the literature to define a *reduced* Raman spectrum as

$$I_{\mathrm{p}}^{\mathrm{R}}(\omega)=\omega(\omega_{\mathrm{L}}-\omega)^{-4}[1+n(\omega,T)]^{-1}I_{\mathrm{p}}(\omega), \tag{3.7}$$

which, according to Shuker and Gammon, should be proportional to the density of states within each band. For the anti-Stokes spectrum the factors $[1+n(\omega,T)]$ and $(\omega_{\mathrm{L}}-\omega)$ are replaced by $n(\omega,T)$ and $(\omega_{\mathrm{L}}+\omega)$, respectively. It is assumed that the Raman scattering intensity $I_{\mathrm{p}}(\omega)$, and hence $I_{\mathrm{p}}^{\mathrm{R}}(\omega)$, have been corrected for any instrumental artifacts such as a frequency dependence of the system through-put or instrumental transfer function.

In principle, both $I(\omega)$ and $I^{\mathrm{R}}(\omega)$ must be corrected for any differences in the absorption coefficient α or single surface reflectance R at the frequencies of the incident and scattered photons (ω_{L} and $\omega_{\mathrm{L}}\pm\omega$, respectively). In practice these corrections are unimportant unless $\hbar\omega_{\mathrm{L}}$ falls in the region of the exponential absorption edge of the solid, in which case one may also observe a resonant enhancement of the Raman effect (Kobliska and Solin 1972).

In single crystals one can use the dependence of the Raman spectra for various crystal orientations on the polarisation of the incident and scattered photons to determine the symmetries of the vibrational modes. However, in powdered crystalline solids one cannot usually determine mode symmetries from polarisation measurements because the individual scattering sites are randomly oriented with respect to both the incident and scattered photons. In this case the symmetry properties of a given mode can only be characterised by a *depolarisation ratio* p which is defined as the ratio of the intensity of scattered light polarised in the scattering plane to that polarised perpendicular to the scattering plane. For amorphous solids p is a continuous function of ω.

Experimentally one can determine $p(\omega)$ from either the Raman spectrum or the reduced Raman spectrum as follows:

$$p(\omega)=I_{\mathrm{HV}}(\omega,T)/I_{\mathrm{VV}}(\omega,T)=I_{\mathrm{HV}}^{\mathrm{R}}(\omega)/I_{\mathrm{VV}}^{\mathrm{R}}(\omega), \tag{3.8}$$

where the two indices (H = horizontal and V = vertical) refer to the polarisation of the incident and scattered photon, respectively.

3.4. Data reduction in neutron scattering

In inelastic neutron scattering the one-phonon contribution to the spectrum is proportional to

$$S(Q,E)=\frac{n(E)-1}{E}\mathrm{e}^{-W(Q)}\sum_n \delta(E_n-E)I_n(Q), \qquad (3.9)$$

where E is the energy and Q the momentum transfer vector, n is the phonon occupation factor, W the Debye–Waller factor and the sum is over all vibrational eigenstates n. The factor $I_n(Q)$ is given by eq. (2.6) (§2.10) for coherent scattering. As mentioned in §2.10, this factor is independent of n in the limit of high Q. It follows that, if the prefactors in eq. (3.9) are estimated and divided out, the experimental spectrum yields the *density of states* in this limit.

In practice multiphonon processes also become important at high Q and some procedure must be adopted for subtracting them out (Axe et al. 1974). Alternatively one may try to work at intermediate values of Q and attempt to account for departures from the phonon density of states through a model calculation (Carpenter and Pelizzari 1975a, b).

There are several experimental difficulties involved in high-Q measurements (Egelstaff 1965). For example, to maintain the validity of the incoherent approximation (see the discussion following eq. (2.6)) at low frequencies requires the measurement of small energy changes at sufficiently high Q. This means that high incident neutron energies must be employed. If the energies are not sufficiently high then oscillations in the peak structure with Q are obtained. Resolution can also be a problem in neutron scattering measurements. In practice the experiments which are most easily interpretable have so far been performed only on elemental amorphous solids such as Se, As and Ge (Axmann et al. 1969, Axe et al. 1974, Leadbetter et al. 1976).

3.5. Electron tunnelling

In electron tunnelling measurements the amorphous solid is used as the insulating layer in a tunnel junction and the vibrational modes are excited directly by the tunnelling electrons (Giaever and Zeller 1968, Ladan and Zylbersztejn 1972). Measurements are made of the derivative of the voltage with respect to the current as a function of the voltage. The difficulties in relating such measurements to a phonon density of states are significant and the technique has not been widely used to date. Among the experimental difficulties are the separation of excitations in the barrier itself from those in the electrodes or in an oxidised layer on the electrodes, the subtraction of various background signals, and the differences in the data observed for the two different polarities of the electrodes. Nonetheless the resulting plot of σ^{-1} $(\mathrm{d}\sigma/\mathrm{d}V)$, where σ is the conductivity across the junction, for amorphous Ge, resembles the phonon density of states (see §4.2). Bennett et al. (1968) have calculated that for crystalline barriers containing impurities the appropriate matrix element is inversely proportional to the energy.

3.6. Special problems at low frequencies

At low frequencies both Raman and infrared measurements require extra precautions. In addition several other experimental techniques (see §3.3) become useful for reduced frequencies $\bar{\nu}$ below about 10 cm^{-1} (300 GHz, 14 K, 81 meV).

3.6.1. *Far infrared absorption*

The normal infrared analyses described in §3.2 are of course also applicable in the far infrared. Most amorphous solids absorb only weakly in this spectral range ($k<1$) so that both the single surface reflectance R and the real part of the refractive index n are essentially constant (cf. eqs. (3.1) and (3.2)). This fact means that the absorption coefficient α and the conductivity σ are simply related by a constant. At particularly low absorption levels ($\alpha \sim 1$ cm^{-1}) thick samples are required (so that $\alpha d \sim 1$) and one must be careful that scattering processes within the sample volume are not mistaken for absorption.

The *tunnelling modes* which have been postulated to exist in amorphous solids at far infrared energies are highly anharmonic and should exhibit the phenomenon of saturation at low temperature as one increases the infrared power (Anderson et al. 1972, Phillips 1972). Measurements of this type are important because they provide the only clear test of whether the absorption results from phonons similar to those present in crystalline solids or from the highly anharmonic tunnelling modes. Unfortnately such measurements have not yet been performed in the far infrared because they require rather high powers (von Schickfus and Hunklinger 1977). Far infrared laser sources coupled with cavity techniques borrowed from standard microwave procedures should be capable of observing saturation if it exists in the far infrared.

3.6.2. *Raman scattering*

At low frequencies ($\bar{\nu}<40$ cm^{-1}) the Raman spectra must either be taken using very high resolution ($\sim$0.5 cm^{-1}) or some appropriate procedure must be adopted to correct for the effects of stray background light and perhaps also the Rayleigh tail of the incident laser line. In some cases a triple monochromator has been used in place of the ordinary double monochromator in order to get reliable data at low frequencies.

3.6.3. *Microwave measurements*

In the microwave region there are two standard techniques for measuring conductivity or dielectric loss. In the first method, which works when small samples can be used (perturbation technique), the field distribution inside the microwave cavity is assumed to remain unchanged upon introduction of the sample. If ν and ν_0 are the perturbed and unperturbed cavity resonant frequencies, respectively, and Q is the quality factor of the perturbed cavity, then the necessary conditions for the validity of the perturbative approach are $(\nu-\nu_0)/\nu \ll 1$ and $Q \ll 1$. The complex dielectric constant for a homogeneous, isotropic sample in the perturbative case is given by

$$\varepsilon' = 1 + \frac{\nu-\nu_0}{2\nu}\frac{V}{v}, \tag{3.10}$$

$$\varepsilon'' = \frac{1}{4}\frac{Q-Q_0}{Q_0 Q}\frac{V}{v}, \tag{3.11}$$

where V and v are the volumes of the cavity and the sample, respectively.

For most dielectrics or semiconductors in the microwave region the condition $Q \gg 1$ is easily satisfied, but for some materials one cannot obtain the second necessary condition for the perturbation technique to be appropriate. In this case a non-perturbative technique is necessary in which the field distribution both before and after insertion of the sample into the cavity is calculated. Experimentally the sample is moulded to fit the cavity cross section and cut to a length of $\lambda_s/2$, where λ_s is the microwave wavelength in the sample. Both ε' and ε'' are then determined by solving some rather complicated expressions which depend on details of the cavity configuration (Altschuler 1963). Some iterative procedure may be necessary because λ_s depends on ε'. This method may involve some complicated algebra, but it is accurate provided the samples are homogeneous and isotropic. If ε is known (from far infrared reflectivity measurements, for example), then λ_s is known accurately and the procedure is somewhat simplified.

Measurements of power saturation at low temperatures can in principle be performed using either of these two procedures and standard microwave attenuation techniques. To date only one saturation measurement has been reported (von Schickfus and Hunklinger 1977), and these results are discussed in §5.2.

3.6.4. *Pulse techniques (ultrasonic attenuation, phonon echoes, polarisation echoes, magnetic resonance)*

Several rather specialised techniques which involve pulses of acoustic waves, electromagnetic waves, or both, have been employed recently to investigate tunnelling modes in amorphous solids. These techniques only yield effective densities of tunnelling modes because the coupling depends on the details of the experiment just as it does in infrared or Raman measurements. For example, in a phonon echo experiment (Golding and Graebner 1976) the coupling involves acoustic excitations (ultrasonic waves) while in a polarisation echo measurement (Shiren et al. 1977) the coupling involves both a travelling ultrasonic wave and a standing-wave electric field. Unfortunately the matrix elements are even more difficult to estimate for the tunnelling modes than they are for the modes which dominate at higher frequencies. Perhaps because of this difficulty there has been an unfortunate tendency to ignore the matrix elements and to equate densities of states determined by different experimental techniques. In our view, this tendency should be resisted.

Because the simplest approximation for the tunnelling modes involves only two energy levels, these modes can be treated using much the same formalism developed for spin-$\frac{1}{2}$ states in magnetic resonance (NMR and ESR). The eigenstates can be described via a Pauli spin matrix. The lifetime T_1 and phase memory time T_2 of the excited state can be characterised by analogy with the corresponding spin–lattice and spin–spin relaxation phenomena in magnetic resonance. Measurements of T_1 and T_2 will also vary depending on the experiment.

In magnetic resonance the coupling to the tunnelling modes is indirect and one infers properties of these modes by observing their effect on the relaxation of excited magnetic spins via the spin–lattice relaxation (Szeftel and Alloul 1975, Rubinstein et al. 1975, Reinecke and Ngai 1975). Ultrasonic attenuation experiments can distinguish between travelling sound waves and localized tunnelling modes by measuring the attenuation as a function of incident power (Hunklinger et al. 1972, Golding et al. 1973). At low temperatures the saturable component of this attenuation is a measure of the influence of the tunnelling modes.

3.6.5. *Specific heat and thermal conductivity*

Specific heat and thermal conductivity data provided the first evidence for tunnelling modes (Zeller and Pohl 1971). In many amorphous solids the low-temperature specific heat is proportional to T (amorphous As is currently the only known exception). In addition the cubic term in amorphous solids is enhanced over that which one would predict using the velocity of sound as determined from the Debye approximation (Zeller and Pohl 1971, Stephens 1973).

The thermal conductivities of amorphous solids are all very similar (Zeller and Pohl 1971), while those of crystalline solids vary considerably. Below 1 K the conductivities in amorphous solids all conform to a power law

$$\kappa(T) \propto T^{\delta}, \tag{3.12}$$

where δ lies between 1.8 and 2.0 for all amorphous materials studied (Pohl et al. 1974). The understanding of these thermal conductivity results depends upon the assumptions made concerning the heat transport mechanisms. Most authors have assumed, probably correctly, that the tunnelling modes do not transport heat so that the temperature dependence of the thermal conductivity results entirely from changes in the phonon mean free path with temperature (Pohl et al. 1974) which may be due to scattering from the tunnelling modes (Anderson et al. 1972).

4. *Some examples*

4.1. Group IV elements

The Group IV elements were among the first materials to be subjected to detailed spectroscopic studies in their amorphous form and our understanding of their properties has benefited from a continued debate among various groups. Initially, the concentrated interest in these materials could be justified only by academic arguments. Interesting prospects of device applications have emerged more recently (Wronski et al. 1976, Spear 1977, Wilson and Weaire 1978), particularly for a-Si prepared by the r.f. glow discharge deposition technique. Some of the most recent vibrational studies are aimed at the characterisation of this type of material by the analysis of vibrational bands associated with hydrogen atoms incorporated in the structure.

Earlier work was mostly devoted to samples prepared by sputtering or evaporation. While electronic properties displayed a good deal of sensitivity to the method and conditions of preparation, no clear dependence on these was demonstrated for vibrational properties.

Si and Ge have so much in common that we shall discuss them together in what follows. Of the other Group IV elements, Sn has not been prepared in amorphous form (except possibly in a metallically bonded structure) while C exhibits an interesting variety of amorphous forms with tetrahedral sp^3 and/or planar sp^2 bonding. These have not yet been fully sorted out and we shall confine attention here to the more straightforward case of Si and Ge.

The experimental facts were made very plain at an early stage by Brodsky and Lurio (1974), as shown in fig. 9, from which they drew two conclusions, as follows:

(1) Both infrared and Raman spectra appear to reflect all the main features of the density of states.

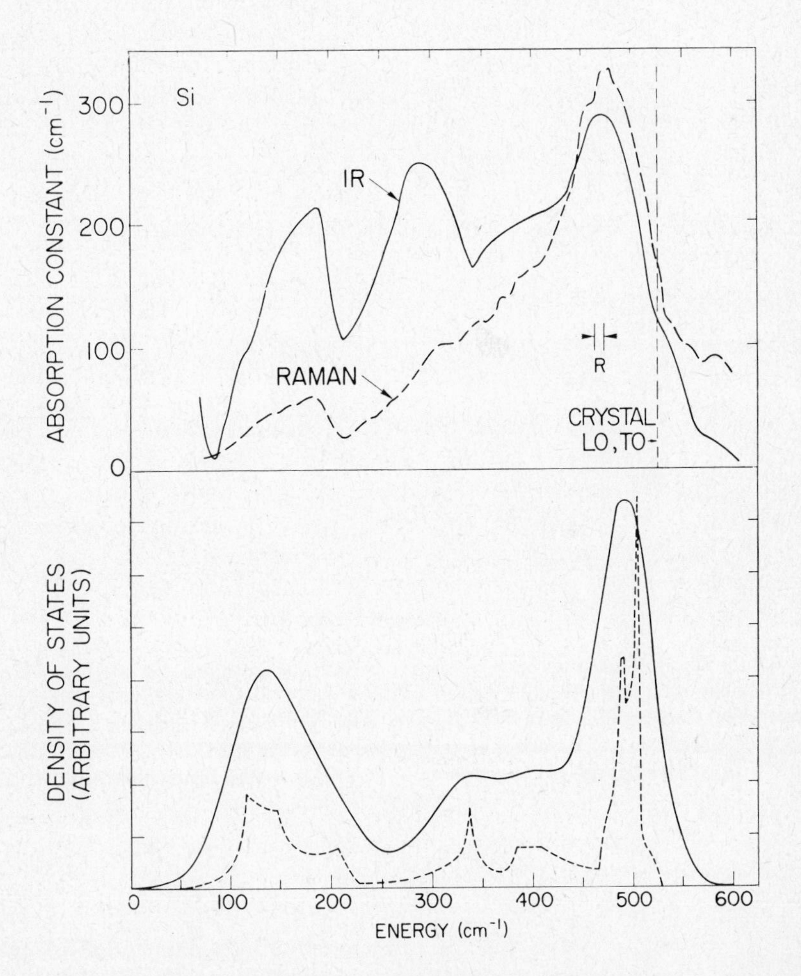

Fig. 9. (a) Top: Infrared absorption coefficient at 300 K for amorphous Si as a function of wave number (solid line). Reduced Raman spectrum (see §3.5) at 300 K for amorphous Si (dashed line) from Smith et al. (1971a). Bottom: Density of states of crystalline Si (dashed line) as obtained from a fit to neutron scattering data of Dolling and Cowley (1966). The solid line is a broadened density of states.

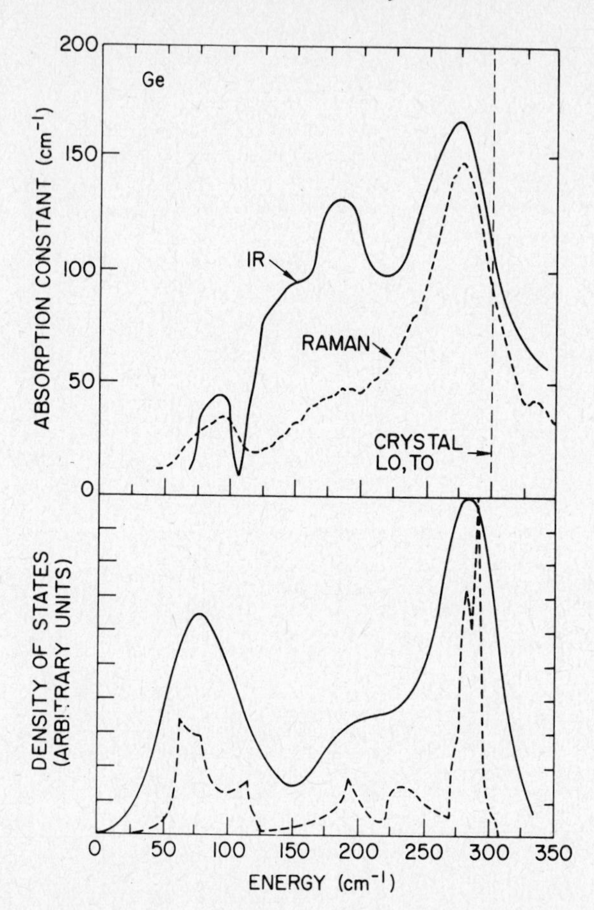

Fig. 9. (b) Top: Infrared absorption coefficient at 300 K for amorphous Ge as a function of wave number (solid line). Reduced Raman spectrum (see §3.5) at 300 K for amorphous Ge (dashed line) from Smith et al. (1971b). Bottom: Density of states of crystalline Ge (dashed line) as obtained from a fit to neutron scattering data of Dolling and Cowley (1966). The solid line is a broadened density of states (Smith et al. 1971a). Entire figure from Alben et al. (1975).

(2) The density of states appears to be remarkably similar to that of the crystalline form, although the latter must be considerably broadened.

At first sight, this resemblance would appear to call into question the random network model. Can a structure so very different from the diamond cubic crystal really produce much the same vibrational spectrum? The answer is yes, and this can now be seen from a variety of viewpoints.

Firstly, there is the analysis of Weaire and Alben (1972), already summarised in §2.9. This shows that in a much simplified central force model, the gross features of the density of states are largely independent of structure. Specifically, as shown in fig. 7, there are structure independent delta functions at the top and bottom of the spectrum and a band bounded by them. At this skeletal level only the structure of the middle band depends on the (topological) structure. Intriguing possibilities emerge for the relationship of this part of the vibrational spectrum with the topological structure. We shall return to this point shortly.

Two ingredients remain to be added – non-central forces and local distortions. The effect of the former is to broaden the two delta functions into peaks, while the distortions mainly have the effect of broadening the *upper* peak, as shown in fig. 10 (Weaire and Alben 1972).

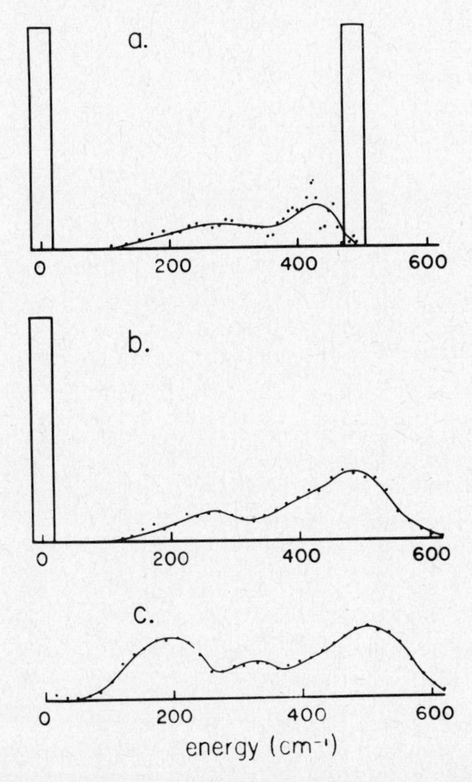

Fig. 10. Phonon density of states for a random network of 61 atoms with periodic boundary conditions: (a) as derived from a one-band Hamiltonian (§2.9); (b) as calculated for the Keating model (see §2.3) with $\beta=0$; and (c) as calculated for the Keating model with $\beta=0.2\alpha$. In each case $\alpha=0.495\times10^5$ dyn/cm (Weaire and Alben 1972).

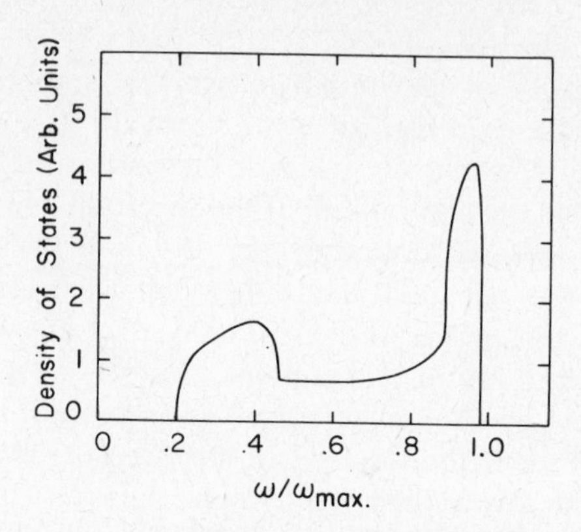

Fig. 11. Phonon density of states derived for a single tetrahedral unit with the boundary conditions described in §2.7. The maximum frequency is the frequency of the zone centre optic mode in the diamond cubic structure. The ratio $\beta/\alpha = 0.6$, which is appropriate to Ge (Thorpe 1973).

Another calculation which is helpful at this stage is that of Thorpe (1973), described in §2.7. In fig. 11 we see that this self-consistent formulation, derived from a *single tetrahedral cluster* of five atoms, yields a density of states having the same general features mentioned above!

Alben et al. (1975) incorporated simple semi-empirical forms for the matrix elements appropriate to the infrared and Raman spectra in their numerical calculations. In this way, the different relative weightings of the two spectra could be at least partially understood (fig. 12).

The success of this analysis reinforced confidence in the random network model, but the question remained – could it be used to refine the characterisation of the model? Renewed efforts to calculate the vibrational spectra of larger random network models have largely been in pursuit of this goal.

To cut a long story mercifully short, the answer to the above question would appear to be *no*. It is true that calculations by Beeman and Alben (1977) and Meek (1976a), using the two methods of §2.6 and Born or Keating force constants, revealed various interesting correlations of structure and spectra (figs. 13, 14). However, further calculations by Meek (1976b) caused a revision of his earlier conclusions. In these later calculations he incorporated a simplified form of Weber's bond charge force model (§2.3) and the additional forces had the effect of washing out most of the interesting dependences of the calculated spectrum on the details of structure! It can hardly be doubted that this is an improved model and its disappointing consequences must therefore be faced.

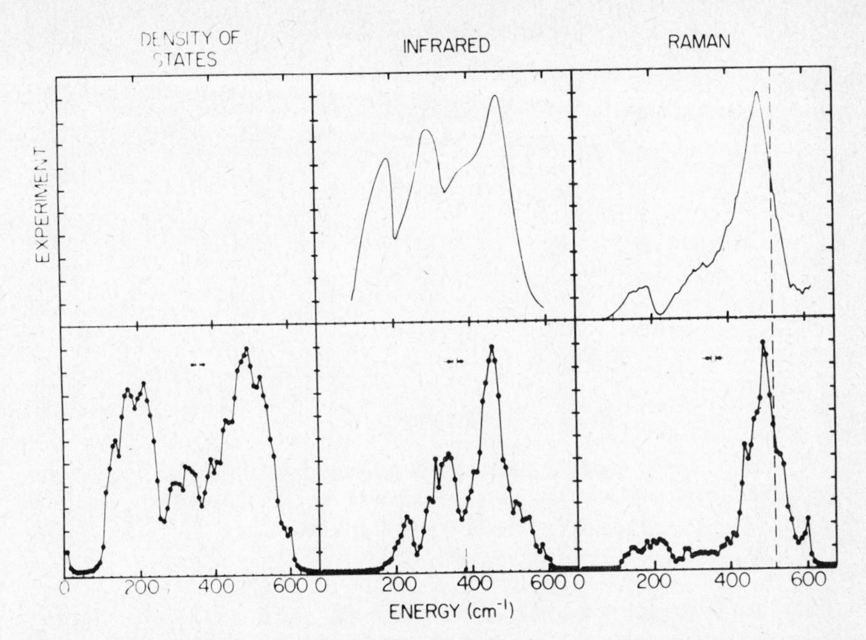

Fig. 12. Theoretical results for the phonon density of states, infrared absorption and Raman scattering for 61- and 62-atom periodic models compared with the experimental data of Brodsky and Lurio (1974) on amorphous Si. The specific infrared absorption and Raman mechanisms used in the theory are those of Alben et al. (1975). The position of the maximum frequency mode for diamond cubic (Raman active TO mode) is indicated by the vertical dashed line on the Raman graphs. The position of the density-of-states minimum near the centre of the spectrum for diamond cubic with the force constants used in these calculations is indicated by the arrow at 380 cm^{-1} on the infrared theory results. The spectra are formed by adding weighted Lorentzian contributions from $k=0$ modes and are normalised to the same maximum value (Alben et al. 1975).

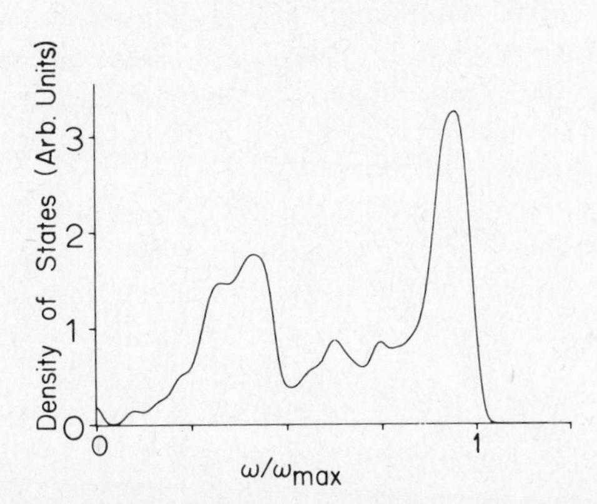

Fig. 13. Phonon density of states calculated for a 201-atom model (Steinhardt et al. 1974) of a tetrahedrally-coordinated amorphous solid using Born force constants and both central and non-central forces (Beeman and Alben 1977).

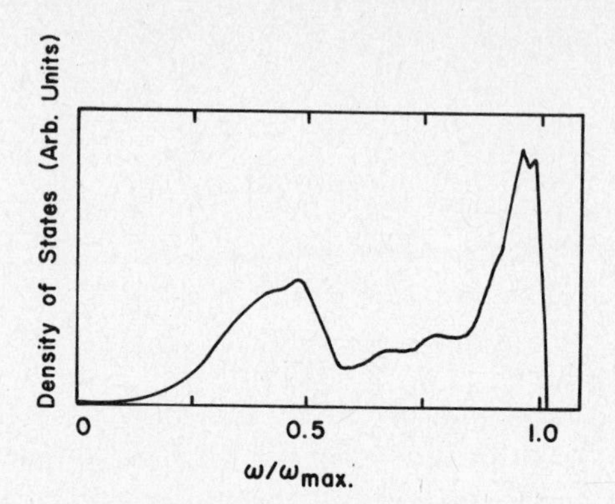

Fig. 14. Phonon density of states calculated for a 201-atom model (Steinhardt et al. 1974) of a tetrahedrally-coordinated amorphous solid using Born force constants and both central and non-central forces (Meek 1976a).

Thus we have achieved, at the expense of much analysis (including alarms and diversions not mentioned here), little more than the reconciliation of the random network model with the conclusions of Brodsky and Lurio with which we began.

Finally, we turn to the recent studies of glow discharge a-Si. These focus on the role of hydrogen, which appears to be responsible for the favourable electronic properties of this material. Indeed, rather similar properties can be induced in sputtered specimens by depositing them in the presence of gaseous hydrogen (Connell and Pawlik 1976).

Brodsky et al. (1977) and Lucovsky et al. (1979) have published an extensive analysis of the vibrational spectra of glow discharge a-Si, using infrared and Raman techniques. They were able to identify Si–H bond and stretch bands and various bands of lower frequency associated with

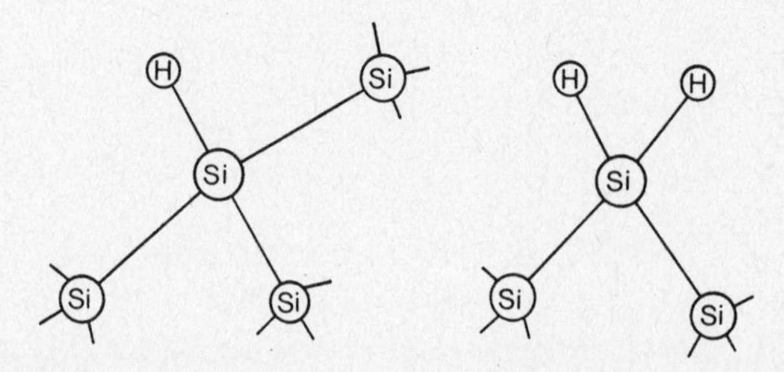

Fig. 15. Schematic diagram of possible bonding configurations for hydrogen incorporated into amorphous silicon: at left, an Si–H complex; at right, an $Si–H_2$ complex.

the “bending” or “wagging” of Si–H bonds. The details of these bands were interpreted in terms of Si–H, $Si–H_2$ and $Si–H_3$ complexes, that is, silicon atoms bonded to a number of hydrogen atoms as well as silicon atoms in the random network (fig. 15).

One may use the strength of the absorption bands to monitor hydrogen content in these films. However, while this is straightforward if only relative concentrations are needed, there is some uncertainty concerning the absolute scale involved, and calibration by comparison with other techniques is essential.

Finally, mention should be made of experimental studies of metastable polymorphs of Si and Ge. Si III and Ge III have, respectively, 8 and 12 atoms per unit cell and potentially offer much detailed information on the effect of distortion and band topology on interatomic forces. However, the infrared and Raman data so far available are rather preliminary in nature, and only a crude force constant analysis has been undertaken (Kobliska et al. 1972). The difficulty here lies in the quality of the available samples. Given better material, these fascinating allotropic forms would be worthy of much more detailed attention.

4.2. Group V elements

Much of what we have said about the Group IV amorphous semiconductors is precisely paralleled by their Group V counterparts. Tetrahedral coordination is replaced by a trigonal arrangement of bonds. The random network has a local topology which is somewhat planar in nature in this case (Greaves and Davis 1974) and reference is often made to its essentially two-dimensional character. The mean value of the interbond angle is not determined by symmetry and for As, on which most work has been done, it is thought to be about 97° (Smith et al. 1975).

The most elementary idealisation of the local structure and forces is obtained by considering bonds oriented at 90° to one another and associated with central nearest-neighbour forces. Trivially, the degrees of freedom associated with each bond are entirely decoupled. Each bond contributes modes at $\omega^2=0$ and α/m, where α is the force constant. Thus the density of states consists of two delta functions. This structure is found in the observed density of states of each form of As, since it consists of two separate bands.

These features are illustrated in the high-Q neutron data of Leadbetter et al. (1976) shown in fig. 16, which deserves pride of place here since it is probably the most impressive application to amorphous solids of the technique so far made. In addition to measuring the density of states of amorphous As, Leadbetter et al. made measurements for the rhombo-

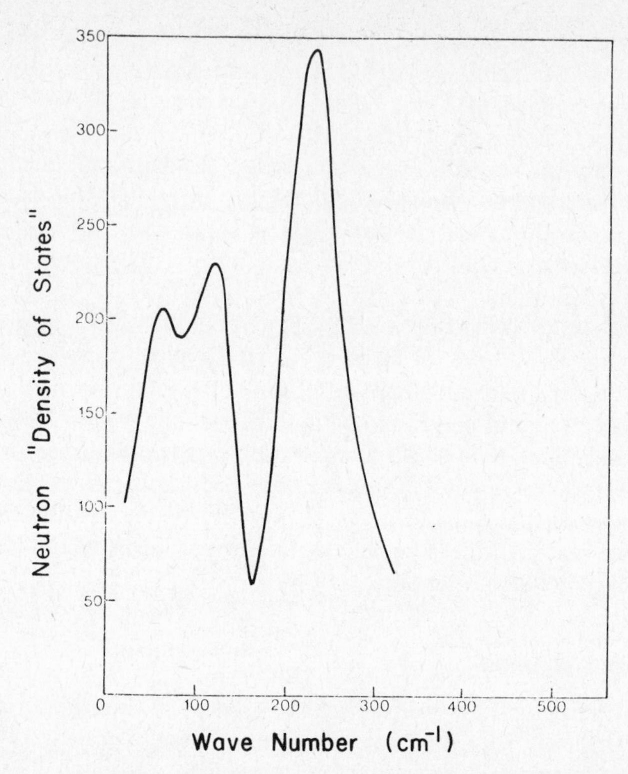

Fig. 16. An estimate of the phonon density of states for amorphous As from neutron scattering results (Leadbetter et al. 1976). The curve represents an average of cold neutron results for scattering angles $2\theta \gtrsim 36°$. Such an average tends to eliminate the coherence (interference) effects which are important because arsenic is predominantly a coherent scatterer. Corrections have been made for multiphonon contributions (see §§2 and 3.4).

hedral crystalline form and the relatively rare orthorhombic form. They noted that the latter gives results closer to those of the amorphous phase.

Infrared (Lucovsky and Knights 1974) and Raman spectra (Lannin 1977) show similar structure, with significant modulation due to matrix element effects (fig. 17). Figs. 18 and 19 show the comparison of these data with the calculations of Beeman and Alben (1977). These calculations were carried out for the Greaves–Davis random network model using Born forces (§2.3) and the equation-of-motion method (§2.6). The major success of this calculation is the reproduction of the polarisation dependence of the shape of the upper peak of the spectrum.

An exhaustive analysis of the dependence of the density of states on the distortions and topological statistics of the random network has been given

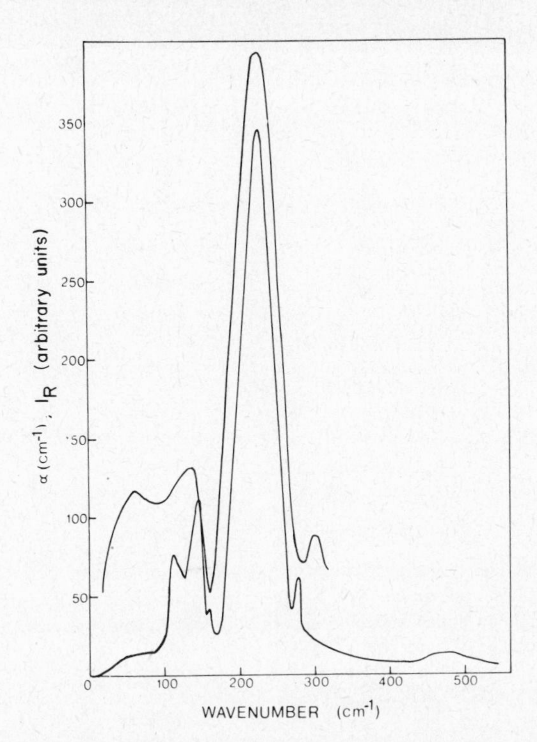

Fig. 17. Reduced Raman spectrum (top; Lannin 1977) and infrared absorption (bottom; Lucovsky and Knights 1974) as functions of wave number for amorphous As. The Raman results are for films and the infrared results for bulk material. Note that some "extra" features appear in the infrared results, for example the peaks near 150 and 280 cm^{-1}.

by Pollard and Joannopoulos (1978), using Bethe (§2.8) and cluster–Bethe calculations. These authors point out an interesting isomorphism between the electronic and vibrational densities of states for pyramidally bonded amorphous solids.

There are some additional sharp features in the infrared spectrum of bulk amorphous As (fig. 17) and in the Raman spectrum of bulk amorphous As (Nemanich et al. 1978). Two of these weak features near 165 and 280 cm^{-1} have been attributed by Nemanich et al. (1978) to specific defects in the As structure. These authors used a model calculation based on the cluster–Bethe-lattice method (§2.8) and Born force constants (§2.3) to calculate the vibrational density of states for one-fold- and two-fold-coordinated and tetrahedrally-coordinated defects. The results indicate that tetrahedrally-coordinated defects or perhaps nearest-neighbour pairs of four-fold and two-fold-coordinated defects can account for the extra features near 165 and 280 cm^{-1}.

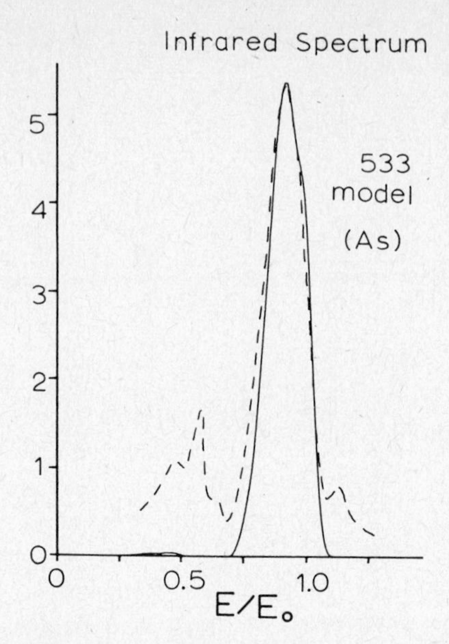

Fig. 18. Calculated (solid line) and observed (broken line; Lucovsky and Knights 1974) infrared absorption for amorphous As. Model parameters have been chosen so that the position of the calculated upper peak is in accord with experiment (Beeman and Alben 1977).

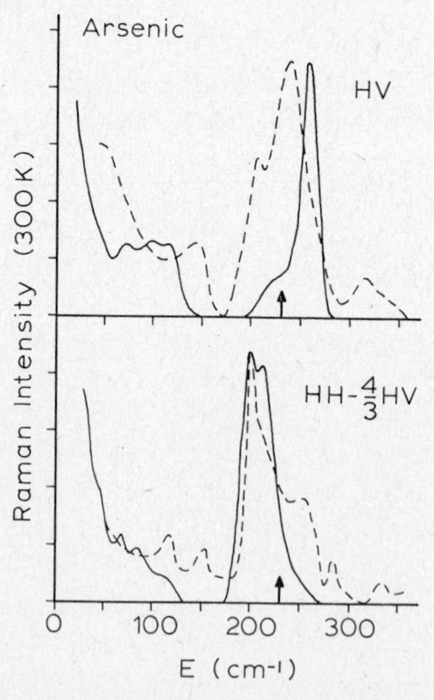

Fig. 19. Calculated (solid line) and observed (broken line; Lannin 1977) Raman spectra (*unreduced*, 300 K) for bulk amorphous As. The adjustment of the energy scale for the model calculation is identical with that imposed on the infrared spectrum (fig. 18). The vertical arrow marks the position of the infrared peak. In accord with experiment, the calculated polarized Raman spectrum peaks below, and the depolarized Raman spectrum peaks above the infrared peak (Beeman and Alben 1977).

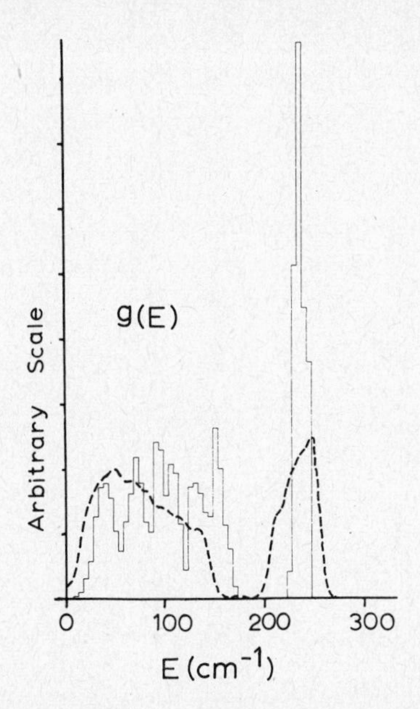

Fig. 20. Phonon density of states for a three-parameter model for trigonal crystalline Se (histogram) and for a 539-atom model of amorphous Se (broken line) (Beeman et al. 1976).

4.3. Group VI elements

Beeman et al. (1976) have analyzed the vibrational properties of a model for amorphous Se, consisting of convoluted chains (Long et al. 1976). They used the equation-of-motion method (§2.6) and a simple valence force potential (§2.3) for intrachain forces and a 6–12 potential ($V \propto ar^{-6} + br^{-12}$) for interchain forces. Results are shown in fig. 20. Again the spectrum breaks into two parts, easily understood in terms of the dominant central nearest-neighbour forces which are common to the crystalline and amorphous phases. Similar results have been obtained by Meek (1976b) who explored the spectra of an isolated Se chain with various choices of dihedral angles, using the recursion method.

The molecular type of analysis has also been applied to this case by Lucovsky and Knights (1974) who gave an interpretation in terms of the modes of a bent diatomic Se molecule. The limitations of this approach have already been discussed (§2.7).

4.4. IV–VI and related compositions

Vitreous silica is the most familiar of amorphous materials and many of the questions of previous sections were posed for it long before amorphous solids became of general interest. Thus the first calculations of the vibrational density of states of a random network were undertaken by Bell and Dean (Bell and Dean (1970), Bell 1972, Dean 1972) for silica, as already mentioned in §2.4. Some of their results are shown in fig. 21. These authors were also among the first to stress the importance of not using terminology appropriate only to periodic systems. Instead they defined quantities indicative of the general character of the modes. Thus a fractional "bond

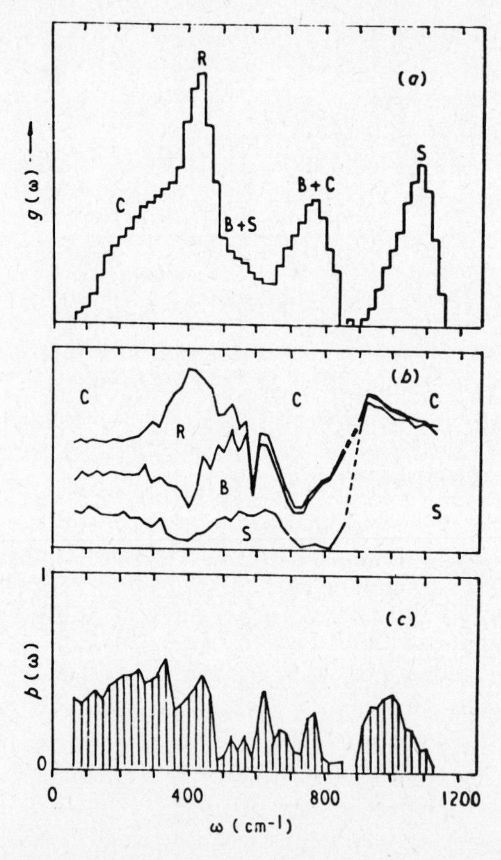

Fig. 21. (a) Phonon density of states for a random network model of glassy SiO_2 containing 101 silicon atoms and 233 oxygen atoms. Rigid boundary conditions for the 62 surface oxygen atoms were imposed. The symbols S, B, R, C refer to atomic motions which are primarily bond-stretching, bond-bending, bond-rocking, and those involving silicon atoms only, respectively. (b) Mode assignment diagrams. Vertical distances between adjacent curves give the proportion of energy arising from the various types of atomic motions, S, B, R, C. (c) Mode localisation pattern. $p(\omega)$ is an indication of the number of silicon atoms which are involved in the vibrational motion at a particular frequency (Bell and Dean 1970).

stretching" or "bond bending" character may be calculated, or the "participation ratio", which is the effective fraction of atoms contributing to states at a given frequency and is indicative of their degree of localisation.

What is not at all clear in fig. 21, but emerges from the Bethe lattice studies of Sen and Thorpe (1977), is the extent to which the features of interest are attributable to the short-range order alone. Sen and Thorpe showed that most of these features (and the character of the associated modes) are well reproduced by simple Bethe lattice calculations. Similar but less simplified calculations were performed by Laughlin and Joannopoulos (1977), including the evaluation of intensities appropriate to Raman and infrared spectra. The calculated infrared spectrum, based on dynamic charges (§2.10) inferred from data on crystalline quartz, was found to be in excellent agreement with the data of Galeener and Lucovsky (1976). Agreement with the Raman spectrum was less satisfactory.

Galeener and Lucovsky have posed a formidable theoretical puzzle by asserting that there is an LO–TO splitting in silica (Galeener and Lucovsky 1976). No one, as yet, has given a cogent explanation (or refutation!) of this point of view.

There are a number of similar compounds for which these studies of SiO_2 can serve as a guide to interpretation. For example, the Raman and infrared spectra of amorphous GeS_2, $GeSe_2$ and BeF_2 have been determined recently (Galeener et al. 1978, Lucovsky and Galeener 1976).

4.5. Other compounds, particularly As_2S_3

There has been a good deal of scattered and often inconclusive work on amorphous compounds. For example, it was thought at one time that the question of the existence of "wrong bonds" (III–III or V–V) in amorphous III–V compounds (Shevchik 1974) might be settled by vibrational spectroscopy, but this has proved difficult.

Another family of compounds which has attracted attention is that of the chalcogenide glasses, typified by As_2S_3. The only known crystalline form of As_2S_3 has 20 atoms per unit cell and consequently has a rich infrared and Raman spectrum. In fact essentially all optic modes are either infrared or Raman active in this solid (Zallen et al. 1971). We can thus construct broadened averages of the infrared and Raman spectra of the crystalline phase in the high-frequency regime and compare them with the spectra of the amorphous solid, as in fig. 22. In fig. 22 are shown the absorption coefficient α for glassy As_2S_3 (solid line) and a broadened average of crystalline As_2S_3 (dashed and dotted lines) which was obtained

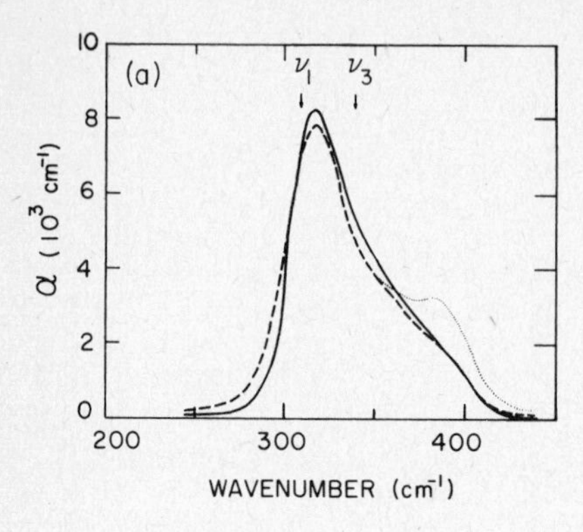

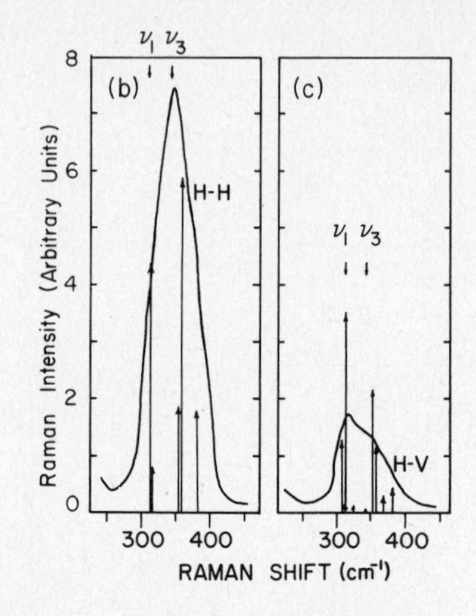

Fig. 22. (a) Comparison of the experimental infrared absorption spectrum of glassy As_2S_3 (solid line) with a broadened average of the absorption in crystalline As_2S_3 including only modes for which the electric field is parallel to the crystalline layer phase (dashed line) and including all modes (dotted line). The vertical scale applies to all curves. (b) Comparison of the polarised (H–H) Raman spectrum of glassy As_2S_3 with a histogram representing the average of the corresponding polarised crystalline As_2S_3 spectrum. (c) Comparison of the depolarised (H–V) Raman spectrum of glassy As_2S_3 with a histogram representing the average of the corresponding depolarised crystalline As_2S_3 spectrum. The vertical arrows labelled ν_1, ν_3 refer to "stretching" modes of an isolated AsS_3 "molecule" (Lucovsky 1972). The same scaling parameter is employed to compare the crystalline and glassy spectra in both (b) and (c) (Klein et al. 1977).

by convoluting the crystalline spectrum with a Gaussian distribution function of width $\sim 10\ cm^{-1}$ (Klein et al. 1977). The dotted line represents the additional contribution to α from out-of-plane modes ($E \| b$) in crystalline As_2S_3. In constructing the averages, account has been taken of differences in local field corrections between the crystal for which ε is anisotropic and the glass for which ε is isotropic. Detailed averages were not performed for the polarised Raman spectra, but figs. 22b and 22c indicate the positions and strengths of the crystalline modes (as adjusted for local field differences between the two phases) by arrows and the analogous glass spectra by solid lines. From these three diagrams it can be concluded that (1) the infrared and Raman vibrational spectra of crystalline and glassy As_2S_3 are similar (Klein et al. 1977) and (2) the infrared and Raman matrix elements are not slowly varying functions of the frequency as they are in amorphous Si but that (3) there is a certain amount of spectral complementarity between the two processes (Lucovsky 1972).

4.6. Amorphous metals

There is relatively little known about the vibrational properties of amorphous metals. This is largely because Raman or infrared absorption spectroscopy is impracticable for metals (see §3.3). However, a growing interest in the structural, mechanical and magnetic properties of these materials has stimulated some recent work on the vibrational density of states. A rather featureless spectrum, obtained from neutron spectroscopy on a Co–P alloy, has been reported (Moss et al. 1974) but it was admittedly preliminary in nature. It presumably represented that part of the density of states which is associated with the (majority) Co constituent. Von Heimendahl and Thorpe (1975) used a nearest-neighbour central force model to calculate the density of states of a cluster of atoms having the dense random packed structure characteristic of an amorphous metal. They used a somewhat unusual version of this model structure (Connell

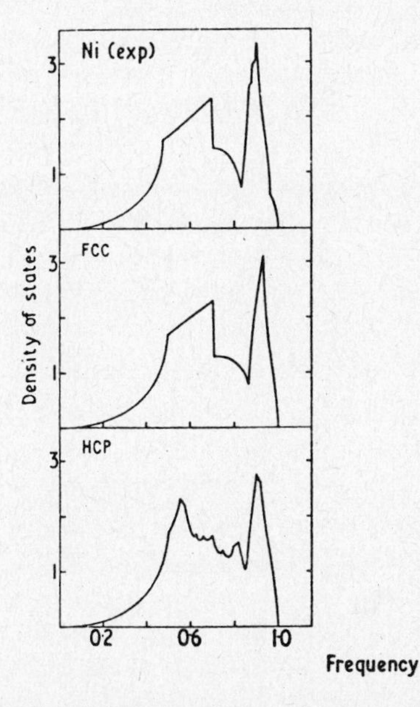

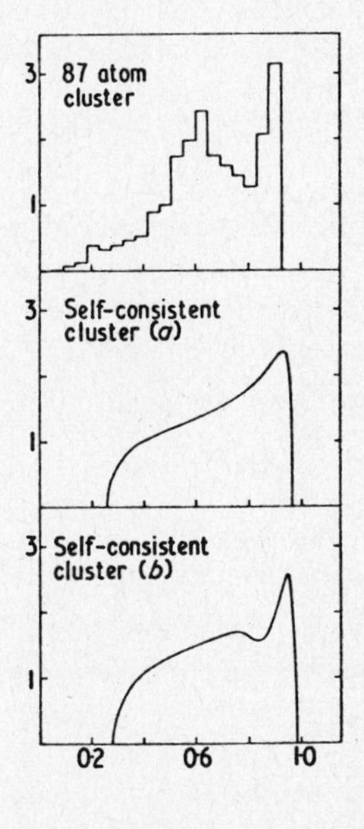

Fig. 23. At left: Phonon density of states for close-packed crystals determined by Brillouin zone integration. The frequency scale is in units of the maximum frequency. The f.c.c. Ni spectrum was computed from experimental dispersion curves (Birgeneau et al. 1964). The f.c.c. and h.c.p. spectra are for a model with only a nearest-neighbour central force constant. At right: Phonon density of states for clusters. The 87-atom cluster has free surface atoms and the spectrum is projected onto the 20 fully bonded atoms at the centre. The "self-consistent" cluster (a) has no triangles of near-neighbour bonds, but (b) has 12 such triangles (von Heimendahl and Thorpe 1975).

and Temkin 1974) but this point is not particularly significant. Their calculations are shown in fig. 23, together with a comparison with an experimental spectrum for crystalline Ni (another element which readily forms amorphous metallic alloys). The calculation on the random cluster was of the "brute force" variety (§2.5), projecting out the density of states associated with 20 atoms in the centre of an 87-atom cluster. Also shown in fig. 23 are calculations for a 12 coordinated Bethe lattice (§2.8), and a similar "self-consistent" calculation with some inclusion of rings of bonds, which are seen to be essential to the appearance of the sharp peak at the top of the spectrum.

Similar but more detailed calculations were performed by Rehr and Alben (1977) using the equation-of-motion method.

5. *Low frequency modes*

A complete and consistent picture of the low-frequency/low-temperature properties of amorphous solids has yet to emerge. One can distinguish two areas of current activity. Firstly, the conventional spectroscopic techniques of §3 may be extended to low frequencies. The question then is – what asymptotic form does the infrared absorption or the Raman scattering intensity take at low frequency? Most interpretations have been based on a Debye density of states. On the other hand, specific heat and related measurements at very low temperatures have been widely interpreted in terms of additional modes, analogous to impurity tunnelling states, such as those in doped alkali halides (Narayanamurti and Pohl 1970).

5.1. Spectroscopic measurements

Historically, SiO_2 was the first amorphous material in which the low-frequency behaviour was investigated in any detail. In 1965 Hadni et al. (1965a, b) suggested that because of its temperature independence, the far infrared absorption in glassy SiO_2 must be due to a single quantum process. In crystalline SiO_2 (α-quartz) the absorption in this spectral region is highly temperature dependent and its magnitude is considerably less than in the glassy phase. Stolen (1970) first showed that, at least in SiO_2, both the Raman and infrared spectra result from a coupling to the same low-frequency modes.

In fig. 24 are plotted infrared and Raman data for three well-studied glasses: SiO_2, As_2Se_3 and Se. The solid lines represent far infrared and microwave conductivity data at 300 K and the dashed lines indicate the data at low temperatures as extracted from several sources (Hadni et al. 1965b, Amrhein and Mueller 1968a, Bagdade and Stolen 1968, Strom et al.

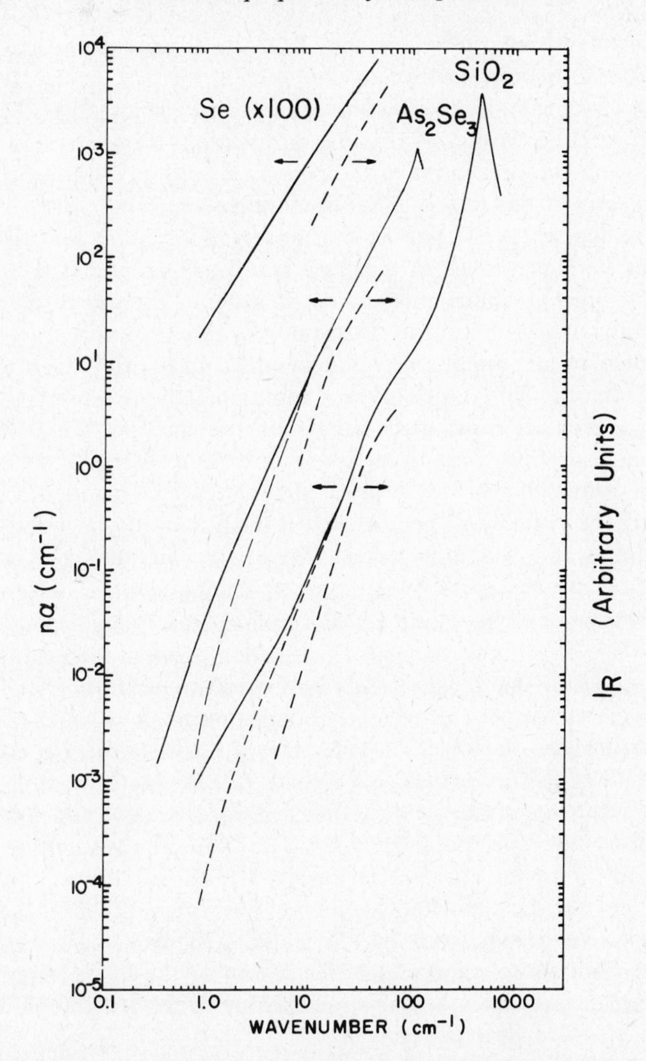

Fig. 24. Infrared absorption (solid lines at 300 K; dashed lines at 10 K) and reduced Raman spectra (dotted lines at $T \lesssim 30$ K) for amorphous Se, As_2Se_3 and SiO_2 at low frequencies. Sources of data are indicated in the text.

1974, Mon et al. 1975, Strom and Taylor 1977). It can be seen from fig. 24 that below about 10–20 cm^{-1} the infrared absorption is temperature dependent, a fact which was first pointed out for SiO_2 by Wong and Whalley (1971) from far infrared data and by Amrhein and Mueller (1968a, b) from microwave data. The dotted lines in fig. 24 denote low-temperature reduced Raman spectra $I_R(\omega)$ as defined in §3.5 (Stolen

1970, Gorman and Solin 1976, Nemanich 1977). Because of the experimental difficulties in measuring Raman scattering at low frequencies (cf. §4.1) the Raman data generally cover a less extended frequency range. Nonetheless, it is apparent from fig. 24 that the infrared and Raman results exhibit similar frequency dependences in the region of overlap. Although it is not shown explicitly in this figure, the lowest-frequency Raman data exhibit a temperature dependence similar to that observed in the far infrared.

The selected data of fig. 24 illustrate several experimental facts which are characteristic of amorphous solids in general. Firstly, there exists a spectral range where both the absorption coefficient and the reduced Raman intensity are temperature independent and scale roughly as the frequency squared. This range covers approximately an order of magnitude in frequency for most materials ($\sim$10–100 cm^{-1}) although in some materials such as SiO_2 underlying peaks are also observed. At reduced frequencies below about 5 to 20 cm^{-1} the frequency dependence of both the infrared absorption and the reduced Raman intensity becomes greater than quadratic at low temperatures. There is in addition a component which is temperature dependent in this frequency range.

In the frequency range where α scales approximately quadratically with frequency there appears to be a good correlation between the magnitude of the absorption and the density of states as determined from specific heat experiments in a number of covalently bonded amorphous solids (Strom et al. 1974, Strom and Taylor 1977). This fact is illustrated in fig. 25 where the magnitude of the far infrared absorption K_0 ($n\alpha = K_0(\hbar\omega)^2$) is plotted as a function of the magnitude of the density of states as determined from the cubic term in the specific heat c_3 ($C_V = c_3 T^3$). The additional factor $K^2 = [(n^2+2)/3]^2$ in the abscissa represents a local field correction and is important only for the glass with the highest absorption ($Tl_2SeAs_2Te_3$). The conclusion to be drawn from fig. 25 is that the matrix element which couples the photons to phonons in this region of the far infrared is not only constant in frequency but also remarkably constant from material to material over a wide variety of glasses.

Bagdade and Stolen (1968) first proposed that local fluctuations in the electronic charge, as considered by Schlömann (1964) and Vinogradov (1960) for disordered crystals, could explain the general features of the temperature-independent (low-temperature) far infrared absorption. These authors suggested that charged defects, such as an oxygen which is not bonded to two silicons, were responsible for the absorption, while others (Wong and Whalley 1971, Whalley 1972) emphasized that the lack of long-range order could provide the coupling dynamically even without specific charged defects.

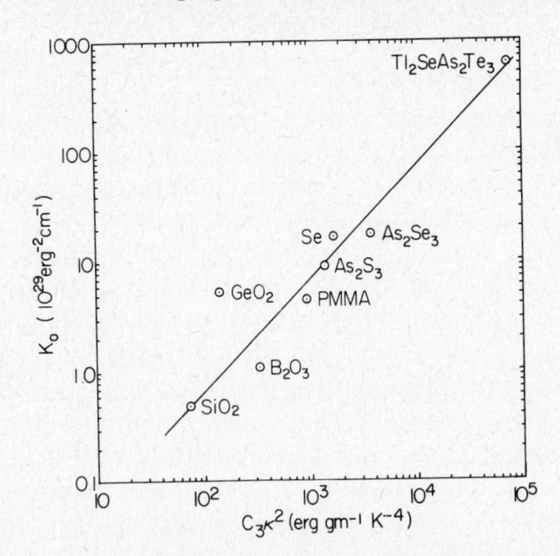

Fig. 25. Magnitude of the temperature-independent component of the far infrared absorption ($n\alpha = K_0(\hbar\omega)^2$ in some amorphous solids as a function of the magnitude of the low-frequency phonon density of states determined from the cubic term in the specific heat ($C_V = C_3T^3$). The factor K^2 is defined in the text (Strom and Taylor 1977).

The frequency dependence of the matrix element which expresses the photon–phonon coupling of the temperature-independent components of α and I_R is still a matter of some debate. The simplest approximation is to assume that all modes are equally effective and that the matrix element is frequency independent as suggested by Shukar and Gammon (1970). More detailed calculations, such as those of Alben et al. (1975) mentioned in §4.1 indicate that the matrix element scales in general as the frequency squared. The "charge fluctuation" model (Whalley and Bertie 1967, Whalley 1972) and the "elastic continuum" model (Martin and Brenig 1974) both predict an ω^2 dependence for the matrix element at low frequencies, but the definition of "low frequency" can depend on specific values for the model parameters. It has also been suggested (Strom et al. 1974, Strom and Taylor 1977) that the procedures of Vinogradov (1960) and Schlömann (1964) can be applied in general to charge fluctuations and not necessarily just to specific charged defects, in which case the matrix element is constant at higher frequencies (10 cm$^{-1} \lesssim \bar{\nu} \lesssim 100$ cm^{-1}) but scales quadratically in frequency at the lowest frequencies ($\bar{\nu} \lesssim 10$ cm^{-1}) where the charge fluctuations interact coherently. Finally, a theoretical argument which suggests an ω^4 dependence for the matrix element at low frequencies has also been presented by Prettl et al. (1973). Although the details of the coupling mechanism remain uncertain, it appears likely from very general considerations (Alben et al. 1975) that at low frequencies the matrix element scales roughly as ω^2, not as ω^4 as suggested by Prettl et al. (1973).

That portion of α and I_R at the lowest frequencies ($\bar{\nu} \lesssim 10\ cm^{-1}$) which is temperature dependent is not at all well understood. Suggested explanations range from highly damped phonon modes (Winterling 1975, and Strom and Taylor 1977, using concepts originally proposed by Fulde and Wagner 1971) and multiphonon processes (Amrhein and Mueller 1968a, b) to relaxation processes involving localized atomic tunnelling modes (Jäckle 1972, Theodorakopoulos and Jäckle 1976). For several reasons the understanding in this area is likely to remain unsatisfactory for some time. Firstly, it is experimentally difficult to map out complete frequency and temperature dependences in the relatively inaccessible spectral range between the microwave and far infrared regions or between typical Brillouin and Raman scattering regimes. Secondly, the absorption (and perhaps also the scattering) is strongly influenced by impurities in this region (Amrhein and Mueller 1968a, b). Finally, the complete frequency and temperature dependences are complex and difficult to characterise by simple theoretical models.

We conclude this discussion with a brief mention of similar low-frequency absorption and scattering observed in amorphous Ge. In fig. 26 are plotted room temperature data for the absorption coefficient $\alpha(\omega)$ (Taylor et al. 1976), the reduced Raman intensity $I_R(\omega)$ (Lannin 1972) and the density of states $n(\omega)$ as determined from neutron scattering measure-

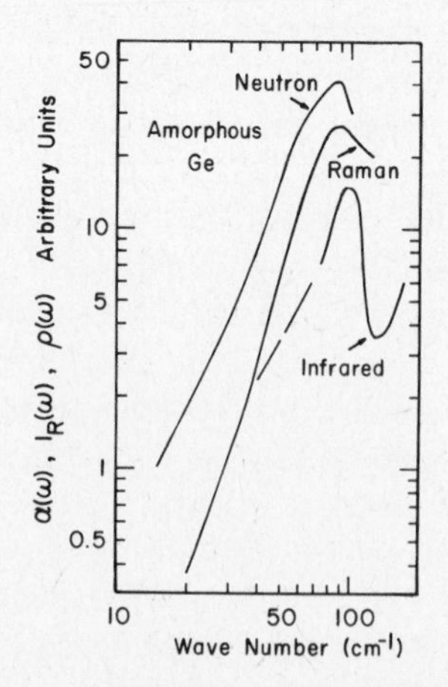

Fig. 26. Phonon density of states as determined from neutron scattering (Axe et al. 1974), reduced Raman spectrum (Lannin 1972) and far infrared absorption (Taylor et al. 1976) for amorphous Ge at 300 K as functions of wave number.

ments (Axe et al. 1974). The most striking feature of the data in fig. 26 is the degree to which both the Raman and infrared results follow the density of states. It has been suggested that the matrix element goes as ω^2 (Lannin 1972, 1973) or as ω^4 (Connell 1974) in amorphous Ge and Si at low frequencies. One should note that the density of states for amorphous Ge scales approximately as ω^4 (and not as ω^2!) in the range 30–60 cm^{-1}, so that comparisons which assume a Debye density of states in this frequency range are bound to be misleading. Because of the significant experimental difficulties involved in measurements of the sort displayed in fig. 26, the behaviour of the matrix element at low frequencies ($<30\ \mathrm{cm}^{-1}$) in amorphous Ge will probably remain a matter of theoretical concern which is unencumbered by experimental verification.

The magnitude of the far infrared absorption in amorphous Ge has been observed (Taylor et al. 1976) to vary with the sample preparation conditions in such a manner that voids have been suggested as responsible for the absorption.

5.2. Low-frequency localized tunnelling modes

The specific heats of amorphous solids at low temperatures are characterised by a term which is linear in the temperature and of nearly constant magnitude from material to material. Because of the universal appearance of this low-temperature specific heat term in amorphous solids it has become commonly accepted that this term represents an intrinsic property of glasses and is not due to impurities.

This behaviour is not understandable in terms of the Debye model, which predicts $C_V \propto T^3$. The "linear term" as it is sometimes called, was explained by Anderson et al. (1972) and by Phillips (1972) using a statistical distribution of localized atomic tunnelling levels. In this model it is assumed that there exist in any glass atoms or groups of atoms which have two alternative equilibrium configurations. There is also assumed to be a broad distribution of energy differences between the two metastable positions.

A number of alternative models have been suggested to explain the low-temperature thermal properties of amorphous solids. In an early attempt Klemens (1965) suggested that inelastic scattering of phonons from spatial fluctuations in the elastic constants could explain the low-temperature behaviour of glasses. More recently electronic states have also been invoked (Kaplan et al. 1971, Redfield 1971). Rosenstock (1972) postulated that atoms trapped in voids or pores of a certain distribution of sizes could account for the experimental evidence. Two additional models rely on postulated peculiarites of the phonon dispersion curves which cause states with small energies to have large wave vectors. Fulde and Wagner (1971) suggested a damping of low-frequency modes which they described macroscopically via an expression for the viscosity with which is associated a characteristic relaxation time. Takeno and Goda (1972) proposed that the dispersion of phonon excitations contains a roton-like part which gives rise to an additional density of low-frequency modes. However, the most successful and most commonly accepted explanation remains the tunnelling model whose essential features we shall now attempt to describe.

We illustrate the essential features of the tunnelling model in fig. 27 where we show schematically the potential energy in terms of an appropriate generalized coordinate r for an atom or group of atoms capable of tunnelling between two metastable equilibrium positions. A plethora of synonyms for these excitations has evolved – tunnelling modes, highly anharmonic modes, configurational rearrangement modes, disorder modes, two-level states or two-level systems, softons, softarons, and bisofterons! With apologies to proponents of the alternative descriptions, we call these excitations "tunnelling modes".

Those tunnelling modes for which the energy barrier Δ of fig. 27 is great enough so that resonant tunnelling between the two minima does not occur, but small enough so that thermal equilibrium can occur on a laboratory time scale, will contribute to the specific heat. Thus for the low-temperature specific heat ($T \lesssim 1$ K) those tunnelling modes for which E

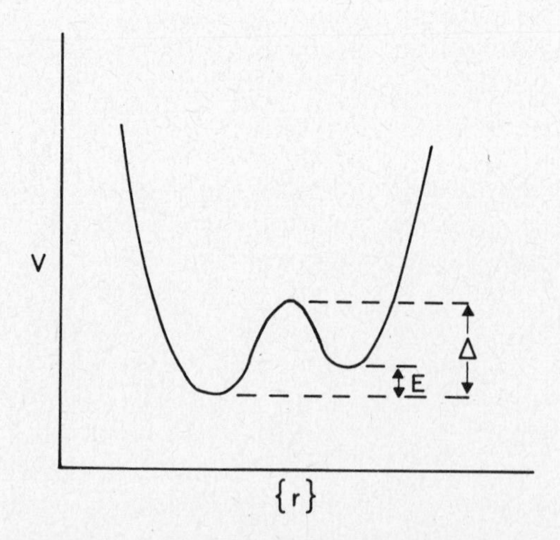

Fig. 27. Schematic diagram of potential energy of a tunnelling mode as a function of generalised coordinate. E is the excitation energy splitting and D is the potential barrier (Reinecke and Ngai 1975).

is approximately equal to kT and Δ is within the required range will provide the dominant contribution. If the distribution of energy differences between the two levels is smooth on a scale of kT, then the specific heat is proportional to T (because the density of tunnelling modes with $E \lesssim kT$ is proportional to T). Thus,

$$C_V = k\int_0^\infty n(E)\left[\left(\frac{E}{kT}\right)^2 \frac{\exp(-E/kT)}{1-\exp(-E/kT)^2}\right] dE$$

$$\propto n(0)\,T,$$

where E is defined as in fig. 27 and $n(0)$ is the assumed constant energy density of the tunnelling modes.

We note there are many modes with small E which do not contribute to the linear term in the specific heat because their energy barriers are too big for tunnelling to occur or because they require the cooperative motion of too many atoms. There are also many modes with larger E whose contributions to the specific heat are masked by the rise of the usual lattice term ($C_V \sim T^3$). In fact, although specific heat may be the most direct measurement of these tunnelling modes, the T^3 term dominates above about 1 K.

The tunnelling mode model also explains the thermal conductivity behaviour described in §3.6. At low temperatures the thermal conductivity varies as T^2 which implies that the phonon mean free path is proportional to ω^{-1}. In disordered systems one expects a mean free path due to phonon scattering which is proportional to ω^{-4} (Zeller and Pohl 1971, Anderson et al. 1972), and this behaviour is actually observed at higher temperatures. The behaviour at lower temperatures (mean free path proportional to ω^{-1}) can be explained in terms of resonant scattering of acoustic phonons off some of the same tunnelling modes invoked to account for the linear term in the specific heat.

At present, the most definitive experimental test of the applicability of the tunnelling model to amorphous solids is an observed saturation in the low-temperature ultrasonic attenuation in glasses (Hunklinger et al. 1972, Golding et al. 1973). The attenuation of ultrasonic phonon pulses due to resonant absorption by the tunnelling modes should saturate at high powers because of the equalization of the occupation of the two levels. Furthermore, the decay length due to resonant absorption (Anderson et al. 1972, Phillips 1972), is essentially proportional to $\omega^{-1}\coth(\hbar\omega/2kT)$ which for phonons of frequency small compared to kT predicts an attenuation proportional to ω^2/T.

In fig. 28 we show the temperature and intensity dependence of the ultrasonic attenuation in glassy SiO_2 at 1GHz (Hunklinger et al. 1973, Dransfeld and Hunklinger 1977). Both the decrease in the attenuation at higher powers and the dependence on T^{-1} at low temperatures and low powers are evident from the figure.

Analogous measurements of the saturation and temperature dependences of the microwave dielectric absorption in SiO_2 have also been performed (von Schickfus and Hunklinger 1977). Fig. 29 shows the low-temperature dielectric loss data for glassy SiO_2 at two different incident powers. The qualitative agreement between the power and temperature dependences of the data of figs. 28 and 29 is apparent. However, in contrast to the constant acoustic absorption observed in samples of differing purities, the coupling to electromagnetic waves is strongly dependent on the impurity content. Von Schickfus and Hunklinger (1977) observed a linear dependence of the electrical coupling with the OH^- content with the known dipole moment for OH^- ions, and this led these authors to suggest that the spatial extent of the tunnelling modes to which the electromagnetic waves couple is $\lesssim 20$ Å. It appears from these measurements that infrared activity is not solely due to the localised tunnelling modes, at least in SiO_2, and that the presence of impurities may be necessary to create infrared activity.

Measurements of thermal expansion in glassy SiO_2 at low temperatures have also been explained in terms of tunnelling modes. Phillips (1972)

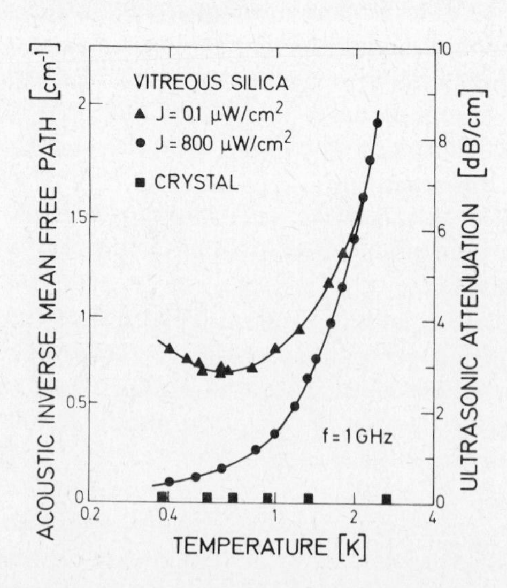

Fig. 28. Temperature dependence of the attenuation of longitudinal acoustic waves of two different intensities in glassy SiO_2 (Hunklinger et al. 1973).

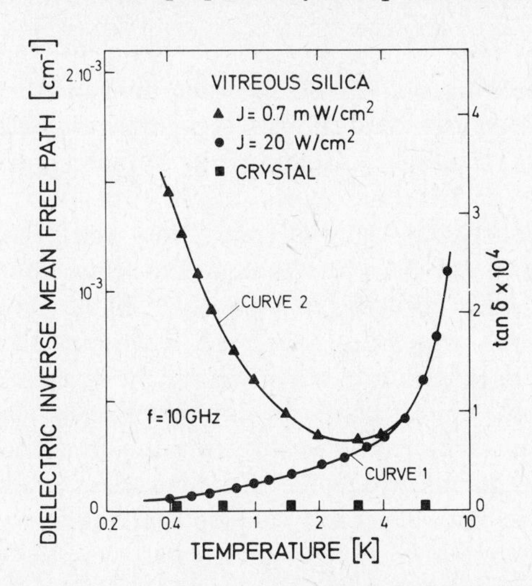

Fig. 29. **Microwave dielectric absorption in glassy SiO_2 as a function of temperature for two different microwave intensities (von Schickfus and Hunklinger 1977).**

initially suggested that tunnelling should give rise to an anomalously large thermal expansion and associated Grüneisen constant γ. Experimental values of γ in SiO_2 reach -40 below 2 K (White 1975) and are comparable to those obtained for known tunnelling impurities in alkali halides.

At higher temperatures ($T > 1$ K) the ultrasonic attenuation in glasses has been ascribed to relaxation processes in two-level systems (Jäckle 1972). The temperature dependences of ultrasonic absorption in a number of glasses have been used in attempts to estimate the density of tunnelling modes above 10 K (Ng and Sladek 1975). The difficulty with this procedure is that some very specific model parameters must be assumed, and the conclusions are very dependent on these assumptions.

The presence of tunnelling modes in some chalcogenide and oxide glasses has also been inferred, albeit rather indirectly, from measurements of nuclear spin–lattice relaxation rates (Rubinstein and Taylor 1974, Rubinstein et al. 1975, Szeftel and Alloul 1975, Rubinstein and Resing 1976). In crystalline solids with nuclei which possess quadrupole moments, nuclear spin–lattice relaxation is well understood and results from first-order Raman processes involving the inelastic scattering of a phonon by the nuclear spin system. Since nuclear spin–lattice relaxation proceeds via vibrational modes, one might expect greater low-temperature spin–lattice relaxation rates in glasses over those in comparable crystalline solids, and in fact increased relaxation rates do appear to be a general feature of disordered materials.

In our view a more promising, although experimentally complex, research area for investigating the detailed nature of the tunnelling modes is the application of pulsed acoustic waves to measure such coherent phenomena as phonon echoes (Golding and Graebner 1976, Black and Halperin 1977) or polarization echoes (Shiren et al. 1977). Phonon echoes, which are the acoustic analogue of spin echoes and photon echoes, have been observed in glassy SiO_2 at temperatures below about 50 mK. Such echoes have not been observed in crystalline solids. On the other hand, polarization echoes, which are observed when an ultrasonic pulse is followed by a pulsed standing wave electric field, have been correlated with OH^- impurities in a number of different glasses. These findings are consistent with our earlier comments concerning ultrasonic and microwave saturation measurements where the ultrasonic data imply a coupling to intrinsic modes which only exhibit a dipole coupling when associated with OH^- impurities.

Fig. 30 shows the dependence of the polarization echo amplitudes on OH^- content in a number of amorphous solids. It is clear from this figure that the amplitude is proportional to the OH^- concentration over a range of two orders of magnitude. The OH^- concentration was extracted from

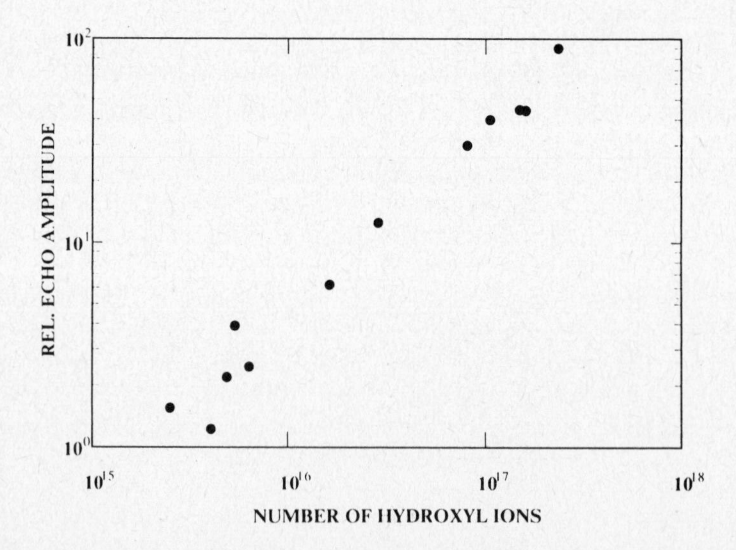

Fig. 30. Polarisation echo amplitudes as a function of number of hydroxyl ions (OH^-) in various amorphous solids. In order of increasing OH^- content these glasses are: As_2S_3 (Servo), As_2S_3 (Ealing), fused SiO_2, As_2S_3 (Ealing), As_2S_3 (Servo), glassy SiO_2, glassy SiO_2, Pyrex, glassy SiO_2 (Suprasil 2), Pyrex, Pyrex, BK-7 (Shiren et al. 1977).

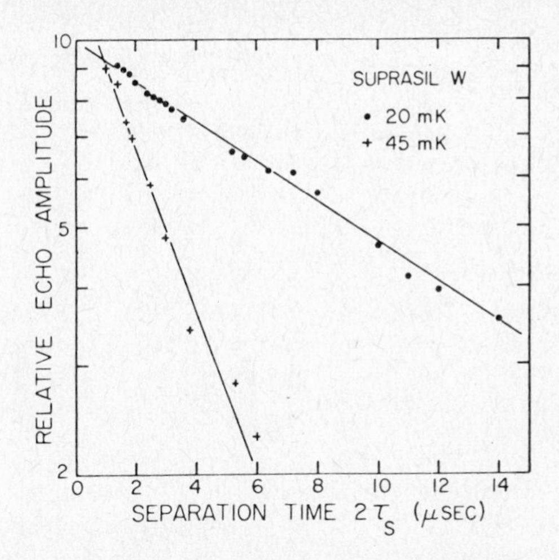

Fig. 31. Decay of the stimulated phonon echoes at 20 and 45 mK as functions of time of separation between the two ultrasonic pulses. The decay time T_2' is 14 μsec and 3 μsec at 20 and 45 mK, respectively (Golding and Graebner 1976).

infrared absorption data (Shiren et al. 1977). An analysis of these data led these authors to postulate the existence of two different two-level systems, only one of which couples to electromagnetic fields, in the amorphous solids studied. Whether or not these two systems are idential to those invoked by Black and Halperin (1977) is not clear.

In the polarization echo experiments the "spin–lattice" and "spin–spin" relaxation times T_1' and T_2' are short compared with the experimentally-necessary pulse widths so that stimulated echoes were not observed. However, stimulated phonon echoes can be observed (Golding and Graebner 1976) from which one can study the decay processes of the tunnelling modes. As an example we show in fig. 31 the decay of a stimulated phonon echo as a function of the time of separation between the two ultrasonic pulses. From these decay curves at 20 and 50 mK one obtains values for T_2'. A detailed interpretation of T_2' is beyond the scope of this discussion (see Black and Halperin (1977) for a detailed theoretical treatment), but qualitatively this quantity represents a dephasing time (or a transverse homogeneous lifetime) for the excited two-level systems which occurs because of an interaction between them. One can also obtain values for the longitudinal relaxation time T_1' from the lifetime of the stimulated echo. This quantity provides a measure of the decay of the excited tunnelling mode via interactions with phonons (see Black and Halperin (1977) and references contained therein).

References

Alben, R., D. Weaire, J. E. Smith and M. H. Brodsky (1975). Phys. Rev. B **11**, 2271.
Altschuler, H. M. (1963), Ed. by M. Sucher and J. Fox (Wiley, New York), p. 330.
Amrhein, E. M. and F. H. Mueller (1968a), J. Am. Chem. Soc. **90**, 3146.
Amrhein, E. M. and F. H. Mueller (1968b), Trans. Faraday Soc. **64**, 666.
Anderson, A. (1971), *The Raman effect* (Marcel Dekker, New York).
Anderson, P. W., B. I. Halperin and C. Varma (1972), Phil. Mag. **25**, 1.
Axe, J. D., D. T. Keating, G. S. Cargill III and R. Alben (1974), in *Proceedings of international conference on tetrahedrally bonded amorphous solids*, Ed. by M. H. Brodsky, S. Kirkpatrick and D. Weaire (American Institute of Physics, New York), pp. 279–283.
Axmann, A., W. Gissler and T. Springer (1969), in *The physics of selenium and tellurium*, Ed. by W. C. Cooper (Pergamon, London), pp. 299–308.
Bagdade, W. and R. Stolen (1968), J. Phys. Chem. Solids **29**, 2001.
Beeman, D. and R. Alben (1977), Adv. in Physics **26**, 339.
Beeman, D., D. Shasha and R. Alben (1976), *Structure and excitations of amorphous solids*, Ed. by G. Lucovsky and F. L. Galeener (AIP, New York), p. 245.
Bell, R. J. (1972), Rep. Prog. Phys. **35**, 1315.
Bell, R. J. and P. Dean (1970), Disc. Faraday Soc. **50**, 55.
Bennett, A. J., C. B. Duke and S. O. Silverstein (1968), Phys. Rev. **176**, 969.
Bernal, J. D. and J. Mason (1960), Nature **188**, 910.
Birgeneau, R. J., J. Cordes, G. Dolling and A. D. B. Woods (1964), Phys. Rev. **136**, A1359.
Black, J. L. and B. I. Halperin (1977), Phys. Rev. B **16**, 2879.
Böttger, H. (1974), Phys. Stat. Sol. (b) **62**, 9.
Brockhouse, B. N. (1961), in *Inelastic scattering of neutrons in solids and liquids* (International Atomic Energy Agency, Vienna).
Brodsky, M. H. (1975), in *Topics in applied physics*, Ed. by M. Cardona (Springer-Verlag, New York), Vol. 8, pp. 205–251.
Brodsky, M. H. and A. Lurio (1974), Phys. Rev. B **9**, 1646.
Brodsky, M. H., M. Cardona and J. J. Cuomo (1977), Phys. Rev. B **16**, 3556.
Carpenter, J. M. and C. A. Pelizzari (1975a), Phys. Rev. **12**, 2391.
Carpenter, J. M. and C. A. Pelizzari (1975b), Phys. Rev. **12**, 2397.
Connell, G. A. N. (1974), in *Proceedings of the twelfth international conference on the physics of semiconductors*, Ed. by M. H. Pilkuhn (Teubuer, Stuttgart), p. 1003.
Connell, G. A. N. and J. R. Pawlik (1976), Phys. Rev. B **13**, 787.
Connell, G. A. N. and R. J. Temkin (1974), in *Proceedings of international conference on tetrahedrally bonded amorphous solids*, Ed. by M. H. Brodsky, S. Kirkpatrick and D. Weaire (American Institute of Physics, New York), p. 192.
Dean, P. (1972), Rev. Mod. Phys. **44**, 127.
Dean, P. and M. D. Bacon (1964), Proc. Roy. Soc. (London) **83**, 1027.
Dolling, G. and R. A. Cowley (1966), Proc. Phys. Soc. (London) **88**, 463.
Donovan, T. M. (1974), in *Proceedings of international conference on tetrahedrally bonded amorphous solids*, Ed. by M. H. Brodsky, S. Kirkpatrick and D. Weaire (American Institute of Physics, New York), pp. 1–17.
Dransfeld, K. and S. Hunklinger (1977), *Proceedings of the 7th international conference on amorphous and liquid semiconductors*, Edinburgh, Ed. by W. E. Spear (CICL, Univ. of Edinburgh).
Edwards, J. T. and D. J. Thouless (1972), J. Phys. C **5**, 807.
Egelstaff, P. H. (1965), *Thermal neutron scattering*, (Academic Press, New York).
Fulde, P. and H. Wagner (1971), Phys. Rev. Lett. **27**, 1280.

Galeener, F. L. and G. Lucovsky (1976), Phys. Rev. Lett. **37**, 1474.
Galeener, F. L., G. Lucovsky and R. H. Geils (1978), Solid State Commun. **25**, 405.
Giaever, I. and H. R. Zeller (1968), Phys. Rev. Lett. **21**, 1385.
Golding, B. and J. E. Graebner (1976), Phys. Rev. Lett. **37**, 852.
Golding, B., J. E. Graebner, B. I. Halperin and R. J. Schultz (1973), Phys. Rev. Lett. **30**, 223.
Gorman, M. and S. A. Solin (1976), Solid State Commun. **18**, 1401.
Greaves, G. N. and E. A. Davis (1974), Phil. Mag. **29**, 1201.
Hadni, A. (1967), *Essentials of modern physics applied to the study of the infrared* (Pergamon, New York).
Hadni, A., J. Claudel, X. Gerbaux, G. Morlot and J. M. Munier (1965a), Appl. Optics **4**, 487.
Hadni, A., G. Morlot, X. Gerbaux, D. Chanal, F. Brehat and P. Strimer (1965b), Compt. Rend. Hebd. Seanc. Acad. Sci. Paris **260**, 4973.
Haydock, R., V. Heine and M. J. Kelly (1972), J. Phys. C **5**, 2845.
Heine, V. and D. Weaire (1970), Solid State Phys. **24**, 249.
Hunklinger, S., W. Arnold, S. Stein, R. Nava and K. Dransfeld (1972), Phys. Lett. **42A**, 253.
Hunklinger, S., W. Arnold and S. Stein (1973), Phys. Lett. **45A**, 311.
Jäckle, J. (1972), Z. Phys. **257**, 212.
Joannopoulos, J. D. and F. Yndurain (1974), Phys. Rev. **10**, 5154.
Kaplan, T. A., S. D. Mahanti and W. M. Hartmann (1971), Phys. Rev. Lett. **27**, 1796.
Keating, P. N. (1966), Phys. Rev. **145**, 637.
Klein, P. B., P. C. Taylor and D. J. Treacy (1977), Phys. Rev. B **16**, 4511.
Klemens, P. G. (1965), in *Physics of non-crystalline solids*, Ed. by J. A. Prins (North-Holland, Amsterdam), p. 162.
Klug, D. D. (1976), in *Structure and excitations of amorphous solids*, Ed. by G. Lucovsky and F. L. Galeener (AIP, New York), p. 229.
Kobliska, R. J. and S. A. Solin (1972), Solid State Commun. **10**, 231.
Kobliska, R. J., S. A. Solin, M. Selders, R. K. Chang, R. Alben, M. F. Thorpe and D. Weaire (1972), Phys. Rev. Lett. **29**, 725.
Ladan, F. R. and A. Zylbersztejn (1972), Phys. Rev. Lett. **28**, 1198.
Lance, A. L. (1964), *Introduction to microwave theory and measurements* (McGraw-Hill, New York).
Lannin, J. S. (1972), Solid State Commun. **11**, 1523.
Lannin, J. S. (1973), Solid State Commun. **12**, 947.
Lannin, J. S. (1977), Phys. Rev. B **15**, 3863.
Laughlin, R. B. and J. D. Joannopoulos (1977), Phys. Rev. B **16**, 2942.
Leadbetter, A. J. (1969), J. Chem. Phys. **51**, 779.
Leadbetter, A. J., D. M. Smith and P. Seyfert (1976), Phil. Mag. **33**, 441.
Long, M., P. Galison, R. Alben and G. A. N. Connell (1976), Phys. Rev. B **13**, 1821.
Lucovsky, G. (1972), Phys. Rev. B **6**, 1490.
Lucovsky, G. (1974), in *Proceedings of the 5th international conference on liquid and amorphous semiconductors*, Ed. by J. Stuke and W. Brenig (Taylor and Francis, London), p. 1099.
Lucovsky, G. and F. L. Galeener (1976), in *Proceedings of the 6th international conference on liquid and amorphous semiconductors*, Ed. by B. T. Kolomiets (Nauka, Leningrad), p. 207.
Lucovsky, G. and J. C. Knights (1974), Phys. Rev. B **10**, 4324.
Lucovsky, G., R. J. Nemanich and J. C. Knights (1979), Phys. Rev. B **19**, 2064.
Martin, A. J. and W. Brenig (1974), Phys. Stat. Sol. (b) **64**, 163.
Martin, D. H. (1967), *Spectroscopic techniques for far infra-red, submillimetre and millimetre waves* (North-Holland, Amsterdam).
Martin, R. M. (1970), Phys. Rev. **1**, 4005.
Meek, P. E. (1976a), Phil. Mag. **33**, 897.
Meek, P. E. (1976b), Phil. Mag. **34**, 767.

Meek, P. E. (1977), J. Phys. C **10**, L59.
Mon, K. K., Y. J. Chabal and A. J. Sievers (1975), Phys. Rev. Lett. **35**, 1352.
Mort, J. (1973), in *Electronic and structural properties of amorphous semiconductors*, Ed. by P. J. LeComber and J. Mort (Academic Press, New York), pp. 475–492.
Moss, S. C., D. L. Price, J. M. Carpender, D. Pan and D. Turnbull (1974), Bull. Am. Phys. Soc. **19**, 321.
Narayanamurti, V. and R. O. Pohl (1970), Rev. Mod. Phys. **42**, 201.
Nemanich, R. J. (1977), Phys. Rev. B **16**, 1655.
Nemanich, R. J., G. Lucovsky, W. Pollard and J. D. Joannopoulos (1978), Solid State Commun. **26**, 137.
Ng, D. and R. J. Sladek (1975), Phys. Rev. B **11**, 4017.
Phillips, W. A. (1972), J. Low Temp. Phys. **7**, 351.
Pohl, R. C., W. F. Love and R. B. Stephens (1974), in *Proceedings of the 5th international conference on amorphous and liquid semiconductors*, Ed. by J. Stuke and W. Brenig (Taylor and Francis, London), Vol. 2, pp. 1121–1132.
Pollard, W. B. and J. D. Joannopoulos (1978), Phys. Rev. B **17**, 1770.
Prettl, W., N. J. Shevchik and M. Cardona (1973), Phys. Stat. Sol. (b) **59**, 241.
Redfield, D. (1971), Phys. Rev. Lett. **27**, 730.
Rehr, J. J. and R. Alben (1977), Phys. Rev. B **16**, 2400.
Reinecke, T. L. and K. L. Ngai (1975), Phys. Rev. B **12**, 3476.
Robinson, L. C. (1973), *Physical principles of far-infrared radiation* (Academic Press, New York).
Rosenstock, H. (1972), J. Non-Cryst. Solids **7**, 123.
Rubinstein, M. and H. A. Resing (1976), Phys. Rev. B **13**, 959.
Rubinstein, M. and P. C. Taylor (1974), Phys. Rev. B **9**, 4258.
Rubinstein, M., H. A. Resing, T. L. Reinecke and K. L. Ngai (1975), Phys. Rev. Lett. **34**, 1444.
Schlömann, E. (1964), Phys. Rev. A **135**, 413.
Sen, P. N. and M. F. Thorpe (1977), Phys. Rev. **15**, 4030.
Shevchik, N. (1974), in *Tetrahedrally bonded amorphous semiconductors*, Ed. by M. H. Brodsky, S. Kirkpatrick and D. Weaire (AIP, New York), p. 72.
Shiren, N. S., W. Arnold and T. G. Kazyaka (1977), Phys. Rev. Lett. **39**, 239.
Shuker, R. and R. W. Gammon (1970), Phys. Rev. Lett. **25**, 222.
Smith, J. E., Jr., M. H. Brodsky, B. L. Crowder, M. I. Nathan and A. Pinczuk (1971a), Phys. Rev. Lett. **26**, 642.
Smith, J. E., Jr., M. H. Brodsky, B. L. Crowder and M. I. Nathan (1971b), in *Proceedings of the second international conference on light scattering in solids*, Ed. by M. Balkanski (Flammarion, Paris), p. 330.
Smith, P. M., A. J. Leadbetter and A. J. Apling (1975), Phil. Mag. **31**, 57.
Solin, S. A. (1977), in *Proceedings of international conference on structure and excitations in amorphous solids*, Ed. by G. Lucovsky and F. L. Galeener (AIP, New York), p. 205.
Spear, W. E. (1977), Advan. Phys. **26**, 811.
Steinhardt, P., R. Alben and D. Weaire (1974), J. Non-Cryst. Solids **15**, 435.
Stephens, R. B. (1973), Phys. Rev. B **8**, 2896.
Stimets, R. W., J. Waldman, J. Lin, T. S. Chang, R. J. Temkin and G. A. N. Connell (1973), Solid State Commun. **13**, 1485.
Stolen, R. H. (1970), Phys. Chem. Glasses **11**, 83.
Strom, U. and P. C. Taylor (1977), Phys. Rev. B **16**, 5512.
Strom, U., J. R. Hendrickson, R. J. Wagner and P. C. Taylor (1974), Solid State Commun. **15**, 1871.
Szeftel, J. and H. Alloul (1975), Phys. Rev. Lett. **34**, 657.

Takeno, S. and M. Goda (1972), Prog. Theor. Phys. (Kyoto) **48**, 1468.
Taylor, P. C., U. Strom, J. R. Hendrickson and S. K. Bahl (1976), Phys. Rev. B **13**, 1711.
Theodorakopoulos, N. and J. Jackle (1976), Phys. Rev. B **14**, 2637.
Thorpe, M. F. (1973), Phys. Rev. B **8**, 5352.
Truell, R., C. Elbaum and B. B. Chick (1968), *Ultrasonic methods in solid state physics* (Academic Press, New York).
Vinogradov, V. S. (1960), Fiz. Tverd. Tela **2**, 2622 (Sov. Phys. – Solid State, **2**, 2338 (1960)).
Von Heimendahl, L. and M. F. Thorpe (1975), J. Phys. F **5**, L87.
Von Schickfus, M. and S. Hunklinger (1977), Phys. Lett. **64A**, 144.
Weaire, D. and R. Alben (1972), Phys. Rev. Lett. **29**, 1505.
Weber, W. (1974), Phys. Rev. Lett. **33**, 371.
Whalley, E. (1972), Trans. Faraday Soc. **68**, 662.
Whalley, E. and J. E. Bertie (1967), J. Chem. Phys. **46**, 1264.
White, G. K. (1975), Phys. Rev. Lett. **34**, 204.
Wilkinson, J. H. (1965), *The algebraic eigenvalue problem* (Clarendon Press, Oxford).
Wilson, J. I. B. and D. Weaire (1978), Nature, **275**, 93.
Winterling, G. (1975), Phys. Rev. B **12**, 2432.
Wong, J. and C. A. Angell (1974), Appl. Spectrosc. Rev. **4**, 155.
Wong, J. and C. A. Angell (1976), *Vitreous state spectroscopy* (Marcell Dekker, New York).
Wong, P. T. T. and E. Whalley (1971), Disc. Faraday Soc. **50**, 94.
Wronski, C. R., D. E. Carlson and R. E. Daniel (1976), Appl. Phys. Lett. **29**, 602.
Yamanaka, M. (1976), Rev. Laser Engng **3**, 253.
Zachariasen, W. H. (1932), J. Am. Chem. Soc. **54**, 3841.
Zallen, R., M. L. Slade and A. T. Ward (1971), Phys. Rev. B **3**, 4257.
Zeller, R. C. and R. O. Pohl (1971), Phys. Rev. B **4**, 2029.

CHAPTER 2

Computer Experiments and Disordered Solids

W. M. VISSCHER and J. E. GUBERNATIS

Theoretical Division
Los Alamos Scientific Laboratory
Los Alamos, New Mexico 87545, USA

Dynamical Properties of Solids, edited by
G. K. Horton and A. A. Maradudin

Contents

1. *Introduction*

The use of large digital computers has played an important role in the development of our current understanding of the dynamics of disordered solids. This use occurs in two basic ways: the first is the evaluation of awkward and complicated expressions resulting from approximate theories; the second is the performance of a "computer experiment" whereby a simple simulation of a disordered solid is generated and its various dynamical properties are directly and exactly controlled and analyzed. In this survey of the dynamical properties of disordered systems, we focus attention on the second usage.

Computer experiments on disordered solids usually have one or more of the following attributes: (1) the physical system is simulated by a small idealized model; (2) the random structure is simulated by the computer; (3) the equations of motion are solved numerically; (4) the problem is formulated so that many repetitions of similar arithmetic operations solve it; (5) the computer performs averages over random ensembles.

A more precise definition is impossible; one man's computer experiment may be another's analytic theory. The boundary is fuzzy; there is a large area where the label is a matter of taste. Fortunately, a precise definition is unimportant. It will become clear from the examples which follow what our tastes are.

Disordered solids themselves have certain distinguishing attributes and are generally grouped into two different classes. One class is the substitutionally disordered solid in which various stochastic parameters (random masses, random transfer integrals, etc.) are in one-to-one correspondence with the points or bonds of some periodic lattice. The second class is a structurally disordered solid where there is no long-range order. In addition to possible disorder in various dynamic parameters, there may be disorder in the topology of the lattice. This type of disorder is called amorphous or glass-like.

The development of an understanding of the dynamics of disordered solids is a challenging task because many familiar concepts and techniques developed to study crystalline solids are inapplicable. The disorder destroys the translational invariance on which such notions as Bloch's theorem, unit cells, Brillouin zones, conservation of crystal momentum, etc. are based. The simplifications provided by these ideas vanish; the complexities of a system with an infinite number of particles reappear.

In the study of such complex systems, computer experiments are sometimes used as substitutes for real, laboratory experiments. Theorists often use simple model systems that correspond only loosely to real systems; calibration of theory by laboratory experiments is seldom possible. One of the important uses of computer experiments is to provide benchmarks for the calibration of analytic theory. This role of computer experiments is exemplified in the development of the various versions of the coherent potential approximation (CPA). Analytic bounds on the accuracy of these approximations are seldom estimated, and computer experiments have provided the only direct assessment of their fidelity.

Additionally, laboratory experiments cannot determine the positions of lattice sites of a structurally disordered solid. Pair distribution functions are determined, but structures with different internal details can have identical pair distribution functions. Model amorphous structures can be constructed by hand, but in recent years the generation of these structures has become a new use of the computer.

Too many computer experiments are reported in the literature to mention in a review of modest length. Our choice of topics is governed by our knowledge, our interests and the contents of other recent reviews. With respect to prior reviews, we mention those by Lieb and Mattis (1966), Hori (1968), Dean (1972), Bell (1972a, 1976), Kirkpatrick (1973), Ishii (1973), Elliott et al. (1974), Thouless (1974), Barker and Sievers (1975), Elliott and Leath (1975), Taylor (1975), Cargill (1975), Visscher (1976), Weaire (1976), and Binder and Stauffer (1979).

In this review well-established and much-used methods are sparingly covered, as are topics which are adequately covered already. Subjectivity ultimately dominates; methods and calculations which are intriguing or familiar to us occupy the bulk of the chapter.

The result of this philosophy is, we hope, a chapter more readable than otherwise. The disadvantage is that the contents are far from being comprehensive; for this we apologize, specifically to the many computer experimenters whose important work we might have slighted.

More specifically, we discuss methods for generating compositionally and structurally disordered solids and numerical techniques for computing the physical properties of a disordered solid. Then various physical properties are discussed, including density of states, localization and electrical and thermal transport. We conclude with comments and suggestions for future areas of study.

2. *Computer generation of random solids*

To do computer experiments on a disordered solid, we must first have a model for the disordered structure. That is, we need information about the positions and orientations of the various constituent atoms or molecules. This information may be obtained by *ansatz*, from experiment, from model building (by hand or by computer) or by some combination of the above. Once we have modeled the disorder, then computer experiments may be performed, yielding electronic, optical, magnetic and vibrational observables.

Disordered solids are generally classified as either *compositionally disordered* (still crystalline or microcrystalline, with long-range order) such as random alloys and doped semiconductors or *structurally disordered* (amorphous) such as the random network of glasses and the randomly packed hard sphere model for metals. Often the classification of a structure is equivocal. Some substances often considered to be amorphous include microcrystalline materials with domain size of atomic dimensions and materials which are crystalline but have a high density of dislocations, voids and other defects (Nakamura et al. 1973). Recent reviews of the structure of amorphous solids are Bell (1972a), Dean (1972), Cargill (1975) and Weaire (1976).

2.1. Compositional disorder

To generate a model for a disordered AB alloy, one needs only to fill the sites on a prescribed lattice sequentially with A or B atoms, according to whether a computer-generated pseudo-random number $0<q\leqslant 1$ is less than or greater than C_A, the fractional concentration of A atoms. This procedure must be appropriately modified when lattice sites are inequivalent or there are more than two constituents, as in mixed crystals (e.g. NaCl–KCl). The model is most suitable at high temperatures when the probability of an A atom occupying a given site becomes independent of the occupation of neighboring sites.

More difficult are the situations in which the probability that atom A will occupy an available site depends on the occupancy of nearby sites. Because the A–A, A–B and B–B interactions are usually different, this is generally the physically interesting situation, but often is not the case studied.

2.1.1. *Percolation and random alloys*

If A and B are isotopes of the same atom, then the simplest method of generating the model lattice is appropriate. For example, solid H with some D mixed in or vice versa would be a meaningful realization of a theoretical ideal, an isotopically disordered lattice. Even for an uncorrelated, compositionally disordered alloy, there is microstructure associated with percolative phenomena that can exert considerable influence on the physical properties of a system. For example, the vibrational spectrum of an isotopically disordered harmonic lattice distinctly changes as the concentration of one component or the other passes through a critical value, identified as the critical percolation probability (Payton and Visscher 1967b).

In the present context percolation studies are concerned with a periodic array on which the probability C_A of an A atom occupying a site is independent of the occupancy of other sites. Two types of percolation problem are studied. The *site* percolation problem is concerned with the statistics of clusters of A atoms. The *bond* percolation problem is concerned with the statistics of lattice sites connected by the same kind of bond. The bond percolation problem is isomorphic to a site percolation problem, generally on a different lattice; the converse is not always true. Two useful quantities are the site percolation probability which is the ratio of the number of sites in the largest cluster to the total number of sites, and the critical site percolation concentration which is the largest value of C_A for which no infinite cluster of A atoms exists, in the limit of an infinite lattice. Similar quantities may be defined for the bond percolation problem.

Many studies of percolative systems investigate the percolation probability, critical percolation concentration and cluster statistics. A variety of computer experiments have been done starting with the studies of Dean and Bird (1967) on cluster statistics on various lattices. Analog computations of conduction in percolative systems were done by Last and Thouless (1971) who punched holes in conductive paper to simulate hopping conductivity in 2d, by Watson and Leath (1974, 1975) who randomly cut wires in a 137×137 metal screen and by Adler et al. (1973) who connected points on a $16\times16\times16$ cubic lattice with 10 K resistors, omitting a random set. Recent computer studies include one by Kirkpatrick (1976) who found the site percolation probability for cubic lattices in 3, 4, 5 and 6 dimensions. Some of Kirkpatrick's results are shown in fig. 1. The critical percolation concentrations are estimated for infinite N by extrapolation. Sur et al. (1976) performed a Monte Carlo calculation on a $50\times50\times50$ simple cubic lattice and found that the critical site percolation concentration is 0.3115 ± 0.0005. Similar computer experiments have been performed by Onizuka (1975).

The nature of clusters in the percolative system is also important. The shape of clusters, called "lattice animals" by Domb (1976), was studied by

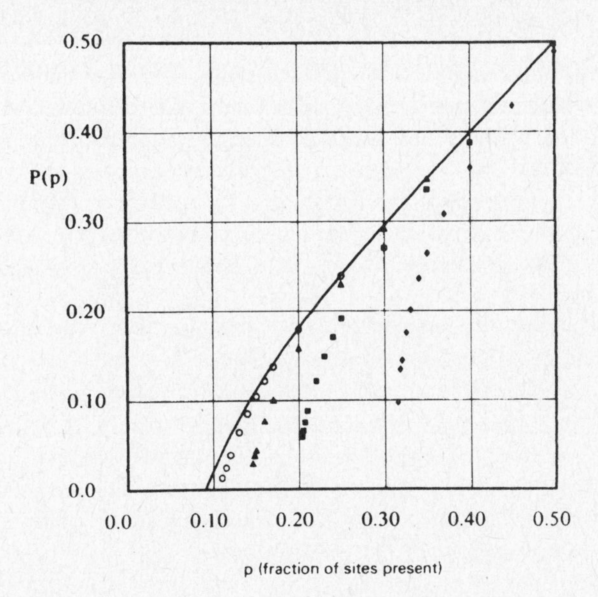

Fig. 1. Site percolation fractions $P(p)$ computed by a computer experiment on simple cubic lattices in 3d (diamonds), 4d (squares), 5d (triangles) and 6d (circles). Each point is an average of 10 samples each containing up to 3×10^6 sites. The solid line is the exact result for percolation on a Bethe lattice (see fig. 9) with the same coordination number, 12, as the 6d simple cubic lattice (Kirkpatrick 1976).

computer experiment by Domb et al. (1975), by Sur et al. (1976) and by Leath (1976). Leath performed a computer experiment on a 2d lattice and concluded that, close to percolation threshold, the "fractal dimensionality" of the boundaries of the largest clusters is the same as the cluster dimensionality. The evidence for this statement was that the ratio of number of boundary sites to number of cluster sites is empirically gaussian-distributed with the width of the gaussian vanishing for very large clusters and the mean approaching a constant independent of cluster size.

Probably the most efficient existing algorithm for keeping track of cluster statistics in a random AB lattice is that given by Hoshen and Kopelman (1976b) who applied it to exciton migration and other problems (Hoshen and Kopelman 1976b, Hoshen et al. 1977, Argyrakis and Kopelman 1977). They experimented on systems with about 10^7 sites, producing the statistics in less than 30 sec on an Ahmdal 470 computer. Their procedure has the computer make a single pass, assigning cluster labels to each A atom in a way that keeps account of cluster coagulation, i.e. clusters which are unconnected early in the computer's pass through the lattice can develop connections later on.

Percolation calculations, thus performed, yield uncorrelated lattices and cluster statistics, which can be used for dynamical numerical experiments. For $T \to \infty$ these models are special cases of the disordered arrays which we will discuss below. They are also of special interest to the experimenter concerned with economy because they can be generated so quickly and easily.

Computer experiments have been performed on a variety of 2d and 3d lattices by Pike and Seager (1974) with applications to conductivity and diffusion reported by Seager and Pike (1974) and Pike et al. (1974). Wintle and Williams (1977) have considered percolation in random networks, both analytically and computer-experimentally.

There is an immense literature concerned with percolation theory, applications and numerical experiments. Electric conduction in inhomogeneous media, hopping conduction in disordered lattices, fluid flow in porous materials, exciton migration and vibrational localization are some of its physical applications. (See Kirkpatrick 1973, and Böttger and Bryskin 1976a for reviews.)

2.1.2. *Non-equilibrium configurations and quenched alloys*

The situation is more complicated if the AA, AB and BB interactions are different, and $T < \infty$. Then the structure depends, in general, on the history of the system as well as the interactions and the temperature, and perhaps the only reliable way to determine it is by a full-fledged numerical experiment. Examples have been reported in a series of papers (Bortz et al. 1974, Mano et al. 1975, Rao et al. 1976, Sur et al. 1977) for 2d and 3d. They start ($t < 0$) with a randomly ($T = \infty$) disordered lattice of A and B atoms which interact with their nearest neighbors with an energy $U_{ij} = Jn_i n_j$, where $n_i = \pm 1$ according to whether the atom at site i is an A or a B atom. J may be either positive or negative. The system is quenched at $t = 0$ to some finite temperature $\beta^{-1} = kT$, which may be greater or less than the critical temperature T_c for this lattice gas (an Ising system). The time evolution is determined by stipulating that the rate of exchange of atoms between nearest-neighbor sites i and j is

$$P_{ij} = \alpha \exp(-\beta \Delta U_{ij})[1 + \exp(-\beta \Delta U_{ij})]^{-1},$$

where ΔU_{ij} is the change in U_{ij} which would result from the exchange and α is an attempt rate. The quenched system is in an unstable state, relaxing toward its equilibrium state, which for $T < T_c$, has separate A and B phases and for $T > T_c$ has some distribution of cluster sizes. The relaxation time is proportional to α^{-1}. Because many alloys are metastable, the structures at intermediate times are as interesting as the equilibrium structures. An example of a structure for $J<0$ for a two-dimensional 80×80 lattice with 20% A atoms is shown in fig. 2, from Rao et al. (1976). With a $T=\infty$ random array, quenched to $T=0.59\,T_c$ at the start, this configuration was realized in 1.5×10^6 exchanges of nearest neighbors. Obviously far from equilibrium, the system is changing slowly. The existence of large A islands reflects the low temperature; for $T>T_c$ there would be none at this concentration.

The rate of approach to equilibrium in these calculations is critically dependent, of course, on proximity to T_c, and the nature of the intermediate states depends on concentration and on the sign and magnitude of J. For $J<0$ the initial random percolation clusters grow. For $J>0$ they shrink.

A number of similar calculations are in the literature. Ogita et al. (1969) made a computer experimental calculation (§2.2.1) of a 2d Ising system (isomorphic to the AB alloy problem) and found configurations like that in

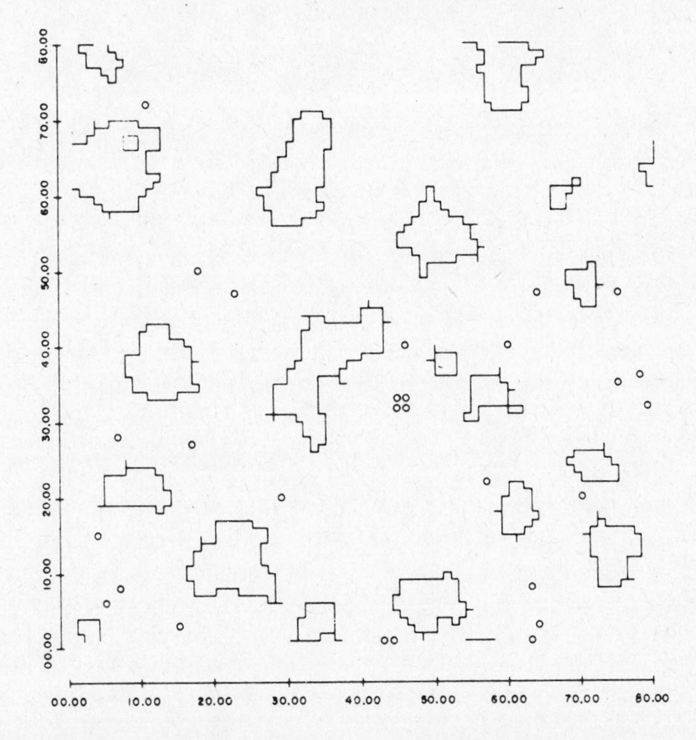

Fig. 2. Snapshot of a quenched 80×80 lattice at $T=0.59\,T_c$ with 20% A atoms after 1.5×10^6 exchanges. A possible structure of a 2d binary alloy (Rao et al. 1976).

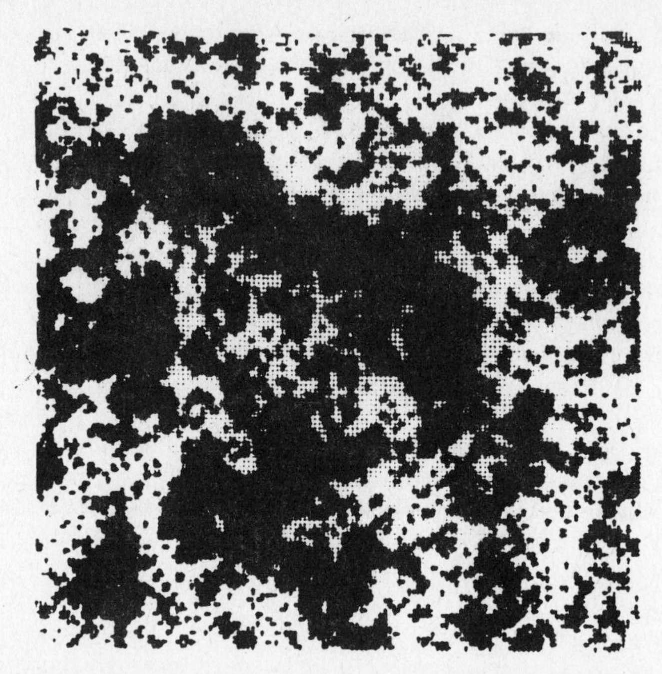

Fig. 3. Spin configurations in a 128 × 128 Ising system for T slightly higher than T_c. There is no net polarization; there are as many up spins (bright spots) as down spins (dark spots). The clusters grow as $T \to T_c$ (Ogita et al. 1969).

fig. 3. Other studies include Flinn (1974) who displays configurations and pair correlation functions as functions of time, Schneider and Stoll (1976) who find evidence for the existence of traveling cluster waves, and Binder (1974a) who treats a square lattice with a gradient in the chemical potential which produces a macroscopic phase separation at late times.

2.1.3. *Equilibrium configurations and annealed alloys*

By carrying the numerical experiment just described to long enough times, one can generate equilibrium configurations. We note that the quenching calculation just described is computationally equivalent to the Monte Carlo method for performing numerical equilibrium averages (see §2.2.1), and when ergodicity is assumed, is also physically equivalent.

This equivalence is underlined in an instructive way in a paper of Binder (1974b), where 2d and 3d binary alloys (and the Heisenberg model) are studied near the phase transition point. As one approaches a critical point, the rate of convergence of the Monte Carlo procedure becomes very slow, physically corresponding to the long-time-scale of critical fluctuations. If T is appreciably larger than T_c, the number and size distribution of clusters is not dominated by cooperative effects, and the probability of a given site being occupied by an A atom depends mostly on the population of the nearest-neighbor sites or perhaps (especially in the case of mixed crystals) on the occupation of next-nearest neighbors.

In one dimension these correlations are of the same nature as the correlations between the successive elements of a Markoff chain. A first-order Markoff chain of A and B atoms is generated according to the rule that the probability that the atom at site $n+1$ is an A atom depends only on the identity of the atom at site n and on the concentration of A atoms. In a second-order Markoff chain it also depends on the identity of the atom at site $n-1$. Such chains are generated on a computer quickly and easily; they have been discussed by Painter and Hartmann (1974) who calculated vibrational spectra and pair-correlation functions for such systems. An interesting result they obtained is that two second-order Markoff chains with identical pair-correlation functions may have quite different spectra. First-order Markoff chains of A and B atoms have been studied by Payton and Visscher (1967a, b). For a 50–50 binary chain, they showed, for example, that, as one goes from the chain in which unlike atoms are completely inhibited from being neighbors through the completely disordered case with uncorrelated neighbors to the ordered binary chain with unlike nearest neighbors, the vibrational spectrum changes from the superposition of two monatomic spectra to the disordered spectrum to the two-branched spectrum characteristic of the ...ABABA... lattice.

Another method which goes a step beyond an alloy generator is one used recently by Walker and Walstedt (1977) to generate the ground state of a spin glass. It has similarities both to the Monte Carlo method and to the molecular dynamics method, but is designed to produce a single state rather than an equilibrium ensemble. A classic example of a spin glass is a dilute alloy of Mn in Cu, the magnetic atom being embedded in a non-magnetic host lattice. The interaction between magnetic moments is ferro or antiferromagnetic depending on separation. Walker and Walstedt generate the structure by first randomly distributing the Mn atoms on the lattice and then randomly orienting their spins. Next the impurity Mn spins are twisted one by one so they become parallel to the local field generated by all the other impurity spins. It is clear that the energy of the system decreases at each step; Walker and Walstedt find convergence to a minimum is speeded by over-twisting each spin at each step by a certain fraction. The minimum they find, incidentally, is very soft. A spin rotated from its equilibrium position feels only a small restoring torque.

2.2. Structural disorder

By a structurally disordered material, an amorphous solid or a glass (terms often used interchangeably), we mean a material which lacks significant long-range crystalline structure. Atomic structure and the nature of the chemical bonds (ionic or covalent) determine the nature of the amorphous structure. If the substance is a metal, then its amorphous form might be expected to be similar to that generated by a random close-packing of hard spheres, because after the valence electrons are stripped off into the extended conduction bands, what is left is usually a core or closed-shell of tightly bound electrons. Experimentally measured radial distribution functions (RDF's) support this view; namely, the interatomic interactions are central two-body forces (along the line connecting centers) and can in many cases be approximated by hard spheres.

If, on the other hand, the solid has elements with strong asymmetries like a glass (SiO_2 being the basic structural unit) or a semiconductor such as Si or Ge which favor tetrahedral bonds, then clearly a hard sphere simulation is inappropriate, and the situation is considerably more complicated.

2.2.1. Dense random-packed hard sphere (DRPHS) simulations

Although random packing of hard spheres has a long history and has been the subject of much study by pure and applied mathematicians, ceramists, physicists and others (mostly not in the context of solid state physics), computer simulations are not yet completely adequate. DRPHS simulations have been reviewed by Finney (1975, 1977), by Cargill (1975), and succinctly by a correspondent in Nature (Anonymous 1972). The generation and description of the properties of a compacted random array of hard spheres has an intuitive appeal, perhaps because it is familiar, but it has successfully resisted the efforts of analysts and geometrical probabilists.

Compacted arrays have been produced in the laboratory by shaking beakers of ball-bearings, kneading wads of them inside balloons, or sticking them together with wax or paint and then by laboriously measuring coordinates and contacts as well as their overall density ρ. For an optimally compacted pile of monosize ball-bearings (Scott and Kilgour 1969), the measured density, when extrapolated to infinite size to eliminate boundary effects, is $\rho_0 = 0.6366$. The density of an ordered close-packed array (either h.c.p. or f.c.c.) is $\pi/\sqrt{18} = 0.74048\ldots$. C. A. Rogers (1958) has lamented the impotence of the analyst vis-à-vis the packing problem: "While many mathematicians believe and all physicists know that density cannot exceed $\pi/\sqrt{18} = 0.74048\ldots$," the best rigorous limit that has been derived is $\sqrt{18}\ (\arccos 1/3 - \pi/3) = 0.77963\ldots$.

Computer experiments have been designed to simulate DRPHS in several ways. The simplest, and to the present time the most successful, is sequential addition of spheres to a random cluster of spheres according to some simple rule. Several calculations have been performed. Some (for example, Bennett 1972, Matheson 1974) start with a seed of three or more spheres and build the DRPHS according to the rule that each added sphere is placed at the three-point contact ("pocket") closest to the center of gravity of the original seed. This algorithm results in packings which are not homogeneous, as one might expect, but have density higher near the center than elsewhere. The asymptotic density (extrapolated from arrays of several thousand spheres) is 4 or 5% lower than the experimental value ρ_0. The fact that each sphere, once it settles in its pocket, is not moved again underlines the serious limitation of this kind of numerical simulation. Collective rearrangements, by which the piles might be compacted, once formed (by jiggling the container, squeezing and kneading the balloon), are not simulated by these programs. We mention below some calculations which do, to a limited extent and in a very different way, allow some rearrangements to compact the system.

Bennett's algorithm and modifications of it have been used by many workers including Sadoc et al. (1973) and Cargill and Kirkpatrick (1976). In these references the effect of binary mixtures of slightly different atomic sizes is considered. Such a structure has been relaxed (i.e. the spheres softened) by Von Heimendahl (1975), improving agreement with experiment.

Another computer experiment with a similar addition algorithm, but with a undirectional, rather than a central, force to hold the spheres to their three-point contacts, was performed by Visscher and Bolsterli (1972). Their program was general enough to handle arbitrary distributions of sphere sizes and allow small spheres to filter down into the interstices between larger ones and was implemented in two as well as three dimensions. They started with a horizontal base divided into periodic bins with each bin containing a certain number of balls fixed in position. The total number could be zero or be enough to form a complete monolayer; the spheres could be either all the same size or a random distribution of sizes. Balls were then added by choosing one at random from the given distribution, dropping it from a random position above the bin and allowing it to roll under the influence of the undirectional gravity until it stopped at the first stable three-point contact (two points in 2d). Because a given ball will seldom find the lowest pocket in the bin, this algorithm, for the case of monosize spheres, gave a loose array ($\rho = 0.582$). In another simulation the density was increased to 0.60 by dropping several balls at each trial and keeping only the lowest position. This program was used to simulate structures of binary alloys, composed of mixtures of spheres of two different radii. One interesting result they obtained in the 2d simulation was that if a random heap was built up on an uneven base, the resulting structure was divided into crystalline domains. Fig. 4 shows a typical pile. A search for evidence of corresponding effects in 3d (mostly by examination of surfaces of section and of RDF's) was negative.

Some avenues less thoroughly explored, but perhaps more promising than the sequential addition algorithm, could be called Monte Carlo and molecular dynamical approaches. These methods exploit the fact that the structures sought are limiting cases of liquid states with $kT/pV \to 0$.

By Monte Carlo (MC) methods we mean application of the scheme invented in 1953 (Metropolis et al.) for calculating many-dimensional phase space integrals (specifically, equilibrium averages of dynamical variables) by sampling the value of the integrand at points along a random walk in phase space, with the steps in the random walk being generated according to the following rule: Choose a sphere in the array (at random, or according to sequential labels) and tentatively move it a distance d in an

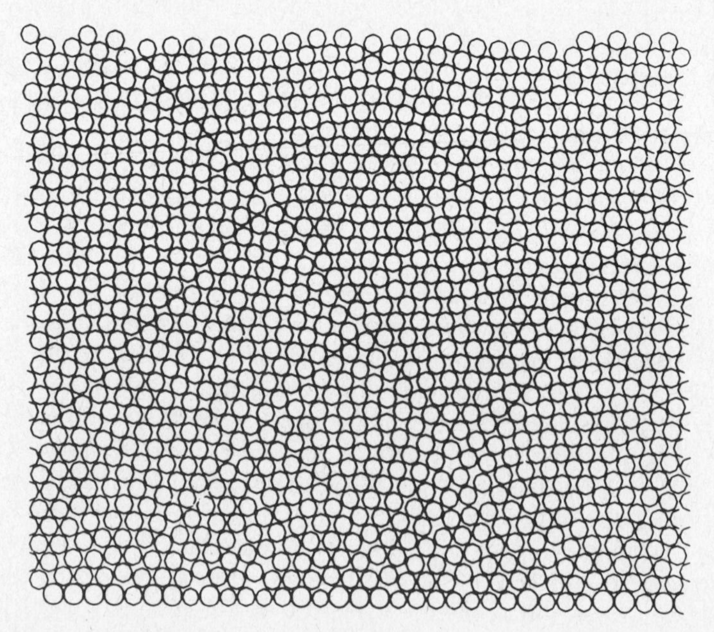

Fig. 4. Two-dimensional DRPHS array, illustrating domain structure (Visscher and Bolsterli 1972).

arbitrary direction, determined by a pseudo-random number generator. Calculate $P=\exp(-\beta\Delta V)$, where $\beta=(kT)^{-1}$ and $\Delta V=V(d)-V(0)$ is the change of the potential energy of the system caused by the move (in the hard sphere system, of course, ΔV is either infinity or zero). If $P\geqslant 1$, accept the move. If $P<1$, choose a random number $0<q\leqslant 1$, and accept the move if $q<P$, otherwise reject it. Repeat the procedure for the next sphere, etc. The parameter d is generally chosen to be smaller than the average separation, so that a reasonably large fraction of trials is accepted. Metropolis et al. (1953) proved that the average of these samples converges to the equilibrium average. The rate of convergence is discussed by Wood (1975) and recently by Valleau and Whittington (1977).

A calculation like this must be done for finite pressure and temperature and therefore for densities less than the DRPHS value. But it can be done for a sequence of densities; then $x=kT/pV$ may then be plotted against density and extrapolated to $x=0$. This analysis was carried out by LeFevre (1972) using both MC and molecular dynamics data. He found that the compacted density at which $p\rightarrow\infty$ is, within uncertainties, equal to 0.64, the Scott–Kilgour value! Other properties of the amorphous ensemble such as RDF can be extrapolated in this way also. MC calculations for hard spheres have been performed by Wood and coworkers (see Wood 1968), by Hoover and Ree (1968) and others. Kincaid and Weis (1977) recently analyzed MC computer experiment data of Weis (1974) to characterize the RDF of a hard sphere solid. Their MC procedure started with a crystalline array; however, they gave results for densities almost up to the f.c.c. value, but their aim was not to study the amorphous solid. Their method is in principle incapable of giving information about an amorphous solid in a reasonable amount of computer time. As would be expected, their RDF for $\rho=0.6366$, when the DRPHS structure is possible, look instead like smeared-out f.c.c. RDF's.

By molecular dynamics (MD) is meant the solution by computer of Newton's equations of motion, which in the present context, would be for a large set of spheres. For hard sphere interactions the spheres move in straight lines between collisions, which makes the computer's task of finding collision times mostly bookkeeping. Equilibrium properties can be determined by MD and the nature of the DRPHS array deduced by extrapolation. Computing time expenditures in the MC and MD approaches are comparable.

MD calculations have been done by many workers for a variety of systems: hard spheres (references may be found in Wood and Erpenbeck 1976) and 6–12 (Lennard–Jones) potentials (see also Hoover et al. 1971, Streett et al. 1974, Raveché et al. 1974). Many of these works make manifest the fact that the structure of the amorphous solid phase depends on its history. When the system is initialized in a crystalline form it will not change, in any reasonable time, to the glassy form or vice versa, unless it is heated, cooled or compressed. The high-density equation of state computed by Hiwatari et al. (1974), in fact, showed three distinct branches, one crystalline and two amorphous. Rahman and his collaborators (Rahman et al. 1975, Mandell et al. 1977), using 500-particle systems with 6–12 interactions, have observed crystal nucleation. Damgaard Kristensen (1976) has recently reported computer experiments on a 336-particle 6–12 system of the thermodynamic and structural properties of non-crystalline arrays and their dependence on quenching rate.

Another method for studying DRPHS has been devised by Stillinger et al. (1964). It bears some resemblance to MC and MD calculations, but is specifically aimed at the compacted system and hence should be expected to be more efficient. Their method, which has been applied only to a 2d system, starts with a random distribution of points in the plane, and lets each point expand into a circle whose radius a increases until two circles contact. Further expansion is allowed by moving the centers of the two circles apart, keeping the center of gravity fixed. When more than two circles are in contact, the rule for expansion is generalized in a simple way. When no further expansion is permitted by the boundaries of the system, then either all the circles are being squeezed by the walls or by each other, or some are caught in cages formed by a compact ring of circles. The number in the latter condition will depend on the initial choice of random position; some will lead to no loose discs and these may be DRPHS configurations. Some choices, conversely, will lead to regular hexagonal arrays.

Although the method of Stillinger et al. is as yet untried in 3d, because it has no intrinsic inhomogeneity or anisotropy (as all of the sequential addition algorithms do), it is a promising candidate for a DRPHS computer algorithm.

Similar to the Stillinger prescription, but perhaps easier to apply, is one invented by Bernal (1959) and developed by Mason (1967). Instead of continuously increasing the sphere radii, this algorithm stipulates that the radii increase by discrete steps. Each increase may generate overlaps; the overlaps are removed by moving each member of a pair of overlapping spheres apart, whether or not further overlaps are generated or aggravated. This process is repeated for the whole system, which has free boundaries, until all overlaps are removed, whereupon the radii are increased another step. Only small arrays (500 spheres) have been generated this way; the results seem promising.

Some workers (Kaplow et al. 1968, Renninger et al. 1974, Rechtin and Averback 1975) have used still a different criterion for acceptance of

Monte Carlo steps. They reject a step if it does not lead to an improvement in the agreement of the computed RDF with the experimentally observed one. Acceptable agreement is obtained after about 1000 steps per atom.

The arrays generated by each of the DRPHS algorithms just described are significantly different from each other. Even the RDF's, the similarity of which is a necessary, but not a sufficient, condition for the similarity of other characteristics of a structure, are often quite dissimilar. To choose from the various algorithms the most appropriate one to apply in a given physical situation, one must, we think, appeal to an ontogenetic criterion. Namely, the prescription should be chosen so that it simulates, as closely as practicable on the computer, the actual physical growth process. For example, in the event that amorphous clusters are grown from a supercooled gas or liquid, the Bennett or Adams–Matheson rule might be applicable. For surface deposition the Visscher–Bolsterli method might be most appropriate, and the MD, MC, Bernal–Mason, and Stillinger prescriptions may be closest to the physics of a quenched melt. (For a discussion of some of the physics involved in amorphous film deposition, see Shevchik 1973b.) Much closer study of amorphous structures, both in the real world and in the computer models, is needed.

2.2.2. *Coordinated random networks (CRN)*

In the pioneering paper of Zachariasen (1932) on the structure of glass, it was made clear that there is a large class of amorphous materials in which the local atomic arrangement is strictly tetrahedrally coordinated, yet without long-range order. Since then many attempts have been made by diverse methods to construct arrays which simulate glass-like structure.

Although these structures are of great current interest for use in calculations of vibrational, electrical and magnetic properties of tetrahedrally bonded amorphous materials, the computer experiment has been less useful for their generation than it has been for the DRPHS arrays because the building blocks are far from spherically symmetric. On the atomic level, the reason for the structural difference between DRPHS and coordinated random network (CRN) is the bonding mechanism determined by the electronic structure of the atoms. If the bonds are highly directional (either covalent or ionic), then CRN's may be formed if crystalline structures are, for some reason, not favored. In the case of many metals, where the valence electrons form extended bands and the closed-shell cores are spherical, the amorphous form is well represented by DRPHS. Clearly the construction, either by computer or otherwise, of CRN's is much harder than DRPHS arrays.

It is known from X-ray diffraction measurements that the tetrahedral bond lengths vary little from their crystalline values and the bond angles are fairly sharply distributed in many amorphous solids. This suggests that analog computer modelling of CRN's might start with sticks and tetrahedrally drilled balls to replace the bubble rafts and bags of ball bearings which were useful in DRPHS. This was done, with considerable success, by Bell and Dean (1972) and others and has been repeated with many variations in constituents, construction algorithms and size (number of atoms). Polk (1971), Polk and Boudreaux (1973), Greaves and Davis (1974) and Long et al. (1976) contain examples of four-, three- and two-coordinated hand-built structures.

These hand-built models have usually contained about 500 units. This number is probably a compromise among several factors, including human patience, gravitational distortion of the model, unwieldiness and expense of parts. Besides, the study of a smaller array would not bear significantly on the bulk properties of amorphous substances because even in a 500-atom model the number of surface atoms, those with exterior dangling bonds, nearly equals the number of interior atoms. The effects of finite size can be minimized in a number of ways: One is to impose periodicity. A second is to cut out smaller clusters from a large one and to average the results (Tong and Choo 1976). A third is to attach Bethe lattices to each unsatisfied external bond (Joannopoulos and Cohen 1976).

It has been suggested by Little (1973), but apparently not yet implemented, that model-building by hand could be computer-aided with the help of interactive graphics techniques whereby models could be built without balls and sticks, without physical existence outside of the computer memory and the cathode ray tube display, but would have coordinates available for inspection, alteration and calculation. This contrasts with the awkwardness of the situation with real models where coordinates must be laboriously measured one by one with laser beams (Polk and Boudreaux 1973) or more pedestrian instruments.

The basic procedure followed by a model builder is to seed a few atoms and with certain rules or construction algorithms build up the CRN. The rules are based on observation of real structures, assumed temperatures and some physical insight (see Shevchik 1974, for a discussion of relevant factors), but ultimately each model builder will get a different model. Almost by definition, the random network cannot be uniquely specified.

Once the model is built and its coordinates are stored in the computer memory (if they are not there already), it is relatively easy to improve it (reduce its energy) by a relaxation program. Without changing the bond topology, one performs a computer minimization of an expression for the potential energy, for example, a bilinear form (Keating 1966) in coordinate differences. (An example of a procedure like this for a two-coordinated convoluted chain is Long et al. 1976; see Steinhardt et al. 1974 for a four-coordinated model.)

The utility of any structure is determined by how well it agrees with structure-dependent observables, the simplest of which are the RDF and DOS. Some comparisons of RDF's of different hand-built models with and without computer relaxation with each other and with experiment are made by Steinhardt et al. (1974). The peaks in the RDF are sharpened by relaxation, but it is not clear that agreement with experiment is improved.

A step beyond computer relaxation has been taken by Beeman and Bobbs (1975). They start with a hand-built model, randomize its coordinates, reassign bonding neighbors, relax it, then reiterate. Procedures which use rules for successive addition of atoms to a cluster that eliminate the need for intervention of human model-builders beyond the seed construction have been described by Duffy et al. (1974), by Shevchik (1973a), and by Shevchik and Paul (1972). Guttman (1975) and Henderson and Herman (1972) have generated CRN's automatically by systematically restructuring crystalline and random arrangement, respectively, with periodic boundary conditions. Cargill and Kirkpatrick (1976) have proposed a method for generating CRN's by supplementing a program for DRPHS (using sequential addition) with a "tetrahedral perfection" criterion.

A comparison of the predictions of some of the models described here has been made by Alben et al. (1975b). They conclude with a caution against taking quantitative details of the numerical experiments too seriously. RDF's do not uniquely specify a CRN. They must be supplemented by other characterizations, such as numbers of 4, 5, 6,... fold rings, which affect intermediate-ranged order.

A computer experiment that closely simulated the physical formation of a CRN would be ideal if it could be done well enough. The sequential deposition algorithms and those which restructure random or crystalline periodic arrays to form CRN's cannot be justified from first principles to the extent that some of their counterparts in the DRPHS problem can. Also molecular dynamics and Monte Carlo experiments are much more difficult when the elementary units (atoms or molecules) are asymmetric. Stillinger (1975) in his review of water theories gives an account of accomplishments in that field. Nothing comparable has been done for four-coordinated random networks, which should have much in common with amorphous ice (Alben and Boutron 1975).

We have not yet mentioned the possibility of simulating CRN's by ld models. It might appear that a harmonic chain with nearest-neighbor force constants chosen according to a continuous probability distribution could have glassy properties. Lu et al. (1974), however, have shown that a simple transformation casts the equation of motion for this system into those for the chain with disorder only in the mass.

Topologically inequivalent to the nearest-neighbor chain, however, is the model invented by Bell (1972b). He simulates a CRN in ld by allowing each atom in the chain to interact with four others, chosen randomly. These chains possess disorder similar to that present in real glasses.

3. *Numerical methods*

Determination of the structure of a disordered solid is sometimes an end in itself, as in the analysis of elastic neutron and X-ray scattering, but more often it is a first step in the determination of dynamical properties. Frequently, in those calculations one is led to sets of coupled differential equations which describe the time evolution of the system. In the simplest cases these equations are linear and first- or second-order in time derivatives; so after Fourier analysis they reduce to a coupled set of linear algebraic equations.

As discussed by Elliott et al. (1974), various physical systems are described by coupled linear equations of the same form. Coupled linear equations are clearly obtained in lattice dynamical problems in the harmonic approximation. They also arise in tight-binding approximations to electronic wave functions, in exciton dynamics and in spin excitations. For example, in a disordered harmonic lattice, one has $-m_i\omega^2u_i = \Sigma_{j\neq i}k_{ij}(u_j - u_i)$, where m_i and u_i are the mass and displacement about site i and k_{ij} is the force constant between atoms i and j. In the tight-binding approximation for an s-band, one has $(E-\varepsilon_i)a_i = \Sigma_{j\neq i}V_{ij}a_j$, where ε_i and a_i are the atomic level (site energy) and wave function amplitude at site i, and V_{ij} is the overlap (sometimes called hopping or transfer) integral between sites i and j.

We now review several tools which are available to the computer experimentalist who needs to cope with these systems of linear equations. A vast amount of work has been done; our selection cannot be comprehensive.

The solution of each class of problem listed above is equivalent to the diagonalization (finding eigenvectors and eigenvalues) of matrices (almost always hermitian). For most purposes knowledge of all eigenvalues and eigenvectors is superfluous; often only certain averages over matrix elements or just the density of states (DOS) is needed. Each method which we describe is better suited to certain problems than others; the observables which are sought often determine which method is best.

The first method, the negative eigenvalue method of Dean and Martin, has been adequately covered in recent reviews (Dean 1972, Bell 1972a); we will discuss it only briefly. The second is the recursion method (Haydock et al. 1972). The third is the equation-of-motion method (Alben and Thorpe 1975, Alben et al. 1975c).

The negative eigenvalue method, sometimes called the negative factor-counting method (NFC), is the most used simply because it is by far the oldest. For certain problems the others give results, especially in two and three dimensions, with less labor and fewer demands on computer hardware.

All three methods are applicable to the analysis of various properties of matrices. Hermiticity is not always required, but for simplicity and because physically interesting matrices are usually hermitian, we assume it.

3.1. The negative factor-counting method

Suppose we have an $N \times N$ matrix M, and we wish to find its eigenvalue spectrum $G(\lambda)$, which is the number of eigenvalues α,

$$M\chi_\alpha = \alpha\chi_\alpha$$

less than λ. The eigenvalues are the roots of the determinantal equation

$$\mathrm{Det}(M-\alpha) \equiv \mathrm{Det}\, L = 0.$$

In terms of the spectral density $g(\lambda)$, one has

$$G(\lambda) = \int_{-\infty}^{\lambda} g(\alpha)\, \mathrm{d}\alpha.$$

The NFC follows from the fact that if L is partitioned

$$L^{(N)} = \begin{bmatrix} L^{(n)} & Y \\ Y^{\mathrm{T}} & Z^{(m)} \end{bmatrix},$$

where $L^{(n)}$ is $n \times n, Z^{(m)}$ is $m \times m$ with $m+n=N$, then

$$\mathrm{Det}\, L^{(N)} = \mathrm{Det}\, L^{(n)}\, \mathrm{Det}\, L^{(m)}, \tag{3.1}$$

where

$$L^{(m)} = Z^{(m)} - Y^{\mathrm{T}}\left[L^{(n)}\right]^{-1} Y. \tag{3.2}$$

In particular, if we take $n=1$, (3.1) reduces the rank of L by one, and it can be applied $N-1$ times to obtain

$$\mathrm{Det}\, L^{(N)} = \prod_{m=1}^{N} l_{11}^{(m)}(\alpha),$$

where $l_{11}^{(m)}(\alpha)$ is the (1,1) element of $L^{(m)}$. The diagonal elements of $Z^{(m)}$ each contain a term $-\alpha$, which for large α dominates other contributions, so as $|\alpha|\to\infty, l_{11}^{(m)}(\alpha)\to-\alpha$ for all m. Thus exactly one of the $l_{11}^{(m)}(\alpha)$'s must change sign at each eigenvalue because (3.1) must change sign N times between $\alpha=-\infty$ and $\alpha=+\infty$, and therefore $G(\lambda)=$number of negative $l_{11}^{(m)}(\alpha)$'s, $m=1,2,\ldots,N$.

In the ld case of a harmonic chain with nearest-neighbor forces only, the row matrix Y has only one non-vanishing element; so (3.2) becomes trivial, with $L^{(m)}$ and $Z^{(m)}$ different only in the (1,1) element. It is this fact that eliminates almost completely the need for computer storage in the ld problem, and it is the reason very long chains ($\sim 10^6$ atoms) have been used. In the 2d and 3d cases the reduction process (3.2) requires one to store one large matrix (up to $N\times N$ in the general case) and the limit on N depends on the computer hardware available. In most practical problems M is banded, so the storage requirements are less onerous.

This procedure is formally identical to the reduction process of gaussian elimination. An error analysis has been made by Wilkinson (1965), and the method is accurate except if an $l_{11}^{(m)}$ becomes very small. (They are the divisors in (3.2).) If this happens, one can simply choose a slightly different value of α.

Examples of the application of NFC abound in the literature. Pre-1972 computations are reviewed in Dean (1972) and Bell (1972a).

Fig. 5 shows the DOS obtained by NFC for a disordered simple cubic lattice with $6\times6\times40=1440$ atoms (Payton and Visscher 1967a), of which ten percent, randomly selected, have one-third the host mass. This figure is an example of the detail the NFC can yield.

Fig. 6 shows the vibrational spectrum of an SiO_2 model of about 330 atoms calculated by Bell et al. (1968) by the NFC method. This DOS is in qualitative agreement with a variety of neutron scattering and optical measurements (Bell 1976).

The NFC method has been applied to disordered systems other than lattice dynamical ones. For example, a number of calculations have been done on disordered magnets wherein the DOS has been obtained by NFC. (See Huber 1974a, b and references contained therein.)

NFC is pre-eminent in the application to which it was first applied, the isotopically disordered harmonic chain. The length of the chain is limited only by one's computer budget, not by memory capacity. Fig. 7, for example, shows the DOS computed by Dean (1972) using the NFC for randomly disordered chains composed of equal parts of two different masses. The lower histogram is the spectrum of a chain of 250000 atoms; the upper one, 32000 atoms. The differences are only statistical. Dean's review, as well as that of Payton and Visscher (1967b) contain discussions

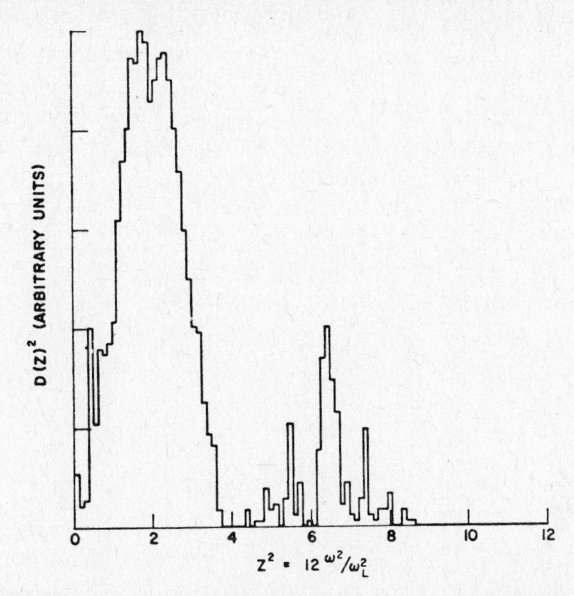

Fig. 5. The DOS for a 6×6×40 simple cubic isotopically disordered lattice, with a 10% admixture of light atoms one-third as heavy as the host. Periodic boundary conditions were used in the short dimensions; fixed in the long dimension (Payton and Visscher 1967a).

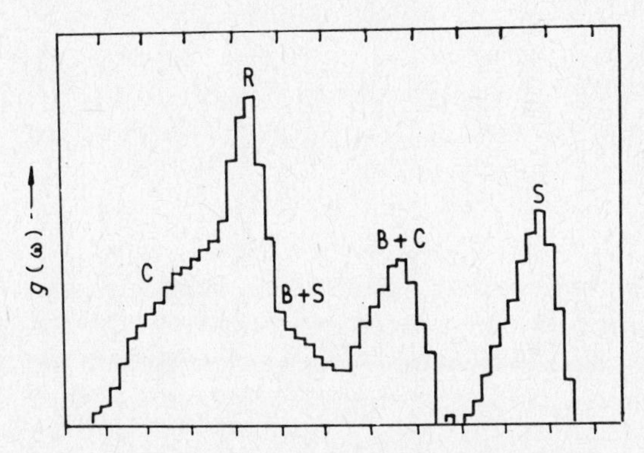

Fig. 6. Vibrational spectrum of a 330-atom model for SiO_2, computed by NFC with fixed boundary conditions (Bell et al. 1968).

of these spectra and an identification of a number of the peaks in the impurity band with local structures in the chain. (See §4.1.)

NFC does not directly give information about eigenvectors. Eigenvalues, however, can be obtained approximately by successive bisection of intervals in α or by simple refinements of that method. Given an approximation to an eigenvalue, the corresponding eigenvector can be generated with any

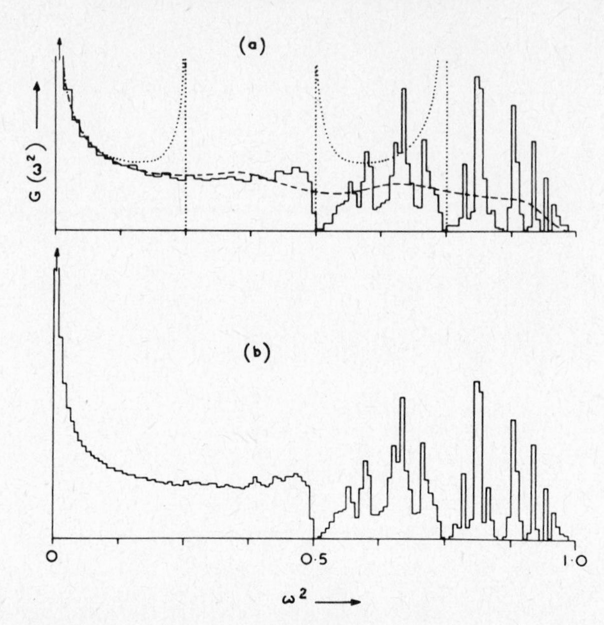

Fig. 7. Computed DOS for binary isotopically disordered chains with equal parts of mass 1 and 2. The chain (a) has 32000 atoms; (b) has 250000. The length of the chain is not limited by computer memory; a 2d or 3d system does require matrix storage and will be limited in size (Dean 1972).

desired accuracy by inverse iteration. That is, if α is an approximate eigenvalue (closer to the desired eigenvalue than to any other), the sequence $w_0, w_1, w_2, \ldots$ generated by

$$w_\nu = (M-\alpha)^{-1} w_{\nu-1},$$

with w_0 arbitrary, will generally converge to the eigenvector corresponding to the eigenvalue to which α is an approximation. The closer α is to the eigenvalue and the better choice one makes for w_0, the quicker the convergence. (See Bell 1976 for a more sophisticated exposition.)

3.2. The recursion method

A new method for finding densities of states, as well as other matrix elements of the resolvents and wave functions, has been developed in the past few years by Haydock et al. (1972, 1975). Its motivation was that since many observable properties of amorphous and disordered solids depend mostly on local structure, it ought to be possible to formulate a theory in which local effects are explicitly separated from bulk averages. The theory may be presented in the following way. It is closely related to the Lanczos method (Wilkinson 1965).

3.2.1. *The spectral density*

Take again our hermitian matrix M, but now consider its resolvent operator

$$R(\lambda)=(M-\lambda)^{-1},$$

where λ has an infinitesimal positive imaginary part. Among the properties of M which may be determined by analyzing R are the spectral density

$$g(\lambda)=\pi^{-1}\operatorname{Im}\operatorname{Tr}R(\lambda)$$

and the "local" or "effective" density of states

$$g_\chi(\lambda)=\pi^{-1}\operatorname{Im}\langle\chi|R(\lambda)|\chi\rangle \qquad (3.3)$$

with respect to some vector $|\chi\rangle$. We will assume $|\chi\rangle$ is normalized, so that

$$g(\lambda)=\sum_\chi g_\chi(\lambda)$$

if the sum is over a complete orthogonal set.

Now suppose that in some representation M is block-tridiagonal,

$$M=\begin{bmatrix} A_1 & B_2 & 0 & \dots & & \\ C_1 & A_2 & B_3 & 0 & & \\ 0 & & & \ddots & B_n \\ & & & C_{n-1} & A_n \end{bmatrix}. \qquad (3.4)$$

The A_k's are $n_k\times n_k$ square matrices and $n_1+n_2+\dots+n_n=N$. Then B_k is $n_{k-1}\times n_k$, C_{k-1} is $n_k\times n_{k-1}$, and since we have assumed a hermitian M,

$$B_{k+1}=C_k^\dagger.$$

There is a matrix identity (Butler 1973) for inverting block matrices of the form (3.4). For example, the upper left diagonal $n_1\times n_1$ block of $R(\lambda)$ is

$$\begin{aligned} R_1(\lambda)&=(A_1-\lambda-B_2D_1C_1)^{-1},\\ D_1&=(A_2-\lambda-B_3D_2C_2)^{-1},\\ D_2&=(A_3-\lambda-B_4D_3C_3)^{-1},\\ &\ \vdots\\ D_{n-1}&=(A_n-\lambda)^{-1}. \end{aligned} \qquad (3.5)$$

This relation can be verified by repeated application of the identity

$$P^{-1}=\begin{bmatrix}(P/U)^{-1} & -R^{-1}S(P/R)^{-1}\\ -U^{-1}T(P/U)^{-1} & (P/R)^{-1}\end{bmatrix} \tag{3.6}$$

for the inverse of the block matrix

$$P=\begin{bmatrix}R & S\\ T & U\end{bmatrix}, \tag{3.7}$$

where

$$(P/U)=R-SU^{-1}T$$

is the Schur complement of U in P. Eq. (3.6) can be checked by multiplying it by (3.7).

If all the blocks in (3.4) happen to be 1×1, then (3.5) says that $R_{11}(\lambda)$ is an Nth order continued fraction, which is easy to calculate. To exploit this fact we must find a tridiagonal representation for M, in other words a basis set $\{|\chi_j\rangle\}, j=1,\ldots,N$ such that

$$M|\chi_j\rangle=a_j|\chi_j\rangle+b_{j+1}|\chi_{j+1}\rangle+c_{j-1}|\chi_{j-1}\rangle. \tag{3.8}$$

Any hermitian matrix can be reduced by a unitary transformation to a tridiagonal one, and this is the first step in many eigenvalue–eigenvector algorithms. The reduction is not unique; in the recursion method it is done as follows:

Start with (3.8), set $|\chi_0\rangle=0$, assume an arbitrary $|\chi_1\rangle$, then $|\chi_2\rangle$ satisfies

$$b_2|\chi_2\rangle=(M-a_1)|\chi_1\rangle.$$

If we require $\{|\chi_j\rangle\}$ to be an orthonormal set, i.e. $\langle\chi_i|\chi_j\rangle=\delta_{ij}$, then (3.8) implies that for $j\geqslant 1$

$$\begin{aligned}a_j&=M_{jj},\\ b_{j+1}&=c_j^*=M_{j+1,j},\end{aligned} \tag{3.9}$$

where

$$M_{ij}=\langle\chi_i|M|\chi_j\rangle.$$

So given M and an arbitrary vector $|\chi_1\rangle$, one can generate the basis of a tridiagonal representation $\{|\chi_j\rangle\}$ and the corresponding tridiagonal matrix elements $\{a_i, b_i, c_i\}$. An orthogonal basis set in N dimensions has only N members; therefore, the recursion (3.8) is guaranteed to terminate with $|\chi_{N+1}\rangle = 0$.

Once $\{a_i, b_i, c_i\}$ is generated recursively, the local density of states (3.3) can be directly evaluated from eq. (3.5),

$$R_{11}(\lambda) = \cfrac{1}{a_1 - \lambda - \cfrac{|b_2|^2}{a_2 - \lambda - \cfrac{|b_3|^2}{\ddots \; a_{N-1} - \lambda - \cfrac{|b_N|^2}{a_N - \lambda}}}}. \tag{3.10}$$

Eq. (3.10) is exact and is conveniently evaluated on a computer, but if N is large, it is unnecessarily cumbersome. One can, instead of evaluating the entire set $\{a_i, b_i, c_i\}$, consider M to be tridiagonal in the first n elements, the remainder M' being a single $(N-n) \times (N-n)$ block:

$$M = \begin{bmatrix} a_1 & b_2 & 0 & \cdots & & \\ c_1 & a_2 & b_3 & 0 & \cdots & \\ 0 & & \ddots & a_n & b_{n+1} \cdots & \\ \vdots & & & c_n & \boxed{} & \\ & & & \vdots & \boxed{M'} & \end{bmatrix} \tag{3.11}$$

The $(1, 1)$ element of the resolvent of (3.11) is, from (3.5), exactly

$$R_{11}(\lambda) = \cfrac{1}{a_1 - \lambda - \cfrac{|b_2|^2}{a_2 - \ddots \; - \cfrac{|b_n|^2}{a_n - \lambda - |b_{n+1}|^2 R'_{11}(\lambda)}}}, \tag{3.12}$$

where $R'(\lambda)$ is the resolvent $(M' - \lambda)^{-1}$. Thus the effect of all further iterations of the recursion relation (3.8) can be absorbed into $R'_{11}(\lambda)$, which is the first diagonal element of $R'(\lambda)$, i.e. $R'_{11}(\lambda) = \langle \chi_{n+1} | R'(\lambda) | \chi_{n+1} \rangle$. Computationally this is attractive because n can be much less than N.

Physically, it is attractive because M is usually a sparse local matrix. For example, in a lattice-dynamical problem or a tight-binding electronic problem in the site-representation, only matrix elements of M corresponding to near neighbors are non-zero. This means that the number of sites represented in $|\chi_n\rangle$ increases like n^3 in a three-dimensional system. If $|\chi_1\rangle$ is a localized basis vector, then it is intuitively clear that the properties of the local environment will be described by the first few $|\chi_j\rangle$ and that it would be reasonable to terminate the recursion (3.8) at some $j=n\gg 1$ with the contributions from iterates with $j\geqslant n+1$ replaced by an "effective medium" self-energy-like term such as $R'_{11}(\lambda)$ in (3.12).

Eq. (3.12) is still exact; it becomes approximate only when we guess, on physically or mathematically convenient grounds, an expression for $R'_{11}(\lambda)$. If, for example, the $\{a_i, b_i, c_i\}$ settle down, i.e.

$$a_j\approx a, \qquad b_j\approx b, \qquad c_j\approx c, \qquad j>n, \tag{3.13}$$

then for $N\to\infty$, $R'_{11}(\lambda)$ is the infinite continued fraction

$$R'_{11}(\lambda)=\cfrac{1}{a-\lambda-\cfrac{|b|^2}{a-\lambda-\cfrac{|b|^2}{\ddots}}}$$

$$=\frac{1}{a-\lambda-|b|^2R'_{11}(\lambda)},$$

whose solution is

$$|b|R'_{11}(\lambda)=x-\sqrt{x^2-1}\,, \tag{3.14}$$

with $x=(a-\lambda)/2|b|$. So (3.12) affords a physically appealing and numerically tractable way to separate the local "cluster" effects $j\leqslant n$ from those of the "medium" or "environment" for $n<j<\infty$. $R'_{11}(\lambda)$ is real if $|x|\geqslant 1$. Then $R_{11}(\lambda)$ is real also, except at its n singularities which will have imaginary δ-function parts corresponding to discrete localized states. On the other hand, if $|x|<1$, then $R'_{11}(\lambda)$ is complex, as is $R_{11}(\lambda)$. $R_{11}(\lambda)$ is non-singular for λ real, but can have a resonance structure. Physically, this means that resonances lie within the band of the imbedding medium, so they can decay by exciting it. The rate of decay, the width of the in-band resonance mode, is determined by the imaginary part of $R'_{11}(\lambda)$. These are generalizations of the single-impurity resonance modes, the simplest examples of which are the modes associated with a heavy (or weakly-bound) impurity in a lattice which were predicted many years ago by Brout and Visscher (1962) (see also Kagan and Iosilevskii 1963). Their ubiquity in disordered crystals has since been manifested through optical, thermal and vibrational properties. (See Barker and Sievers 1975 for a recent review.)

It is intuitively clear and is borne out by numerical results that the larger n, the less effect a particular choice of $R'_{11}(\lambda)$ has on results for any local observable, including the density of states. If n is big enough, the choice of $R'_{11}(\lambda)$ is unimportant. But other possibilities besides (3.13) exist. For example, the $\{a_j, b_j, c_j\}$ might converge to a limit cycle in which one set of values is approached when j is even, another set when j is odd. It is easy to show that this is the case for the alternating binary chain, and it has been shown numerically that this also happens in the closely related method of moments approach to a two-banded electronic density of states problem (Gaspard and Cyrot-Lackmann 1973).

It is also possible that a higher-order limit cycle might be attained if the medium has a multiple-banded spectrum, with the order equal to the number of bands. If the positions of the band edges are known, as is often the case for systems with dilute impurities, the imaginary part of $R'_{11}(\lambda)$ can be chosen so its edges coincide with them to speed convergence of the recursion method.

An example of calculations of a DOS calculated with the recursion method is shown in fig. 8. Here are compared the results of an exact calculation with recursion method calculations with $n=5$ and $n=20$ with the terminating $\{a_j, b_j, c_j\}$ equal to the last calculated ones in each case.

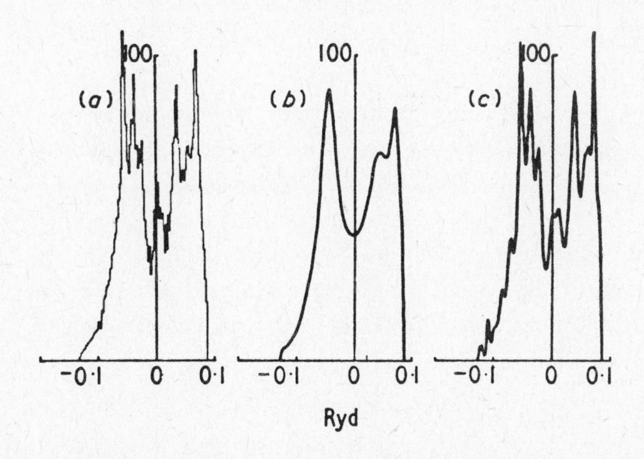

Fig. 8. Illustration of the convergence of the recursion method for a periodic system, the f.c.c. d-band DOS. (a) is an exact histogram for the infinite lattice. (b) and (c) are from the recursion method with $n=5$ and $n=20$, respectively. The terminating (a, b, c) is the last one calculated in each case (Haydock et al. 1975).

Although this system is an ordered f.c.c. d-band, with 2000 atoms in the cluster, the cluster boundary is reached in less than 10 iterations and the results do seem to improve for $n=20$. Curve (a) in fig. 8 is the DOS for the infinite periodic lattice; the recursion method seems to converge to it nicely.

3.2.2. *Eigenvectors and eigenvalues*

Once $\{a_j, b_j, c_j\}$ is determined from the recursion (3.8), it is easy to find the eigenvectors. Let $|\psi_\lambda\rangle$ be an eigenvector of M with eigenvalue λ; expanded in the complete set $\{|\chi_j\rangle\}$ it is

$$|\psi_\lambda\rangle = \sum_{j=1}^{N} P_j(\lambda)|\chi_j\rangle. \tag{3.15}$$

Now, using (3.8) it follows that

$$\lambda P_j(\lambda) = a_j P_j(\lambda) + b_j P_{j-1}(\lambda) + c_j P_{j+1}(\lambda) \tag{3.16}$$

with

$$i \leqslant j \leqslant N \qquad \text{and} \qquad P_0 = P_{N+1} = 0.$$

The secular equation is

$$\mathrm{Det}\begin{bmatrix} a_1-\lambda & c_1 & 0 & \cdots & \\ b_2 & a_2-\lambda & c_2 & & \\ 0 & b_3 & & \ddots & c_{N-1} \\ \vdots & & & b_N & a_N-\lambda \end{bmatrix} = 0. \tag{3.17}$$

Computationally the eigenvalues and eigenvectors can be straightforwardly, but tediously, found as follows: Start with $P_0=0$ and $P_1=1$, and generate succeeding polynomials $P_j(\lambda)$ (of order $j-1$) with (3.16). $P_N(\lambda)$ is determined by (3.16) with $j=N-1$; the equation with $j=N$,

$$(\lambda - a_N)P_N(\lambda) = b_N P_{N-1}(\lambda),$$

is an Nth order algebraic equation whose roots are the solutions of (3.17). In the process of generating $\{P_j(\lambda)\}$ we have, with the eigenvalues, determined the eigenvectors $|\psi_\lambda\rangle$ which can be normalized by replacing P_j with $P_j/(\Sigma P_j^2)^{1/2}$.

But one would like to be able to do something for the eigenvectors analogous to what was done for the resolvent, that is, to calculate the $P_j(\lambda)$ explicitly only for $1 \leqslant j \leqslant n$ and approximate the rest (Visscher 1977). Thus, if (3.13) holds with $c=b$, a solution to (3.16) for $j \geqslant n$ is

$$P_j(\lambda) = \exp[\mathrm{i}\alpha(\lambda)j], \tag{3.18}$$

where

$$\lambda - a = 2b\cos\alpha(\lambda). \tag{3.19}$$

If $\alpha(\lambda)$ is real, λ is inside the allowed band of the effective medium, and incoming, outgoing or standing waves can be constructed, depending on the boundary conditions imposed. If $\alpha(\lambda)$ is complex ($|\cos\alpha(\lambda)| > 1$), then (3.18) is exponentially damped as j increases and describes a localized mode.

Eq. (3.16) can now be used to connect the $P_j(\lambda)$ for $j \geqslant n$ to those with $j < n$:

$$\begin{aligned}\lambda P_n(\lambda) &= a_n P_n(\lambda) + b_n P_{n-1}(\lambda) + c_n P_{n+1}(\lambda)\\ &= [a_n + c_n \exp(\mathrm{i}\alpha)] P_n(\lambda) + b_n P_{n-1}(\lambda).\end{aligned}$$

Then, if $\hat{M}(\lambda)$ denotes the transpose of the upper left $n \times n$ block of (3.11), with a_n replaced by $a_n + c_n \exp[\mathrm{i}\alpha(\lambda)]$, the eigenvalue equation becomes

$$[\hat{M}(\lambda) - \lambda]|\hat{\psi}_\lambda\rangle = 0, \tag{3.20}$$

where

$$|\hat{\psi}_\lambda\rangle = \sum_{j=1}^{n} P_j(\lambda)|\chi_j\rangle$$

is an eigenvector of $\hat{M}(\lambda)$ on the truncated space $j \leqslant n$. Notice from (3.14), (3.19) that

$$|b| R'_{11}(\lambda) = -\exp[\mathrm{i}\alpha(\lambda)], \tag{3.21}$$

so the resolvent of $\hat{M}(\lambda)$ is identical with the upper left block of the resolvent of M when $N \to \infty$.

Finding eigenvectors and eigenvalues when λ is outside the band ($|\cos\alpha(\lambda)| > 1$ and $\alpha(\lambda)$ complex) is straightforward, because as $N \to \infty, \alpha(\lambda)$ approaches a limit smoothly. The eigenvalues are the roots of

$$\mathrm{Det}[\hat{M}(\lambda) - \lambda] = 0 \tag{3.22}$$

and the eigenvectors can be generated from (3.16). If $\alpha(\lambda)$ is real, the situation changes because a boundary condition at $j=N$ must be satisfied by the oscillating function (3.18) so that $\hat{M}(\lambda)$ does not lose its dependence on N for large N. Thus, the $P_j(\lambda)$'s do not smoothly approach a limit, and one cannot solve (3.20) for large N. Instead, we consider the inhomogeneous analog of (3.20) for a system which is driven, say, at site 1:

$$[\hat{M}(\lambda)-\lambda]|\hat{\psi}_\lambda\rangle = f|\chi_1\rangle$$

or

$$|\hat{\psi}_\lambda\rangle = f\hat{R}(\lambda)|\chi_1\rangle, \tag{3.23}$$

where

$$\hat{R}(\lambda)=[\hat{M}(\lambda)-\lambda]^{-1}. \tag{3.24}$$

Eq. (3.23) gives the response of the system driven at site 1 with amplitude f at frequency λ, in lattice-dynamical language. In this form we *can* let $N\to\infty$ because it contains no products of $P_j(\lambda)$, whose limiting behavior is unknown, with $\exp[i\alpha(\lambda)]$, whose limiting behavior is pathological but known in principle. Matrix elements of $\hat{R}(\lambda)$ have poles for real λ outside the band, and inside the band they have resonance structure, as will be illustrated later in a simple case.

From eq. (3.23) one finds, for $f=1$,

$$\begin{aligned} P_1(\lambda) &= \hat{R}_{11}(\lambda), \\ P_2(\lambda) &= \hat{R}_{21}(\lambda), \\ &\vdots \\ P_n(\lambda) &= \hat{R}_{n1}(\lambda). \end{aligned} \tag{3.25}$$

Since $\hat{M}(\lambda)$ is tridiagonal, one can write

$$\hat{R}_{n1}(\lambda)=b_2b_3\ldots b_n/\mathrm{Det}[\hat{M}(\lambda)-\lambda] \tag{3.26}$$

because the $(n,1)$ cofactor is just the product of the b's. The rule for generating the $(j-1,1)$ cofactor from the $(j,1)$ is (3.16), with a_n replaced by $a_n-b^2R'_{11}(\lambda)$. The ratio of the amplitude at the "center" to that at the "edge" of the cluster is the "response ratio"

$$P_1(\lambda)/P_n(\lambda)=\mathrm{Cof}_{11}[\hat{M}(\lambda)-\lambda]/b_2\ldots b_n. \tag{3.27}$$

To summarize, if M is a hermitian operator, for example, representing a tight-binding hamiltonian system with finite range interactions, then the recursion (3.8) can efficiently generate $\{a_j, b_j, c_j\}$. The method is intuitively appealing because it emphasizes the local environment which is expected to dominate the local physical observables. If, by direct calculation or other means, one can justify replacement of $\{a_j, b_j, c_j\}$ with a, b, c (or a limit cycle) for $j>n$, where n is manageably small, then the resolvent and important eigenvalues and eigenvectors can be calculated from the first n recursions.

3.2.3. *The linear chain with a single impurity*

As a simple example of the application of the recursion method, consider the system illustrated in fig. 9, an infinite harmonic chain with an impurity at its center which can either be weakly or strongly bound, relative to the other force constants. The equations of motion are

$$\begin{aligned} \ddot{x}_i &= x_{i+1} - 2x_i + x_{i-1}, \qquad |i| \geqslant 2, \\ \ddot{x}_{\pm 1} &= x_{\pm 2} - x_{\pm 1} + \gamma(x_0 - x_{\pm 1}), \\ \ddot{x}_0 &= \gamma(x_1 - 2x_0 + x_{-1}), \end{aligned} \tag{3.28}$$

where all masses and the host force constants are unity. As usual we will take $x_i(t) = x_i(0)\exp(-\mathrm{i}\omega t)$, so $\ddot{x}_i = -\omega^2 x_i$.

The site basis representation of M can be read from (3.28):

$$M = \begin{bmatrix} \ddots & & & & & & 0 \\ -1 & 2 & -1 & & & & \\ & -1 & 1+\gamma & -\gamma & & & \\ & & -\gamma & 2\gamma & -\gamma & & \\ & & & -\gamma & 1+\gamma & -1 & \\ & & & & -1 & 2 & -1 \\ 0 & & & & & & \ddots \end{bmatrix}.$$

After taking $|\chi_1\rangle$ to be a site basis state for $i=0$, $|\chi_1\rangle = (\ldots 0, 1, 0, \ldots)$, (3.8)

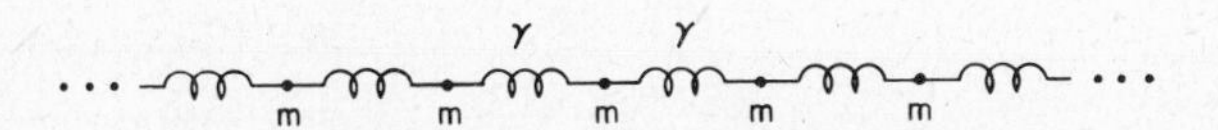

Fig. 9. Harmonic chain with a single weakly-bound impurity at its center.

gives for M in the χ-basis

$$M=\begin{bmatrix} a_1 & b_2 & 0 & \dots \\ c_1 & a_2 & b_3 & \dots \\ 0 & c_2 & a_3 & b_4 \dots \\ \vdots \end{bmatrix}=\begin{bmatrix} 2\gamma & -\sqrt{2}\,\gamma & 0 & \dots & \\ -\sqrt{2}\,\gamma & 1+\gamma & -1 & 0 & \dots \\ 0 & -1 & 2 & -1 & \dots \\ \vdots & & & \dots & \end{bmatrix}. \tag{3.29}$$

So we can choose $n=2$, and with $\lambda=\omega^2$ and

$$\begin{aligned} |\chi_2\rangle &= 2^{-1/2}(\dots 0,0,1,0,1,0,0,\dots), \\ |\chi_3\rangle &= 2^{-1/2}(\dots 0,1,0,0,0,1,0,\dots), \\ &\vdots \end{aligned} \tag{3.30}$$

we find from (3.10)

$$R_{11}(\lambda)=\cfrac{1}{2\gamma-\omega^2-\cfrac{2\gamma^2}{1+\gamma-\omega^2-\cfrac{1}{2-\omega^2-\cfrac{1}{2-\omega^2-\cfrac{1}{\ddots}}}}}$$

$$=\cfrac{1}{2\gamma-\omega^2-\cfrac{2\gamma^2}{1+\gamma-\omega^2-R'_{11}(\lambda)}},$$

where

$$R'_{11}(\lambda)=1-\frac{\omega^2}{2}+(\omega^2-4)^{1/2}\frac{\omega}{2}.$$

Some algebra yields

$$R_{11}(\lambda)=\frac{\gamma^2(\omega^2-4)^{1/2}}{\omega(1-\gamma)\left[\left(\omega^2-\omega_0^2\right)^2+(\Gamma/2)^2\right]}, \tag{3.31}$$

with

$$\omega_0^2 = 2\gamma + \frac{1}{2}\gamma^2/(1-\gamma),$$

$$(\Gamma/2)^2 = 2\gamma^3/(1-\gamma) - \frac{1}{4}\gamma^4/(1-\gamma)^2.$$

The matrix $\hat{M}(\lambda)$ is

$$\hat{M}(\lambda) = \begin{bmatrix} 2\gamma & -\sqrt{2}\,\gamma \\ -\sqrt{2}\,\gamma & 1+\gamma-R'_{11}(\lambda) \end{bmatrix}$$

and the response ratio is

$$\begin{aligned} \frac{P_1(\lambda)}{P_2(\lambda)} &= \frac{\hat{R}_{11}(\lambda)}{\hat{R}_{21}(\lambda)} \\ &= \frac{1+\gamma-\omega^2-R'_{11}(\lambda)}{\sqrt{2}\,\gamma} \\ &= \frac{\gamma-\omega^2/2+(\omega^2-4)^{1/2}\omega/2}{\sqrt{2}\,\gamma} \\ &= \frac{\gamma-1+\exp[-\mathrm{i}\alpha(\lambda)]}{\sqrt{2}\,\gamma}, \end{aligned}$$

where $\omega^2 = 2 - 2\cos\alpha(\lambda)$. For $\omega^2 = \omega_0^2$ and $\gamma \ll 1$, $|P_1(\lambda)/P_2(\lambda)|^2 \approx 1/\gamma$, so the resonance mode has a large central amplitude with a small outgoing wave. In this limit $\Gamma/2\omega_0^2 \approx (\gamma/2)^{1/2}$, so the rate of decay is small compared to the frequency. In the other limit $\gamma \gg 1$ a localized mode with complex $\alpha(\lambda)$ appears at $\omega^2 = 3\gamma$, in which the three central atoms oscillate thus: $(\rightarrow\leftarrow\rightarrow)$. One might think that a resonance mode would appear also corresponding to $(\rightarrow\rightarrow\rightarrow)$, equivalent to the vibration of an impurity of three times the host mass, but it is a peculiarity of the ld system that this mode disappears into the singularity at zero frequency. The fact that M is tridiagonal in the site-representation in this example is a consequence of the simple system we chose to illustrate the method which works for general M.

3.2.4. Related techniques

The recursion method is intimately connected with the method of moments (Gaspard and Cyrot-Lackmann 1973, Wheeler et al. 1974) in which the density of states is calculated by manipulating expectation values of powers of the hamiltonian. Work utilizing Padé approximants to moment expansions is also closely related (Nickel 1974), and there are obvious formal similarities to CPA expressions (Elliott et al. 1974). As emphasized by Haydock et al. (1975), the recursion method has the great advantage over the moment techniques in that it is numerically stable, i.e. an increase in n will yield a better answer. The interested reader is referred to their paper for mathematical references.

Yndurian et al. (1974) have recently developed a method similar in spirit to the recursion method, but different in detail. Joannopoulos and Cohen (1976) give a review and bibliography. The local structure is again regarded as the factor which determines the local dynamics; accordingly, they start with a relatively small cluster, a CRN because they are concerned with this type of amorphous solid. The dangling bonds of the cluster (the surface) are then satisfied by attaching to each a Bethe lattice with the same coordination number as the CRN. A Bethe lattice is a structure, specified by the number of nearest neighbors, for which the path between any two points is unique. For three nearest neighbors, such a structure is illustrated in fig. 10. An attractive feature of a Bethe lattice is that its dynamics are exactly solvable in a simple closed form. The rationale for believing that it leads to a good approximation is, again, the faith that local properties should depend mostly on local cluster structure, and only weakly on the distant boundaries.

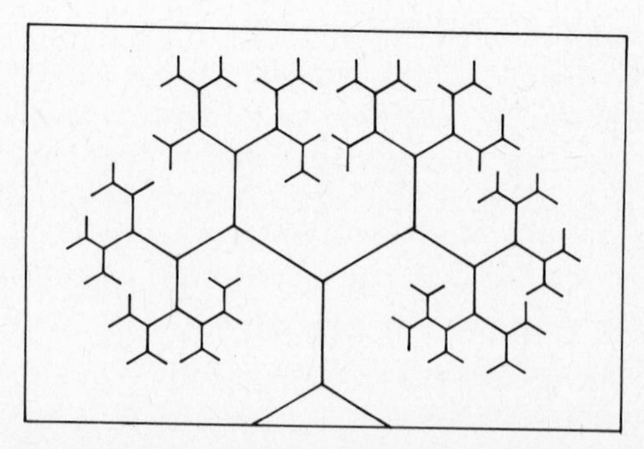

Fig. 10. Bethe lattice (or Cayley tree) with coordination number 3. The trunk divides into limbs, limbs into branches, branches into twigs, ad infinitum.

3.3. The equation-of-motion method

Closest to the spirit of computer experimentation is the equation-of -motion method (Alben et al. 1975c) because it actually calculates physical observables by following the time evolution of dynamical quantities or wave functions. Moreover, it can be applied to find the spectral density and other properties of a wide class of matrices.

As an illustration of the equation-of-motion (EOM) method, we will compute the resolvent of a hermitian matrix M:

$$R(\lambda)=(M-\lambda)^{-1}.$$

We begin by defining a state vector $|\chi(t)\rangle$ which satisfies

$$\left(\mathrm{i}\frac{\mathrm{d}}{\mathrm{d}t}-M\right)|\chi(t)\rangle=0. \tag{3.32}$$

Next we let

$$|\chi(t)\rangle=|\tilde{\chi}(t)\rangle\exp(-\mathrm{i}\lambda t),$$

where, with (3.32), we have formally

$$|\tilde{\chi}(t)\rangle=\exp[\mathrm{i}(\lambda-M)t]|\chi\rangle$$

in which $|\chi\rangle=|\chi(0)\rangle=|\tilde{\chi}(0)\rangle$ is some suitably chosen initial state.

Now we consider the damped and truncated time integral

$$|I(\chi,\lambda)\rangle=\int_0^T\mathrm{d}t\,\mathrm{e}^{-\varepsilon t}|\chi(t)\rangle. \tag{3.33}$$

If $T\to\infty$ and $\varepsilon\to 0$, (3.33) becomes

$$|I(\chi,\lambda)\rangle=-\mathrm{i}R(\lambda)|\chi\rangle, \tag{3.34}$$

so that any elements of the matrix $R(\lambda)$ can be computed approximately by numerically solving (3.32) and integrating the result in (3.33). The numerical schemes may be simple first-order ones (Alben et al. 1975c) or more elaborate Runge–Kutta methods (Weaire and Williams 1976).

There are two parameters in (3.33) which smear out the λ-dependence in $|I(\chi,\lambda)\rangle$. If $T\to\infty$, then

$$|I(\chi,\lambda)\rangle=[\varepsilon-\mathrm{i}(\lambda-M)]^{-1}|\chi\rangle \tag{3.35}$$

corresponds to a lorentzian resolution function of width ε. If $\varepsilon=0$,

$$I(\chi,\lambda)=\exp[\mathrm{i}(\lambda-M)T/2]\frac{\sin(\lambda-M)T/2}{(\lambda-M)/2}|\chi\rangle,$$

whose real part is another approximation to a δ-function. In any numerical calculation one should choose a combination of ε and T which optimizes accuracy for given computing expenditure. The more computing time, the smaller ε and the larger T one can use to get higher resolution in λ.

In practice the lorentzian broadening function is sometimes found to be not as optimal in producing details of spectra as is a gaussian broadening (Alben et al. 1977a, b). That is, instead of using (3.33), one damps the time integral with a gaussian in t:

$$|I(\chi,\lambda)\rangle=\int_0^T \mathrm{d}t\,\mathrm{e}^{-\varepsilon^2 t^2}|\chi(t)\rangle. \tag{3.36}$$

Just as (3.33) is equivalent to folding (3.34) with a lorentzian, (3.36) corresponds to folding it with a gaussian if $T\to\infty$. A version of EOM much like (3.36) has been used by Joannopoulos and Mele (1976).

The choice of (3.36) or (3.33) depends on the particular application, as do the choices of ε and T. For example, unless one is interested in finite size effects, one should choose ε much larger than the average level spacing.

We will now discuss several examples of the use of EOM, which we hope will illustrate its simplicity, economy and versatility. Suppose M is the dynamical matrix for a disordered system. We can choose $\{|\chi_j\rangle\}$ to be site vectors with only one non-zero component: $|\chi_j\rangle\equiv|j\rangle$, where j is a site label. Then if $|\chi(0)\rangle=|i\rangle, |\chi(t)\rangle$ can be obtained by numerical integration of (3.32), (3.33) may be integrated numerically, and the site matrix elements of $R(\lambda)$ are

$$R_{ji}(\lambda)=\langle j|I(i,\lambda)\rangle\equiv I_j(i,\lambda). \tag{3.37}$$

For the large systems which we would like to calculate ($N\sim 10^4$), such complete information is neither desirable nor obtainable. Certain linear combinations of $R_{ji}(\lambda)$ are easy to calculate directly or to approximate.

The spectral density, for example, is

$$g(\lambda)=\pi^{-1}\,\mathrm{Im}\,\mathrm{Tr}\,R(\lambda).$$

We can approximate this without actually computing N diagonal elements from (3.37): Take

$$|\chi\rangle = N^{-1/2} \sum_{j=1}^{N} \exp(\mathrm{i}\phi_j)|j\rangle.$$

Then compute $|I(\chi,\lambda)\rangle$ and take the inner product

$$\langle\chi|I(\chi,\lambda)\rangle = N^{-1} \sum_{ij=1}^{N} \exp[\mathrm{i}(\phi_i - \phi_j)] R_{ij}(\lambda). \tag{3.38}$$

If $(\phi_1,\ldots,\phi_N)$ are chosen randomly and independently on $0<\phi_i \leqslant 2\pi$ and if an average of (3.38) is taken over many such realizations, then clearly only the terms for which $i=j$ will survive, i.e. the trace remains. In practice only a few realizations of $(\phi_1,\ldots,\phi_N)$ need be generated and summed over to perform the average; results good to a few percent have been obtained with only four realizations for $N=8000$ (Alben et al. 1975c).

Neutron scattering amplitudes involve matrix elements between phonon states

$$S(\boldsymbol{k}',\boldsymbol{k},\lambda) = \langle\chi_{\boldsymbol{k}'}|(M-\lambda)^{-1}|\chi_{\boldsymbol{k}}\rangle.$$

This can be generated by letting $|\chi(t)\rangle$ evolve according to (3.32) from

$$|\chi_k\rangle = N^{-1/2} \sum_{j=1}^{N} \exp(\mathrm{i}\boldsymbol{k}\cdot\boldsymbol{r}_j)|j\rangle$$

doing the time integral to get $|I(\chi,\lambda)\rangle$ and using (3.34) to obtain

$$S(\boldsymbol{k}',\boldsymbol{k},\lambda) = \langle\chi_{\boldsymbol{k}}|I(\boldsymbol{k},\lambda)\rangle.$$

Calculations of DOS for a simple cubic disordered alloy (Alben et al. 1975) are shown in fig. 11. Although this system is an electronic one with non-comparable parameters, one can compare this DOS with that shown in fig. 5 for a disordered harmonic lattice with 1440 atoms, calculated by NFC. The $c=0$ curve in fig. 11 is not an EOM result; it is an exact histogram and illustrates the desirability of having $\lambda>0$ in EOM. The fluctuations are caused by the finiteness of the system. The dashed line is the DOS for the infinite s.c. lattice. For the EOM calculations here $\lambda=0.03$ and only partly damps these fluctuations. This figure shows the results using the lorentzian broadening function (3.33). Alben et al. (1977b) have repeated this calculation with the gaussian function (3.36), which they find to be distinctly preferable for this problem.

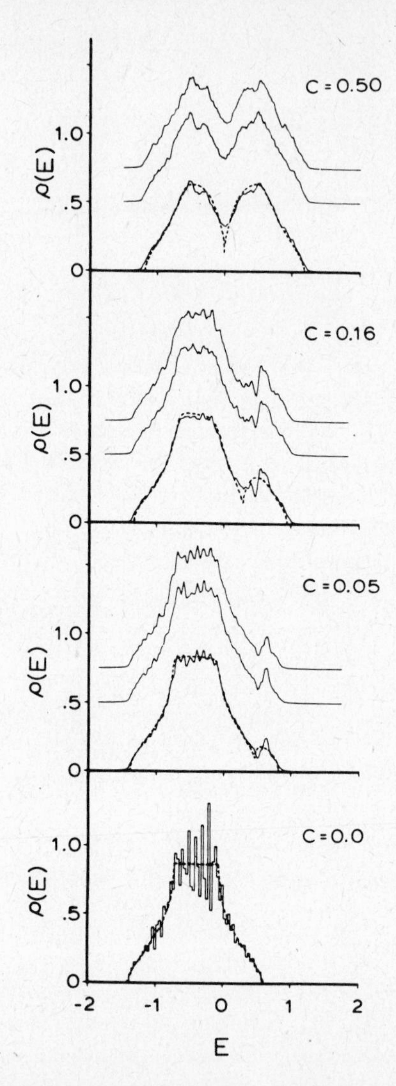

Fig. 11. EOM results for the tight-binding DOS for a simple cubic alloy. Four concentrations of impurity are shown for a system with 7980 atoms (19×20×21). The $c = 0$ case is an exact histogram for this model. The lorentzian broadening has a full width of 0.03 energy units, the same as the bar width of the histogram. For each value of c the upper two curves are results averaged over two statistically independent realizations of $(\phi_1, \ldots, \phi_N)$ and two distributions of impurities. The lower curve is the average of the upper curves (Alben et al. 1975).

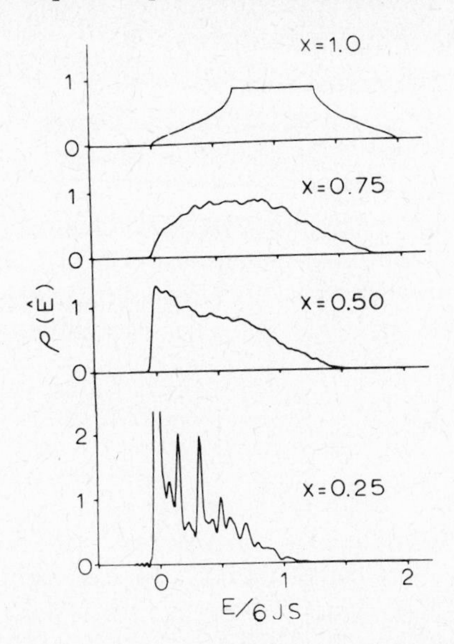

Fig. 12. DOS by EOM for simple cubic ferromagnet, with non-magnetic substitutionally random impurities. The interactions are nearest-neighbor, and the spectrum changes character when the density of spins x is about equal to the critical site-percolation concentration (Alben et al. 1977a).

Many other applications of EOM have been made, mostly by Alben and his collaborators. Quite different physical systems are, as we have previously emphasized, handled by exactly the same computational methods, and the results look similar. For example, fig. 12, taken from Alben et al. (1977a), shows the DOS for diluted s.c. ferromagnets. The concentration of magnetic atoms decreases from top to bottom. The top is the familiar s.c. spectrum with four van Hove singularities. Dilution shifts the bulk of the states down; when the critical percolation concentration is passed, the nature of the spectrum changes to one characteristic of isolated islands of impurity spins.

4. *Density of states*

Because of the periodicity of crystalline solids, the wavevector $\boldsymbol{k}$ is a good quantum number and is used to label the eigenvalues $E(\boldsymbol{k})$ of the system under study. Detailed information about the eigenvalues is often presented in elaborate graphs, band structures, of $E(\boldsymbol{k})$ versus $\boldsymbol{k}$ for special directions of $\boldsymbol{k}$ and polarizations labeled by symbols representing different symmetries inherent to the crystal.

In disordered solids one has to settle for considerably less information. Instead of a band structure, one computes a density of states, that is, the distribution of eigenvalues. The effect of disorder on the DOS is one of many topics concerned with the properties of disordered solids that is studied.

We now discuss computer experimental calculations of the DOS in one dimension and in two- and three-dimensions. The discussion of ld studies is brief because comprehensive reviews already exist. The discussion of 2d and 3d studies considers the density of states of substitutionally disordered alloys and amorphous solids.

4.1. One dimension

Most early investigations into the properties of disordered systems dealt with one-dimensional systems. A short history of these investigations begins with Dyson's exact calculation of the frequency spectrum of a one-dimensional glass-like chain (Dyson 1953). A more tractable exact analysis was performed later by Schmidt (1957). In relation to computer experiments, there were two notable developments. James and Ginzbarg (1953) exploited properties of solutions in one dimension and demonstrated how to calculate the density of states by counting the nodes of a Sturm sequence, and Dean and Martin (1960) introduced a general counting technique known as the negative eigenvalue theorem (the NFC method of §3.1), which became the computational basis for most density of states calculations in one-, two- and three-dimensions.

In 1960 Dean used NFC to calculate the frequency spectrum of an isotopically disordered linear chain and surprisingly found that the spectrum had fine structure with many peaks. At first there was some doubt about Dean's results since they were based on a numerical calculation using a counting method for a finite chain. As the chain became longer, it was expected that the spectrum would become smooth. Dean and Bacon (1963), however, removed these doubts by giving a physical explanation of the peaks. They were able to relate numerically the frequencies of distinct peaks in the spectrum to impurity frequencies associated with several kinds of clusters of atoms surrounded by the host atoms. They also investigated the concentration dependence of these peaks and showed that the strength of each peak is proportional to the probability of occurrence of the corresponding cluster in the disordered lattice. An additional explanation of the peaks was given by Hori and Matsuda (1964) and Hori and Fukushima (1963); Dean's discovery later received further numerical verification by Agacy (1964) who used the exact formalism of Schmidt (1957).

The fine structure noted by Dean (1961) is associated principally with localized states in the impurity band, that is, with states confined to regions of light impurities embedded in regions of heavier host masses. For example, the local arrangement (H, heavy mass, L, light mass)

...HHLHH...

in a disordered chain behaves similarly to a single light atom in an otherwise monatomic chain of heavy atoms.

In a series of papers, Wu and coworkers showed that for a system diluted with light mass impurities, the main band also has fine structure (Gubernatis and Wu 1974, Wu and Chao 1975) at the high-frequency end. This fine structure may be associated with resonance modes (Wu 1975, Chen et al. 1977) identified with such configurations as

$$\ldots\mathrm{HLH}\underbrace{\ldots}_{n}\mathrm{HLH}\ldots .$$

Physically, Wu (1975) argues that the presence of the two light impurities in a sea of heavy hosts introduces discontinuities that result in the interference of reflected and transmitted waves in the region between the light impurities so that resonance modes can be set up within the particular local arrangement.

The fine structure caused by either light mass impurity or resonant modes is associated with clustering phenomena resulting from fluctuations that must exist in the system because of the disorder, and thus is not restricted to isotopically disordered systems. Fine structure associated with impurity modes is found in electronic systems (Agacy 1964); however, structure resulting from resonant modes has not been studied.

Another effect of fluctuations, tailing in the density of states, is expected to be present in isotopically disordered systems, but has been studied only in electronic systems. Tailing arises in the following manner: Consider an electron in a substitutional alloy of A and B atoms, and let V_B, the potential of a B atom at any given lattice site, be less than V_A. Because of the fluctuations, the density of B atoms will vary from place to place in the crystal. Depending on the concentration c of B, there is a certain probability that a group of B atoms will be surrounded by enough A atoms that one can associate a potential well with the island of B atoms. If the cluster is wide enough or the potential strong enough, then at least one electronic bound state can be formed in the well. It is this bound state that one associates with the peaks in the spectrum. The states associated with the widest and deepest wells are the farthest away from the center of the band. They also are of the lowest energy, and, more significantly, they are the least likely to occur. As a result they figure less prominently in the peaky nature of the spectrum but give rise to the tails in the density of states (Lifshitz 1964); that is, the density of states $g(E)$ is of the form

$$g(E) \propto \exp\left[-\alpha/(E-E_0)^{1/2}\right], \tag{4.1}$$

where E_0 is a band edge. Using Schmidt's formalism, Gubernatis and Taylor (1971) found this behavior numerically. A rigorous derivation is given by Friedberg and Luttinger (1975), not only for one dimension, but also for three dimensions. Near band edges the tailing causes small values of the density of states, and in computer experiments is generally observed as a shrinkage of the band edges from values predicted by band gap theorems (Saxon and Hutner 1949, Luttinger 1951, Matsuda 1964, Hori 1964).

An additional feature of the spectra of one-dimensional disordered lattice deserves mention. In Dean's spectra (1960) there were valleys at which the density of states appeared to vanish. Borland (1964), Matsuda (1964) and Hori (1964) showed that for a large enough mass ratio there exist special frequencies at which the density of states must vanish. These special points exist for a variety of one-dimensional models (Wada 1966, Hori 1968, Gubernatis and Taylor 1973) and can in a given band be infinite in number. Consequently, the existence of the special frequencies (or energies) itself can cause considerable fine structure in the spectra.

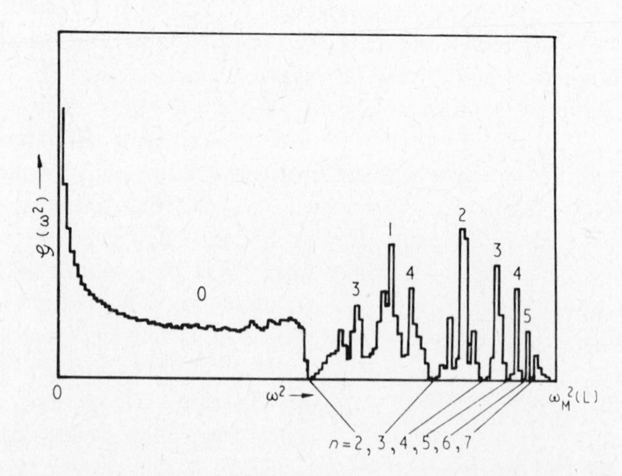

Fig. 13. Squared frequency density of states for a binary isotopically disordered chain of 250000 atoms with mass ratio 2:1 and containing equal numbers of light and heavy atoms (Dean 1972). The numbers above the peaks indicate the size of the clusters of light masses which form the localized modes responsible for those peaks, and n labels the special frequencies (Bell 1972). An indication of band tailing is seen near $\omega_M^2(L)$, which is the maximum frequency in the pure light mass chain.

The density of states in fig. 13 exhibits many of the features just discussed. Fine structure is evident in the impurity band on the left-hand side of the figure. Peak 1 is associated with one L atom surrounded by two or more H atoms on either side. Peak 2 stems from a cluster of two L atoms; peak 3 stems from a cluster of three L atoms; etc. Zeroes in $g(\omega^2)$ labeled $n=2,3,\ldots$ are special frequencies, and at the extreme right of the figure, tailing in the density of states is suggested by the steady decrease in $g(\omega^2)$.

We mentioned the principal features of spectra of one-dimensional disordered lattices. Extensive reviews by Hori (1968), Dean (1972) and Bell (1972a) discuss individual calculations for a variety of models and types of disorder.

4.2. Two and three dimensions

The computer experimentalist can study systems with many different kinds of disorder, and can consider the lattice dynamics, electronic or magnetic properties of each. Fig. 14 shows 2d representations of several types of disorder one might have. In the following pages we cite examples of computer experiments on some of them.

4.2.1. *Disordered alloys*

The easiest densities of states to calculate are those of disordered crystalline alloys, partly because the structures of disordered alloys are

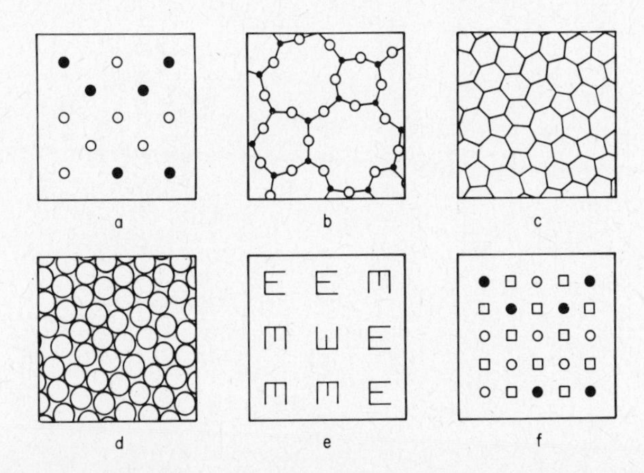

Fig. 14. 2d representations of several structures for which one might compute DOS and other observables: (a) a substitutionally disordered alloy, (b) an A_2B_3 glassy network, (c) a monatomic CRN, (d) a DRPHS, (e) an orientationally disordered solid like ice, (f) a mixed crystal like NaCl–KCl.

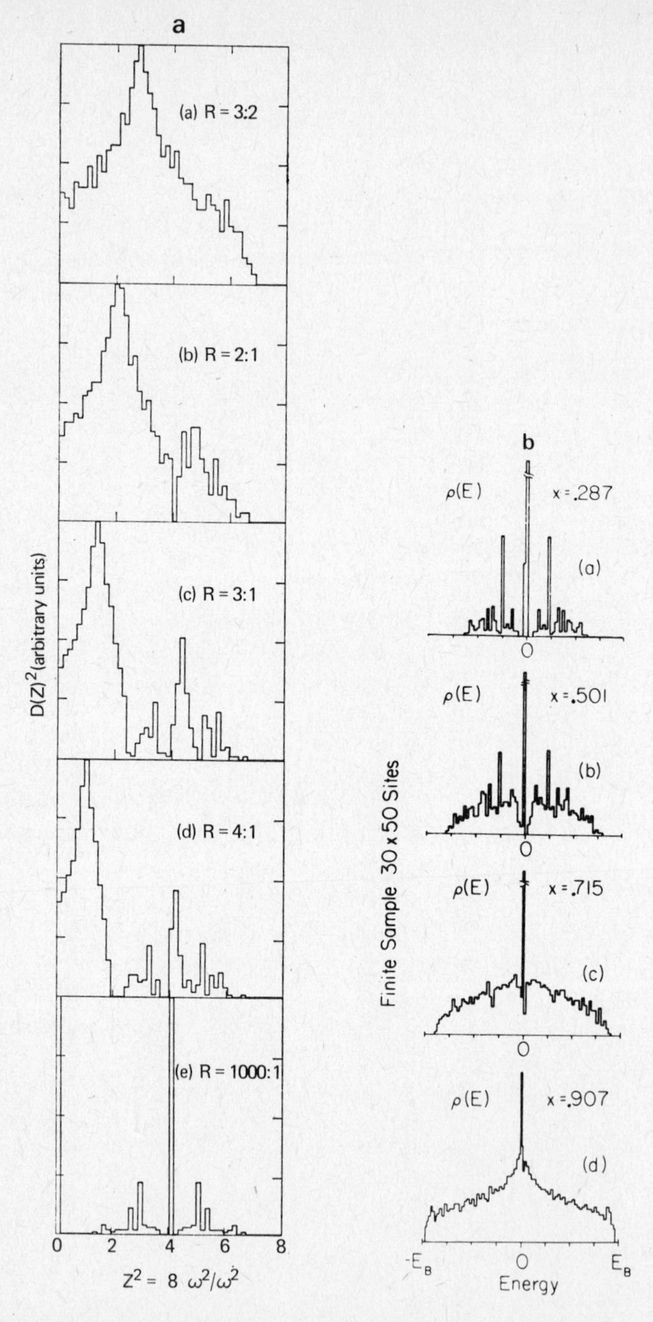

Fig. 15. (a) DOS of a 900-site square lattice (calculated by NFC) which is a randomly disordered mixture of 75% heavy and 25% light masses with mass ratio R. Special frequencies seem to begin to appear at $R=2$ (Payton and Visscher 1967a). (b) DOS of the A subband of a binary alloy on a 1500-site square lattice. The concentration of A atoms is x. The similarity of the DOS for $x=0.287$ with that for $R=1000:1$ in (a) is obvious. As x increases the A subband approaches the familiar perfect square lattice DOS (Kirkpatrick and Eggarter 1972).

easy to generate computationally. The original binary mass-disorder calculations of Dean and his collaborators were on the ld analogs of fig. 14a, and it is natural to ask whether the features they found (fine structure of the DOS and special frequencies) are also present in 2d and 3d.

In fact the DOS for 2d and 3d disordered alloys exhibit much structure in the impurity bands, which often is correlated with the occurrence of islands of impurities in the host lattice. This correlation is discussed by Payton and Visscher (1967b) and also by Bell (1972a) and Dean (1972). Some special frequencies, discussed above in relation to the ld case, seem to persist in the 2d case, and may also exist in 3d. (See, however, Hori and Wada 1970 and Hori 1968.) Fig. 15a shows a set of DOS of lattice vibrations of the 2d square lattice, for a binary mixture of 75% heavy and 25% light atoms with all force constants equal (diagonal disorder; or isotopically disordered). For mass ratios of 2 : 1 and larger the DOS dips to zero at $\omega^2 = \frac{1}{2}\omega_0^2$. Empirically this is a special frequency in that the fraction of states above it is the same in all cases with mass ratio greater than 2 : 1 (Payton and Visscher 1967a). The ld special frequencies mentioned in §4.1 also had this property. The fact that the DOS vanishes at each special (or "forbidden") frequency means that modes are trapped *between* special frequencies while parameters (mass ratios, force constants) are varied, and the integrated density of states at those points is invariant.

Some results of an NFC calculation by Kirkpatrick and Eggarter (1972) are shown in fig. 15b. They considered an electronic problem: a tight-binding approximation to an AB alloy on a square lattice. The figure shows the A subband when the B energies are very large. (As previously noted, the electronic tight-binding problem is isomorphic to this lattice-dynamical one, so the similarities between the DOS are expected.) Kirkpatrick and Eggarter remark that the spike in the middle of their spectra is due to isolated A sites (atoms); thus it represents localized modes. But as the concentration of B atoms increases the band of extended states seems to envelop them. To have a localized mode in the midst of a continuous spectrum is thought to be impossible because nothing prevents the localized and extended modes from mixing. The fact that the local modes occur at a special frequency removes the objection because the spectrum vanishes there.

Fig. 15 illustrates the DOS of systems with diagonal disorder; i.e. in the lattice-dynamical (electronic) case, different masses (site energies) but equal force constants (transfer integrals). In fig. 16 an example (Alben et al. 1976) of the converse is illustrated. The system is a simple cubic binary alloy with nearest-neighbor transfer integrals dependent on the identities of the neighbors but with all site energies (diagonal elements) the same.

The results, for two different values of the disorder parameter δ, completely lack the fine structure characteristic of the corresponding binary

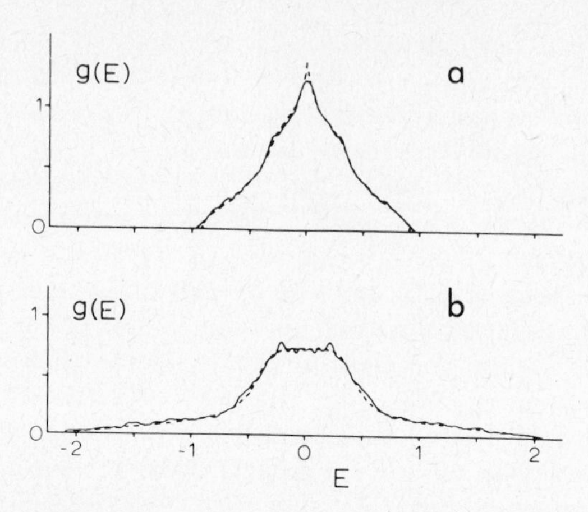

Fig. 16. Electronic tight-binding DOS for a simple cubic AB alloy with 15% A atoms, and different AA, AB and BB transfer integrals t_{ij}, but all diagonal matrix elements equal. If $t_{AB}=\frac{1}{2}(t_{AA}+t_{BB})$ and $t_{BB}=1/6$ then in (a) $\delta=6t_{AB}-1=-0.5$; in (b) $\delta=1.5$. Compared to a similar case with diagonal disorder (fig. 11) these DOS are very smooth (Alben et al. 1976).

diagonal disordered problem. It is generally true that off-diagonal disorder in 2d and 3d (not only in the substitutional alloy problem, but also structurally disordered systems, which may be considered to have a kind of off-diagonal disorder) does less violence to the DOS than diagonal disorder does.

An example of the application of EOM to a magnetic system is shown in fig. 17. Alben et al. (1977a) linearized the equations of motion for the spins and calculated the local densities of states for a mixture of two different ferromagnetic atoms, randomly placed on a b.c.c. lattice and interacting with nearest neighbors. The top graph is for pure B and illustrates the three van Hove singularities of the b.c.c. spectra. As the A concentration increases the logarithmic singularity at $E=1$ disappears, and a peak corresponding to isolated A atoms begins to build up. It will become a van Hove singularity at $E=\frac{1}{2}$ for $x=0$. Decomposition of the DOS into components in this way shows that the A and B bands seem to repel each other. When one is anomalously high, the other is anomalously low, leading to a pronounced dip, like a band gap or incipient special frequency, between peaks.

Although often mathematically identical to lattice-dynamical or electronic calculations, the computation of magnetic excitations (magnons) involves basically very different physical ideas. Consider, for example, the dilute ferromagnet (Huber 1974a, b). If only nearest-neighbor exchange

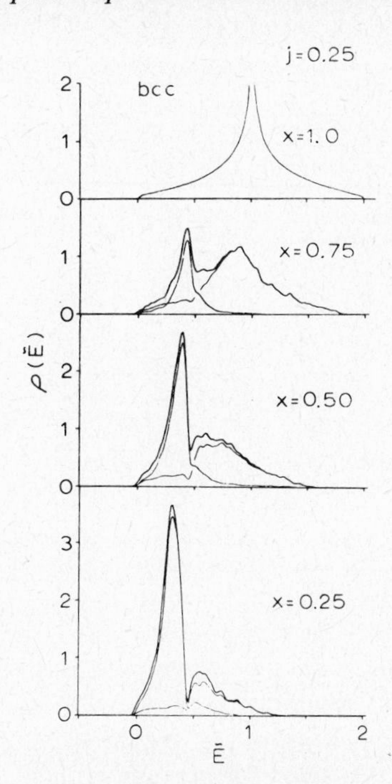

Fig. 17. DOS for b.c.c. random binary ferromagnetic alloy calculated by EOM. The spin–spin interaction energies of the two components differ by a factor 4; concentrations are x and $1-x$. Total DOS is the heavy line; component DOS are the light lines (Alben et al. 1977).

integrals are considered, then the system obviously cannot be ferromagnetic or have a continuous spectrum of magnon modes until the concentration of magnetic atoms exceeds the critical site percolation concentration. For more dilute systems the excitation spectrum will be discrete, corresponding to modes of the many isolated finite magnetic islands. This corresponds in lattice dynamics to the doping of a heavy host lattice with a light impurity; there too the spectrum undergoes a qualitative change as the percolation concentration is passed.

Another magnetic system with a rich potential for physical and computer experiments is the spin glass. (See Binder and Stauffer 1979 for a recent review.) The archetypical spin glass, Fe in Au, involves a few percent of substitutional magnetic impurities in a non-magnetic host. The interaction between magnetic atoms, however, has an oscillating sign

$$J(r) \sim r^{-3}\cos(2k_F r),$$

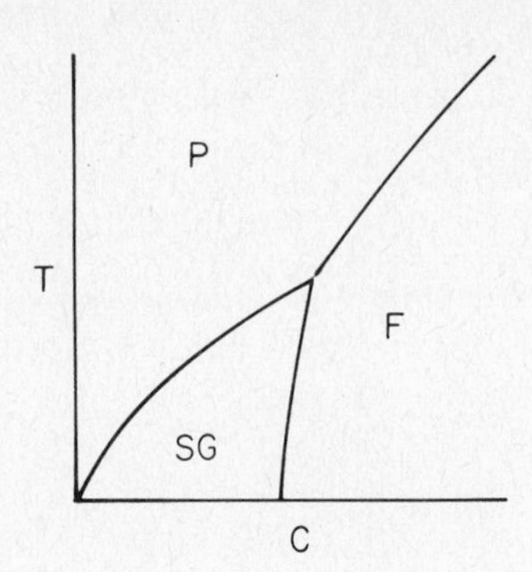

Fig. 18. Temperature vs concentration of magnetic atoms phase diagram for a spin glass (SG). F = ferromagnetic, P = paramagnetic.

so that a pair of spins interact ferromagnetically or antiferromagnetically, depending on their separation. $J(r)$ is called the RKKY interaction. With such a system one cannot define magnetic excitations until one knows the lowest energy state about which the system oscillates. Walker and Walstedt (1977) performed such a computer experiment. (See also §2.1.3.) Fig. 18 shows a schematic phase diagram for a spin glass. For $T=0$ and high concentration the near-neighbor ferromagnetic $J(r)$ dominates, and the system is a ferromagnet. But below some concentration c (related to a percolation concentration) the antiferromagnetic longer-range interactions begin to compete successfully, destroying the ferromagnetism and creating the spin-glass phase, characterized by no long-range magnetic order. Computer experiments on them are being actively pursued.

4.2.2. *Structural disorder*

Several examples of computer experiments to determine DOS of structurally disordered systems are discussed in this subsection. One feature common to such systems is that each has short-range order similar to some regular lattice. As a result, the DOS and other observables in structurally disordered systems often resemble those of a related periodic lattice. This remark is illustrated in fig. 19, which shows DOS calculated by Meek (1976) for several tetrahedrally bonded CRN's, and for a finite and infinite crystalline structure. The two-peaked nature of all of these spectra is characteristic of structures in which each atom is a member of several closed rings of bonds. It can be shown (Alben et al. 1975a) that under certain conditions a perfectly tetrahedrally bonded CRN will have localized modes, in which the atoms in the rings are displaced as shown in fig. 20, that form a low-frequency (ideally zero) band corresponding to

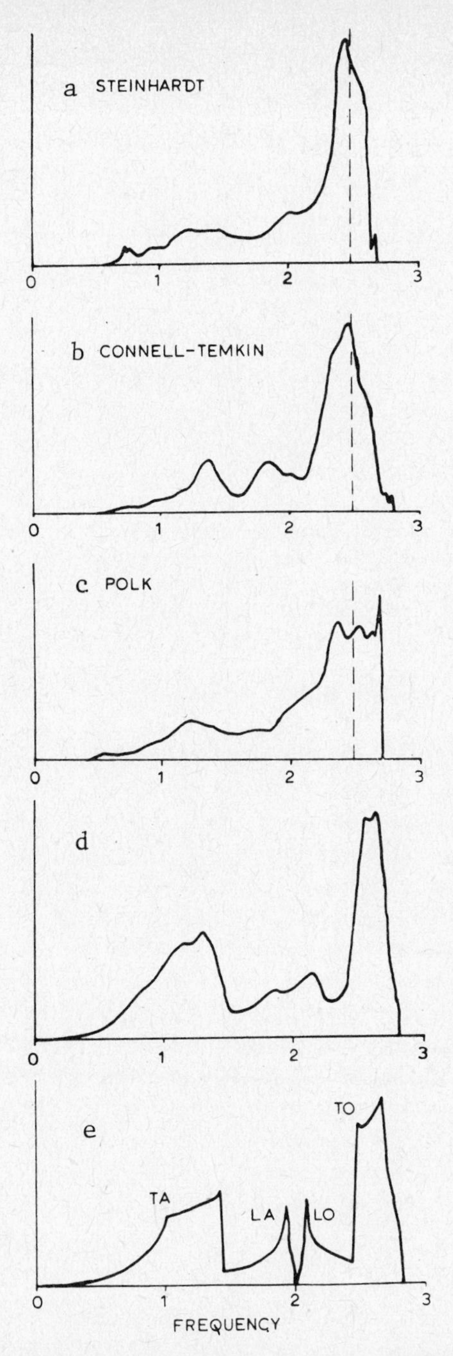

Fig. 19. Vibrational DOS for several CRN models: (a) Steinhardt, 201-atom model, (b) Connell–Temkin, 238-atom model, (c) Polk, 519-atom model, (d) 344-atom diamond structure micro-crystallite, (e) infinite diamond structure. Except for (e), the DOS were calculated by the recursion method with $n=20$ (Meek 1976).

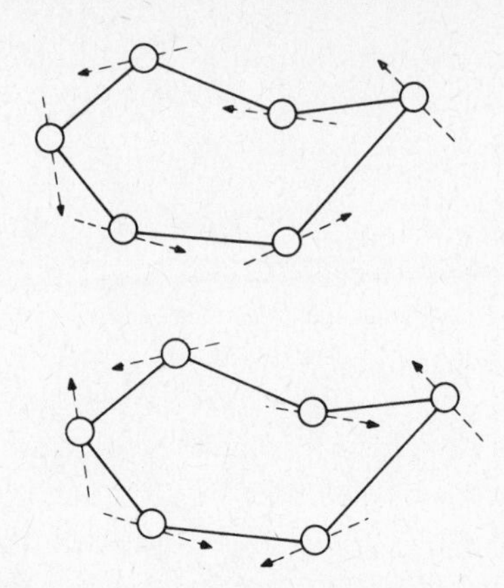

Fig. 20. Examples of localized modes which may be formed on a closed ring of bonds if there is perfect local tetrahedral symmetry and central nearest-neighbor forces. The displacement vectors are in the planes of the bonds in the ring and perpendicular to their bisector (Alben et al. 1975a).

displacements in the same direction around the ring and a high-frequency band corresponding to alternating phases. The CRN spectra contain contributions from rings of different orders lying at different frequencies which broaden the peaks.

The relation of ring statistics to vibrational spectrum has received considerable study. One correlation which seems generally established is that the steep drop at the upper limit of frequency in fig. 19 is enhanced by the presence of odd-membered (in this case five-fold) rings. The Connell–Temkin structure, alone of those illustrated here, has only even-membered rings, and its DOS declines much more gradually than the others.

Already shown (fig. 6) was a spectrum computed by NFC by Bell et al. (1968). (Also see §5.2.) They studied the nature of the modes contributing to each peak in this DOS of an SiO_2 structure. Each Si atom has four oxygens arranged tetrahedrally around it: each O atom forms a bridge between two silicons. The spectrum can be largely analyzed in terms of the roles of the bridging bonds between the oxygens and their neighbors. The bonds may be stretched, rocked or bent; this is the meaning of the notations S, R, and B in fig. 6.

5. *Localization*

Important questions about the dynamics of disordered solids include some about the nature of localized eigenstates and the conditions for which localization occurs. Considerable effort in which computer experiments have played an important role has been directed at providing solid answers to these questions. Answers are beginning to emerge for simple model systems.

The extent and degree of localization depends on the "amount" of disorder in the solid. Apparently, it also depends on the dimensionality of the system since in one dimension, almost all eigenstates are localized even for the slightest amount of disorder, while in higher dimensions, for certain systems, all eigenstates are localized only when the disorder is sufficiently strong. In addition to this physical difference, there are several practical differences between localization in one and higher dimensions. In contrast to work on higher-dimensional systems, several exact results have been proved for 1d systems and computer experiments are more easily performed. Below we discuss in separate subsections localization in 1d and higher-dimensional systems.

5.1. One dimension

In 1961, Mott and Twose investigated the nature of electron wave functions in a one-dimensional disordered lattice and conjectured that essentially all eigenstates are localized. By localization was meant an exponential decay of the wave function envelope away from some point in the lattice. This surprising proposition applied for all strengths of the potential and types of disorder and had profound implications for transport phenomena. For example, for independent electrons it suggested the vanishing of the d.c. electrical conductivity. The argument of Mott and Twose (see also Mott 1967 and Mott and Davis 1971) was intuitive. More rigorous arguments exist, and many are discussed in the review by Ishii (1973). Since our emphasis is on computer experiments, we will discuss only special points about analytic studies.

It is rigorously known that for certain models, e.g. the Anderson and isotopic mass (diagonally) disordered systems, all eigenstates are localized* and that for other models, e.g. the one-dimensional liquid of δ-function potentials, some states are extended. These conclusions hold under several different definitions of localization. One definition (Borland 1963, Matsuda and Ishii 1970, Ishii 1973) is the exponential growth of the eigenstate envelope as the equation of motion is integrated (or iterated) along the lattice from one end. Another definition (Rubin 1968) frequently used is concerned with the transmission of a plane wave through a barrier consisting of a random array of scatterers. As the length of the barrier

*Except the mode with zero frequency (Ishii 1973).

becomes infinite, localization is indicated by the barrier becoming perfectly reflecting. From the transmission problem viewpoint, one sees why some states may not be localized, since if each scatterer has a transmission resonance, in analogy with the Ramsauer–Townsend effect, then the wave function can be extended. (See also Tong 1970.)

Most studies treat diagonal disorder. However, considering the off-diagonally disordered single band, tight-binding model, Theodorou and Cohen (1976) showed that one eigenstate is always extended in contrast to the diagonal disorder problem where all states are localized. Consequently, in one dimension, without specifying the model or type of disorder it is generally incorrect to say that all eigenstates are localized, although almost all are.

Computer experiments on localization differ from analytic studies in several ways. First, they deal with single realizations of possible configurations. Second, they deal with systems of finite length. The second difference is the most serious because the various proofs of localization (Ishii 1973) are difficult to apply to the interpretation of the results of computer experiments. Extended states exist in systems which, if infinite, would have only localized states.

In a computer experiment, one way to determine if a state is localized is to compute numerically the eigenstate, plot the amplitude (its square, its logarithm or its envelope) as a function of position along the chain, and observe directly to what extent the amplitude is appreciable. Such studies include those by Rosenstock and McGill (1961), Roberts and Mackinson (1962), Borland (1963), Dean and Bacon (1963), Dean (1964), Hertzenberg and Modinos (1964), Payton and Visscher (1967b), Wu (1975) and Chen et al. (1977).

Other studies use a variety of computable parameters as indicators of localization. In barrier transmission studies (Rubin 1968, Erdös and Herndon 1973) transmission and reflection coefficients are computed. Several other workers (Tong and Wong 1972, Bush 1972, Frisch et al. 1973) computed the rate of exponential growth of the eigenstate as the equation of motion is integrated starting at one end of the lattice. For example, Bush used the Schrödinger equation for the Anderson model (see §5)

$$a_{n+1} = [(E-\varepsilon_n)/V]a_n - a_{n-1}$$

to generate successive values of a_n as functions of E. Values of n_i and a_{n_i} defined by $|a_{n_1}| > e^4, |a_{n_2}| > e^6, \ldots, |a_{n_{20}}| > e^{42}$ were collected, and the slope of a linear least-squares fit of $\ln|a_{n_i}|$ vs n_i determined an exponential growth length for a given configuration. Averaging this length over several configurations defined L, the localization length. In addition to L, Bush (1972) also computed transmission and reflection ratios and found close agreement in their predictions. Instead of the amplitudes at the sites, Hori (1968) and Minami and Hori (1970) studied the phases of the displacements of vibrating masses. An equivalence between these phase-theoretic studies and the barrier transmission work of Rubin is demonstrated by Minami and Hori (1970).

Still other definitions of localization are used. We mention Bell's use (1976) of the participation ratio and Moore's use (1973) of the standard deviation of the wave function. The participation ratio is a quantity proposed by Bell et al. (1970) and Visscher (1972) to describe the proportion of the total number N of atoms in a system which effectively contribute to the eigenstate amplitude. If a_i is the eigenstate amplitude at site i, the participation ratio is

$$P=\left[\sum_{i=1}^{N}|a_i|^2\right]^2 / N\sum_{i=1}^{N}|a_i|^4. \tag{5.1}$$

If all sites participate equally, $P=1$; if only one site, $P=1/N$. Moore computed

$$S=\sum_{i=1}^{N}\left[i^2|a_i|^2-\left(\sum_{j=1}^{N}j|a_j|^2\right)^2\right] / N\sigma, \tag{5.2}$$

where

$$\sigma=\left[(N^2-1)/12\right]^{1/2}$$

is the standard deviation of a state distributed uniformly over all sites. This ratio is a measure of the spatial extent of the state. We note that Painter and Hartmann (1976) computed both L, as defined by Bush, and P for a disordered binary chain and found that these measures, while agreeing at low frequencies, disagreed near peaks in the impurity band. They traced the discrepancy to modes localized in more than one region of the chain. Albers and Gubernatis (1977) computed both S and P for the Anderson model and found good agreement between these measures. An example of their results is shown in fig. 21.

Summarizing the results of all the studies, one finds that some states are clearly localized while others extend over the length of the system. The demarcation between localization and extension is not sharp, and the degree of localization is a function of the amount of disorder. For example, using the isotopic mass model, Dean and Bacon (1963), Dean (1964) and

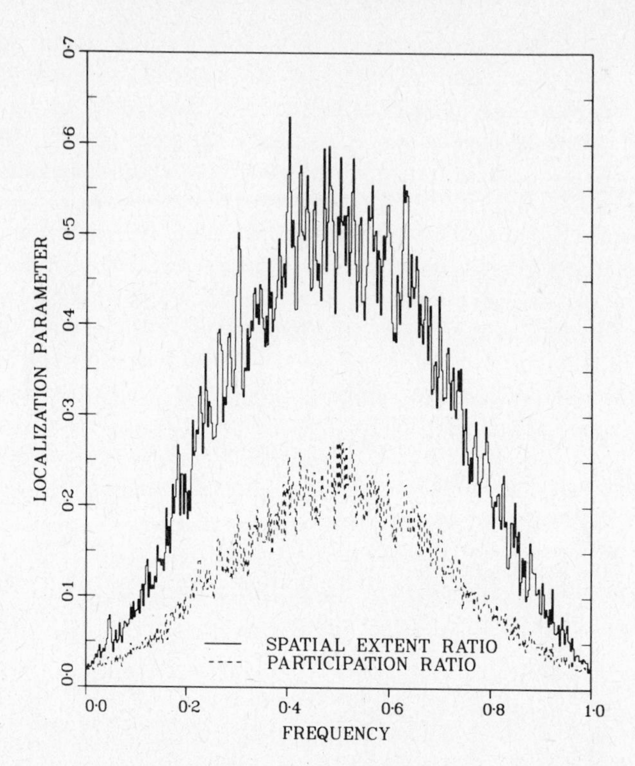

Fig. 21. A comparison between two measures of localization for a 1d 400-atom Anderson model with $W/V = 1$. The participation ratio is defined by (5.1); the spatial extent ratio by (5.2) (Albers and Gubernatis 1977).

Payton and Visscher (1967b) find that the high-frequency modes (especially the impurity band) are highly localized and low-frequency (long-wavelength) modes are extended (fig. 22). Additionally, Payton and Visscher (1967b) show that the degree of localization differs if the lattice is light mass diluted as opposed to heavy mass diluted. These facts can be correlated with the features of the density of states discussed previously.

In connection with these generalizations, we mention two exact results. Herbert and Jones (1971) and Thouless (1972) derive an expression which relates the decay length α^{-1} of a localized state at a given energy E to a moment of the density of states. (See also Papatriantafillou 1973 and Sen 1973.) It is found for the Anderson model with diagonal disorder that

$$\alpha = \int \mathrm{d}x \, \rho(x) |E - x| - \ln|V|,$$

where $\rho(x)$ is the density of states and V is the overlap integral. Hirota (1973) derives expressions for several hamiltonians relating a decay length to the variance of the disorder.

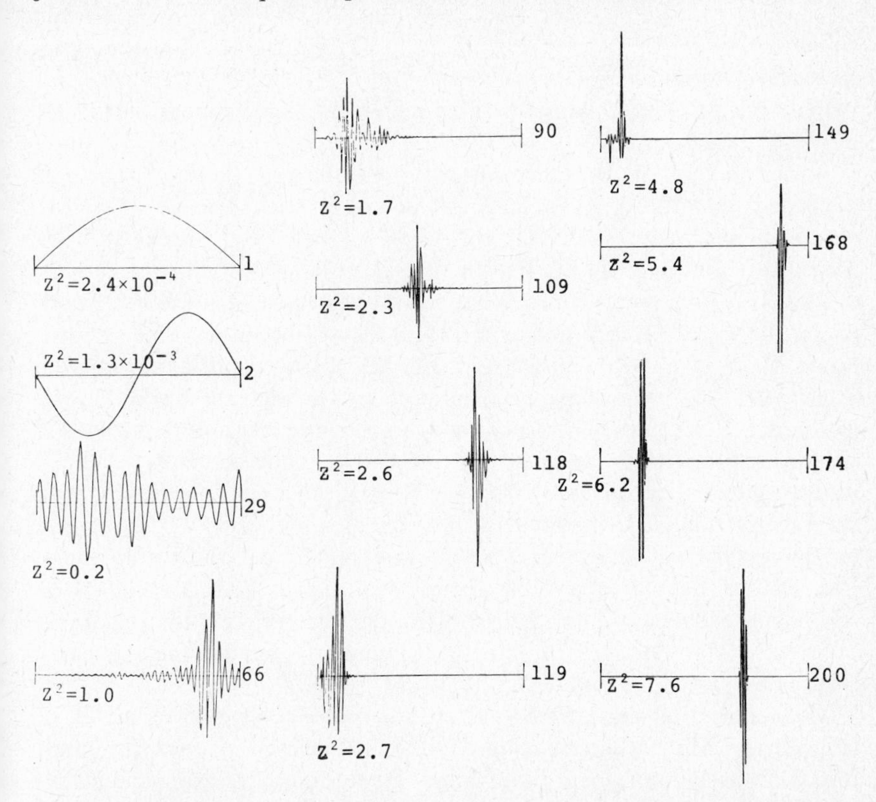

Fig. 22. Normal modes of an isotopically disordered chain containing 100 light atoms and 100 heavy atoms with a mass ratio 3 : 1. Squared frequencies are given by $Z^2/8$ in units of the maximum squared frequency of the pure light chain. Mode numbers are indicated to the right of each graph (Payton and Visscher 1967b).

5.2. Higher dimensions

Anderson (1958) suggested that under some circumstances the exact eigenstates in a random lattice might be localized rather than extended throughout the lattice and presented a mathematical estimate of the condition for localization according to a specific definition of localization. He studied the diffusion of an electron through a random lattice whose dynamics are described by a single-band, tight-binding hamiltonian in a representation of basis states fully localized at each lattice site. In this representation, energies ε_i are assigned to each lattice site i, and tunneling between sites i and j is governed by the overlap integral V_{ij}. Anderson mimicked disorder in the actual electron potential by assigning to the ε_i a random probability distribution. One choice for this distribution was a rectangular distribution of width W centered around zero. He further simplified the problem by restricting V_{ij} to nearest neighbors and ignoring its random variations by setting it equal to a constant V. This particular model of a disordered lattice is often called the Anderson model; it has only diagonal disorder.

Anderson took W/V as a measure of the disorder in the lattice and suggested that for a given lattice structure a critical value of W/V, $(W/V)_c$, may exist such that for $W/V \geqslant (W/V)_c$ all eigenstates are localized. He estimated $(W/V)_c$ for several lattices.

Mott (1967), in particular, emphasized the relevance of Anderson's work to the theory of electron transport in a variety of non-crystalline systems. Since Mott's work, key questions in the theory of electron transport in non-crystalline systems have become when and to what extent does localization occur. Numerous investigations, both theoretical and numerical, of these questions have examined the Anderson model or closely related models. From these investigations a physical picture of localization with several commonly accepted features has emerged. For $W/V < (W/V)_c$ the energy spectrum is separated at critical energies, E_c, called mobility edges, into distinct regions containing either localized or extended states. As W/V approaches $(W/V)_c$ the mobility edges merge, and the regions of extended states disappear, a transition popularly termed the Anderson transition.

The details of electron transport near the mobility edges are a matter of controversy. Mott argues that the $T=0$ d.c. electrical conductivity jumps discontinuously from some finite value (the minimum metallic conductivity) to zero as the Fermi energy E_F is varied across E_c into the region of localized states. For an exponentially localized state of energy E (the wave function envelope proportional to $\exp(-\alpha r)$) near the mobility edge, Mott (1976) also claims that the decay length α^{-1} is proportional to $(E_c - E)^{-s}$, and if s is below certain critical values, 1 and 2/3 in two and three dimensions, there is no minimum metallic conductivity. He maintains that experimental evidence (Pollitt 1976) supports the existence of a minimum metallic conductivity.

On the other hand, Cohen and Jortner (Cohen 1970, Cohen and Jortner 1973, 1974) claim that long-ranged variations in the electron potential energy always exist and interpose near the mobility edge an inhomogeneous regime, where extended and localized states coexist, to which classical percolation theory can be applied so that the electrical conductivity can rise continuously from zero. This conflicts with Mott's arguments about the decay length in which he claims that the existence of a minimum metallic conductivity precludes such long-ranged fluctuations. Recently, Licciardello and Thouless (1977) argue that these fluctuations probably occur, but do not prevent a sharp mobility edge. Other theorists have estimated values of s that lie below Mott's critical values for the existence of minimum metallic conductivity (Anderson 1972, Abram and Edwards 1972, Freed 1972, Abram 1973).

Recently, Economou and Antoniou (1977) have questioned a basic feature of the picture of localization. They suggest that when off-diagonal (restricted to nearest neighbors) disorder is added to the Anderson model, a region of extended states around the middle of the band always exists: That is, the mobility edges do not merge, and there is no Anderson transition. Many theorists have implicitly assumed that the effects of off-diagonal disorder are qualitatively similar to those of diagonal disorder. Generalizations about the nature of localization drawn from calculations involving only diagonal disorder may need some re-evaluation.

Various theoretical efforts have been directed at accurate estimates of $(W/V)_c$. These efforts are reviewed by Thouless (1974) and Licciardello and Economou (1975); however, the work on the spatial extent of the wave functions near mobility edges is less extensive and unreviewed. The analytic details of all this work are involved; we will not discuss them. We only comment that the values of $(W/V)_c$ for various lattices are not precisely known, the more important questions about the spatial extent of electron wave functions near mobility edges are not fully answered, and consequences of off-diagonal disorder are essentially unexplored.

Many computer experiments have been performed to complement and test a number of the concepts and theoretical developments discussed above. These 2d and 3d calculations are more difficult to execute than those in one dimension. One reason is that there are no exact results to provide a calibration. Another reason is that the finite amount of available computer storage limits the number of lattice sites that can be studied. This number is approximately independent of the dimensionality of the system. Consequently for a fixed number of sites, the fraction of sites on the boundary of the system rapidly increases with dimensionality. Care is thus required to distinguish features of the eigenstates caused by the disorder from those associated with boundary effects. Most investigators study 2d systems because the fraction of boundary sites is much less than in 3d.

For a square lattice of 10^4 sites, Yoshino and Okazaki (1976) studied the localization of eigenstates in the Anderson model. They observed directly the behavior of eigenstates as a function of W/V and estimated $(W/V)_c$. Partly because of the size of their system, nearly two orders of magnitudes larger than most others, their work is to date the most convincing numerical demonstration of the existence of the Anderson transition, and their estimate of $(W/V)_c = 6.5$ may be the best available from computer experiments.

They studied localization in the following manner: Because of the symmetry in the Anderson model for a square lattice, the eigenvalues near $E=0$ lie in the center of the band, and the eigenstates near the band center are expected to be among the last to remain extended as W/V approaches $(W/V)_c$ from below. They computed just a few eigenstates near the band center and from their behavior assessed the onset of the Anderson transition. They examined the quantities

$$\ln A_x = \ln\left(\sum_y |a_{xy}|^2\right),$$

$$\ln A_y = \ln\left(\sum_x |a_{xy}|^2\right),$$

as a function of x and y, the positions along columns and rows of the lattice. a_{xy} is the computed eigenstate amplitude at position (x,y).

Exponential localization suggests that A_x behaves approximately as $\exp(-2\beta|x-x_0|)$ and A_y as $\exp(-2\beta|y-y_0|)$, where the position (x_0,y_0) is the center of localization. Their computed results show a distinct dependence of $\ln A_x$ and $\ln A_y$ on W/V with quite different behavior for small (extended states) and large (localized states) values of W/V, and the published results are shown in fig. 23. For large W/V, plots of $\ln A_x$ and $\ln A_y$ as functions of x and y show regions of nearly linear behavior from which they estimated values of β. The value of $(W/V)_c$ was estimated from a curve (see fig. 24) of β versus W/V by extrapolating β to zero. Their estimate of $(W/V)_c=6.5$ is consistent with numerical work of Edwards and Thouless (1972) and Licciardello and Thouless (1975a, b). This value is significantly smaller than the theoretical estimate of Anderson (1958), implying that localization in a disordered system, at least for a square lattice, occurs more easily than he originally suggested.

Thouless and coworkers use a specific localization criterion. For the Anderson model they (Edwards and Thouless 1972) investigate the shifts of individual eigenvalues caused by changes in boundary conditions, arguing that extended states are more sensitive to boundary conditions than localized states and thus exhibit larger energy shifts. For exponentially localized states the energy shift is expected to decrease exponentially with the distance between the localization center and boundary; for extended states they argue that the shift is of the order $\hbar/\tau$, where τ, dependent on the size of the system, is the time an electron takes to diffuse to the boundary.

In practice they examined systems with at most several hundred lattice sites. So to reduce problems with surface states they used periodic boundary conditions and then replaced conditions along one pair of

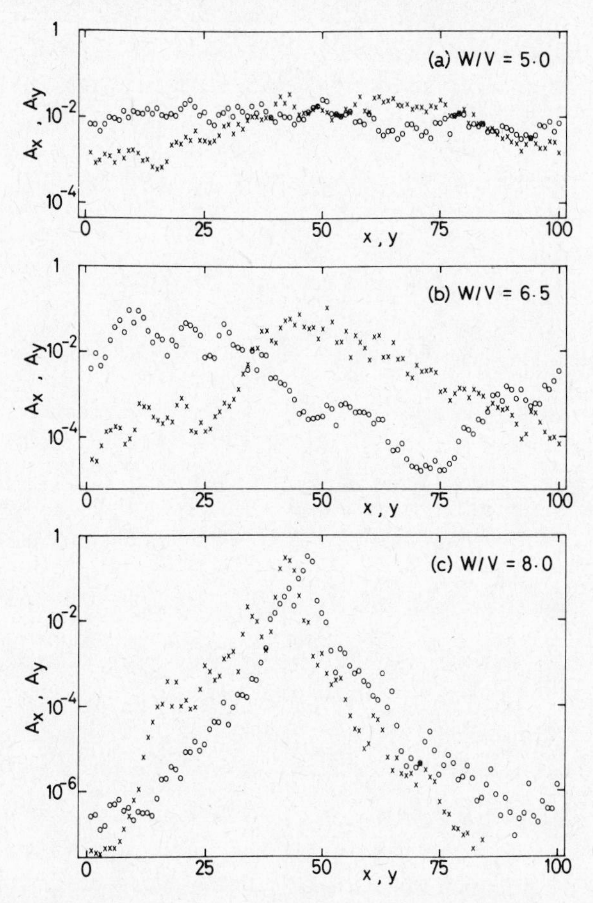

Fig. 23. $\ln A_x$ vs x (circles) and $\ln A_y$ vs y (crosses) for an eigenstate near the band center for different values of disorder (Yoshino and Okazaki 1976).

boundaries with antiperiodic boundary conditions. In this scheme all states are effectively equidistant from the boundaries. For each boundary condition, all the eigenvalues were computed and individual differences ΔE determined.

Edwards and Thouless found that eigenstates in a square lattice localized more easily than eigenstates in a diamond lattice. These lattices have the same coordination number, but, of course, exist in spaces of different dimensionality. Furthermore, the Anderson transition was not as clearly defined for the diamond as for the square lattice. This behavior of the diamond lattice led them to suggest the existence of states intermediate between fully localized and fully extended. This possibility was studied by

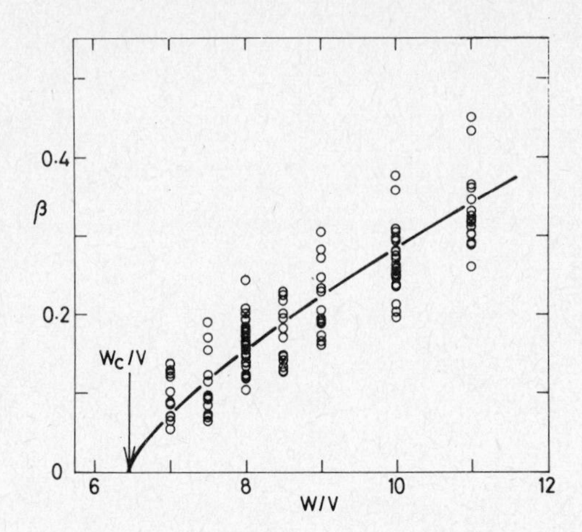

Fig. 24. The spatial extent β versus disorder. A circle denotes a value of β estimated from the $\ln A_x(\ln A_y)$ vs $x(y)$. Different circles for the same value of W/V correspond to different eigenstates near the band center (Yoshino and Okazaki 1976).

Last and Thouless (1974) who suggest a power law, rather than an exponential, fall-off of the eigenstate amplitude envelope from its maximum.

Licciardello and Thouless (1975a, b) use a scaling argument which suggests for 2d disordered systems that the value of the minimum metallic conducitivity is a universal constant. They find that for extended states

$$N\,\overline{\Delta E}\,\rho(E)=2\sigma\hbar/e^2, \tag{5.3}$$

where N is the number of lattice sites, $\overline{\Delta E}$ is the average energy shift, $\rho(E)$ is density of states per site at the Fermi energy and σ is the $T=0$ d.c. conductivity. They classify states as localized if $N\overline{\Delta E}\rho(E)$ decreases steadily with increasing N (a conclusion based on an expected exponential decrease of $\overline{\Delta E}$ with the number of lattice sites); otherwise, the states are extended. From computations of $N\overline{\Delta E}\rho(E)$ as a function of E for systems of different N, they estimate E_c, the value of E at which the N dependence of $N\overline{\Delta E}\rho(E)$ changes. Then from (5.3) they find the minimum metallic conductivity. Since this number is supposed to be a universal constant, they computed its value for different lattice structures and find it equals $(0.12\pm0.03)e^2/\hbar$. Their results for the different lattice structures and different values of W/V are shown in fig. 25.

Licciardello and Thouless (1977) reinterpreted their conductivity results to investigate Mott's (1976) predictions about the spatial extent of the wave function near the mobility edge. They find that the decay rate α^{-1} is

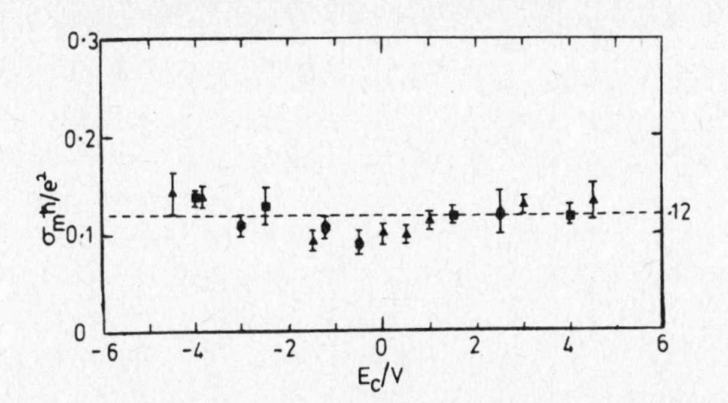

Fig. 25. The minimum metallic conductivity for the honeycomb, square, and triangular lattice (indicated by the appropriate symbol) for various values of W/V. The error bars indicate the spread in results as the system size was varied over 64, 100, 144 and 196 atoms (Licciardello and Thouless 1975a, b).

proportional to $(E_c - E)^{-s}$ with $s \geqslant 1$, which indicates that a minimum metallic conductivity exists in two dimensions.

Kikuchi (1974) has used the Edwards–Thouless localization indicator to study localization in a two-dimensional structurally disordered lattice. He modeled structural disorder by randomly placing a small number (one percent of total number of sites) of "impurities" on a large, in some cases over 10^4 sites, *empty* square array. The interactions among these impurities are described by a single-band, tight-binding hamiltonian that has for the overlap integral between two impurities separated by a distance R_{ij} (off-diagonal disorder)

$$V_{ij} = -V_0(1 + R_{ij}/a)\exp(-R_{ij}/a).$$

This interaction, characterized by a range a, was truncated at 50 lattice spacings. He considered cases where all the ε_i were zero or specified according to the rectangular distribution and also varied a. He found that increasing a increases the region of the energy band for which extended states exist. He did not identify any mobility edges.

Whereas Yoshino and Okazaki (1976) computed both eigenvalues and eigenvectors and Thouless and coworkers computed just eigenvalues, Weaire and Williams (1976, 1977) and Weaire and Srivastava (1977) computed neither. They studied localization in the Anderson model by using the equation-of-motion method of Alben et al. (1975c) which we described in §3.3. With this method they computed the average inverse participation ratio and examined it as a measure of localization for lattices of several hundred sites of dimension one through four. Weaire and Williams could not identify the Anderson transition, but they obtain suggestive results and indicate various refinements and generalization of the method as areas for future study. Their result for a four-dimensional lattice, for example, supports the contention of Thouless (1976) that Anderson localization occurs in four dimensions. Weaire and Srivastava refined the method used by Weaire and Williams (1977) by incorporating a more careful extrapolation to infinite time and were able to more accurately pinpoint the Anderson transition in a diamond lattice. They find $(W/V)_c \simeq 8$, about half the numerical estimate of Edwards and Thouless (1972), but nearly equal to the theoretical prediction of Licciardello and Economou (1975).

A number of other workers have examined related aspects of localization by various techniques and for various model systems. Khor and Smith (1971a, b) studied the localization of eigenstates in the band tails for 2d and 3d structurally and site disordered lattices and determined localization by direct examination of contour plots of the absolute value of the wave functions or $\ln|\psi(\boldsymbol{r})|$. Their study was not limited to the Anderson model as they also did a finite difference solution of the Schrödinger equation with spherically symmetric potentials. One of their conclusions was that states in the band tails localize around impurity clusters.

Schönhammer and Brenig (1973) infer results about the position of the Anderson transition and mobility edges from the behavior of the $T=0$ d.c. conductivity. More specifically, they compute a sum related to the variance of the wave function. A finite value of this sum is expected only if the wave function is localized and is connected with a vanishing d.c. conductivity. Thus, the convergence or divergence of their sum indicates the localization or extension of the wave functions. They compute the sum for various values of W/V for $E_F=0$ to determine $(W/V)_c$ and for various values of E_F at fixed W/V to determine E_c. Their value of $(W/V)_c$ for a square lattice is higher than that obtained by Khor and Smith (1971a), Edwards and Thouless (1972) and Yoshino and Okasaki (1976), but their value for a diamond lattice is in reasonable agreement with that of Edwards and Thouless (1972) and the behavior of their values of E_c as a function of W/V is in qualitative agreement with theory (Bishop 1974).

A variety of disordered systems were studied by Visscher (1972). He considered the single-band, tight-binding hamiltonian not only for a rectangular distribution of site energies (the Anderson choice), but also for gaussian and binary distributions. The overlap was variable ranged

$$V_{ij} = V_0 \exp(-R_{ij}/R),$$

where R_{ij} is the distance between sites i and j, and R is the range of the interaction. Both structurally (off-diagonal disorder) and site disordered square lattices were studied. His localization criterion was the participation ratio, and his qualitative conclusions included the observations that increasing the range of interaction in a site disordered system increased localization, but in a structurally disordered system decreased localization. Most of the systems exhibited neither sharp mobility edges nor Anderson transitions.

Work discussed so far has been for the Anderson model or related tight-binding hamiltonians. A number of studies of localization in harmonic lattices also exist. In fact, the earliest numerical demonstrations of localization in two and three dimensions were for isotopically disordered, binary, harmonic lattices (Dean and Bacon 1962, 1965, Payton 1966, Payton and Visscher 1967b). Localization was determined by direct observation of the eigenmodes.

In general, calculations of localization in harmonic lattices have not addressed the theoretical questions raised by Mott (1967) and others. The interest has been to relate features (peaks) of frequency spectrum to identifiable types of localized modes. Some peaks are identified with modes localized about islands of light impurity clusters. The observation of this type of localization is reminiscent of similar observations for one-dimensional systems. The higher-dimensional systems, however, exhibit features not present in one dimension. For example, Payton and Visscher (1967b) found that the peaky structure of the density of states is much smoothed after the critical percolation concentration for light masses is passed. Localization associated with internal degrees of freedom is also possible. Studying a model of SiO_2 glass, Bell et al. (1970, 1971) used the participation ratio as a criterion for localization and identified features in the frequency spectrum from local normal modes associated with the bending, rocking and stretching of molecular bonds. Dean (1972) and Bell (1972a, 1976) have amply reviewed other work on localization in site and structurally disordered harmonic lattices.

6. *Electrical transport*

As mentioned in the previous section, eigenstate localization has significant consequences for electrical transport in a disordered system. Intuitively, for non-interacting electrons, no d.c. conductivity is expected when the Fermi energy is in the midst of localized states. The localization makes all transitions except those to nearby states improbable, and since these neighboring states must have at least slightly different energies, energy must be supplied for transitions to occur. With an a.c. electric field, the energy of a photon is available; at finite temperatures, the energy of a phonon is available. Conductivity by phonon-assisted tunneling is often called hopping conductivity.

Below we consider electrical transport in disordered solids in the presence of weak d.c. and a.c. electric fields. For each type of field, conduction with and without phonon assistance ("phononless") is discussed.

6.1. d.c. conductivity

6.1.1. *Phononless conduction*

There are several formal arguments for a zero d.c. conductivity in the absence of an electron–phonon interaction (Halperin 1967, Schönhammer and Brenig 1973, Thouless 1974). For example, applying the Kubo–Greenwood formula to the Anderson model, Schönhammer and Brenig showed that the minus second moment of the real part of the conductivity is

$$\int_0^\infty \omega^{-2}\sigma(\omega)\,\mathrm{d}\omega = \frac{e^2}{\Omega}\sum_{ij}(x_i - x_j)^2 R_{ij}, \tag{6.1}$$

with x_i the position of the ith site and

$$R_{ij} = \sum_{\alpha\beta} \frac{\langle\alpha|i\rangle\langle i|\beta\rangle\langle\beta|j\rangle\langle j|\alpha\rangle}{E_\alpha - E_\beta}\left[f(E_\alpha) - f(E_\beta)\right],$$

where $|\alpha\rangle$ is an eigenstate with eigenvalue E_α and Ω is the volume and f is the Fermi–Dirac distribution function. They then argued that if E_F lies in a region of localized states, states with finite variance, the sum on the right-hand side converges and bounds the integration on the left-hand side. However, for this integral to be finite, $\sigma(\omega)$ must go to zero faster than ω. (Mott 1970 argues that $\sigma(\omega) \sim \omega^2(\ln \omega)^4$.)

Schönhammer and Brenig (also §5.2) supported their argument by a computer experiment. More precisely, what Schönhammer and Brenig investigated at $T = 0$ was the convergence of

$$S_{ij} = \sum_{k=j}^{i} (x_k - x_j)^2 R_{kj}.$$

If S_{ij} converges as a function of $|x_i - x_j|$, localization is suggested. As their results are basically consistent with other criteria for localization, the d.c. conductivity in a static lattice does appear to vanish. Additional support for a vanishing d.c. conductivity is found in the localization studies of Licciardello and Thouless (1975a, b), also described in the previous section.

A strictly one-dimensional argument for a vanishing $T=0$ d.c. conductivity is given by Landauer (1970) in a work reminiscent of barrier transmission studies discussed in §5.1. He showed that the average d.c. electrical resistance of an array of scatterers grows exponentially as the length of the array increases.

6.1.2. *Phonon-assisted conduction*

The hopping conductivity problem has received considerable attention especially because of its connection to electrical and other transport properties of amorphous semiconductors. Most studies of d.c. hopping conductivity are based on the work of Miller and Abrahams (1960) who reduced the problem of hopping conduction among localized states with the assistance of a single phonon to a problem of current flow through a network of random conductors. The conductors link randomly positioned localized sites, and the net conductance of the network can be found by applying Kirchoff's law. The conductor between sites i and j has conductance (Ambegaokar et al. 1971)

$$G_{ij}=\frac{e^2}{kT}\gamma_0\exp\left(-\alpha R_{ij}-\frac{|E_i-E_j|+|E_i|+|E_j|}{2kT}\right) \tag{6.2}$$

$$\equiv\frac{e^2}{kT}\gamma_0\exp(-s_{ij}), \tag{6.3}$$

where γ_0 is a material-dependent constant depending on the electron–phonon coupling constant, the phonon density of states, etc. R_{ij} is the distance between sites i and j; and the site energies E_i are measured with respect to the Fermi energy. This expression was originally derived for transport among impurity states in a doped semiconductor. The ith impurity state was regarded as a hydrogenic ground state (exponentially localized) with an energy E_i and an effective Bohr radius α^{-1}. Hence α measures the degree of localization.

It is expected that the net conductance of such a network of random conductors is determined by paths where the average hop is one of high conductance (least resistance). As Mott (1969) observed, at low temperatures such paths are not simply ones dominated by nearest-neighbor hops, but are paths with hops over variable distances between sites of nearly equal energy. That is, what is important is on the average to minimize s_{ij}, not simply to replace R_{ij} by an average nearest-neighbor distance. For this reason hopping conductivity is often called a variable-range process.

The variable-range hopping has led to several interesting predictions for the electrical conductivity: One is that at sufficiently low temperatures the (real) d.c. conductivity behaves like (Mott 1968, Brenig et al. 1973a)

$$\sigma(0)=\sigma_0 \exp\left[-\left(T_0^{(d)}/T\right)^{-1/(d+1)}\right], \tag{6.4}$$

where d is the dimensionality of the system and σ_0 and $T_0^{(d)}$ are material-dependent constants. A second is that the (complex) a.c. conductivity at a given low T is of the form (Pollack and Geballe 1961)

$$\sigma(\omega) \sim \omega^{-s}, \tag{6.5}$$

where $s<1$.

The parameters σ_0 and $T_0^{(d)}$ and the form (6.4) have been the subjects of diverse investigations. Mott (1969), making various physical assumptions, minimized the term s_{ij} in (6.3) and found that

$$kT_0^{(3)}=\lambda\alpha^3/g(E_{\mathrm{F}}),$$

where λ is a dimensionless constant, α the localization length, and $g(E_{\mathrm{F}})$ the density of states at the Fermi level. Ambegaokar et al. (1971) refined Mott's estimate through the reduction of the problem of finding the effective conductance of a random network of Miller–Abraham conductors (6.2) to a problem in percolation theory. They were able to relate $T_0^{(3)}$ to a critical percolation density ν_{c}, to be defined below,

$$kT_0^{(3)}=4\nu_{\mathrm{c}}\alpha^3/g(E_{\mathrm{F}}).$$

The relevance of percolation theory to this problem can be understood as follows: They define a critical percolation conductance G_{c} to be the largest value of G such that if only those conductances in the random network for which $G_{ij}>G_{\mathrm{c}}$ are kept, and the others set to zero, then a conductance path exists through the system. They note that the conductance of the random network is determined mostly by the conductances with values in the neighborhood of G_{c}, because regions with $G_{ij} \ll G_{\mathrm{c}}$ are isolated from each other and shorted by the percolative network with $G_{ij} \sim G_{\mathrm{c}}$, whereas regions with $G_{ij} \gg G_{\mathrm{c}}$ might as well have $G_{ij}=\infty$, because their effect is limited by the percolative network which connects them to one another. Ambegoakar et al. call the network with $G_{ij} \gtrsim G_{\mathrm{c}}$ the critical subnetwork.

The importance of percolation in conduction through a random network was investigated in a computer experiment by Ambegaokar et al. (1973). They considered a number N ($\sim$1000) of random points enclosed in a cube and periodically extended in all directions. The points were connected by conductances $G_{ij} = G_0 \exp(-\alpha R_{ij})$, with R_{ij} limited to about the first eight nearest neighbors. (Note that this is not a variable-range hopping network because the energy dependence of (6.2) is absent.) The conductivity of the system was computed by assigning voltages to each site and solving Kirchoff's equations iteratively until a potential difference U_0 existed between every point and its periodic image in the z-direction. The current I was computed and was found to behave like

$$I = G_0 U_0 \mathrm{e}^{-r_c}. \tag{6.6}$$

The link to theory is made in the following manner: If r_c is connected to percolation, then it is the distance up to which randomly distributed points would have to be connected pairwise to obtain a non-zero probability of connected points extending across the system. The critical percolation concentration is

$$p_c^{(3)} = \frac{4\pi}{3} \rho \left(\frac{r_c}{2} \right)^3, \tag{6.7}$$

where ρ is the density of sites. If a sphere of radius $\frac{1}{2} r_c$ is drawn about each "bonded" site, then $p_c^{(3)}$ is the ratio of the volume enclosed by these spheres to the total volume. When r_c determined from (6.6) was substituted into (6.7), Ambegaokar et al. found $p_c^{(3)} = 0.30 \pm 0.015$ in the limit $N \to \infty$ (fig. 26). This value is in close agreement with critical percolation probabilities found in percolation theory, e.g. Holcomb and Rehr (1969).

One should note that the percolation problem of Ambegaokar et al. (1973) is neither the site nor the bond percolation problem discussed in §2.1.1. (Pike and Seager 1974 discuss classification schemes for generalized percolation problems and present numerical results for a variety of 2d and 3d models.)

The prefactor to the exponential in (6.4) has been investigated by Mott (1972), Pollack (1972), Ambegaokar et al. (1973), Brenig et al. (1973b), Kurkijärvi (1974), Levinshtein et al. (1976) and Shklovskii and Efros (1976). When the energy dependence between sites is ignored, i.e. temperature dependences are ignored, Ambegaokar et al. and Kurkijärvi argue that

$$\sigma_0 \propto G_c \propto (r_c \alpha)^{-\nu} \mathrm{e}^{-\alpha r_c},$$

and Levinstein et al. (1976) and Shklovskii and Efros (1976) show that

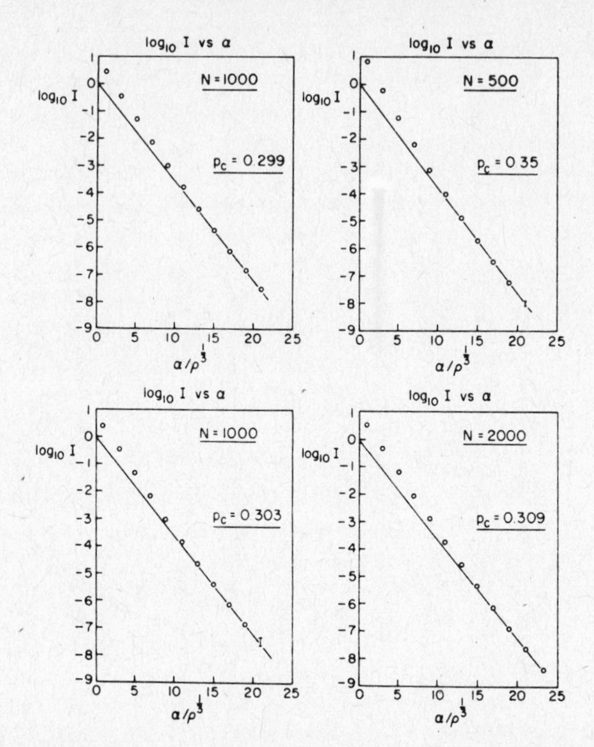

Fig. 26. Logarithm of the current I for a network of random conductors vs $\alpha/\rho^{1/3}$ for systems of different sizes. The critical percolation constant p_c is calculated from the slope of the asymptote (Ambegaokar et al. 1973).

$\nu = 1 + \nu'$ is related to the critical exponent ν' for the correlation length defined for the phase transition between an unpercolated and a percolated system. When the energy dependences of the sites are included, Ambegaokar et al. (1973) suggest that the temperature dependence of the prefactor is $[T_0^{(d)}/T]^{-\nu/4}$. In a computer experiment that used the same model as Ambegaokar et al. (1973), Kurkijärvi (1974) found ν to be 0.60 ± 0.25.

In review, Ambegaokar et al. (1971, 1973) take a Miller–Abrahams random network of conductors and assume percolation theory describes its effective behavior. When the effective behavior is evaluated, (6.4) follows with σ_0 and $T_0^{(d)}$ depending on percolation parameters. The applicability of percolation theory is then demonstrated by using a computer experiment to evaluate a random network of conductors closely related to the Miller–Abrahams network.

Others have taken a direct approach and have used the computer to evaluate the effective conductance of a Miller–Abrahams random network as a function of temperature to find that their results are described by (6.4). These workers include Seager and Pike (1974), Maschke et al. (1974b) and Rycerz and Mościński (1976a, b, c). The results of Maschke et al. for a 2d system are shown in fig. 27. $\ln\sigma$ is proportional to $T^{-1/3}$ over a large temperature range.

In addition to finding $\ln\sigma \propto T^{-1/(d+1)}$, Seager and Pike (1974) and Maschke et al. (1974b) emphasize the importance of the conductances varying by many orders of magnitude for the critical subnetwork ideas of Ambegaokar et al. (1971) to be valid. In particular, Seager and Pike computed the conductance of a simple cubic network of conductors G_{ij} whose logarithm is uniformly distributed between $-\log A$ and $\log A$ and compared this conductance with $\log G_c = \frac{1}{2}\log A$ predicted by simple percolation models for this distribution of G_{ij}. Their result, shown in fig. 28, indicates that for $A < 10^3$ the percolation model is not particularly useful in characterizing the network, but for $A > 10^3$ the percolation model fits the computed results.

Legions have investigated network problems for which percolation theory is inapplicable. In the interpretation of the results of these works, percolation theory often is replaced by a self-consistent effective medium approximation (Bruggemann 1935, Landauer 1952, Krumhansl 1973, Kirkpatrick 1973, Elliott et al. 1974, Stroud 1975).

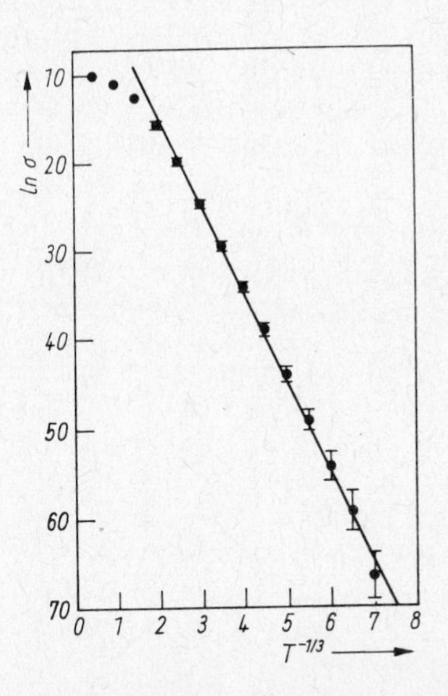

Fig. 27. Logarithm of the conductivity vs $T^{-1/3}$ for a 2d variable-range hopping network (Maschke et al. 1974b).

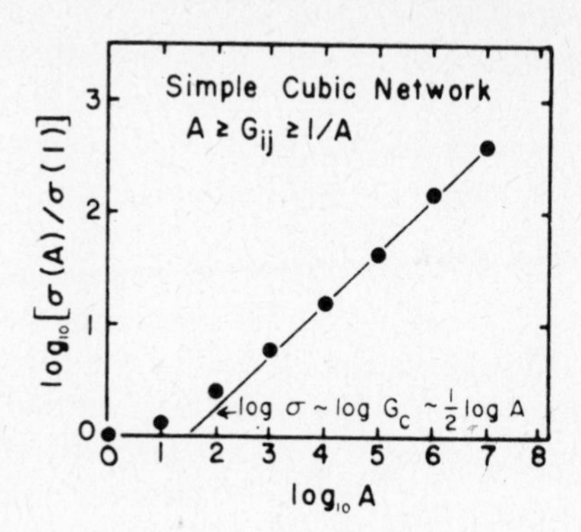

Fig. 28. Logarithm of the conductivity of a simple cubic network of random conductors vs the logarithm of the conductance spread. The solid line is the variation predicted from percolation theory (Seager and Pike 1974).

Among the investigations in this conduction regime are the computer experiments of Kirkpatrick (1971, 1973) and analog experiments of Last and Thouless (1971) and Adler et al. (1973). Their principal result is that the $T=0$ conductivity $G(p)$ is not proportional to the percolation probability $P(p)$. This is illustrated in fig. 29 (Kirkpatrick 1973) where the computed conductivity and the theoretical percolation probability are compared for site and bond percolation in a 3d simple cubic lattice. The difference between $G(p)$ and $P(p)$ is generally accounted for by the observation that when p is near p_c much of the material contributes very little to the conductivity because the percolation channels through the material are tortuous and constricted (Last and Thouless 1971, Adler et al. 1973). Also shown in fig. 29 are the predictions of the self-consistent effective medium approximation.

Hopping conduction in one-dimensional systems requires special attention. A network with random conductors between nearest neighbors is a trivial problem: The net resistance is the sum of individual resistances. On the other hand, variable-range hopping is a subtle problem: The expected temperature dependence (from 6.4), $\ln\sigma \propto -T^{-1/2}$, is not explained by the critical subnetwork concept of Ambegaokar et al. (1971) since a 1d system cannot percolate. The conductance of the network is dominated by high-resistance hops called blockages or hard hops, and as Kurkijärvi (1973) showed, these blockages in a system of infinite length lead to

$$\ln\sigma \propto -T^{-1}. \tag{6.8}$$

On the other hand, Shante et al. (1973) argued that real materials in which charge is transported effectively in only one dimension actually consist of

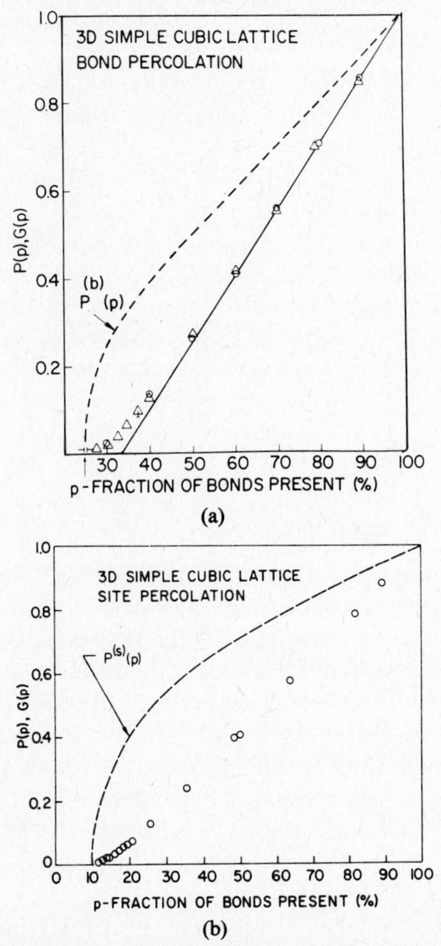

Fig. 29. Percolation probability $P(p)$ (dashed line) and conductance $G(p)$ (data points) for a 3d simple cubic network of random conductors. (a) Bond percolation. The solid line is the prediction of the self-consistent effective medium approximation. (b) Site percolation (Kirkpatrick 1973).

a large number of parallel chains of finite length. Some of these chains will not have blockages and, being of higher conductance, dominate the conductivity of a bundle of parallel, finite chains. This critical subnetwork provides continuous paths for the charge transport and the form

$$\ln \sigma \propto -T^{-1/2} \tag{6.9}$$

is recovered. Computer experiments by Shante et al. (1973), Maschke et al.

(1974a) and Rycerz and Mościński (1976a) verify the temperature dependences in (6.8) and (6.9).

Brenig et al. (1973a), Kurkijärvi (1973) and Shante et al. (1973) discuss fine points about the applicability of hopping conductivity to one dimension. The computed temperature dependence (6.9) is in qualitative agreement with measured conductivities of quasi-one-dimensional materials like TCNQ-compounds and Krogman salts (Schegolev 1972). The applicability of hopping conductivity to these materials is discussed by Bloch et al. (1972) and Bloch and Varma (1973).

In closing this section, we note that variable-range hopping is not the only process that leads to $\ln\sigma \propto T^{-1/4}$. For example, we refer the reader to Adler et al. (1973), Kirkpatrick (1974) and Emin (1976).

6.2. a.c. conductivity

6.2.1. *Phononless conduction*

For one-dimensional systems Halperin (1967) has derived an exact formalism for the ensemble averaged a.c. conductivity which is related to Schmidt's (1957) exact formalism for the integrated density of states. His equations apparently await solution. (Hirsch and Eggarter 1976, 1977 have solved Halperin's equations for the spectral density.) Hirsch and Eggarter (1976) suggest $\sigma(\omega)$ may be calculated by use of a new numerical approach to computer experiments in one dimension with continuous disorder.

The $T=0$ a.c. conductivity has been studied in two computer experiments on one-dimensional systems (Penchina and Mitchell 1972, Albers and Gubernatis 1977). These experiments are based on the direct evaluation of the Kubo–Greenwood formulas

$$\sigma(\omega)=\frac{2e^2\pi}{L\omega m^2}\sum_{\alpha}\sum_{\beta\neq\alpha}\int_{E_F-\hbar\omega}^{E_F}\mathrm{d}E\,|p_{\alpha\beta}|^2\delta(E-E_\alpha)\delta(E-E_\beta+\hbar\omega) \quad (6.10)$$

$$=\frac{2e^2\pi\omega}{L}\sum_{\alpha}\sum_{\beta\neq\alpha}\int_{E_F-\hbar\omega}^{E_F}\mathrm{d}E\,|x_{\alpha\beta}|^2\delta(E-E_\alpha)\delta(E-E_\beta+\hbar\omega), \quad (6.11)$$

where x and p are the position and momentum operators for an electron in a system of length L. $x_{\alpha\beta}$ and $p_{\alpha\beta}$ equal $\langle\alpha|x|\beta\rangle$ and $\langle\alpha|p|\beta\rangle$. Clearly, the principal numerical task is the evaluation of the eigenvalues and eigenvectors.

Penchina and Mitchell studied a single electron in an array of randomly positioned δ-functions. This single electron occupied $E_1=E_F$, the lowest energy level of the system. In (6.10) β is thus restricted to $\beta=1$ as the only contributions to the conductivity are transitions from this lowest level to all other levels. Even when averaged over ten configurations and subjected to a smoothing operation, their results varied drastically from configuration to configuration, and they concluded that the conductivity spectrum is not well defined. However, they do claim that the overall behavior of the results is consistent with a vanishing d.c. conductivity.

Albers and Gubernatis (1977) studied the $T=0$ a.c. conductivity of the Anderson model. For several different values of W/V they computed the eigenvalues and eigenvectors and used them in (6.11). With E_F at the middle of the band, they found for weak disorder, $W/V=1$, that $\sigma(\omega)$ resembles conductivity of an ordered system by having a strong peak near $\omega=0$. As W/V increases, this peak shrinks and moves to higher frequencies, and the overall value of the conductivity is lowered. Relative to the system length (400 sites), eigenstates near the center of the band for the weak disorder case were quite extended compared to those near the band tails. When the Fermi energy was placed in each transition region between the relatively extended and quite localized states, the conductivities were quite similar to conductivity when the Fermi energy was in the middle of the band with relatively extended states immediately above and below. Their conductivity spectrum, averaged over 12 configurations, was well defined, and the structure in their spectrum was consistent with a system of small length averaged over a small number of configurations. The computed conductivity for strong disorder, $W/V=10$, is shown in fig. 30 alongside the conductivity computed for the perfectly ordered lattice. The effect of the disorder is clearly to decrease the conductivity and to shift the peak near $\omega=0$ that is present in the ordered lattice to higher frequencies.

6.2.2. *Phonon-assisted conduction*

A widely studied consequence of phonon-assisted conduction in the presence of an a.c. electric field is the finding of Pollack and Geballe (1961) that for a fixed temperature and a large number of materials the conductivity behaves as

$$\sigma(\omega) \propto \omega^{0.8}. \tag{6.12}$$

Starting with a rate equation, they showed

$$\sigma(\omega) = \int \frac{P(\tau)(\omega\tau)^2}{1+(\omega\tau)^2} \mathrm{d}\tau,$$

where $P(\tau)$ is the distribution of relaxation times τ. Basically, this equation

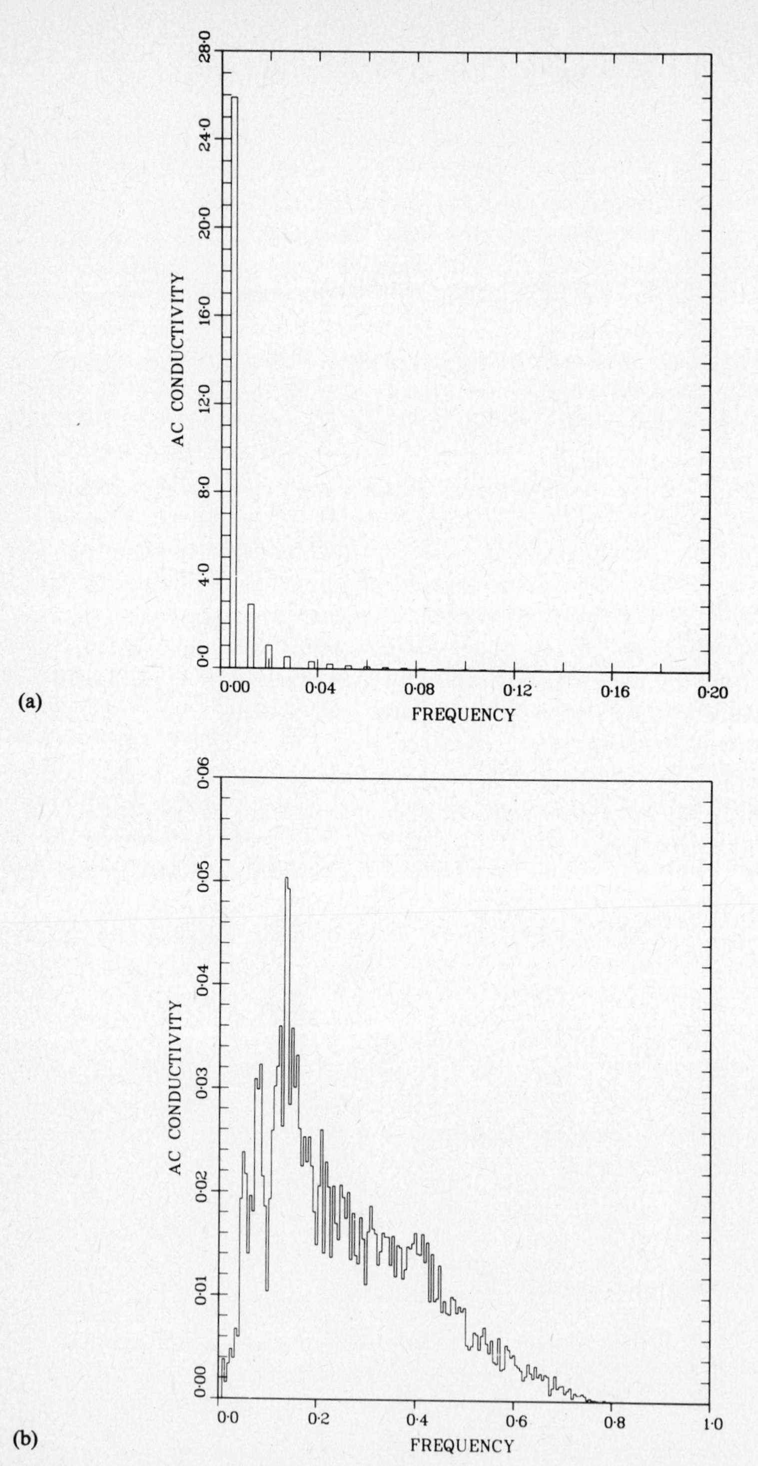

Fig. 30. Electrical conductivity of a 400 site s-band tight-binding model vs angular frequency. The conductivity is in units of $2\pi ae^2/\hbar$; $\hbar\omega$ is in units of the band width. (a) The ordered lattice. (b) The Anderson model with $W/V = 5$ (Albers et al. 1977).

represents the average of independent Debye loss processes each characterized by a random relaxation time τ. When the relaxation time τ_{ij} for hops between sites i and j with an energy difference ΔE_{ij} was taken from the theory of Miller and Abrahams (1960),

$$\tau_{ij} \propto (\alpha R_{ij})^{-3/2} \exp(2\alpha R_{ij}) \tanh(\Delta E_{ij}/2kT) \tag{6.13}$$

and when $P(\tau)$ was for a uniform distribution of hopping sites and temperatures comparable to ΔE_{ij}, they showed that the computed frequency dependence was fitted by (6.12). Their result was developed for impurity band conduction in doped semiconductors, but was carried over, as were the results of Miller and Abrahams, to more general classes of disordered solids with subsequent refinements and extensions.

A basic approximation in their analysis is that hopping takes place only between a pair of sites and each pair is independent of other pairs. A number of theoretical papers are based on the pair approximation. (See Böttger and Biryskin 1976b for a discussion and references.) In particular, Butcher and Morys (1973, 1974) showed that this pair approximation provides an exact solution to the rate equation formulation of the hopping conductivity in the limit of low site densities. Additionally, $\sigma(\omega) \propto \omega^{0.8}$ if $\omega\tau_0 < 10^{-1}$, where τ_0 is the minimum relaxation time.

As discussed in §6.1.2, d.c. hopping between pairs is equivalent to a network of random conductors. Pollak (1974) and Kirkpatrick (1973) have suggested a generalized random impedance network to describe a.c. hopping. Their model is the Miller–Abrahams network in the d.c. limit and corresponds to the Pollack–Geballe theory at high frequencies. The modification of the Miller–Abrahams network is simple: Between ground and each node of the conductors (the localized sites) is placed a capacitor C_i in series with a voltage generator V_i. The capacitors depend on temperature and the site energy, and the voltage generators depend on the applied field. Although there are a number of computer experiments on a network of random resistors, we are unaware of any computer experiments on this generalized impedance network.

Scher and Lax (1972, 1973a, b) have presented a theory, generalized by Moore (1974a, b), designed to go beyond the pair approximation. For systems with a large number of randomly distributed localized sites, they regard the electrical charge as a classical particle undergoing a continuous time random walk (Montroll and Weiss 1965) with the probability per unit time of a hop determined quantum mechanically. Simply stated, the probability function in the classical random walk problem is multiplied by a site- and time-dependent function $Q(i, t-t')$ called the waiting time probability, which is the probability that a carrier at site i at time t' makes no hops by the time t. $Q(i,t-t')$ is guessed or determined from first principles.

The basic equation in the model (Moore 1974a), with $t-t'$ Fourier transformed to ω, is

$$P(i,j;\omega)=Q(i;\omega)\delta_{ij}+\sum_{k=1}^{N}Q(i;\omega)W(i,k)P(k,j;\omega), \tag{6.14a}$$

where

$$Q(i;\omega)=\left[\mathrm{i}\omega+\sum_{j=1}^{N}W(i,j)\right]^{-1} \tag{6.14b}$$

and $P(i,j;t-t')$ is the probability of a charge carrier at site j at time t' hopping (by any number of hops) to site i by time t and (6.14b) relates $W(i,j)$, the transition rate for hops between sites i and j, to $Q(i,t-t')$.

Assuming the Miller–Abrahams transition rate (6.12) for $W(i,j)$, Moore (1974b) solved (6.14) for a 1d lattice of 50 sites for four cases: a regular lattice of equal energy sites, a random lattice of equal site energies, a regular lattice with random site energies and a random lattice with random site energies. He found that both spatial and energy randomness can lead to an a.c. conductivity proportional to ω^s, where $s<1$ and variations in s with α or T are controlled entirely by multiple hopping processes. The pair approximation yielded the correct qualitative dependence of the conductivity with frequency for the case of spatial randomness, but gave an incorrect dependence for the case of energy randomness. It also appeared to become more exact as the degree of localization increased or as the temperature decreased.

7. *Thermal transport*

The simple classical methods which account satisfactorily for many physical observables fail miserably when they are applied to thermal conductivity. The reason for this failure is not fully understood, but much has been learned from computer experimentation. In the following paragraphs we will review what is known from both theoretical and numerical studies of such systems. We discuss only 1d and 2d systems because we are unaware of any computer experiments on thermal transport in 3d disordered systems.

It has been realized for many years that an ordered harmonic lattice will offer no resistance to heat flow, because the normal modes are phonons and a superposition of phonons with random phases cannot describe a temperature gradient ∇T. So, although the heat flow J in such a lattice is finite, the conductivity K, defined by Fourier's law,

$$J = -K\nabla T \tag{7.1}$$

is infinite. A question must studied, but not yet fully answered, is, What characteristics must be added to a harmonic system to force it to obey Fourier's law?

Analytic solutions have been obtained for the stationary non-equilibrium density matrix for ordered harmonic lattices both in one (Rieder et al. 1967, Nakazawa 1970) and two (Helleman 1971) dimensions. In these calculations a harmonic lattice interacted at one end with a reservoir of temperature T_1, at the other end with a reservoir at temperature T_N, and expressions were obtained for the kinetic temperatures of each atom in the lattice as well as expectation values of all other bilinear forms in the coordinates and momenta. It was found that for long lattices the temperature gradient vanished except very near the ends of the lattice, and that the heat current was proportional to $T_1 - T_N$ and was asymptotically independent of N. This means in eq. (7.1), depending upon whether one takes ∇T to be the observed internal temperature gradient (zero) or the overall gradient $(T_1 - T_N)/N$, one gets $K = \infty$ or $K \propto N$. So, as expected, the ordered harmonic chain is pathological.

Thermal conductivity might be limited by anharmonic forces, impurities, structural disorder, phonon scattering from boundaries, or interactions of phonons with other excitations such as electrons or magnons. Computer experimental studies have been performed to investigate some of these. As discussed below, disorder alone is not sufficient to force a system to obey Fourier's law.

Much analytic work has been done on the harmonic chain with isotopic (diagonal) disorder by Lebowitz and his collaborators (Casher and Lebowitz 1971, O'Connor and Lebowitz 1974, O'Connor 1975, Spohn and Lebowitz 1976) who obtained a variety of interesting rigorous results, especially regarding the asymptotic behavior of J as $N \to \infty$ and its relation to the nature of the spectrum of the chain.

Less rigorous, partly because it treats the interactions of the chain with the end reservoirs perturbatively in the chain–reservoir coupling λ, is the work of Matsuda and Ishii (1970) and Ishii (1975). They find, to first order in λ, a simple explicit formula for the heat current through the isotopically disordered harmonic chain

$$J = \lambda k(T_1 - T_N) \sum_{\nu=1}^{N} \left(\chi_{\nu,1}^{-2} + \chi_{\nu,N}^{-2}\right). \tag{7.2}$$

The sum here is over the normal modes ν of the disordered chain, and $\chi_{\nu,j} = \sqrt{m_j}\, u_j^{(\nu)}$, where $u_j^{(\nu)}$ is the displacement of the jth atom in the νth mode and $\Sigma_j \chi_{\nu,j}\chi_{\mu,j} = \delta_{\mu\nu}$. Eq. (7.2) exhibits the dependence of the heat current on the displacements of the end atoms. One can make some interesting conclusions by inspection of (7.2). First, if the chain is ordered and monatomic, then for large N

$$\chi_{\nu,1} = \sqrt{2/N}\ \sin\,(\cos)\ k_\nu, \tag{7.3}$$

where k_ν is the wave number of the νth mode, and the boundary conditions are fixed or free according to whether the sin or cos appears. So

$$J = \lambda k(T_1 - T_N)N^{-1}\sum_\nu \sin^2(\cos^2)\,k_\nu \sim \lambda k(T_1 - T_N)/2$$

in agreement with Rieder et al. (1967). The thermal conductivity $K \sim N\lambda k/2$, by the definition (7.1), with $\nabla T = \Delta T/N$.

One of the principal results of Matsuda and Ishii is that in an isotopically disordered chain with a unit force constant the rate of exponential fall-off of the amplitude of a normal mode about its center of localization j_0 (defined loosely by the relation $|u_j| \lesssim \exp[-\gamma|j - j_0|]$) for low frequencies ω is

$$\gamma = \langle (m - \langle m\rangle)^2\rangle \omega^2/8\langle m\rangle. \tag{7.4}$$

Thus only very-low-frequency modes can extend from one end $(j - 1)$ of the lattice to the other $(j - N)$. (Note that in order for a mode ν to contribute to (7.2) *both* $\chi_{\nu,1}$ and $\chi_{\nu,N}$ must be appreciable.) If we arbitrarily say that only modes for which $\gamma N/2 < 1$ contribute, then the number of contributing terms in the sum of (7.2) is

$$\nu_c = \frac{4}{\pi c}\left[\frac{\langle m\rangle N}{\langle (m - \langle m\rangle)^2\rangle}\right]^{1/2}, \tag{7.5}$$

where c is the sound velocity. These low-frequency extended modes are phonon-like and still are approximately given by (7.3). One can now estimate (7.2) for the disordered chain, getting

$$J \sim \lambda k(T_1 - T_N)\frac{64}{3\pi c^3}\left[\frac{\langle m\rangle}{\langle \delta m^2\rangle N}\right]^{3/2} \quad \text{(fixed ends)}, \tag{7.6}$$

$$J \sim \lambda k(T_1 - T_N)\frac{4}{\pi c}\left[\frac{\langle m\rangle}{\langle \delta m^2\rangle N}\right]^{1/2} \quad \text{(free ends)}, \tag{7.7}$$

where $\delta m = m - \langle m\rangle$.

These approximate predictions have been shown by a computer experiment to be quite accurate (Visscher 1971, 1976). The eigenvectors for isotopically disordered harmonic chains with $N=1000$ were calculated and used to directly evaluate (7.2). The results approximately confirmed eqs. (7.5), (7.6) and (7.7). But these equations imply that $K \sim N^{1/2}$ for fixed end boundary conditions and $K \sim N^{-1/2}$ for free ends; this is an unpleasant result because K should be an intrinsic property of the chain, not dependent on the details of the reservoir interactions or the termination of the chain.

Although the Matsuda–Ishii development is a perturbation theory in λ, and there are reasons to believe that $K(\lambda)$ should be singular at $\lambda=0$ (O'Connor and Lebowitz 1974, Visscher 1971), the fact that the formulae are partly computer-experimentally verified gives one some confidence in them.

A different termination was studied by Rubin (1963, 1968, 1970) and by Rubin and Greer (1971). They considered an isotopically disordered chain of length N imbedded in an infinite monatomic one and studied the steady state ultimately attained when the left-hand reservoir (a semi-infinite harmonic chain) initially has a different temperature from that on the right. Using this steady-state heat current and the temperature gradient within the disordered region, they showed analytically that K diverges as $N \to \infty$ *at least* as fast as $N^{1/2}$. They also performed a computer experiment, and found that in fact $K \sim N^{1/2}$. This conclusion is supported by the rigorous work of Keller et al. (1978). Thus this termination gives results similar to the free end boundary condition.

Something more than disorder is needed to cause the harmonic chain to obey Fourier's law. Jackson et al. (1968) performed a computer experiment on a chain with disorder in the anharmonic force constants between neighboring particles. They computed the autocorrelation function of the total heat current and tried to obtain the conductivity by the Kubo formula

$$K = \frac{1}{NT^2} \int_0^\infty \langle J(\tau) J(0) \rangle \, d\tau.$$

Their lattice was too short for this to be successful.

A numerical experiment on the isotopically disordered anharmonic chain was performed by Payton et al. (1967), who used an interaction between nearest neighbors of the form

$$V(x_i) = \varepsilon_0 + \tfrac{1}{2}\gamma x_i^2 - \tfrac{1}{3}\mu x_i^3 + \tfrac{1}{4}\nu x_i^4,$$

where $x_i = u_i - u_{i-1}$. They used a 100-atom chain which had the first particle in contact with a hot reservoir and the last with a cold reservoir. Reservoir interactions were simulated by periodically allowing elastic collisions between the end particles and reservoir particles chosen from hot and cold maxwellian velocity distributions. Various degrees of mass-disorder and various amounts of anharmonicity were used.

K was determined by molecular-dynamics simulations by dividing the observed steady-state heat current by the observed gradient in kinetic temperature ∇T near the middle of the chain. Some of the results are shown in fig. 31. Most of the error bars are attributable to inaccuracies in estimation of ∇T.

For harmonic lattices K is U-shaped as a function of c, the concentration of one of the two components. $K \to \infty$ for $c \to 0$ and for $c \to 1$ as expected, but when anharmonicity is introduced in the disordered case, K increases! This is in conflict with Mathiesson's rule, which would lead one to expect two independent scattering mechanisms to *add* and increase the resistivity. The fact that the opposite occurs means that the mechanisms

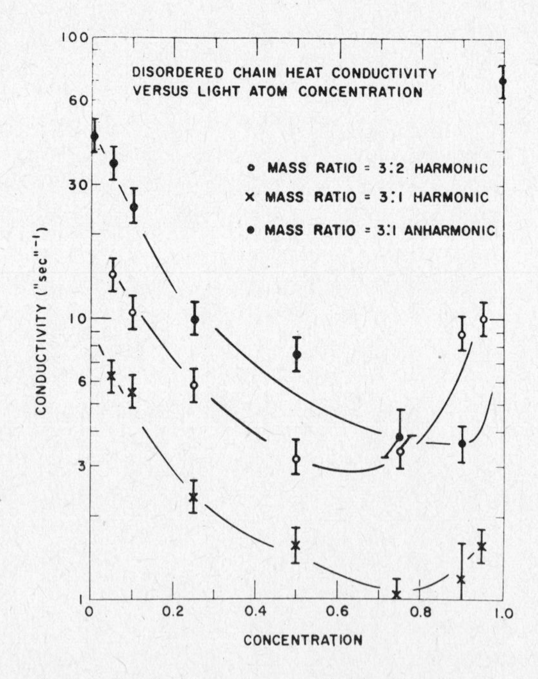

Fig. 31. Heat conductivity for the binary isotopically disordered linear chain with harmonic and anharmonic nearest-neighbor interactions. The error bars are estimates of the accuracy of the temperature gradient: thermal conductivity = average heat current/average temperature gradient. The anharmonic force parameters approximated those appropriate to noble gas solids (Payton et al. 1967).

are not independent, which can be intuitively understood in the following way:

Most of the normal modes of the disordered chain are localized and cannot contribute to energy transport from hot reservoir to cold reservoir. Non-linearities in the equations of motion (anharmonicities) allow localized modes to decay either into other localized modes, which can be localized at another point in the chain, or into extended modes, which can themselves transfer heat or excite other localized modes. Both of these decay processes increase the heat current, so the conductivity too should rise. No quantitative theory of this effect exists.

The experiment was also performed for a two-dimensional system (Payton et al. 1967, Rich et al. 1971). It was thought that because of the extra degrees of freedom in the multi-dimensional lattice (energy–momentum conservation in phonon collision processes is more easily satisfied) and because transverse phonons then exist, the results might differ from those in one dimension. But they were qualitatively similar.

In connection with these computer experiments on two-dimensional systems, Payton et al. (1968) made a computer-generated movie. Sample frames are shown in fig. 32 illustrating the interplay between anharmonicity and disorder. These are equal time plots of kinetic energy of particles in disordered two-dimensional lattices which were excited by giving each atom along the end row an equal impulse at $t=0$. In the harmonic disordered cases the wave front is severely attenuated. Both the speed and amplitude of the wave front are enhanced by the presence of anharmonicity. Different verbal descriptions of the physics involved must be equivalent. One is, as above, that the anharmonicity allows the energy to hop along from localized mode to localized mode, with the phases of the

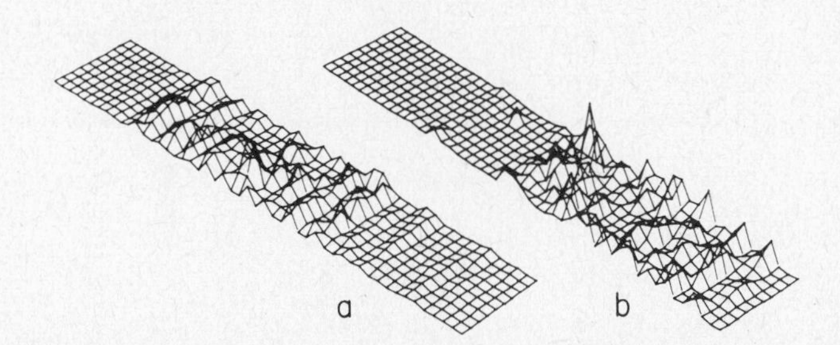

Fig. 32. Comparison at identical times of the energy penetration into a region of 15% heavy impurity for an anharmonic (a) and a harmonic (b) 10×50 lattice. The wave started at $t=0$ with all atoms at rest except the 10 along the right-hand edge which each received an equal impulse. Displacement out of the plane is proportional to kinetic energy. Periodic boundary conditions in the short dimension were used, fixed boundary conditions in the long dimension. There were no impurities in the first 16 rows (Payton et al. 1968).

different modes locked together to form a coherent wavefront. Another is that the wave front is in fact a soliton or solitary wave and is little attenuated by the impurities.

Other mechanisms besides anharmonicity can cause exchange of energy between normal modes and hence give rise to a normal thermal conductivity. In order to simulate all of them the "self-consistent-reservoir model" was invented in 1970 by Bolsterli et al., applied to the classical disordered lattice by Rich and Visscher (1975) and to the quantum-mechanical ordered lattice by Visscher and Rich (1975). The model is illustrated in fig. 33. Each particle in the system is coupled with strength λ to its own thermal reservoir, which can emit and absorb phonons. The rationale for the reservoirs is that they simulate non-harmonic forces, such as anharmonicities and electron–phonon interactions, in an analytically tractable way. The Fokker–Planck equation for this system can be solved exactly numerically with each reservoir at a different temperature. A self-consistent condition is then imposed, so that the temperature of each reservoir, except those interacting with the end atoms, is required to be equal to the kinetic temperature of the corresponding atom.

Some of the results are shown in fig. 34 for self-consistent disordered chains with fixed and free boundary conditions. K is computed from the internal temperature gradient, and for large N is *independent* of boundary conditions. It is known (eqs. (7.6), (7.7)) that for large N, $K_N(\lambda=0) \sim N^{1/2}$ or $N^{-1/2}$ for free or fixed ends. But as N increases, the minimum value of λ for which $K_N(\lambda)$ is independent of the boundary conditions decreases! This suggests that

$$\lim_{N\to\infty} \lim_{\lambda\to 0+} K_N(\lambda)$$

exists, where the order of the limits is crucial. If this conjecture should be

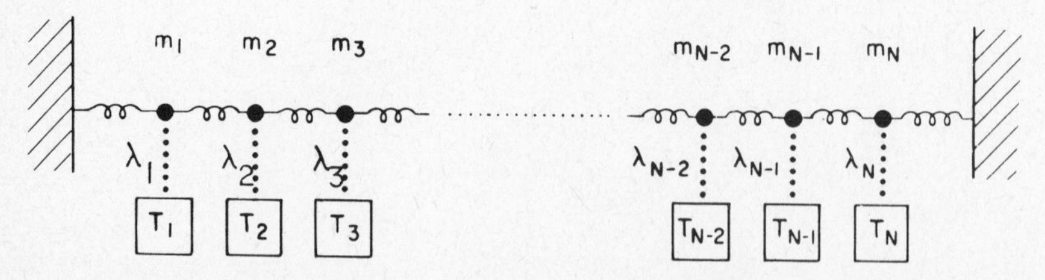

Fig. 33. Model for the disordered harmonic chain with self-consistent reservoirs. Each atom interacts with its own independent reservoir, which may be thought of as being a viscous fluid in random motion surrounding the atom. The temperatures of the end reservoirs T_1 and T_N are prescribed; the other T_i are determined in the steady state by the self-consistency condition that they exchange no heat with the lattice (Rich and Visscher 1975).

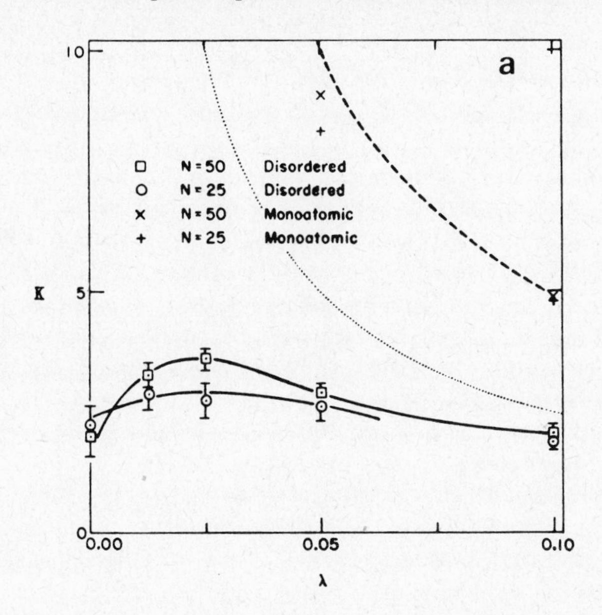

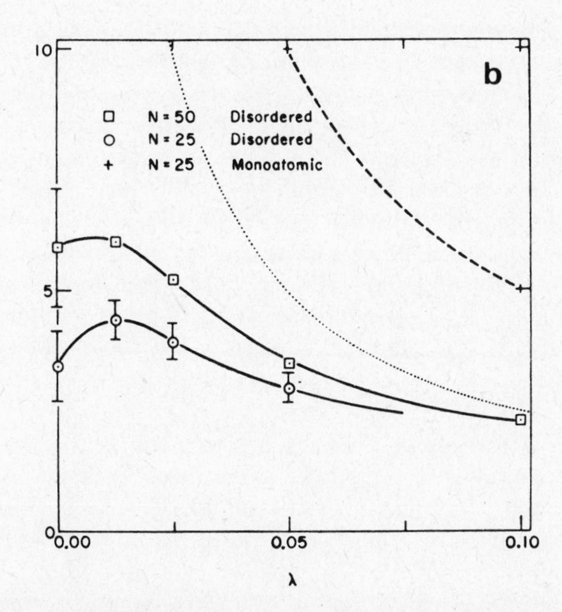

Fig. 34. Thermal conductivities for self-consistent chains with fixed (a) and free (b) boundary conditions. The chains are randomly disordered mixtures of masses 1 and 2 in equal numbers. The dotted and dashed lines are analytic results for monatomic infinite chains of mass 1 and 2, respectively (Rich and Visscher 1975).

true, it would be another instance in which the existence of an ergodogenic mechanism (interaction with heat baths), not the magnitude of λ, that stabilizes a transport coefficient. (Another example is cited by Visscher 1974.)

The conditions which a dynamical system must satisfy in order that it should have a normal thermal conductivity are not known. The experiments of Payton et al. (1967) were performed only for $N = 100$ so they say nothing about the N-dependence. Fig. 32 leads one to suspect that the self-reinforcement (i.e. formation of solitons or solitary waves) of non-linearities, generating long-lived, non-interacting excitations, might well cause many non-linear disordered systems to flout Fourier's law. To attack these problems by perturbation techniques is probably futile even in the absence of disorder, but is certainly so in its presence. The computer experiment will surely be useful here.

The characteristics which a dynamical many-body system must have to obey Fourier's law probably include stochasticity. In this context "stochastic" means that two orbits initially close together in phase space will diverge exponentially and their phase points after a short time will be uncorrelated. Systems which diverge only linearly in time are "integrable". (See Ford 1975 for a general discussion.) The harmonic system is integrable; it has as many constants of the motion as it has degrees of freedom. Some non-linear systems are also integrable, notably the Toda lattice (Toda 1975). The integrability of the Toda chain was discovered, in fact, by computer experiment (Ford et al. 1973) before it was demonstrated mathematically. Other systems have been shown by numerical experimentation to be stochastic in certain regions of their phase space. Still other non-linear systems, particularly the Fermi–Pasta–Ulam chain (Fermi et al. 1955) have been studied by computer experiment for many years, but their stochasticity properties are still uncertain. Much work remains to be done.

8. *Concluding remarks*

We have tried to present a survey of the methods used in computer experiments on disordered solids and a sampling of results of their application. Unfortunately, the survey is not comprehensive, nor is the sampling up-to-date. The subject of computer experiments on disordered solids is simply too large to be reviewed in an article of modest length. It is also a very active area of research, and new papers continually appear. For example, in recent days journals arrived in our library with papers by Yoshino and Okazaki (1977) giving details and results to supplement their 1976 paper (see §5.2), by Aoki (1977) presenting a study of electrons in random potentials in strong magnetic fields, and by Sakata et al. (1977) investigating spin glasses. In addition to preprints, we included work from journals in our library at the end of August 1977.

A natural question to ask is: What is the future of computer experiments on disordered solids? In regard to computer experiments themselves, as computer hardware continues to improve, so will computer experiments. In some scientific and technological areas such as fluid dynamics, plasma physics and classical statistical mechanics, computer experiments have developed in scope and reliability to the point that they are replacing physical experiments. Disordered solids is not yet one of these areas, but many solid state physicists are beginning to recognize the importance of computer experiments to investigations of many features of the dynamics of disordered solids.

New computer experimental techniques must be developed to enable studies on the dynamics of disordered, anharmonic, quantum lattices, and on electronic properties of random solids in which the tight-binding hamiltonian cannot be applied. The growing technological and industrial importance of alloys and amorphous solids seems to insure that computer studies of them will continue to be vigorously pursued.

Acknowledgements

We thank all who sent us preprints and reprints of their work. In this regard, we especially thank Drs. S. Kirkpatrick, K. Binder and D. Stauffer. We also thank Drs. A. R. Bishop and R. Albers for their helpful comments and suggestions about the manuscript.

References

Abram, R. A. (1973), J. Phys. C **6**, L379.
Abram, R. A. and S. F. Edwards (1972), J. Phys. C **5**, 1183.
Adler, D., L. P. Flora and S. D. Senturia (1973), Solid State Commun. **12**, 9.
Agacy, R. L. (1964), Proc. Phys. Soc. **83**, 591.
Alben, R. and M. F. Thorpe (1975), J. Phys. C **8**, L275.
Alben, R. and P. Boutron (1975), Science **187**, 430.
Alben, R., D. Weaire, J. E. Smith, Jr. and M. H. Brodsky (1975a), Phys. Rev. B **11**, 2271.
Alben, R., L. von Heimendahl, P. Galison and M. Long (1975b), J. Phys. C **8**, L468.
Alben, R., M. Blume, H. Krakauer and L. Schwartz (1975c), Phys. Rev. B **12**, 4090.
Alben, R., H. Krakauer and L. Schwartz (1976), Phys. Rev. B **14**, 1510.
Alben, R., S. Kirkpatrick and D. Beeman (1977a), Phys. Rev. B **15**, 346.
Alben, R., M. Blume and M. McKeown (1977b), Phys. Rev. B **16**, 3829.
Albers, R. and J. E. Gubernatis (1977), Phys. Rev. B **17**, 4487.
Ambegaokar, V., B. I. Halperin and J. S. Langer (1971), Phys. Rev. B **4**, 2612.

Ambegoakar, V., S. Cochran and J. Kurkijarvi (1973), Phys. Rev. B **8**, 3682.
Anderson, P. W. (1958), Phys. Rev. **109**, 1492.
Anderson, P. W. (1972), Proc. Nat. Acad. Sci. **69**, 1097.
Anonymous (1972), Nature **239**, 488.
Aoki, H. (1977), J. Phys. C **10**, 2583.
Argyrakis, P. and R. Kopelman (1977), J. Chem. Phys. **66**, 3301.
Barker, Jr., A. S. and A. J. Sievers (1975), Rev. Mod. Phys. **47**, Suppl. 2.
Beeman, D. and B. L. Bobbs (1975), Phys. Rev. B **12**, 1399.
Bell, R. J. (1972a), Rep. Prog. Phys. **35**, 1315.
Bell, R. J. (1972b), J. Phys. C **5**, L315.
Bell, R. J. (1976), in *Methods in computational physics*, Ed. by B. Alder, S. Fernbock and M. Rotenberg (Academic Press, New York), Vol. 15, p. 216.
Bell, R. J., and P. Dean (1972), Phil. Mag. **25**, 1381.
Bell, R. J., N. F. Bird, and P. Dean (1968), J. Phys. C **1**, 299.
Bell, R. J., P. Dean and D. C. Hibbins-Butler (1970), J. Phys. C **3**, 2111.
Bell, R. J., P. Dean and D. C. Hibbins-Butler (1971), J. Phys. C **4**, 1214.
Bennett, C. H. (1972), J. Appl. Phys. **43**, 2727.
Bernal, J. D. (1959), Nature **183**, 141.
Binder, K. (1974a), Z. Phys. **267**, 313.
Binder, K. (1974b), Advan. Phys. **23**, 917.
Binder, K. and D. Stauffer (1979), in *Monte Carlo methods in statistical physics*, Ed. by K. Binder (Springer-Verlag, Berlin), unpublished.
Bishop, A. R. (1974), Phys. Lett. **49A**, 5.
Bloch, A. N. and C. M. Varma (1973), J. Phys. C **6**, 1849.
Bloch, A. N., R. B. Weisman and C. M. Varma (1972), Phys. Rev. Lett. **28**, 753.
Bolsterli, M., M. Rich and W. M. Visscher (1970), Phys. Rev. A **1**, 1086.
Borland, R. E. (1963), Proc. Roy. Soc. **A274**, 529.
Borland, R. E. (1964), Proc. Phys. Soc. (London) **83**, 1027.
Bortz, A. B., M. H. Kalos, J. L. Lebowitz and M. A. Zendejas (1974), Phys. Rev. B **10**, 535.
Böttger, H. and V. V. Bryksin (1976a), Phys. Stat. Sol. (b) **78**, 9.
Böttger, H. and V. V. Bryksin (1976b), Phys. Stat. Sol. (b) **78**, 415.
Brenig, W., G. H. Döhler and H. Heyszenau (1973a), Phil. Mag. **27**, 1093.
Brenig, W., G. Döhler and P. Wölfle (1973b), Z. Phys. **258**, 381.
Brout, R. and W. M. Visscher (1962), Phys. Rev. Lett. **9**, 54.
Bruggeman, D. A. G. (1935), Annln. Phys. **24**, 636.
Bush, R. L. (1972), Phys. Rev. B **6**, 1182.
Butcher, P. N. and P. L. Morys (1973), J. Phys. C **6**, 2147.
Butcher, P. N. and P. L. Morys (1974), in *Amorphous and liquid semiconductors*, Ed. by J. Stuke and W. Brenig (Taylor and Francis, London), p. 153.
Butler, W. H. (1973), Phys. Rev. B **8**, 4499.
Cargill, III, G. S. (1975), in *Solid state physics*, Ed. by H. Ehrenreich, F. Seitz and D. Turnbull (Academic Press, New York), Vol. 30, p. 227.
Cargill, III, G. S., and S. Kirkpatrick (1976), in *Structure and excitations of amorphous solids*, AIP Conference **20**, Ed. by G. Lucovsky and F. L. Galeener (American Institute of Physics, New York), p. 339.
Casher, A. and J. L. Lebowitz (1971), J. Math. Phys. **12**, 1701.
Chen, G. L., S. Y. Wu and M. Schwartz (1977), Phys. Stat. Sol. (b) **83**, 115.
Cohen, M. H. (1970), Phys. Rev. Lett. **35**, 1475.
Cohen, M. H. and J. Jortner (1973), Phys. Rev. Lett. **30**, 699.
Cohen, M. H. and J. Jortner (1974), Phys. Rev. A **10**, 978.

Damgaard Kristensen, W. (1976), J. Non-Cryst. Solids **21**, 303.
Dean, P. (1960), Proc. Phys. Soc. **254**, 507.
Dean, P. (1961), Proc. Phys. Soc. **260**, 263.
Dean, P. (1964), Proc. Phys. Soc. London **84**, 727.
Dean, P. (1972), Rev. Mod. Phys. **44**, 127.
Dean, P. and M. D. Bacon (1962), Nature **194**, 541.
Dean, P. and M. D. Bacon (1963), Proc. Phys. Soc. **81**, 642.
Dean, P. and M. D. Bacon (1965), Proc. Roy. Soc. **A283**, 64.
Dean, P. and N. F. Bird (1967), Proc. Camb. Philos. Soc. **63**, 477.
Dean, P. and J. L. Martin (1960), Proc. Roy. Soc. **A259**, 409.
Domb, C. (1976), J. Phys. A **9**, L141.
Domb, C., T. Schneider and E. Stoll (1975), J. Phys. A **8**, L90.
Duffy, M. G., D. S. Boudreaux and D. E. Polk (1974), J. Non-Cryst. Solids **15**, 435.
Dyson, F. J. (1953), Phys. Rev. **92**, 1331.
Economou, E. N. and P. D. Antoniou (1977), Solid State Commun. **21**, 285.
Edwards, J. T. and D. J. Thouless (1972), J. Phys. C **5**, 807.
Elliott, R. J. and P. L. Leath (1975), in *Dynamical properties of solids*, Ed. by G. K. Horton and A. A. Maradudin (North-Holland, Amsterdam), Vol. 2, p. 385.
Elliott, R. J., J. A. Krumhansl and P. L. Leath (1974), Rev. Mod. Phys. **46**, 465.
Emin, D. (1976), in *Physics of structurally disordered solids*, Ed. by S. S. Mitra (Plenum, New York), p. 461.
Erdös, P. and R. C. Herndon (1973), in *Computational methods for large molecules and localized states in solids*, Ed. by F. Herrman, A. D. McLean and R. K. Nesbet (Plenum, New York), p. 275.
Fermi, E., J. Pasta and S. Ulam (1955), Report, LA-1940, Los Alamos Scientific Laboratory.
Finney, J. L. (1975), J. de Physique **36**, Suppl. 4, C2-1.
Finney, J. L. (1977), Nature **266**, 309.
Flinn, P. A. (1974), J. Stat. Phys. **10**, 89.
Ford, J. (1975), in *Fundamental problems in statistical mechanics*, Ed. by E. G. D. Cohen (North-Holland, Amsterdam), Vol. 3, p. 215.
Ford, J., S. D. Stoddard and J. S. Turner (1973), Prog. Theor. Phys. **50**, 1547.
Freed, K. (1972), Phys. Rev. B **5**, 4802.
Friedberg, R. and J. M. Luttinger (1975), Phys. Rev. B **12**, 4460.
Frisch, U., C. Froeschle, J.-P. Scheidecker and P.-L. Sulem (1973), Phys. Rev. A **8**, 1416.
Gaspard, J. P. and F. Cyrot-Lackmann (1973), J. Phys. C **6**, 3077.
Greaves, G. N. and E. A. Davis (1974), Phil. Mag. **29**, 1201.
Gubernatis, J. E. and P. L. Taylor (1971), J. Phys. C **4**, L94.
Gubernatis, J. E. and P. L. Taylor (1973), J. Phys. C **6**, 1889.
Gubernatis, J. E. and S.-Y. Wu (1974), Solid State Commun. **15**, 1915.
Guttman, L. (1975), Phys. Rev. B **11**, 764.
Halperin, B. I. (1967), Adv. Chem. Phys. **13**, 123.
Haydock, R., V. Heine and M. J. Kelly (1972), J. Phys. C **5**, 2845.
Haydock, R., V. Heine and M. J. Kelly (1975), J. Phys. C **8**, 2591.
Helleman, R. H. G. (1971), Ph.D. thesis, Yeshiva University, (University Microfilms Inc., Ann Arbor).
Henderson, D. and F. Herman (1972), J. Non-Cryst. Solids **8–10**, 359.
Herbert, D. C. and R. Jones (1971), J. Phys. C **4**, 1145.
Herzenberg, A. and A. Modinos (1964), Biopolymers **2**, 561.
Hirota, T. (1973), Prog. Theor. Phys. **50**, 1240.
Hirsch, J. E. and T. P. Eggarter (1976), Phys. Rev. B **14**, 2433.

Hirsch, J. E. and T. P. Eggarter (1977), Phys. Rev. B **15**, 779.
Hiwatari, Y., H. Matsuda, T. Ogawa, N. Ogita and A. Ueda (1974), Prog. Theor. Phys. **52**, 1105.
Holcomb, D. F. and J. J. Rehr (1969), Phys. Rev. **183**, 773.
Hoover, W. G. and F. H. Ree (1968), J. Chem. Phys. **49**, 3609.
Hoover, W. G., S. G. Gray and K. W. Johnson (1971), J. Chem. Phys. **55**, 1128.
Hori, J. (1964), Prog. Theor. Phys. (Kyoto) **31**, 940.
Hori, J. (1968), *Spectral properties of disordered chains and lattices*, (Pergamon, Oxford).
Hori, J. and M. Fukushima (1963), J. Phys. Soc. Japan **19**, 296.
Hori, J. and H. Matsuda (1964), Prog. Theor. Phys. **32**, 183.
Hori, J. and K. Wada (1970), Prog. Theor. Phys. Suppl. **45**, 36.
Hoshen, J. and R. Kopelman (1976a), Phys. Rev. B **14**, 3438.
Hoshen, J. and R. Kopelman (1976b), J. Chem. Phys. **65**, 2817.
Hoshen, J., R. Kopelman and E. M. Monberg (1977), J. Stat. Phys. **19**, 219.
Huber, D. L. (1974a), Phys. Rev. B **10**, 4621.
Huber, D. L. (1974b), Solid State Commun. **14**, 1153.
Ishii, K. (1973), Prog. Theor. Phys. Suppl. **53**, 77.
Jackson, E. A., J. R. Pasta and J. F. Waters (1968), J. Comp. Phys. **2**, 207.
James, H. and A. S. Ginzbarg (1953), J. Phys. Chem. **57**, 840.
Joannopoulos, J. D. and M. L. Cohen (1976), in *Solid state physics*, Ed. by H. Ehrenreich, F. Seitz and D. Turnbull (Academic Press, New York), Vol. 31, p. 71.
Joannopoulos, J. D. and E. J. Mele (1976), Solid State Commun. **20**, 729.
Kagan, Yu. and Ya. Iosilevskii (1963), Sov. Phys.–JETP **17**, 195.
Kaplow, R., T. A. Rowe and B. L. Averbach (1968), Phys. Rev. **168**, 1068.
Keating, P. N. (1966), Phys. Rev. **145**, 637.
Keller, J. B., G. C. Papanicolaou and J. Weilenmann (1978), Commun. Pure Appl. Math. **32**, 583.
Khor, K. E. and P. V. Smith (1971a), J. Phys. C **4**, 2029.
Khor, K. E. and P. V. Smith (1971b), J. Phys. C **4**, 2041.
Kikuchi, M. (1974), J. Phys. Soc. Japan **37**, 904.
Kincaid, J. M. and J. J. Weis (1977), Molec. Phys. **34**, 931.
Kirkpatrick, S. (1971), Phys. Rev. Lett. **27**, 1722.
Kirkpatrick, S. (1973), Rev. Mod. Phys. **45**, 574.
Kirkpatrick, S. (1974), in *Amorphous and liquid semiconductors*, Ed. by J. Stuke and W. Brenig (Taylor and Francis, London), p. 183.
Kirkpatrick, S. (1976), Phys. Rev. Lett. **36**, 69.
Kirkpatrick, S. and T. P. Eggarter (1972), Phys. Rev. B **6**, 3598.
Krumhansl, J. A. (1973), in *Proceedings of the international symposium on amorphous magnetism*, Ed. by H. O. Hooper and A. M. de Graaf (Plenum, New York), p. 15.
Kurkijärvi, J. (1973), Phys. Rev. B **8**, 922.
Kurkijärvi, J. (1974), Phys. Rev. B **9**, 770.
Landauer, R. A. (1952), J. Appl. Phys. **23**, 779.
Landauer, R. (1970), Phil. Mag. **21**, 863.
Last, B. J. and D. J. Thouless (1971), Phys. Rev. Lett. **27**, 1719.
Last, B. J. and D. J. Thouless (1974), J. Phys. C **7**, 699.
Leath, P. L. (1976), Phys. Rev. B **14**, 5046.
LeFevre, E. J. (1972), Nature **235**, 20.
Levinshtein, M. E., B. I. Shklovskii, M. S. Shur and A. L. Efros (1976), Sov. Phys.–JETP **42**, 197.
Licciardello, D. C. and E. N. Economou (1975), Phys. Rev. B **11**, 3697.
Licciardello, D. C. and D. J. Thouless (1975a), Phys. Rev. Lett. **21**, 1475.

Licciardello, D. C. and D. J. Thouless (1975b), J. Phys. C **8**, 4157.
Licciardello, D. C. and D. J. Thouless (1977), Commun. Phys. **2**, 7.
Lieb, E. H. and D. C. Mattis (1966), *Mathematical physics in one dimension*, (Academic Press, New York), Chs. 2 and 3.
Lifshitz, I. M. (1964), Adv. Phys. **13**, 483.
Little, W. A. (1973), in *Computational methods for large molecules and localized states in solids*, Ed. by F. Herman, A. D. McLean and R. K. Nesbet (Plenum Press, New York), p. 49.
Long, M., P. Galison, R. Alben and G. A. N. Connell (1976), Phys. Rev. B **13**, 181.
Lu, M.-S., M. Nelkin and M. Arita (1974), Phys. Rev. B **10**, 2315.
Luttinger, J. M. (1951), Philips Res. Rep. **6**, 303.
Mandell, M. J., J. P. McTague and A. Rahman (1977), Bull. A.P.S. **22**, 297.
Mano, J., A. B. Bortz, M. H. Kalos and J. L. Lebowitz (1975), Phys. Rev. B **12**, 2000.
Maschke, K., H. Overhof and P. Thomas (1974a), Phys. Stat. Sol. (b) **61**, 621.
Maschke, K., H. Overhof and P. Thomas (1974b), Phys. Stat. Sol. (b) **62**, 113.
Mason, G. (1967), Disc. Faraday Soc. **43**, 75.
Matheson, A. J. (1974), J. Phys. C **7**, 2569.
Matsuda, H. (1964), Prog. Theor. Phys. **31**, 161.
Matsuda, H. and K. Ishii (1970), Prog. Theor. Phys. Suppl. **45**, 56.
Meek, P. E. (1976), Phil. Mag. **33**, 897.
Metropolis, N., A. W. Rosenbluth, A. H. Teller and E. Teller (1953), J. Chem. Phys. **21**, 1087.
Miller, A. and E. Abrahams (1960), Phys. Rev. **120**, 745.
Minami, S. and J. Hori (1970), Prog. Theor. Phys. Suppl. **45**, 87.
Montroll, E. W. and G. H. Weiss (1965), J. Math. Phys. **6**, 167.
Moore, E. J. (1973), J. Phys. C **6**, 1551.
Moore, E. J. (1974a), J. Phys. C **7**, 339.
Moore, E. J. (1974b), J. Phys. C **7**, 1840.
Mott, N. F. (1967), Advan. Phys. **16**, 49.
Mott, N. F. (1968), J. Non-Cryst. Solids **1**, 1.
Mott, N. F. (1969), Phil. Mag. **19**, 825.
Mott, N. F. (1970), Phil. Mag. **22**, 7.
Mott, N. F. (1972), J. Non-Cryst. Solids **8–10**, 1.
Mott, N. F. (1976), Commun. Phys. **1**, 203.
Mott, N. F. and E. A. Davis (1971), *Electronic processes in non-crystalline materials* (Oxford University Press, Oxford).
Mott, N. F. and W. D. Twose (1961), Advan. Phys. **10**, 107.
Nakamura, F., J. Takamura and M. Chikasaki (1973), J. Mater. Sci. **8**, 385.
Nakazawa, H. (1970), Prog. Theor. Phys. Suppl. **45**, 231.
Nickel, B. G. (1974), J. Phys. C **7**, 1719.
O'Connor, A. J. (1975), Commun. Math. Phys. **45**, 63.
O'Connor, A. J. and J. L. Lebowitz (1974), J. Math. Phys. **15**, 692.
Ogita, N., A. Ueda, T. Matsubara, H. Matsuda and F. Yonezawa (1969), J. Phys. Soc. Japan Suppl. **26**, 145.
Onizuka, K. (1975), J. Phys. Soc. Japan **39**, 527.
Painter, R. D. and W. M. Hartmann (1974), Phys. Rev. B **10**, 2159.
Painter, R. D. and W. M. Hartmann (1976), Phys. Rev. B **13**, 479.
Papatriantafillou, C. T. (1973), Phys. Rev. B **7**, 5386.
Payton, D. N., III, (1966), LA-3471-MS, Los Alamos Scientific Laboratory.
Payton, D. N., III, and W. M. Visscher (1967a), Phys. Rev. **154**, 802.
Payton, D. N., III and W. M. Visscher (1967b), Phys. Rev. **156**, 1032.
Payton, D. N., III and W. M. Visscher (1968), Phys. Rev. **175**, 1201.

Payton, D. N., III, M. Rich and W. M. Visscher (1967), Phys. Rev. **160**, 706.
Payton, D. N., III, M. Rich and W. M. Visscher (1968), in *Localized excitations in solids*, Ed. by R. F. Wallis (Plenum, New York), p. 657.
Penchina, C. M. and D. L. Mitchell (1972), J. Non-Cryst. Solids **7**, 127.
Pike, G. E. and C. H. Seager (1974), Phys. Rev. B **10**, 1421.
Pike, G. E., W. J. Camp, C. H. Seager and G. L. McVay (1974), Phys. Rev. B **10**, 4909.
Polk, D. E. (1971), J. Non-Cryst. Solids **5**, 365.
Polk, D. E. and D. S. Boudreaux (1973), Phys. Rev. Lett. **31**, 92.
Pollak, M. (1972), J. Non-Cryst. Solids **11**, 1.
Pollak, M. (1974), in *Amorphous and liquid semiconductors*, Ed. by J. Stuke and W. Brenig (Taylor and Francis, London), p. 127.
Pollak, M. and T. H. Geballe (1961), Phys. Rev. **122**, 1742.
Pollitt, S. (1976), Commun. Phys. **1**, 207.
Rahman, A., M. J. Mandell and J. P. McTague (1975), J. Chem. Phys. **64**, 1564.
Rao, M., M. H. Kalos, J. L. Lebowitz and J. Mano (1976), Phys. Rev. B **13**, 4328.
Raveché, H. J., R. D. Mountain and W. B. Streett (1974), J. Chem. Phys. **61**, 1970.
Rechtin, M. D. and B. L. Averbach (1975), Phys. Stat. Sol. (a) **28**, 283.
Renninger, A. L., M. D. Rechtin and B. L. Averbach (1974), J. Non-Cryst. Solids **16**, 1.
Rich, M. and W. M. Visscher (1975), Phys. Rev. B **11**, 2164.
Rich, M., W. M. Visscher and D. N. Payton III (1971), Phys. Rev. A **4**, 1682.
Rieder, Z., J. L. Lebowitz and E. Lieb (1967), J. Math. Phys. **8**, 1073.
Roberts, A. P. and R. E. B. Makinson (1962), Proc. Phys. Soc. **79**, 630.
Rogers, C. A. (1958), Proc. London Math. Soc. **8**, 609.
Rosenstock, H. B. and R. E. McGill (1968), Phys. Rev. **176**, 1004.
Rubin, R. J. (1963), Phys. Rev. **131**, 964.
Rubin, R. J. (1968), J. Math. Phys. **9**, 2252.
Rubin, R. J. (1970), J. Math. Phys. **11**, 1857.
Rubin, R. J. and W. L. Greer (1971), J. Math. Phys. **12**, 1686.
Rycerz, Z. and J. Mościński (1976a), Comp. Phys. Commun. **11**, 163.
Rycerz, Z. and J. Mościński (1976b), Comp. Phys. Commun. **11**, 169.
Rycerz, Z. and J. Mościński (1976c), Comp. Phys. Commun. **11**, 177.
Sadoc, J. F., J. Dixmier and A. Guinier (1973), J. Non-Cryst. Solids **12**, 46.
Sakata, M., F. Matsubara, Y. Abe and S. Katsura (1977), J. Phys. C **10**, 2887.
Saxon, D. S. and R. F. Hutner (1949), Philips Res. Rep. **4**, 81.
Scher, H. and M. Lax (1972), J. Non-Cryst. Solids **8–10**, 497.
Scher, H. and M. Lax (1973a), Phys. Rev. B **7**, 4491.
Scher, H. and M. Lax (1973b), Phys. Rev. B **7**, 4502.
Schmidt, H. (1957), Phys. Rev. **105**, 425.
Schneider, T. and E. Stoll (1976), Phys. Rev. B **13**, 1216.
Schönhammer, K. and W. Brenig (1973), Phys. Lett. **42A**, 447.
Scott, G. D. and D. M. Kilgour (1969), J. Phys. D **2**, 863.
Seager, C. H. and G. E. Pike (1974), Phys. Rev. B **10**, 1435.
Sen, P. N. (1973), Solid State Commun. **13**, 1693.
Shante, V. K. S., C. M. Varma and A. N. Bloch (1973), Phys. Rev. B **8**, 4885.
Shchegolev, I. F. (1972), Phys. Stat. Sol. (a) **12**, 9.
Shevchik, N. J. (1973a), Phys. Stal. Sol. (b) **58**, 111.
Shevchik, N. J. (1973b), J. Non-Cryst. Solids **12**, 141.
Shevchik, N. J. (1974), Phys. Stat. Sol. (b) **61**, 589.
Shevchik, N. J. and W. Paul (1972), J. Non-Cryst. Solids **8–10**, 381.
Shklovskii, B. I. and A. L. Efros (1976), Sov. Phys.–Usp. **18**, 845.
Spohn, H. and J. L. Lebowitz (1977), Commun. Math. Phys. **54**, 97.
Steinhardt, P., R. Alben and D. Weaire (1974), J. Non-Cryst. Solids **15**, 199.

Stillinger, F. H. (1975), Advan. Chem. Phys. **33**, 1.

Stillinger, Jr., F. H., E. A. DiMarzio and R. L. Kornegay (1964), J. Chem. Phys. **40**, 1564.

Streett, W. B., H. J. Raveché and R. D. Mountain (1974), J. Chem. Phys. **61**, 1960.

Stroud, D. (1975), Phys. Rev. B **12**, 3368.

Sur, A., J. L. Lebowitz, J. Mano, M. H. Kalos and S. Kirkpatrick (1976), J. Stat. Phys. **15**, 345.

Sur, A., J. L. Lebowitz, J. Mano and M. H. Kalos (1977), Phys. Rev. B **15**, 3014.

Taylor, D. W. (1975), in *Dynamical properties of solids*, Ed. by G. K. Horton and A. A. Maradudin (North-Holland, Amsterdam), Vol. 2, p. 285.

Theodorou, G. and M. H. Cohen (1976), Phys. Rev. B **13**, 4597.

Thouless, D. J. (1972), J. Phys. C **5**, 77.

Thouless, D. J. (1974), Phys. Rep. **13**, 94.

Thouless, D. J. (1976), J. Phys. C **9**, L603.

Toda, M. (1975), Phys. Lett. **18C**, 1.

Tong, B. Y. (1970), Phys. Rev. A **1**, 52.

Tong, B. Y. and F. C. Choo (1976), Solid State Commun. **20**, 957.

Tong, B. Y. and T. C. Wong (1972), Phys. Rev. B **6**, 4482.

Valleau, J. P. and S. G. Whittington (1977), J. Comp. Phys. **24**, 150.

Visscher, W. M. (1971), Prog. Theor. Phys. **46**, 729.

Visscher, W. M. (1972), J. Non-Cryst. Solids **8–10**, 477.

Visscher, W. M. (1974), Phys. Rev. A **10**, 2461.

Visscher, W. M. (1976), in *Methods in computational physics*, Ed. by B. Alder, S. Fernbach and M. Rotenberg (Academic Press, New York), Vol. 15, p. 371.

Visscher, W. M. (1977), unpublished.

Visscher, W. M. and M. Bolsterli (1972), Nature **239**, 504.

Visscher, W. M. and M. Rich (1975), Phys. Rev. A **12**, 675.

Von Heimendahl, L. (1975), J. Phys. F **5**, L141.

Wada, K. (1966), Prog. Theor. Phys. **36**, 726.

Walker, L. R. and R. E. Walstedt (1977), Phys. Rev. Lett. **38**, 514.

Watson, B. P. and P. L. Leath (1974), Phys. Rev. B **9**, 4893.

Watson, B. P. and P. L. Leath (1975), Phys. Rev. B **12**, 498.

Weaire, D. (1976), Contemp. Phys. **17**, 173.

Weaire, D. and V. Srivastava (1977), Commun. Phys. **2**, 73.

Weaire, D. and A. R. Williams (1976), J. Phys. C **9**, L461.

Weaire, D. and A. R. Williams (1977), J. Phys. C **10**, 1239.

Weis, J.-J. (1974), Molec. Phys. **28**, 187.

Wheeler, J. C., M. G. Prais and C. Blumstein (1974), Phys. Rev. B **10**, 2429.

Wilkinson, J. H. (1965), *The algebraic eigenvalue problem* (Clarendon Press, Oxford).

Windsor, C. G. and J. Locke-Wheaton (1976), J. Phys. C **9**, 2749.

Wintle, H. J. and T. P. T. Williams (1977), Can. J. Phys. **55**, 635.

Wood, W. W. (1968), in *Physics of simple liquids*, Ed. by H. N. V. Temperley, J. S. Rowlinson and G. S. Rushbrooke (Wiley Interscience, New York), p. 115.

Wood, W. W. (1975), in *Fundamental problems in statistical mechanics*, Ed. by E. D. G. Cohen (North-Holland, Amsterdam), Vol. 3, p. 331.

Wood, W. W. and J. J. Erpenbeck (1976), Ann. Rev. Phys. Chem. **27**, 319.

Wu, S. Y. (1975), Phys. Stat. Sol. (b) **64**, K75.

Wu, S. Y. and M. Chao (1975), Phys. Stat. Sol. (b) **68**, 349.

Yndurian, F., J. D. Joannopoulos, M. L. Cohen and L. Falicov (1974), Solid State Commun. **15**, 617.

Yoshino, S. and M. Okazaki (1976), Solid State Commun. **20**, 81.

Yoshino, S. and M. Okazaki (1977), J. Phys. Soc. Japan **43**, 415.

Zachariasen, W. H. (1932), J. Am. Chem. Soc. **54**, 3841.

CHAPTER 3

Morphic Effects in Lattice Dynamics

EVANGELOS M. ANASTASSAKIS

Physics Department, Polytechnion
Athens 147, Greece

Dynamical Properties of Solids, edited by
G. K. Horton and A. A. Maradudin

Contents

1. *Introduction*

When an *external force* is applied to a system the symmetry of the system (if any) is lowered in general. Effects which may arise from this reduction of symmetry are called *morphic effects* (Müller 1935). By external force we mean any perturbation which is externally applied to the system and which transforms in a well-defined manner under the space symmetry operations of the system and under the time reversal symmetry operation.

In *Crystal Physics* morphic effects have been known for a long time. They appear as higher-order *macroscopic* effects such as piezoelectricity, photoelasticity, the electro-optic effect, electrostriction, etc. (Nye 1964, Bhagavantam 1966). Most of the technological applications of non-linear optics are in fact based on some kind of a morphic effect.

In *Atomic, Molecular and Solid State Physics*, morphic effects are encountered as *microscopic* effects, that is, only in microscopic scale. Stark and Zeeman shifts and splittings of electronic energy levels are examples of such phenomena, as are the effects of an electric and magnetic field and a uniaxial stress on the energy and degeneracy of impurity and point defect electronic states (Hayes and Macdonald 1967 and references therein). The behavior of trivalent impurity states of semiconducting diamond (type IIb) under such perturbations constitutes a typical example of this type of microscopic morphic effects (Bagguley et al. 1966, Crowther et al. 1967, Anastassakis 1969). In the same spirit one may include among morphic effects the changes in the electronic band structures of solids due to an electromagnetic or elastic perturbation, e.g. electro-piezo-optical effects (Cardona 1969, Seraphin 1972, Balslev 1972, Bir and Pikus 1975, Kondo and Moritani 1976, Chandrasekhar and Pollak 1977) and magneto-optical effects (Aggarwal 1972).

In *Lattice Dynamics*, one way that morphic effects may manifest themselves is from the changes in the selection rules for the interaction of light with the collective excitations which can be sustained by the lattice. The best-known examples of such interactions are the absorption and scattering of light by the vibrational normal modes (phonons) of the crystal. Neutron scattering is another form of interaction which in principle could qualify as an experimental procedure for the observation and study of morphic effects in Lattice Dynamics. In practice, however, the use of neutron scattering is avoided because of its serious limitations of resolution. X-ray spectroscopy also presents a valid possibility, although no systematic experimental work seems to have been done in connection with morphic effects. In this chapter we are concerned only with those morphic effects that can be observed through infrared absorption and Raman scattering techniques. Whether or not a particular vibrational normal mode can produce a scattering and/or absorption process depends critically on the symmetry characteristics of the mode; in other words, it is a matter of *mode property*. The concept of mode properties will be introduced later more quantitatively in terms of well-defined quasi-macroscopic parameters of the crystal. These parameters also contain information about the strength of the corresponding effect. It is clear now that the lowering of symmetry may have an influence on the mode properties, thus leading to related morphic effects. Besides scattering and absorption, external forces may also affect the dispersion curves of the pure lattice. This will appear as a shift of the frequencies of the various normal modes and/or lifting of their degeneracies.

The various types of generalized forces used in this chapter are classified as follows. A force is characterized as a *polar force* if it transforms under the rotation symmetry operations of the system like a polar tensor of appropriate rank (e.g. an electric field, a stress or strain, an electric field gradient). It is an *axial force* if it transforms like an axial tensor under the same operations (e.g. a magnetic field, a uniform rotation). Likewise, it is a *time symmetric* force if its sign remains unchanged under the time reversal symmetry operation (e.g. an electric field, a stress). Otherwise it is a *time antisymmetric* force (e.g. a magnetic field, the gradient of the electric component of an electromagnetic field, a uniform rotation). Hydrostatic pressure and temperature may also be regarded as generalized forces of *scalar character*. They do not affect the symmetry of the system directly. Strictly speaking, the resulting effects do not conform with the definition of morphic effects. On the other hand, these forces may induce phase transitions which are accompanied by changes in the symmetry of the system. Thus, morphic effects due to scalar forces may be produced indirectly. Finally, forces which change their sign under an improper rotation only (= rotation + reflection in a plane perpendicular to the axis of rotation) will be named *pseudoscalar forces*. In principle such forces are capable of producing morphic effects although no physical example of this kind of force is commonly known.

In recent years considerable theoretical and experimental effort has been put into morphic effects in Lattice Dynamics. Examples of such studies are the effects of an electric field and uniaxial stress on first- and second-order infrared absorption and Raman scattering by the long-wavelength optical phonons of certain crystals, and also the effects of a uniaxial stress on the frequencies and degeneracies of such modes (Anastassakis 1972 and references therein, Angress et al. 1975). Similar effects due to a magnetic field were reported more recently (Schaack 1977a, b). Morphic effects on the optical and vibrational properties of disordered solids have been and are the subjects of extensive investigations (Maradudin et al. 1967, Ganesan 1971, Barker and Sievers 1975, Taylor 1975).

All these theoretical and experimental studies have clearly demonstrated that with the use of morphic effects one can predict situations, test the validity of theoretical models and, generally speaking, obtain significant knowledge about the various aspects of lattice-vibrational modes and their interactions. In most cases such knowledge is unaccessible in other ways.

In the present review of morphic effects in Lattice Dynamics every effort has been made to include a reasonable amount of information on relevant methods and experimental facts. Particular emphasis is given to morphic effects due to non-scalar forces. For completeness the effects of hydrostatic pressure and temperature are briefly discussed and also spatial dispersion effects and their role in resonant Raman scattering.

We shall proceed as follows. In §§2–6, a brief summary will be given of those basic definitions and ideas from Lattice Dynamics and group theory which are necessary for the treatment of morphic effects (only basic knowledge of group theory is required throughout this chapter, except for §17 where some knowledge of corepresentation theory may be particularly helpful). This will be followed by applications to specific types of forces, including electric field, stress, magnetic field, hydrostatic pressure, temperature, magnetic ordering, spatial dispersion effects and coupling of normal modes (§§6–19). The chapter is closed with a summary of what has been accomplished and what may be accomplished in the future by the use of morphic effects. The effects of a force on infrared absorption and Raman scattering and on the frequency and degeneracy of optical phonons are discussed in separate (successive) sections for every type of force. At the beginning of each such section the general symmetry analysis and other theoretical discussion are given. This is followed by a review of the corresponding experimental work. The author has tried to include in each case the major and most relevant published theoretical and experimental results up to mid-1977. He expresses his regrets for any accidental omission of important articles.

2. *Normal coordinate formalism*

For a crystal with s atoms per unit cell there exist $3s$ branches for the crystal dispersion curves. Each point on these branches corresponds to a normal mode of vibration with a well-defined wavevector $\boldsymbol{q}$ and frequency

$\omega(\boldsymbol{q}/j)$. The index j designates the branch j with, in general, a degeneracy l_j. An additional index σ, where $\sigma=1,\ldots,l_j$ is needed to complete the designation of a degenerate normal mode. Obviously, for fixed $\boldsymbol{q}$ and j all σ modes have the same eigenfrequency $\omega(\boldsymbol{q}/j)$. The values for $\omega(\boldsymbol{q}/j)$, and the three Cartesian coordinates $e_\alpha(\kappa/\boldsymbol{q}j\sigma)$ of the eigenvector for the κth atom in the unit cell are both found from the solution of the eigenvalue equation

$$\omega^2(\boldsymbol{q}/j)e(/\boldsymbol{q}j\sigma)=\boldsymbol{D}(\boldsymbol{q})\cdot e(/\boldsymbol{q}j\sigma). \tag{2.1}$$

$\boldsymbol{D}(\boldsymbol{q})$ is a $3s\times3s$ Hermitian matrix, the *Fourier-transformed dynamical matrix* for the wavevector $\boldsymbol{q}$ (Born and Huang 1962, Birman 1974a, b). Its components $D_{\alpha\beta}(\boldsymbol{q}/\kappa\kappa')$ are expressed in terms of the components of the real force constant matrix $\Phi_{\alpha\beta}(l\kappa/l'\kappa')$. The latter involves the Cartesian coordinates of the real displacements $\mu_\alpha(l\kappa)$ and $\mu_\beta(l'\kappa')$ from their equilibrium positions of the atoms κ and κ' in cells l and l' respectively. Accordingly, the eigenvector $\boldsymbol{e}(/\boldsymbol{q}j\sigma)$ is represented by a column matrix with $3s$ components $e_\alpha(\kappa/\boldsymbol{q}j\sigma)$, where $\alpha=1,2,3$, and $\kappa=1,\ldots,s$. Three of the solutions $\omega^2(\boldsymbol{q}/j)$ of eq. (2.1) tend to zero as $q\to0$. The corresponding branches are recognized as the *acoustic branches*. The remaining $3s-3$ branches are the *optical branches*. Phonons from the acoustic or optical branches will be characterized as *acoustic or optical phonons* respectively.

It turns out from space group symmetry considerations that the branch index j may be used to designate one of the irreducible representations of the unitary space group $\mathcal{G}(\boldsymbol{q})$ of $\boldsymbol{q}$. More specifically, the eigenvectors $\boldsymbol{e}(/\boldsymbol{q}j\sigma)$ can serve as a basis for the jth irreducible representation of $\mathcal{G}(\boldsymbol{q})$. The degeneracy of the branch is now interpreted as the degeneracy of jth irreducible representation. In other words $\boldsymbol{e}(/\boldsymbol{q}j\sigma)$ transforms as the σth row of the jth irreducible representation of $\mathcal{G}(\boldsymbol{q})$ (Warren 1968, Maradudin and Vosko 1968, Birman 1974a, b).

Since the eigenvectors $\boldsymbol{e}(/\boldsymbol{q}j\sigma)$ form a complete set of basis functions, the time-dependent Cartesian components $\mu_\alpha(l\kappa)$ of the real atomic displacements can be expanded in terms of the basis $\boldsymbol{e}^*(/\boldsymbol{q}j\sigma)$ for all possible values of $\boldsymbol{q}$, j and σ. The coefficients in this expansion represent the time-dependent complex normal coordinates for the corresponding vibrational modes. We designate these by $u(\boldsymbol{q}j\sigma/t)$. The exact relations between $\mu_\alpha(l\kappa)$ and $u(\boldsymbol{q}j\sigma/t)$ are

$$\mu_\alpha(l\kappa)=(NM_\kappa)^{-1/2}\sum_{\boldsymbol{q},j,\sigma}\exp(-\mathrm{i}\boldsymbol{q}\cdot\boldsymbol{R}_{\mathrm{L}})e^*_\alpha(\kappa/\boldsymbol{q}j\sigma)u(\boldsymbol{q}j\sigma/t) \tag{2.2}$$

and

$$u(\boldsymbol{q}j\sigma/t)=N^{-1/2}\sum_{l,\kappa,\alpha}\exp(\mathrm{i}\boldsymbol{q}\cdot\boldsymbol{R}_{\mathrm{L}})e_{\alpha}(\kappa/\boldsymbol{q}j\sigma)\mu_{\alpha}(l\kappa)M_{\kappa}^{1/2}, \tag{2.3}$$

where $\boldsymbol{R}_{\mathrm{L}}$ is a lattice translation vector, M_{κ} is the mass of the atom κ in the lth unit cell and N the number of unit cells in the crystal. The normal coordinate formalism has the advantage that it describes collectively the normal mode j at $\boldsymbol{q}$ and makes no reference to the individual atoms in the unit cell or to their physical displacements associated with the mode j. Furthermore, the normal coordinate $u(\boldsymbol{q}j\sigma/t)$ transforms according to the σth row of the irreducible representation j of $\mathcal{G}(\boldsymbol{q})$, that is, the set of functions $u(\boldsymbol{q}j\sigma/t)$, $\sigma=1,\ldots,l_j$ forms a basis for the irreducible representation j (Birman 1974a, b). The most important property of the normal coordinate formalism is that the eigenvalue equation (2.1) can be replaced by as many independent simple harmonic oscillator equations of motion as the number of normal modes, i.e.

$$\ddot{u}(\boldsymbol{q}j\sigma/t)+\omega^{2}(\boldsymbol{q}/j)u(\boldsymbol{q}j\sigma/t)=0, \qquad \sigma=1,\ldots,l_j. \tag{2.4}$$

This can be easily derived from Hamilton's equations of motion assuming, within the harmonic approximation, a classical Hamiltonian of the form

$$\mathcal{H}=\frac{1}{2}\sum_{\boldsymbol{q},j,\sigma}\left\{\dot{u}(\boldsymbol{q}j\sigma/t)^{*}\dot{u}(\boldsymbol{q}j\sigma/t)+\omega^{2}(\boldsymbol{q}/j)u(\boldsymbol{q}j\sigma/t)^{*}u(\boldsymbol{q}j\sigma/t)\right\}. \tag{2.5}$$

For the remainder of this chapter we will be mostly concerned with optical normal modes of long wavelength (i.e. $\boldsymbol{q}\simeq 0$ optical phonons). The corresponding unitary symmetry group is the point group of the crystal. Its irreducible representations will be designated by $\Gamma(j)$ and the normal coordinates of the jth $\boldsymbol{q}\simeq 0$ optical phonon with a degeneracy l_j by $u_{j\sigma}$, $\sigma=1,\ldots,l_j$. Unless otherwise specified, we will be dealing with the real part of the Hamiltonian (2.5) and the normal coordinate $u_{j\sigma}$. The latter will be regarded from now on as a quasi-macroscopic parameter of the crystal which carries all the symmetry characteristics of the optical phonon. Under these circumstances eq. (2.5) takes the form

$$\mathcal{H}=\frac{1}{2}\sum_{j,\sigma}\left\{\dot{u}_{j\sigma}^{2}+\omega_j^{2}u_{j\sigma}^{2}\right\}. \tag{2.6}$$

Eq. (2.6) will be used later to derive the relationship between ω_j and the effective force constant of the $\boldsymbol{q}\simeq 0$ optical phonon.

3. *Mode properties and related morphic effects*

Infrared absorption and Raman scattering are the most convenient phenomena from which the existence and physical behavior of certain crystal vibrational normal modes can be clearly and directly demonstrated. Neutron scattering of course is an equally important experimental procedure particularly useful in the gross study of all possible dispersion curves of the crystal. However, resolution restrictions disqualify neutron scattering for the observation and study of such fine effects as, for instance, force-induced or spatial-dispersion-induced frequency splittings, etc. As we shall see, Raman scattering is a uniquely sensitive tool for probing phenomena like these. Most, if not all, of the known morphic effects have been studied using infrared absorption or Raman scattering. For the remainder of this chapter we restrict the discussion to such effects only.

In this section, we will first give a brief introduction to infrared absorption and Raman scattering with no external forces present (*intrinsic effects*). This will be followed by the corresponding morphic effects due to the presence of external forces (*induced effects*). The discussion includes the frequency and degeneracy of normal modes with and without external forces present. Both intrinsic and induced effects are introduced only from a phenomenological point of view, since it is the symmetry rather than the quantitative aspects of the problem that we are concerned with.

3.1. First-order infrared absorption

First-order (one-phonon) infrared absorption due to a normal mode j can occur when the mode j is capable of modulating $\boldsymbol{M}$, the crystal electric dipole moment per unit cell, at the frequency of the mode ω_j. If the crystal is exposed to a continuum of electromagnetic radiation, it is well known that the transmission spectrum of the crystal will exhibit an absorption band at ω_j. Since the frequencies ω_j usually fall in the infrared region of the spectrum, the distinct absorption lines described above will also appear in the infrared. We are not concerned with overtone or combination frequency absorption. These effects are easily described by extending the process described above to overtone or combination frequencies of modulation.

For a phenomenological treatment of the effect we assume that the dipole moment $\boldsymbol{M}$ is a function of the normal coordinates $u_{j\sigma}$ associated with a particular mode of symmetry $\Gamma(j)$ and degeneracy l_j. In addition to this *internal perturbation*, the electric moment $\boldsymbol{M}$ may also depend on an *externally imposed perturbation*, due to a generalized applied force $\boldsymbol{F}$. This force transforms under the crystal point group symmetry operations in a specified manner and exhibits a certain number of independent components relative to the system of crystallographic axes. We designate these components by F_ν, with the understanding that the index ν may in fact stand for all indices of a tensor of appropriate rank. We treat $\boldsymbol{F}$ as a time symmetric force. The case of time antisymmetric forces will be examined in §15.

We can now expand the components of $\boldsymbol{M}$ to first order in the variables $u_{j\sigma}$ and F_ν, ignoring for the moment all higher-order terms,

$$M_\mu \simeq M_{0\mu} + \left(\frac{\partial M_\mu}{\partial u_{j\sigma}}\right) u_{j\sigma} + \left(\frac{\partial^2 M_\mu}{\partial F_\nu \, \partial u_{j\sigma}}\right) F_\nu u_{j\sigma}$$

$$\equiv M_{0\mu} + (e_{j,\mu\sigma} + \Delta e_{j,\mu\sigma}) u_{j\sigma}, \tag{3.1a}$$

where

$$e_{j,\mu\sigma} = \left(\frac{\partial M_\mu}{\partial u_{j\sigma}}\right) \tag{3.1b}$$

$$\Delta e_{j,\mu\sigma} = \left(\frac{\partial^2 M_\mu}{\partial F_\nu \, \partial u_{j\sigma}}\right) F_\nu . \tag{3.1c}$$

Summation over repeated indices is implied in eqs. (3.1a, c). The partial derivatives are calculated at the equilibrium position with $\boldsymbol{F}=0$. In what follows we will always assume this to be the case. The term $M_{0\mu}$ corresponds to the μth component of the electric moment per unit cell when no atomic motion is involved and no forces are applied. This term exists only in certain crystals (e.g. ferroelectrics, pyroelectrics). For simplicity we will ignore this term from now on. The remaining two terms in eq. (3.1a) are the phonon modulated contributions to $\boldsymbol{M}$ at the frequency ω_j. The coefficient $e_{j,\mu\sigma}$ in eq. (3.1b) corresponds to the $\mu\sigma$ component of a tensor-like quantity $\boldsymbol{e}_j$ which is defined *as the intrinsic effective charge per unit cell associated with the optical mode j*. It is solely the symmetry of the mode j which determines whether or not $\boldsymbol{e}_j$ exists. As we shall see, for $\boldsymbol{e}_j$ to be non-zero the normal coordinate $u_{j\sigma}$ must transform like a polar vector. This means that $\boldsymbol{e}_j$ transforms like a general $3 \times l_j$ polar tensor. Likewise, the coefficient $\Delta e_{j,\mu\sigma}$ in eq. (3.1c) represents the $\mu\sigma$ component of a tensor-like quantity $\Delta\boldsymbol{e}_j$ which is defined as the linear *force-induced effective charge per unit cell associated with the optical mode j*. Its contribution to $\boldsymbol{M}$ depends linearly on the applied force. Again, it is the symmetry of the mode j which determines whether or not $\Delta\boldsymbol{e}_j$ exists for a given type of applied force. For $\Delta\boldsymbol{e}_j$ to be non-zero, the normal coordinate $u_{j\sigma}$ does not have to transform like a polar vector. This means that $\Delta \boldsymbol{e}_j$ cannot be identified with a second-rank polar tensor in general. It consists of an array of $3\times l_j$ components which carry their own transformation properties under symmetry operations of the crystal point group.

The effective charge of eq. (3.1b) is defined under constant macroscopic electric field. It can also be introduced with different independent parameters kept constant. Thus, the *Szigetti charge* is defined as $\boldsymbol{e}_j^{\mathrm{s}}=(\partial \boldsymbol{M}/\partial u_j)_{E_\ell}$, where $\boldsymbol{E}_\ell$ is the local electric field. If all these independent parameters are taken into consideration one has to include in eq. (3.1a) all the corresponding terms. We omit all these terms, as their inclusion does not add any further information to the symmetry aspects of the problem.

Physical quantities like $\boldsymbol{e}_j$ and $\Delta \boldsymbol{e}_j$ are properties of the optical phonon, in the sense that their existence is determined by the symmetry of the phonon. They are hereafter defined as *mode properties*. Furthermore, partial derivatives like the one in eq. (3.1c) describe the possible linear dependence of a mode property on the external force. They represent component arrays of tensor-like quantities whose existence and transformation properties for a given type of force are also determined by the symmetry of the mode. They are hereafter defined as *mode coefficients*.

In view of all the simplifications introduced above, eq. (3.1a) can be rewritten in a more compact form, i.e.

$$\boldsymbol{M}_{jF}\equiv(\boldsymbol{e}_j+\Delta \boldsymbol{e}_j)\cdot\boldsymbol{u}_j\equiv\boldsymbol{M}_j+\Delta\boldsymbol{M}_j, \tag{3.2}$$

where matrices of appropriate dimensions are implied for all quantities involved.

Suppose that an incident linearly polarized electromagnetic field $\mathcal{E}^0(\omega_j)$ is absorbed by the mode j. The intensity of the absorption is proportional to $|\mathcal{E}^0\cdot\boldsymbol{M}_j|^2$. It is assumed, of course, that $\boldsymbol{M}_j\neq 0$, or $\boldsymbol{e}_j\neq 0$, in other words the mode j is *infrared active*. From the requirement $|\mathcal{E}^0\cdot\boldsymbol{M}_j|\neq 0$ one can determine the *polarization selection rules* for this phonon–photon interaction. Obviously, if any absorption is to occur the polarization of $\mathcal{E}^0$ must not be perpendicular to that of $\boldsymbol{M}_j$. To put it differently, only components of $\mathcal{E}^0$ parallel to $\boldsymbol{M}_j$ can be absorbed.

Similar arguments hold when a force is applied. If the symmetry of the mode j is such that $\boldsymbol{e}_j=0$ (*infrared-inactive* mode) but $\Delta\boldsymbol{e}_j\neq 0$, one may expect force-induced infrared absorption to occur at the mode frequency ω_j. Its intensity will be proportional to $|\mathcal{E}^0\cdot\Delta\boldsymbol{M}_j|^2\propto F^2$. The polarization selection rules are derived from a similar requirement, that is $|\mathcal{E}^0\cdot\Delta\boldsymbol{M}_j|\neq 0$. If, on the other hand, the symmetry of the mode is such that $\boldsymbol{e}_j\neq 0$ and $\Delta\boldsymbol{e}_j\neq 0$, then the intensity of the absorption will be proportional to

$|\mathscr{E}^0 \cdot (\boldsymbol{M}_j + \Delta \boldsymbol{M}_j)|^2$ and the polarization selection rules are in general modified by the applied force. In both cases ($\boldsymbol{e}_j = 0$, $\Delta \boldsymbol{e}_j \neq 0$ and $\boldsymbol{e}_j \neq 0$, $\Delta \boldsymbol{e}_j \neq 0$) the resulting phenomena constitute *morphic effects associated with the mode j*.

3.2. First-order Raman scattering

First-order (one-phonon) Raman scattering due to a normal mode j can occur when $\boldsymbol{\alpha}$, the crystal electric polarizability per unit cell, exhibits a linear dependence on the normal coordinate $u_{j\sigma}$. Physically this means that there exists a contribution $\boldsymbol{\alpha}_j$ in $\boldsymbol{\alpha}$ which is modulated by the mode j at the frequency ω_j. When the crystal is exposed to a monochromatic light beam of frequency ω_0 (usually in the visible region of the spectrum) the associated electric field $\mathscr{E}^0$ sets up an electric moment per unit cell $\boldsymbol{M}_j$, equal to $\boldsymbol{\alpha}_j \cdot \mathscr{E}^0$. Since $\boldsymbol{\alpha}_j$ and $\mathscr{E}^0$ are modulated at the frequencies ω_j and ω_0 respectively, the resulting dipole moment per unit cell oscillates at the frequency $\omega_S \equiv \omega_0 - \omega_j$ or $\omega_A \equiv \omega_0 + \omega_j$. This means that the crystal radiates (scatters) these two frequencies which fall close to ω_0 (since $\omega_j \ll \omega_0$). They are commonly known as the *Stokes* and *anti-Stokes* components respectively of first-order Raman scattering by the mode j. From a microscopic point of view one can also regard the Stokes process as a sequence of the following events, not necessarily in this order: (i) absorption of a photon ω_0 through an electron–photon interaction, (ii) creation of a phonon ω_j through an electron–phonon interaction, and (iii) emission of a photon $\omega_S = \omega_0 - \omega_j$ through an electron–photon interaction. An analogous sequence of events may be invoked to describe an anti-Stokes process. A great deal of theoretical and experimental work has been done in recent years to establish on a microscopic scale the exact nature and sequence of events which lead to Raman scattering. No general and universally applicable theory is available as yet, although a large number of models have been proposed and successfully used to interpret the results from groups of materials. For example, one can mention the work on insulators and semiconductors (Pinczuk and Burstein 1975, Richter 1976 and references therein) and on metals (Mills et al. 1970).

We are not concerned here with the microscopic aspects of light scattering but rather with the symmetry characteristics of the effect under the influence of an external force. In the presence of an optical mode j and an applied force $\boldsymbol{F}$, we can expand $\boldsymbol{\alpha}$ to terms linear in $u_{j\sigma}$ and F_ν, following the same phenomenological approach as in §3.1:

$$\boldsymbol{\alpha} \simeq \boldsymbol{\alpha}_0 + \boldsymbol{\alpha}_{jF} = \boldsymbol{\alpha}_0 + (\boldsymbol{\alpha}_j + \Delta\boldsymbol{\alpha}_j) = \boldsymbol{\alpha}_0 + (\boldsymbol{a}_j + \Delta\boldsymbol{a}_j) \cdot \boldsymbol{u}_j, \tag{3.3a}$$

where, in full notation,

$$a_{j,\mu\lambda\sigma}=\left(\frac{\partial\alpha_{\mu\lambda}}{\partial u_{j\sigma}}\right), \tag{3.3b}$$

$$\Delta a_{j,\mu\lambda\sigma}=\left(\frac{\partial^2\alpha_{\mu\lambda}}{\partial F_\nu\,\partial u_{j\sigma}}\right)F_\nu. \tag{3.3c}$$

$\boldsymbol{\alpha}_0$ is the electric polarizability per unit cell when no atomic motion is involved and no forces are applied. The tensor-like quantities $\boldsymbol{a}_j$ and $\Delta\boldsymbol{a}_j$, correspond respectively to the *intrinsic* and *force-induced Raman tensor associated with the normal mode j*. In most physical situations they are symmetric in the indices μ,λ (see §4). It is the symmetry of the mode j which determines whether or not $\boldsymbol{a}_j\neq 0$, and $\Delta\boldsymbol{a}_j\neq 0$ for a given type of force. Thus $\boldsymbol{a}_j$ and $\Delta\boldsymbol{a}_j$ constitute mode properties.

Suppose that the symmetry of a given mode j is such that $\boldsymbol{a}_j\neq 0$. We say that the mode is *Raman active*. The scattering intensity is proportional to $|\mathcal{E}^S\cdot\boldsymbol{M}_j|^2=|\mathcal{E}^S\cdot\boldsymbol{\alpha}_j\cdot\mathcal{E}^0|^2$. The same quantity also determines the polarization selection rules for the scattering process. In other words it specifies the components $\alpha_{j,\mu\lambda}$ which will produce a scattered field $\mathcal{E}^S_\mu$ or $\mathcal{E}^S_\lambda$ out of an incident field $\mathcal{E}^0_\lambda$ or $\mathcal{E}^0_\mu$ (the indices μ,λ correspond to a convenient system of Cartesian coordinates). In the presence of a force the intensity will be proportional to $|\mathcal{E}^S\cdot(\boldsymbol{\alpha}_j+\Delta\boldsymbol{\alpha}_j)\cdot\mathcal{E}^0|^2$ and the polarization rules will be modified in general*. If $\boldsymbol{a}_j=0$ and $\Delta\boldsymbol{a}_j\neq 0$ we are dealing with force-induced Raman scattering. Its intensity will be proportional to F^2. As before, these phenomena constitute morphic effects.

So far the discussion has been referred to scattering by a single mode j. The summation over σ implied in eqs. (3.3) indicates that each component of the degenerate mode j will contribute independently to the total scattering intensity at the frequency ω_S or ω_A. Since no phase correlation exists among such individual scattering processes the total scattering intensity may be regarded as arising from incoherent superposition of the individual intensities for all degenerate partners $\sigma=1,\ldots,l_j$ and all optical modes j. In other words the total scattering intensity will be proportional to $\Sigma_{j,\sigma}|\mathcal{E}^S\cdot(\boldsymbol{\alpha}_j+\Delta\boldsymbol{\alpha}_j)\cdot\mathcal{E}^0|^2$ where the index σ is contained in the polarizability terms.

*For the description of a specific scattering geometry or configuration, we shall often use the notation of Damen et al. (1966), for example, $\bar{x}(yz)x$. This means that the incident and observed scattered beams are along $-x$ and x respectively, and they are polarized along y and z respectively.

3.3. The frequencies of $q \simeq 0$ optical phonons as mode properties

The normal coordinate formalism presented earlier allows one to look upon the frequency ω_j of the mode j as associated with *an effective force constant* K_j. This is another way of picturing the reduction of the eigenvalue equation (2.1) into a number of independent equations of motion for a single harmonic oscillator. The frequency of each of these oscillators is related to the classical Hamiltonian (or equivalently the crystal potential energy) through appropriate second-order derivatives of eq. (2.6). By definition the frequency ω_j is non-zero for all optical phonons. Therefore, the symmetry of the mode is not important in determining the existence of ω_j. On the other hand, the symmetry is important in deciding whether or not an applied force can shift the frequency ω_j and/or split the degeneracy of the mode. In this sense ω_j may be regarded as a mode property and the related effects of an applied force as morphic effects. The phenomenological development of this type of morphic effects is based on the secular equation associated with the jth block of the dynamical matrix. When no force is applied we can write

$$|K_{j,\alpha\beta} - \omega_j^2 \delta_{\alpha\beta}| = 0. \tag{3.4}$$

The indices α, β designate values of the row index $\sigma = 1, \ldots, l_j$ of $u_{j\sigma}$. For the 32 crystallographic point groups these indices are the same as the Cartesian indices x, y, z which are associated with the crystallographic system of axes. This simplifies the notation and also makes the eigenvectors of secular equations easier to visualize. According to eq. (2.6) the same row index σ appears in each bracket of the summation; also, all l_j brackets with $\sigma = 1, \ldots, l_j$ contain the same frequency factor ω_j^2. This means that $\boldsymbol{K}_j$ is a diagonal second-rank tensor, i.e.

$$K_{j,\alpha\beta} = K_{j,\alpha\alpha} = K_{j,\alpha\beta} \delta_{\alpha\beta} \equiv K_j \delta_{\alpha\beta}, \tag{3.5}$$

where $\delta_{\alpha\beta}$ is the Kronecker delta. From eqs. (3.4) and (3.5) we find

$$K_j = \omega_j^2. \tag{3.6}$$

This, of course, is a trivial result simply stating that each mode is treated like a simple harmonic oscillator.

In the presence of a force the tensor $\boldsymbol{K}_j$ and the frequency ω_j can be expanded to terms linear in F_ν. Thus,

$$\boldsymbol{K}_{jF} \simeq \boldsymbol{K}_j + \Delta \boldsymbol{K}_j, \tag{3.7}$$

$$\Omega_j \simeq \omega_j + \Delta\omega_j, \tag{3.8}$$

where

$$\Delta K_{j,\alpha\beta} = \left(\frac{\partial K_{j,\alpha\beta}}{\partial F_\nu}\right) F_\nu \equiv K_{j,\alpha\beta\nu} F_\nu. \tag{3.9}$$

The new mode coefficient $K_{j,\alpha\beta\nu}$ transforms under the point group symmetry operations in a well-defined manner. The effect of the force is to mix components of different row indices of the originally degenerate mode j. This means that the tensor ΔK_j is not necessarily diagonal when referred to the crystallographic system of axes. The secular equation now takes the form

$$|(\boldsymbol{K}_j + \Delta \boldsymbol{K}_j)_{\alpha\beta} - \Omega_j^2 \delta_{\alpha\beta}| = 0 \tag{3.10}$$

or

$$|\Delta K_{j,\alpha\beta} - \lambda \delta_{\alpha\beta}| = 0, \tag{3.11}$$

where

$$\lambda \equiv \Omega_j^2 - \omega_j^2 \simeq 2\omega_j \Delta\omega_j. \tag{3.12}$$

Thus the original secular equation has been reduced to a new one given by eq. (3.11). Diagonalization of eq. (3.11) will yield the new eigenvectors of the mode and the new value(s) for the eigenfrequency Ω_j. This will also determine to what extent the initial degeneracy is removed. Before these, of course, one has to establish the exact tensor form of $\Delta \boldsymbol{K}_j$, that is, the non-zero components of $\Delta \boldsymbol{K}_j$ for a given type of force.

Thus far we have assumed that the applied force is time symmetric. When regarding the force constant as a Hermitian generalized kinetic coefficient, it turns out that all force constant tensors introduced above are symmetric in the indices α, β. This follows from Onsager's reciprocity theorem for kinetic coefficients (Landau and Lifshitz 1958). Hence,

$$K_{j,\alpha\beta} = K_{j,\beta\alpha}, \tag{3.13}$$

$$K_{j,\alpha\beta\nu} = K_{j,\beta\alpha\nu}. \tag{3.14}$$

Accordingly, $\Delta \boldsymbol{K}_j$ can be treated as a symmetric second-rank tensor. This helps in further reducing the form of the secular equation (3.11). The case of antisymmetric force constants will be treated in §16.

The shift of ω_j and/or lifting of the mode degeneracy can be best traced through experiments involving absorption or scattering of light. Again, proper use of polarization selection rules enables one to follow and observe each component separately. Such experiments will be discussed in detail in later sections.

4. *Potential energy and mode properties*

The main aspect of morphic effects that we are at present concerned with is to be able to anticipate the kind of morphic effects that can occur for a given type of optical mode and applied force. Group theory will play a major role in pursuing this, as will become evident in the next section. Before that, it is necessary to express all mode properties introduced so far in a more fundamental way, so that group theory can be conveniently applied. The starting point is the classical Hamiltonian $\mathcal{H}(u_j, \boldsymbol{F}, \mathcal{E}, \ldots)$. We assume that $\mathcal{H}$ is a function of normal coordinates u_j of an applied force $\boldsymbol{F}$, of a time varying electromagnetic field $\mathcal{E}$ and perhaps of some other independent parameters. Since these variables will affect the term of $\mathcal{H}$ which corresponds to the potential energy Φ of the crystal, it is permissible to continue this discussion on the basis of Φ rather than $\mathcal{H}$. The objective is to express the mode properties and mode coefficients in terms of appropriate derivatives of Φ for which the symmetry transformation properties are well known or can be easily established. This can be done in a straightforward manner if one takes into consideration the fundamental definitions of the electric moment, the electric polarizability and force constant. These are

$$M_\mu = -\left(\frac{\partial \Phi}{\partial \mathcal{E}_\mu}\right), \tag{4.1}$$

$$\alpha_{\mu\lambda} = -\left(\frac{\partial^2 \Phi}{\partial \mathcal{E}_\mu \, \partial \mathcal{E}'_\lambda}\right), \tag{4.2}$$

$$K_j = \left(\frac{\partial^2 \Phi}{\partial u_j \, \partial u_j}\right). \tag{4.3}$$

An expansion of $\Phi(u_j, \boldsymbol{F}, \mathcal{E}, \ldots)$ about the value $\Phi(0,0,0,\ldots) \equiv \Phi_0$ gives

$$\begin{aligned}\Delta\Phi = \Phi - \Phi_0 &= \left(\frac{\partial^2 \Phi}{\partial \mathcal{E} \, \partial u_j}\right)\mathcal{E} u_j + \left(\frac{\partial^3 \Phi}{\partial \mathcal{E} \, \partial F \, \partial u_j}\right)\mathcal{E} F u_j \\ &\quad + \frac{1}{2}\left(\frac{\partial^3 \Phi}{\partial \mathcal{E} \, \partial \mathcal{E}' \, \partial u_j}\right)\mathcal{E}\,\mathcal{E}' u_j + \frac{1}{2}\left(\frac{\partial^4 \Phi}{\partial \mathcal{E} \, \partial \mathcal{E}' \, \partial F \, \partial u_j}\right)\mathcal{E}\,\mathcal{E}' F u_j \\ &\quad + \frac{1}{2}\left(\frac{\partial^2 \Phi}{\partial u_j \, \partial u_j}\right) u_j u_j + \frac{1}{2}\left(\frac{\partial^3 \Phi}{\partial u_j \, \partial u_j \, \partial F}\right) u_j u_j F + \ldots \\ &= -(e_j + \Delta e_j)\mathcal{E} u_j - \frac{1}{2}(a_j + \Delta a_j)\mathcal{E}\,\mathcal{E}' u_j + \frac{1}{2}(K_j + \Delta K_j) u_j u_j + \ldots,\end{aligned} \tag{4.4}$$

where

$$e_j = -\left(\frac{\partial^2 \Phi}{\partial \mathcal{E}\, \partial u_j}\right) = \left(\frac{\partial M}{\partial u_j}\right), \qquad \Delta e_j = -\left(\frac{\partial^3 \Phi}{\partial \mathcal{E}\, \partial F \partial u_j}\right) F = \left(\frac{\partial^2 M}{\partial F \partial u_j}\right) F, \tag{4.5a}$$

$$a_j = -\left(\frac{\partial^3 \Phi}{\partial \mathcal{E}\, \partial \mathcal{E}'\, \partial u_j}\right) = \left(\frac{\partial \alpha}{\partial u_j}\right),$$

$$\Delta a_j = -\left(\frac{\partial^4 \Phi}{\partial \mathcal{E}\, \partial \mathcal{E}'\, \partial F \partial u_j}\right) F = \left(\frac{\partial^2 \alpha}{\partial F \partial u_j}\right) F, \tag{4.5b}$$

$$K_j = \left(\frac{\partial^2 \Phi}{\partial u_j\, \partial u_j}\right), \qquad \Delta K_j = \left(\frac{\partial^3 \Phi}{\partial u_j\, \partial u_j\, \partial F}\right) F. \tag{4.5c}$$

In the Taylor expansion above we have omitted irrelevant terms and also all tensor indices. We notice that the Raman tensors $\boldsymbol{a}_j$ and $\Delta \boldsymbol{a}_j$ are defined in terms of the polarizability of eq. (4.2). The two fields $\mathcal{E}$ and $\mathcal{E}'$ are treated independently to signify the difference in their frequencies, the one being at ω_0 (incident radiation) the other at ω_S (scattered radiation), or vice-versa. The consequence of this differentiation between $\mathcal{E}$ and $\mathcal{E}'$ is that the Raman tensors are not in general symmetric in the two polarizability indices μ and λ. However, they are usually treated as symmetric tensors. This is a very good approximation, consistent with the vast majority of experimental results, provided that $\omega_j \ll \omega_0$ and that ω_0 is not near resonance (Pinczuk and Burstein 1975). In what follows we will always assume this to be the case, unless otherwise specified.

5. *Group-theoretical considerations*

It was pointed out in the last section that the most basic questions regarding morphic effects can be answered through a proper application of group theory. This will be done in three steps. First we will review some general group-theoretical material which concerns the transformation properties of tensors. Then a method will be discussed for determining almost by inspection whether a particular morphic effect can occur for a given type of force, and if so, how many independent components are to be expected for the appropriate mode property or mode coefficient which describes the morphic effect in question. Finally, the exact form of the component array for this mode property or mode coefficient will be established. This will enable us to determine the polarization selection rules which should be applied in order to observe the morphic effect.

5.1. Tensor transformation properties

For a given point group, a Cartesian tensor $\boldsymbol{t}$ of any rank belongs to a reducible representation $\Gamma(\boldsymbol{t})$ of the group, with characters which can be found using standard techniques (Bhagavantam 1966). The reducible representation $\Gamma(\boldsymbol{t})$ can always be reduced to irreducible representations of the point group. The number of times n_A that the totally symmetric irreducible representation of the group A_1 occurs in this reduction indicates the number of independent components which the tensor $\boldsymbol{t}$ exhibits in this particular point group.

We consider structures with O_h and D_3 symmetry. Typical materials which crystallize in these point groups are diamond (centrosymmetric structure) and α-quartz or trigonal tellurium (non-centrosymmetric structure). For tensors $\boldsymbol{t}$ of rank one (T_l), two ($T_{lm}=T_{ml}$ symmetric, $T_lT_m \neq T_mT_l$ non-symmetric), three ($T_{lmn}=T_{lnm} \neq T_{mln}$) and four ($T_{lmnr}=T_{mlnr} \neq T_{nrlm}$), the reductions $\Gamma(\boldsymbol{t})$ are readily found to be

$$O_h: \quad \Gamma(T_l)=F_{1u}, \tag{5.1a}$$

$$\Gamma(T_{lm})=A_{1g}\oplus E_g \oplus F_{2g}, \tag{5.1b}$$

$$\Gamma(T_lT_m)=A_{1g}\oplus E_g \oplus F_{1g} \oplus F_{2g}, \tag{5.1c}$$

$$\Gamma(T_{lmn})=A_{2u}\oplus E_u \oplus 3F_{1u} \oplus 2F_{2u}, \tag{5.1d}$$

$$\Gamma(T_{lmnr})=3A_{1g}\oplus A_{2g} \oplus 4E_g \oplus 3F_{1g} \oplus 5F_{2g}. \tag{5.1e}$$

$$D_3: \quad \Gamma(T_l)=A_2 \oplus E, \tag{5.2a}$$

$$\Gamma(T_{lm})=2A_1 \oplus 2E, \tag{5.2b}$$

$$\Gamma(T_lT_m)=2A_1 \oplus A_2 \oplus 3E, \tag{5.2c}$$

$$\Gamma(T_{lmn})=2A_1 \oplus 4A_2 \oplus 6E, \tag{5.2d}$$

$$\Gamma(T_{lmnr})=8A_1 \oplus 4A_2 \oplus 12E. \tag{5.2e}$$

The crystal dipole moment M_μ and the symmetric electric polarizability $\alpha_{\mu\lambda}=\alpha_{\lambda\mu}$ transform like a vector and a symmetric second-rank tensor respectively, i.e. like T_l and T_{lm}. Each one of their components belongs to at least one of the irreducible representations of the point group. In table 55 of Herzberg (1964) these irreducible representations are listed for each point group. Physical properties transforming like T_{lmn} and T_{lmnr} are the piezoelectric modulus ($d_{lmn}=d_{lnm}$) and piezo-optical coefficient ($\pi_{lmnr} \neq$

π_{nrlm}) respectively (Nye 1964). Very often, here, we will use the notation $\Gamma(\boldsymbol{M})$, $\Gamma(\boldsymbol{\alpha})$, $\Gamma(\boldsymbol{d})$, $\Gamma(\boldsymbol{\pi})$ instead of the general notation of eqs. (5.1) and (5.2). Since the same reduction holds for any other physical tensor with the same transformation properties, it is obvious that there is nothing unique about the notation $\Gamma(\boldsymbol{M})$, $\Gamma(\boldsymbol{\alpha})$.... It is only a convenient choice of notation which reminds one instantly of the physical quantities with which we are concerned. Clearly, an electric field $\mathfrak{E}$ or $\boldsymbol{E}$ reduces like $\boldsymbol{M}$, that is, $\Gamma(\boldsymbol{E})=\Gamma(\mathfrak{E})=\Gamma(\boldsymbol{M})$. We will often take advantage of such possible alternatives, so that the group-theoretical notation be constantly adjusted to the particular physical problem under discussion.

The distinction between a symmetric and a non-symmetric second-rank tensor is reflected in the way that their corresponding reducible representations are reduced. Since the symmetrized product T_{lm} is part of the general non-symmetric product $T_l T_m$ it follows that $\Gamma(\boldsymbol{\alpha})$ is simply the symmetrized part of the double Kronecker product $\Gamma(\boldsymbol{M})\otimes\Gamma(\boldsymbol{M})$. More specifically, we can write

$$\Gamma(\boldsymbol{M})\otimes\Gamma(\boldsymbol{M})=[\Gamma(\boldsymbol{M})\otimes\Gamma(\boldsymbol{M})]_{\mathrm{S}}\oplus[\Gamma(\boldsymbol{M})\otimes\Gamma(\boldsymbol{M})]_{\mathrm{A}} \tag{5.3a}$$

$$=\Gamma(\boldsymbol{\alpha})\oplus\Gamma(\boldsymbol{H}). \tag{5.3b}$$

The two brackets in eq. (5.3a) correspond to the symmetrized and antisymmetrized parts of the double Kronecker product respectively. Group theory provides us with well-known techniques for calculating each reduction independently (Hamermesh 1964). We introduce in eq. (5.3b) the notation $\Gamma(\boldsymbol{H})$ for the antisymmetrized part. $\boldsymbol{H}$ stands for a magnetic field, or *any* other axial vector (e.g. an angular momentum). It transforms like the cross product of *any* two polar vectors, i.e.

$$\bar{x}\equiv y_1z_2-y_2z_1, \tag{5.4a}$$

$$\bar{y}\equiv z_1x_2-z_2x_1, \tag{5.4b}$$

$$\bar{z}\equiv x_1y_2-x_2y_1. \tag{5.4c}$$

This observation facilitates the reduction of $\Gamma(\boldsymbol{H})$, since the irreducible representations to which $\bar{x},\bar{y},\bar{z}$ belong are usually indicated explicitly in the character tables of all point groups. The same remark, of course, holds for $\Gamma(\boldsymbol{E})$ which, as already mentioned, transforms like x,y,z.

From the character tables of our standard examples (O_h and D_3) we reproduce below a few sets of basis functions for some of their irreducible representations. Each set is included in brackets, or is separated by semicolons (Van der Lage and Bethe 1947, Callaway 1965, Callen 1968).

These functions will be widely used in later sections:

$$\mathrm{O_h}: \quad \mathrm{A_{1g}} \quad x^2+y^2+z^2;\ x^4+y^4+z^4;$$

$$x^2y^2+y^2z^2+z^2x^2 \tag{5.5a}$$

$$\mathrm{E_g} \quad [\sqrt{3}\,(x^2-y^2),(2z^2-x^2-y^2)], \tag{5.5b}$$

$$\mathrm{F_{1g}} \quad [\bar{x},\bar{y},\bar{z}], \tag{5.5c}$$

$$\mathrm{F_{2g}} \quad [yz,zx,xy];$$
$$[yzx^2,zxy^2,xyz^2];$$
$$[yz(y^2+z^2),zx(z^2+x^2),xy(x^2+y^2)], \tag{5.5d}$$

$$\mathrm{F_{2u}} \quad [x(y^2-z^2),y(z^2-x^2),z(x^2-y^2)], \tag{5.5e}$$

$$\mathrm{F_{1u}} \quad [x,y,z]. \tag{5.5f}$$

$$\mathrm{D_3}: \quad \mathrm{A_1} \quad x^2+y^2;\ z^2, \tag{5.6a}$$

$$\mathrm{A_2} \quad z;\ \bar{z}, \tag{5.6b}$$

$$\mathrm{E} \quad [x,y];\ [\bar{x},\bar{y}]$$
$$[xz,yz];\ [2xy,x^2-y^2]. \tag{5.6c}$$

Numerical factors such as $\sqrt{3}$, etc., are necessary to ensure that these functions generate unitary irreducible representations (Callen 1968).

Going back to the problem of reducing $\Gamma(\boldsymbol{H})$, we observe that the dimensionality of both $\Gamma(\boldsymbol{E})$ and $\Gamma(\boldsymbol{H})$ is three. This helps in deciding how many times each irreducible representation occurs in either $\Gamma(\boldsymbol{E})$ or $\Gamma(\boldsymbol{H})$. For the point groups $\mathrm{O_h}$ and $\mathrm{D_3}$ the reductions of $\Gamma(\boldsymbol{H})$ take the form

$$\mathrm{O_h}: \quad \Gamma(\boldsymbol{H})=\mathrm{F_{1g}} \tag{5.1f}$$

$$\mathrm{D_3}: \quad \Gamma(\boldsymbol{H})=\mathrm{A_2}\oplus\mathrm{E}, \tag{5.2f}$$

where the functions (5.5c) and (5.6b, c) have been used respectively. It is easy now to verify that the reduction in eq. (5.3b) does indeed apply to both point groups i.e. eq. (5.1b) + eq. (5.1f) = eq. (5.1c), etc.

The difference in the way that the axial vector $\boldsymbol{H}$ and the polar vector $\boldsymbol{E}$ transform can also be visualized as follows. $\boldsymbol{H}$ and $\boldsymbol{E}$ transform alike under proper rotations and differ only in sign under improper rotations. The same distinction holds for all polar and axial tensors (or pseudotensors) of *any* rank, including scalars and pseudoscalars. Thus, an axial vector $\boldsymbol{H}$ may be regarded as transforming like the product of a pseudoscalar and polar vector $\boldsymbol{E}$. In terms of representations this can be written as

$$\Gamma(\boldsymbol{H})=\Gamma_{\mathrm{p}}\otimes\Gamma(\boldsymbol{E}), \tag{5.7}$$

where Γ_{p} stands for the irreducible representation that a pseudoscalar belongs to. It is a well-defined one-dimensional irreducible representation for each point group, with characters $+1$ and -1 for proper and improper rotations respectively. Clearly, $\Gamma_{\mathrm{p}}\otimes\Gamma_{\mathrm{p}}=\mathrm{A}_1$. For centrosymmetric groups Γ_{p} is always of odd parity. When no improper rotations exist in the group, Γ_{p} coincides with A_1. In table A1 of the Appendix we list the pseudoscalar representations Γ_{p} for all 32 point groups, together with the reductions of the representations of the higher-order tensors $\boldsymbol{m}$ and $\boldsymbol{f}$. These will be introduced in §§15.2 and 15.3 in connection with magnetic-field-induced infrared absorption and Raman scattering. As will be shown in §15.3 the reduction of $\Gamma(\boldsymbol{f})$ is identical to that of a polar second-rank non-symmetric tensor, like, for instance, the tensor $T_l T_m$ in eqs. (5.1c) and (5.2c). The notation of Koster et al. (1963) is used in table A1. For convenience we have included in a separate table A2 the correspondence between the standard Mulliken notation and the one used by Koster et al. (1963) for the irreducible representations of all 32 point groups (see also Cracknell 1968, Appendix). Each point group is given in the Schönflies notation and also in the short international notation. For the centrosymmetric point groups an additional superscript "+" or "−" should be added in the Γ notation to distinguish even- from odd-parity representations respectively. Correspondingly, subscripts "g" or "u" should be added in the alternative notation. For example, if $\Gamma_3\rightarrow\mathrm{B}_2$ then $\Gamma_3^+\rightarrow\mathrm{B}_{2\mathrm{g}}$ and $\Gamma_3^-\rightarrow\mathrm{B}_{2\mathrm{u}}$, etc.

For the point groups $\mathrm{O_h}$ and $\mathrm{D_3}$ of our standard examples, the pseudoscalar irreducible representations are $\mathrm{A_{1u}}$ and $\mathrm{A_1}$ respectively. It is easy to check the validity of eq. (5.7) when considering eqs. (5.1a, f) and (5.2a, f) respectively. (On this and other occasions where quick products of irreducible representations are required, the summary tables of Wilson et al. (1955) are particularly helpful.)

The concept of pseudoscalar representations as introduced in eq. (5.7) will be extensively used in later sections of this chapter where pseudotensors of higher order will be considered.

5.2. Criteria for the existence of intrinsic effects and morphic (induced) effects

Any mode property or mode coefficient which involves the normal coordinate u_j of a phonon with symmetry $\Gamma(j)$ in first order, can be treated as a higher-order tensor-like quantity whose reducible representation has the form $\Gamma(\boldsymbol{t})\otimes\Gamma(j)$. The mode property or coefficient is non-zero for a given crystal and mode, provided that the reduction of $\Gamma(\boldsymbol{t})\otimes\Gamma(j)$ includes the totally symmetric irreducible representation of the crystal point group at least once (in general, n_A times). This is equivalent to saying that $\Gamma(j)$ must be included in the reduction of $\Gamma(\boldsymbol{t})$ at least once. Again, it can be shown that n_A gives the number of independent components of the mode property or coefficient and is equal to the number of times that the irreducible representation $\Gamma(j)$ occurs in the reduction of the reducible representation $\Gamma(\boldsymbol{t})$. In all these statements above it is assumed that the irreducible representations are real and therefore $\Gamma=\Gamma^*$ (to formally determine whether $\Gamma=\Gamma^*$ the so-called Frobenius–Schur test should be made, Tinkham 1964). As examples we consider intrinsic first-order infrared absorption and Raman scattering (no forces applied).

(i) Intrinsic first-order infrared absorption. Intrinsic infrared absorption at the fundamental frequency ω_j of the mode j can occur when $(\partial \boldsymbol{M}/\partial u_j)\neq 0$, i.e., when the associated effective charge of the mode is non-zero. The group-theoretical requirement for this is that $\Gamma(\boldsymbol{M})\otimes\Gamma(j)$ include A_1 at least once, or equivalently, that $\Gamma(j)$ be included in $\Gamma(\boldsymbol{M})$. In other words u_j must transform like the components of a polar vector (since $\boldsymbol{M}$ is a polar vector). This is a well-known result; for an infrared-active mode, u_j always transforms like x,y,z. Consider for instance the reduction of $\Gamma(\boldsymbol{M})$ for a structure with D_3 symmetry. According to eq. (5.2a) only the modes A_2 and E are infrared active, with one independent component each for the tensor $\boldsymbol{e}_j$.

(ii) Intrinsic first-order Raman scattering. The effect can occur when $(\partial\alpha/\partial u_j)\neq 0$, namely when $\Gamma(j)$ is included in $\Gamma(\boldsymbol{\alpha})$. This is the requirement for Raman scattering based on a symmetric Raman tensor. As noted in the discussion following eq. (4.5) this is the usual situation. On the other hand, if for some reason the Raman tensor should be treated as a non-symmetric tensor such that $(\partial^3\Phi/\partial\mathcal{E}_\mu\,\partial\mathcal{E}'_\lambda\,\partial u_j)\neq(\partial^3\Phi/\partial\mathcal{E}_\lambda\,\mathcal{E}'_\mu\,\partial u_j)$, first-order Raman scattering may be expected from all modes j for which $\Gamma(j)$ is included in $\Gamma(\boldsymbol{M})\otimes\Gamma(\boldsymbol{M})$ rather than in $\Gamma(\boldsymbol{\alpha})=[\Gamma(\boldsymbol{M})\otimes\Gamma(\boldsymbol{M})]_S$. We notice that the totally symmetric vibrational mode of any structure is always Raman active. For centrosymmetric crystals the irreducible representations in $\Gamma(\boldsymbol{M})$ are of odd parity with respect to inversion, whereas the irreducible representations in $\Gamma(\boldsymbol{\alpha})$ are of even parity. A mode of a centrosymmetric crystal therefore cannot be both Raman and infrared active. This is the so-called *complementary property* of infrared and Raman activity in centrosymmetric crystals. Let us take the following two examples:

(a). For a D_{4h} structure one finds

$$\Gamma(\boldsymbol{M})\otimes\Gamma(\boldsymbol{M})=2A_{1g}\oplus B_{1g}\oplus B_{2g}\oplus E_g\oplus(A_{2g}\oplus E_g),$$

$$\Gamma(\boldsymbol{\alpha})=2A_{1g}\oplus B_{1g}\oplus B_{2g}\oplus E_g.$$

We see that both the symmetric and the non-symmetric Raman tensors will have exactly the same symmetric form for $\Gamma(j)=A_{1g}, B_{1g}, B_{2g}$. The Raman tensor for $\Gamma(j)=E_g$ has two independent components; it is symmetric with respect to the one and antisymmetric with respect to the other. Finally, we note that pure antisymmetric Raman scattering should be expected only from the A_{2g} mode, with one independent component for its Raman tensor. Since Raman activity is usually considered on the basis of the symmetric part of $\Gamma(\boldsymbol{M})\otimes\Gamma(\boldsymbol{M})$, i.e. $\Gamma(\boldsymbol{\alpha})$, the A_{2g} mode is generally considered to be Raman inactive.

(b). For a D_3 structure the reduction forms for $\Gamma(\boldsymbol{M})$ and $\Gamma(\boldsymbol{\alpha})$ of eqs. (5.2a, b) state that the mode E is both infrared and Raman active, the mode A_1 is Raman active only and the mode A_2 is infrared active only. The number of independent components of $\boldsymbol{e}_j$ and $\boldsymbol{a}_j$ for $\Gamma(j)=A_2, E, A_1$ are 1, 1, 0 and 0, 2, 2, respectively.

The detail matrix form of $\boldsymbol{a}_j$ has been worked out for all Raman-active modes and crystal structures by several workers (Ovander 1960, Loudon 1964, Wallis and Maradudin 1971, Claus et al. 1975). Ovander (1960) and Wallis and Maradudin (1971) treat the Raman tensor as a non-symmetric tensor. It has already been mentioned that antisymmetric Raman scattering can be ignored only as long as the effects of the difference in the frequency of the incident and scattered electromagnetic radiation are negligible. However, these effects may be appreciable when the incident radiation is very close to a resonance frequency. It has also been argued that antisymmetric Raman scattering can occur when the initial state of the electronic transitions which are involved in the phonon Raman scattering is degenerate (Childs and Longuet-Higgins 1961, 1962). Antisymmetric Raman scattering by *phonons* in non-magnetic crystals has not yet been observed. Evidence for antisymmetric Raman scattering by phonons in magnetic crystals has been reported in a few cases (Schaack 1975, Merlin et al. 1977). Antisymmetric *electronic* Raman scattering has been predicted theoretically (Placzek 1934) and observed experimentally for several rare-earth ions in crystals (Kiel et al. 1969 and references therein, Dabrowski et al. 1972, Spiro and Strekas 1972).

In the same spirit we now proceed to determine criteria for the existence of morphic effects. If the applied force $\boldsymbol{F}$ belongs to the representation $\Gamma(\boldsymbol{F})$ (not necessarily irreducible), the force-free tensor, representation $\Gamma(\boldsymbol{t})$ changes to $\Gamma(\boldsymbol{t})\otimes\Gamma(\boldsymbol{F})$ to first order in the force. Thus, the conclusions reached above regarding the intrinsic effects are still valid if $\Gamma(\boldsymbol{t})$ is replaced by $\Gamma(\boldsymbol{t})\otimes\Gamma(\boldsymbol{F})$. Thus we have an easy and quick way of checking whether or not a particular morphic effect is possible and how many independent coefficients are expected for the corresponding mode property or mode coefficient. According to the definitions of eqs. (3.1c), (3.3c) and

(4.5c), the morphic effects for a given choice of force $\boldsymbol{F}$ and mode j are determined by the following reductions respectively;

$$\Gamma(\boldsymbol{M})\otimes\Gamma(\boldsymbol{F})=n_j\Gamma(j)\oplus\ldots, \tag{5.8}$$

$$\Gamma(\boldsymbol{\alpha})\otimes\Gamma(\boldsymbol{F})=n_j'\Gamma(j)\oplus\ldots, \tag{5.9}$$

$$[\Gamma(j)\otimes\Gamma(j)]_S\otimes\Gamma(\boldsymbol{F})=n_A A_1\oplus\ldots. \tag{5.10}$$

Notice that the order of multiplication in the left-hand side of eqs. (5.8) and (5.9) can be changed, i.e., $\Gamma(\boldsymbol{M})\otimes\Gamma(\boldsymbol{F})=\Gamma(\boldsymbol{F})\otimes\Gamma(\boldsymbol{M})$, etc. Eq. (5.8) describes the force-induced infrared absorption due to the mode j with n_j independent coefficients for the quantity $\Delta\boldsymbol{e}_j$. Eq. (5.9) corresponds to the force-induced Raman scattering by the mode j with n_j' independent coefficients for $\Delta\boldsymbol{a}_j$. Likewise, eq. (5.10) refers to the force-induced change in the mode frequency ω_j with n_A independent coefficients for the quantity $\Delta\boldsymbol{K}_j$. We note that only the symmetrized part of the double Kronecker product $\Gamma(j)\otimes\Gamma(j)$ enters eq. (5.10), since we have assumed that $\boldsymbol{F}$ is time symmetric and therefore eq. (3.14) is valid. This procedure for determining whether or not a particular intrinsic or induced effect is possible has been extensively used by Anastassakis and Burstein (1971a, b) and by Humphreys and Maradudin (1972) in connection with morphic effects due to polar forces (electric field, symmetric strain) and their gradients respectively.

The following conclusions emerge directly from eqs. (5.8), (5.9) and (5.10):

(i) When the applied force is a symmetric strain $\boldsymbol{\eta}$, eq. (5.8) becomes $\Gamma(\boldsymbol{M})\otimes\Gamma(\boldsymbol{\eta})=\Gamma(\boldsymbol{\alpha})\otimes\Gamma(\boldsymbol{E})$. In other words eq. (5.8) describes both field-induced Raman scattering and strain-induced infrared absorption due to the mode j.

(ii) For a given structure, all modes with the same degree of degeneracy behave alike when a force is applied. This is because $\Gamma(j)\otimes\Gamma(j)$ reduces in the same way for all modes with the same degeneracy. Thus, for instance, for all triply-degenerate modes of the O_h structure we have

$$F_{1u}\otimes F_{1u}=F_{2u}\otimes F_{2u}=F_{1g}\otimes F_{1g}=F_{2g}\otimes F_{2g}=A_{1g}\oplus E_g\oplus F_{1g}\oplus F_{2g},$$

and also,

$$[F_{1u}\otimes F_{1u}]_S=\ldots=A_{1g}+E_g+F_{2g},$$
$$[F_{1u}\otimes F_{1u}]_A=\ldots=F_{1g}.$$

The degeneracies of A, E, F-type modes are 1, 2, and 3 respectively. This observation simplifies the problem, since now one has to work out the shift and splitting of only one type of mode for each degree of degeneracy, preferably the infrared-active one, i.e. the one for which u_j transforms like the components x,y,z, of a polar vector. The same results will then apply for all other types of modes with the same degeneracy, under the same force.

(iii) For totally symmetric modes, that is $\Gamma(j)=\mathrm{A}_1$, eqs. (5.8), (5.9) and (5.10) become

$$\Gamma(\boldsymbol{M})\otimes\Gamma(\boldsymbol{F})=n_{\mathrm{A}}\mathrm{A}_1\oplus\ldots, \tag{5.8a}$$

$$\Gamma(\boldsymbol{\alpha})\otimes\Gamma(\boldsymbol{F})=n'_{\mathrm{A}}\mathrm{A}_1\oplus\ldots, \tag{5.9a}$$

$$\Gamma(\boldsymbol{F})=n''_{\mathrm{A}}\mathrm{A}_1\oplus\ldots. \tag{5.10a}$$

Eq. (5.8a) implies that $\Delta\boldsymbol{e}_j$ (where $\Gamma(j)=\mathrm{A}_1$) has the same matrix form as a macroscopic tensor whose reducible representation is $\Gamma(\boldsymbol{M})\otimes\Gamma(\boldsymbol{F})$. Thus, when $\boldsymbol{F}$ is an electric field or strain, $\Delta\boldsymbol{e}_j$ will have the same matrix form as a non-symmetric second-rank tensor, or the piezoelectric modulus respectively. In the case of eq. (5.9a) the matrix form would be the same as the piezoelectric modulus, or the piezo-optical tensor respectively. Finally, eq. (5.10a) states that only those components of the applied force which transform according to A_1 can cause a shift of the frequency of the totally symmetric mode A_1 to first order in the force.

(iv) The Kronecker product of complex conjugate modes should be treated carefully. Such modes occur in the uniaxial crystal classes $\mathrm{C}_3, \mathrm{C}_4, \mathrm{C}_6, \mathrm{C}_{3h}, \mathrm{C}_{4h}, \mathrm{C}_{6h}$, and in the cubic classes T and T_h. Using the tables and notation of Koster et al. (1963) we find for instance that the stress-induced Raman tensor in class T is described by *four* independent components, according to eq. (5.9), i.e.

$$[\Gamma_1\oplus(\Gamma_2\oplus\Gamma_3)\oplus\Gamma_4]\otimes[\Gamma_1\oplus(\Gamma_2\oplus\Gamma_3)\oplus\Gamma_4]=4\Gamma_1\oplus\ldots$$

The same tensor in class T_d requires only *three* independent components.

Specific applications of eqs. (5.8), (5.9) and (5.10) will be discussed in later sections, in connection with existing experimental results.

5.3. Component array of mode properties and mode coefficients

In this section we discuss a group-theoretical procedure for deducing the non-zero components of mode properties in which u_j is treated as a quasi-macroscopic parameter. The procedure followed is an extension of the one used by Callen (1968) for the derivation of the component form of *macroscopic* crystal properties like polarizability, piezoelectric modulus, etc.

It involves essentially a straightforward manipulation of the basis functions of the point group irreducible representations and requires no deep knowledge of group theory. The usual macroscopic properties are polar tensors whose reducible representations involve an appropriate direct product of the irreducible representations of a polar vector. In the case of mode properties, the reducible representations correspond to direct products of the irreducible representations of polar vectors and the irreducible representation of the quasi-macroscopic parameter u_j, i.e. $\Gamma(j)$. The latter is not necessarily the same as that for a polar vector. This necessitates the use of the basis functions of $\Gamma(j)$ as well (because of the nature of the macroscopic properties, in Callen's treatment only the basis functions of a polar vector, i.e. x,y,z, are involved, as polynomials of the appropriate order). The physical basis for this method is that *a generalized force (or coordinate) of a given symmetry can only induce a generalized coordinate (or force) of the same symmetry*. In group-theoretical language the symmetry identity between force and coordinate is expressed as identity of their reducible representation. We will first apply this method for the force-induced Raman tensor $\Delta \boldsymbol{a}_j$ and give an explicit example. We will then extend the method to the force-induced effective charge $\Delta \boldsymbol{e}_j$ and force constant $\Delta \boldsymbol{K}_j$.

The term in the expansion of the crystal potential energy (eqs. (4.4) and (4.5b)) which leads to the force-induced Raman scattering is given by

$$\delta\Phi = \frac{1}{2}\left(\frac{\partial^4 \Phi}{\partial \mathcal{E}_\mu\, \partial \mathcal{E}_\lambda\, \partial F_\nu\, \partial u_{j\sigma}}\right) \mathcal{E}_\mu \mathcal{E}_\lambda F_\nu u_{j\sigma}$$

$$\equiv \frac{1}{2}\Phi^{(4)}_{j,\mu\lambda\nu\sigma} \mathcal{E}_\mu \mathcal{E}_\lambda F_\nu u_{j\sigma} = -\frac{1}{2}\Delta a_{j,\mu\lambda\sigma} \mathcal{E}_\mu \mathcal{E}_\lambda u_{j\sigma}, \tag{5.11}$$

where summation over repeated indices is implied. The change in the crystal potential energy $\delta\Phi$ is a scalar and transforms according to the totally symmetric irreducible representation A_1 of the point group.

It is important to note that *there are as many independent terms in the summation of eq.* (5.11) *as the number of independent components of the mode coefficient* $\Phi^{(4)}$. The functional form of $\mathcal{E}_\mu \mathcal{E}_\lambda F_\nu u_{j\sigma}$ is constructed from the basis functions of the irreducible representations of the factors $\mathcal{E}_\mu \mathcal{E}_\lambda$, F_ν and $u_{j\sigma}$. Furthermore, $\delta\Phi$ is a scalar and the coefficients $\Phi^{(4)}_{j,\mu\lambda\nu\sigma}$ are scalars. It then follows from eq. (5.11) that the functional form of $\mathcal{E}_\mu \mathcal{E}_\lambda F_\nu u_{j\sigma}$ *must be totally symmetric too*. In practice, one has to construct polynomials which transform as scalars and whose order is the same as that of $\mathcal{E}_\mu \mathcal{E}_\lambda F_\nu u_{j\sigma}$. Alternatively, one may search for a basis of the jth irreducible representation whose order is the same as that of $\mathcal{E}_\mu \mathcal{E}_\lambda F_\nu$. For the construction of symmetric higher-order polynomials as bases for $\mathcal{E}_\mu \mathcal{E}_\lambda F_\nu u_{j\sigma}$, the following generalized theorem of Unsöld may be of help

(Callen 1968, Falikov 1967). If $(X_1 \ldots X_n)$ and $(\Psi_1 \ldots \Psi_n)$ are basis functions for the same n-dimensional irreducible representation, then $\Sigma_m^n X_m \Psi_m^*$ is a totally symmetric function. To apply this theorem one should make sure that X_m and Ψ_m *transform according to the same row of the irreducible representation* in question. This is particularly important inasmuch as the partner functions found in the literature *are not always correctly ordered and signed* to meet this row correspondence requirement. Once the functional form of eq. (5.11) is established, the detailed components of $\Delta \boldsymbol{a}_j$ can be readily found from eqs. (4.5b) and (5.11).

Following similar arguments, it is easy to show that terms in potential energy which lead to force-induced infrared absorption and frequency change take the form

$$\delta\Phi = \left(\frac{\partial^3 \Phi}{\partial \mathcal{E}_\mu \, \partial F_\nu \, \partial u_{j\sigma}} \right) \mathcal{E}_\mu F_\nu u_{j\sigma}$$

$$\equiv \Phi^{(3)}_{j,\mu\nu\sigma} \mathcal{E}_\mu F_\nu u_{j\sigma} = -\Delta e_{j,\mu\sigma} \mathcal{E}_\mu u_{j\sigma}, \qquad (5.12)$$

and

$$\delta\Phi = \frac{1}{2} \left(\frac{\partial^3 \Phi}{\partial u_{j\alpha} \, \partial u_{j\beta} \, \partial F_\nu} \right) u_{j\alpha} u_{j\beta} F_\nu$$

$$\equiv \frac{1}{2} K_{j,\alpha\beta\nu} u_{j\alpha} u_{j\beta} F_\nu = \frac{1}{2} \Delta K_{j,\alpha\beta} u_{j\alpha} u_{j\beta}. \qquad (5.13)$$

At this point we would like to comment on a different procedure often used by other investigators to establish the non-zero coefficients of the force-induced Raman tensor. It is based on the determination of the new symmetry of the crystal in the presence of an external force. The new point group corresponds to the combined symmetry elements of the force-free structure and those of the applied force alone. Obviously, the symmetries of the modes must conform to the point group symmetry of the deformed crystal. In the case of $SrTiO_3$ for instance, the application of an electric field along a cubic (four-fold) crystal axis reduces the symmetry of the crystal from O_h to C_{4v}. The formal way to trace the changes in symmetry of a mode when the structure changes from the cubic to the tetragonal class is to consider the group-theoretical reduction of the mode representations in the original point group into those of the new point group. In this example one finds, $F_{1u} \rightarrow A_1 \oplus E$, $F_{2u} \rightarrow B_1 \oplus E$, etc. The force-induced effective charge and Raman tensor are the same as the effective charge and Raman tensor for the corresponding phonons in the new point group. It should be noted, however, that in this approach one is not concerned to what order of the applied force the force-induced mode properties occur. For instance, as we shall discuss in §6.2 the silent mode F_{2u} of cubic $SrTiO_3$ becomes Raman active to terms linear with the electric field, but infrared active only to terms quadratic with the field. The detailed description of the force-induced effects therefore requires more than just a consideration of the change of the crystal symmetry.

In the present treatment, the change of the mode properties is treated as a perturbation on the force-free property, and it is possible to determine the effect of the applied force to any order. Thus, it allows one to establish the linear, quadratic and higher-order mode coefficients. It is essential to know whether the effect is linear, quadratic, etc. in the force, for experimental purposes. Thus, when field-modulation techniques are employed for the observation of these effects, the type of dependence on the field (linear, quadratic, etc.) determines whether one should synchronously detect the signal at the fundamental or at a higher harmonic of the modulation frequency (Worlock 1969, Anastassakis and Burstein 1970).

In the case of $\Delta K_{j,\alpha\beta}$ it is possible to follow an alternative procedure, depending on the symmetry of the normal mode j. The basis for this is the observation made at the end of §5.2 that all modes of a particular structure with the same degree of degeneracy behave identically in the presence of a force. This alternative procedure is applicable *when the mode j transforms like x,y,z, i.e. when the mode is infrared active*. We emphasize that the same results will apply for all other modes with the same degeneracy. The procedure is based on the simple observation that the tensor $K_{j,\alpha\beta\nu} = (\partial^3\Phi/\partial u_{j\alpha}\partial u_{j\beta}\partial F_\nu)$ has exactly the same component form as some of the well-known macroscopic crystal properties. If F_ν stands for an electric field E_ν then $K_{j,\alpha\beta\nu}$ has the same form as the piezoelectric modulus $d_{\alpha\beta\nu} = d_{\beta\alpha\nu} \neq d_{\alpha\nu\beta}$. If F_ν is a strain $\eta_{\mu\nu}$, then $K_{j,\alpha\beta\mu\nu}$ has the same form as the photoelastic tensor $P_{\alpha\beta\mu\nu} = P_{\beta\alpha\mu\nu} \neq P_{\mu\nu\alpha\beta}$. Since the matrix forms of these properties are tabulated, one can write the secular equation by inspection. Care should be taken in the case of doubly-degenerate modes of uniaxial crystals which transform like x,y (with z missing). This requires elimination of those columns from these matrices which involve z as one of their suppressed indices. As an example we take the piezoelectric tensor for the D_3 structure (Nye 1964),

$$d_{\alpha\beta\nu} = \begin{bmatrix} Q & -Q & . & P & . & . \\ . & . & . & . & -P & -2Q \\ . & . & . & . & . & . \end{bmatrix},$$

where the column index 1,2,...,6 is a suppressed index for $\alpha\beta = xx,yy,zz,yz,xz,xy$, and the row index ν stands for x,y,z. The conversion to the dynamical matrix is easily obtained if one follows the scheme

$$d_{\alpha\beta\nu} \to K_{j,\alpha\beta\nu} \quad \text{and} \quad \Delta K_{j,\alpha\beta} = K_{j,\alpha\beta\nu} E_\nu. \tag{5.14}$$

The change $\Delta \boldsymbol{K}_j$ in the dynamical matrix can be written as

$$\Delta \boldsymbol{K}_j = \begin{pmatrix} KE_x & -KE_y \\ -KE_y & -KE_x \end{pmatrix},$$

where the third (zz), fourth (yz) and fifth (xz) columns have been dropped from the matrix of $d_{\alpha\beta\nu}$ since z is one of the indices and neither α nor β can be z (the mode E transforms like x,y only). Factors of 2 should be carefully handled in passing from the 2- to the 3-index notation of $d_{\alpha\beta\nu}$. The above results agree with those obtained independently by use of eq. (5.13). It is emphasized that this method can only be applied to infrared-active modes. If for a given mode with degeneracy two, no infrared-active mode with the same degeneracy can be found, the method cannot be applied and the general procedure outlined earlier should be followed. For instance, the modes $E_{1g}, E_{2g}, E_{1u}, E_{2u}$ of the O_h structure cannot be treated by this method since none of them is infrared active.

Establishing the exact matrix form of tensors (mode properties) which describe first- and second-order intrinsic Raman scattering and their morphic effects has been the subject of several independent investigations. Birman and Berenson (1974), Berenson and Birman (1976) and Berenson (1978) have shown that it is possible to use Clebsch–Gordan coefficients to express the Raman tensor elements of a morphic effect of order n as a linear combination of Raman tensor elements of an appropriate morphic effect of order $n-1$. The same method can be used to treat morphic effects due to a magnetic field and its gradients and also for Brillouin scattering processes. Recently Kanamori et al. (1976b) and Matsumoto and Kanamori (1977) completed the reduction of tensors up to fourth rank, e.g. expressions similar to those of eqs. (5.1) and (5.2), for all 32 point groups. They have also determined the complete matrix arrays for the Raman tensors, the dipole moments and the mode force constants, in the presence of an electric and magnetic field and a symmetric strain. This tabulation includes all those modes from all 32 point groups which qualify for exhibiting such morphic effects. The various basis functions which are necessary for the expressions of eqs. (5.11), (5.12) and (5.13) are derived by use of projection operator techniques.

Applications of eqs. (5.11)–(5.14) will be undertaken in later sections in connection with specific experiments.

5.4. Second-order effects

Thus far the discussion has been restricted to morphic effects involving infrared absorption and Raman scattering by long-wavelength optical phonons in first order. The symmetry of these phonons is characterized by one of the irreducible representations of the crystal point group. It is well known that intrinsic infrared absorption and Raman scattering may also occur at the overtone and combination frequencies of vibrational modes. These second-order processes are not restricted solely to long-wavelength optical phonons. The various overtones and combinations allowed are determined by the symmetry groups of the phonon wavevectors. The intensity of the absorption and scattering depends strongly on the combined two-phonon density of states. Detailed group-theoretical methods for determining the selection rules for these effects have been worked out by a number of workers (Birman 1962, 1963, 1974b, Maradudin and Vosko 1968). The concept of morphic effects involving phonon normal coordinates is now extended to include force-induced second-order infrared absorption and Raman scattering. The role of the force is to modify the selection rules of the intrinsic second-order effects and/or to induce infrared absorption and Raman scattering by overtones or combinations which are intrinsically inactive.

The comments following eq. (5.13) apply here as well. That is, one valid procedure for handling the symmetry aspects of second-order morphic effects is to follow the gross reduction of crystal symmetry to a new space group and then apply the selection rules for the new structure. Alternatively, the perturbation method may be employed. This allows the force-free crystal symmetry to be used provided the applied forces are not too strong. One can then proceed by determining the corrections to those mode properties which are responsible for the existence and strength of the intrinsic second-order effect.

In the case of infrared absorption, the terms in the crystal potential energy which correspond to second-order effects (intrinsic and induced) are

$$\begin{aligned}\delta\Phi &= \left(\frac{\partial^3\Phi}{\partial\mathcal{E}\,\partial u_j\,\partial u_{j'}}\right)\mathcal{E}u_ju_{j'} + \left(\frac{\partial^4\Phi}{\partial\mathcal{E}\,\partial F\,\partial u_j\,\partial u_{j'}}\right)\mathcal{E}Fu_ju_{j'}\\ &\equiv -(e_{jj'}+\Delta e_{jj'})u_ju_{j'}.\end{aligned} \tag{5.15}$$

Tensor indices have been omitted for simplicity. In general, normal coordinates u_j and $u_{j'}$ belong to irreducible representations of space groups of different phonon wavevectors for combination processes. They belong to the same irreducible representations for overtone processes. Whichever the combination of wavevectors is, the restriction of conservation of momentum must be satisfied. The mode properties $\boldsymbol{e}_{jj'}$ and $\Delta\boldsymbol{e}_{jj'}$ describe respectively the second-order effective charge of the mode pair jj' and its correction linear in the force. We are only concerned with a direct coupling physical mechanism for the second-order infrared absorption. According to this mechanism the incident photon couples to a pair of phonons directly through second-order terms in the electric moment of the crystal (Lax and Burstein 1953, Burstein 1964).

The existence of the mode coefficients which appear in eq. (5.15) is determined group-theoretically by use of the same criteria developed earlier in §5.2. Thus, an intrinsic and an induced second-order infrared absorption process involving the mode pair jj' is allowed, provided the following relations hold respectively:

$$\Gamma(\boldsymbol{M})\otimes[\mathrm{D}(j)\otimes\mathrm{D}(j')]_{\mathrm{S}}=n_1\Gamma_1\oplus\ldots, \tag{5.16}$$

$$\Gamma(\boldsymbol{M})\otimes\Gamma(\boldsymbol{F})\otimes[\mathrm{D}(j)\otimes\mathrm{D}(j')]_{\mathrm{S}}=n_1'\Gamma_1\oplus\ldots. \tag{5.17}$$

The phonon irreducible representations are denoted by D rather than Γ to emphasize the fact that we are not necessarily dealing with $\boldsymbol{q}\simeq 0$ (Γ point) optical phonons. Γ_1 designates as usual the totally symmetric irreducible representation of the crystal point group. Notice that both $\boldsymbol{M}$ and $\boldsymbol{F}$ are treated within the *point group* of the crystal. The symbols $[\ldots]_{\mathrm{S}}$ are to indicate that interchanging the normal coordinates u_j and $u_{j'}$ in eq. (5.15) does not affect the mode coefficients $(\partial^4\Phi/\partial\mathfrak{E}\,\partial F\partial u_j\partial u_{j'})$. The direct products $\mathrm{D}(j)\otimes\mathrm{D}(j')$ can be reduced by following the general procedure of Birman (1962, 1974b). Detailed reductions for diamond- and zincblende-type crystals have been worked out by Birman (1963, 1974b). It is clear that we are interested only in that part of the reduction which includes Γ-type irreducible representations. It will always be assumed that the construction of basis functions appropriate for the product $\mathrm{D}(j)\otimes\mathrm{D}(j')$ is such that the interchangeability of u_j and $u_{j'}$ is preserved. For brevity in what follows we shall omit the symbol $[\ldots]_{\mathrm{S}}$. Once n_1 and n_1', the number of independent components of the quantities $\boldsymbol{e}_{jj'}$ and $\Delta\boldsymbol{e}_{jj'}$, are determined from eqs. (5.16) and (5.17), one can proceed to obtain the component arrays of $\boldsymbol{e}_{jj'}$ and $\Delta\boldsymbol{e}_{jj'}$ following the general method of §5.3.

It is easy to extend the above discussion to second-order Raman scattering, intrinsic or induced. Again we consider only a direct coupling mechanism of the two phonons to the incident and scattered radiation, via second-order terms in the electric polarizability of the crystal (Burstein

1964). Equations analogous to eqs. (5.15), (5.16) and (5.17) are

$$\delta\Phi = \left(\frac{\partial^4\Phi}{\partial\mathcal{E}\,\partial\mathcal{E}\,\partial u_j\,\partial u_{j'}}\right)\mathcal{E}\,\mathcal{E}\,u_j u_{j'} + \left(\frac{\partial^5\Phi}{\partial\mathcal{E}\,\partial\mathcal{E}\,\partial F\,\partial u_j\,\partial u_{j'}}\right)\mathcal{E}\,\mathcal{E}\,F u_j u_{j'}$$

$$\equiv -(a_{jj'} + \Delta a_{jj'})u_j u_j', \tag{5.18}$$

$$\Gamma(\boldsymbol{\alpha})\otimes \mathrm{D}(j)\otimes \mathrm{D}(j') = n_1\Gamma_1 \oplus \ldots, \tag{5.19}$$

$$\Gamma(\boldsymbol{\alpha})\otimes\Gamma(\boldsymbol{F})\otimes \mathrm{D}(j)\otimes \mathrm{D}(j') = n_1'\Gamma_1 \oplus \ldots. \tag{5.20}$$

Again, the strength of second-order Raman scattering by overtones or combination of phonons will depend on the combined two-phonon density of states.

Based on the work of Birman (1962, 1963) for diamond, Solin and Ramdas (1970) have compiled a useful table, which includes those overtones and combinations of phonons from the critical points of the Brillouin zone which are intrinsically Raman- and/or infrared-active. This table is reproduced in the Appendix as table A3. The same table also includes the number of times the Γ-type irreducible representations, which occur in the reductions of $\Gamma(\boldsymbol{M})$ and $\Gamma(\boldsymbol{\alpha})$, appear in the reduction of overtone and combination reducible representations. The use of this table is easily extended to indicate whether or not a force which reduces like $\Gamma(\boldsymbol{M})$ or $\Gamma(\boldsymbol{\alpha})$ (e.g. an electric field or a symmetric strain) can induce infrared absorption or Raman scattering by any of the overtones or combinations which are included in the table. Indeed, the factor $\Gamma(\boldsymbol{M})\otimes\Gamma(\boldsymbol{F})$ in eq. (5.17) becomes $\Gamma(\boldsymbol{M})\otimes\Gamma(\boldsymbol{M})$ or $\Gamma(\boldsymbol{M})\otimes\Gamma(\boldsymbol{\alpha})$ respectively. The reductions for these factors in the case of diamond are given by eqs. (5.1c) and (5.1d) respectively. If there is a common term between these reductions and the Γ-part of the reduction of $\mathrm{D}(j)\otimes\mathrm{D}(j')$ (table A3), then field-induced or strain-induced infrared absorption by the mode pair jj' is allowed to occur respectively. Similarly, the factor $\Gamma(\boldsymbol{\alpha})\otimes\Gamma(\boldsymbol{F})$ in eq. (5.20) becomes $\Gamma(\boldsymbol{\alpha})\otimes \Gamma(\boldsymbol{M})$ and $\Gamma(\boldsymbol{\alpha})\otimes\Gamma(\boldsymbol{\alpha})$. These reduce according to eqs. (5.1d) and (5.1e) respectively. Again any common terms will result in field-induced or strain-induced Raman scattering by the mode pair jj'.

Let us take from table A3 the example of the overtone

$$[{}^*\mathrm{X}^{(4)}]_{(2)} = \Gamma^{(1+)}\oplus\Gamma^{(12+)}\oplus\Gamma^{(25+)} \tag{5.21}$$

and the combination

$${}^*\mathrm{L}^{(3-)}\otimes{}^*\mathrm{L}^{(1+)} = \Gamma^{(15-)}. \tag{5.22}$$

Since $\Gamma(\boldsymbol{M})=\Gamma^{(15-)}$ and $\Gamma(\boldsymbol{\alpha})=\Gamma^{(1+)}\oplus\Gamma^{(12+)}\oplus\Gamma^{(25+)}$, we conclude that the overtone is *intrinsically* Raman active and infrared inactive. Furthermore, in the notation of table A3, eqs. (5.1c)–(5.1e) are written as

$$\Gamma(\boldsymbol{M})\otimes\Gamma(\boldsymbol{M})=\Gamma^{(1+)}\oplus\Gamma^{(12+)}\oplus\Gamma^{(15+)}\oplus\Gamma^{(25+)}, \tag{5.1c}$$

$$\Gamma(\boldsymbol{M})\otimes\Gamma(\boldsymbol{\alpha})=\Gamma^{(2-)}\oplus\Gamma^{(12-)}\oplus3\Gamma^{(15-)}\oplus2\Gamma^{(25-)}, \tag{5.1d}$$

$$\Gamma(\boldsymbol{\alpha})\otimes\Gamma(\boldsymbol{\alpha})=3\Gamma^{(1+)}\oplus2\Gamma^{(2+)}\oplus4\Gamma^{(12+)}\oplus3\Gamma^{(15+)}\oplus5\Gamma^{(25+)}. \tag{5.1e}$$

There are common terms between eqs. (5.21) and (5.1c) and thus one expects to observe field-induced infrared absorption by the phonon overtone 2TO(X) whose reducible representation is $[{}^*X^{(4)}]_{(2)}$ (Birman 1963). In fact we find $n_1'=3$ (eq. (5.17)). The field-induced second-order effective charge $\boldsymbol{e}_{jj}$ (where $j=\mathrm{TO(X)}$) has 3 independent components, all linear in the applied field. A symmetric strain can induce Raman activity by the same overtone with 12 independent components for $\Delta\boldsymbol{a}_{jj}$ (where $j=\mathrm{TO(X)}$). An electric field cannot induce Raman scattering and a symmetric strain cannot induce infrared absorption by this overtone. Likewise, an electric field cannot induce infrared activity by the combination TO(L) + LO(L) (Birman 1963) whose reducible representation is ${}^*L^{(1+)}\otimes{}^*L^{(-3)}$. It can induce Raman activity, with 3 independent components for $\Delta\boldsymbol{a}_{jj'}$ (where $j=\mathrm{TO(L)}$, $j'=\mathrm{LO(L)}$). A symmetric strain cannot induce Raman activity. It can induce infrared absorption, with 3 independent components for $\Delta\boldsymbol{e}_{jj'}$ (where $j=\mathrm{TO(L)}$, $j'=\mathrm{LO(L)}$). The complete reductions of overtone and combination processes in rocksalt-type crystals in the presence of an electrostatic field have been tabulated by Hayashi and Kanamori (1974). By use of these tables one can readily determine the effects of fields along [001], [111] and [110] on the selection rules for second-order infrared absorption and Raman scattering and also on the degeneracy of these two-phonon processes. As expected, the removal of inversion symmetry results in many drastic changes in the selection rules. Thus, for instance, nearly all two-phonon overtones become infrared active when a field is applied, while most of these overtones are infrared inactive in the field-free structure. A related work has been reported by Balslev (1974) who treats by non-conventional group theory the effect of a stress along [001] and [111] on the relative intensities of second-order infrared absorption and Raman scattering by L- and X-type phonons in diamond- and zincblende-type crystals. Without getting into details Balslev also gives the results for the stress-induced degeneracy removal of one-phonon states at the L and X points of the Brillouin zone in the structures mentioned above. We will return to this point in §11.4.

6. *Electric-field-induced infrared absorption*

6.1. General

According to eqs. (3.1c), (4.5a) this effect can occur when the coefficient $(\partial^2 M/\partial E\,\partial u_j) = -(\partial^3\Phi/\partial\mathfrak{E}\,\partial E\,\partial u_j)$ is non-zero, namely, when the field-induced effective charge Δe_j is non-zero. Here again the interchanging of the indices of E (the applied field, usually at low frequencies, or static) and $\mathfrak{E}$ (the radiation electric field) must be treated carefully. In general, this coefficient is not symmetric (non-symmetric) in the fields E and $\mathfrak{E}$. Therefore it is non-zero when $\Gamma(j)$ is included in $\Gamma(\mathfrak{E})\otimes\Gamma(E)$ which reduces in the same way as $\Gamma(M)\otimes\Gamma(M)$ does. But this is the requirement for the mode j to produce non-symmetric Raman scattering (i.e. symmetric and/or antisymmetric Raman scattering). If on the other hand, and for the same physical reasons mentioned earlier, the fields E and $\mathfrak{E}$ can be considered as interchangeable, the coefficient $(\partial^2 M/\partial E\,\partial u_j)$ has the same form as the symmetric Raman tensor. The effect then is possible, provided that $\Gamma(j)$ is included in $\Gamma(\alpha)$. We can summarize as follows: *electric-field-induced infrared absorption by a mode j is possible only when the mode is Raman active in either symmetric or non-symmetric Raman scattering.* A weak non-symmetric Raman scattering intensity does not necessarily imply a weak field-induced infrared absorption, because the corresponding tensors, although identical in form, depend on the frequencies of the fields in different ways. The former involves fields at approximately equal frequencies, the latter involves fields at very high (optical) and very low (or zero) frequencies.

Another consequence of the above conclusion is that silent modes cannot become infrared active with an electric field in first order, provided that they are silent in both symmetric and non-symmetric Raman scattering (e.g. the F_{2u} mode of cubic $BaTiO_3$). If the mode is silent only with regard to symmetric Raman scattering, the mode may still exhibit infrared absorption with an applied electric field, owing to the presence of antisymmetric components in the Raman tensor (i.e. the A_{2g} mode of the D_{4h} structure, see §5.2).

Electric-field-induced infrared absorption has been observed in diamond (Anastassakis et al. 1966, Anastassakis and Burstein 1970). Diamond belongs to the O_h structure and exhibits one $\boldsymbol{q}\simeq 0$ optical mode of symmetry F_{2g}. According to the reduction of $\Gamma(\alpha)$ given in eq. (5.1b), this mode is Raman active (it is also infrared inactive because of the complementary principle). From the discussion above it follows that an electric field can induce infrared absorption by the mode F_{2g} at the frequency ω_j which at room temperature is equal to 1332.5 cm^{-1} (Solin and Ramdas 1970). Such possibilities for diamond-type crystals were first discussed by Burstein and Ganesan (1965) and by Szigeti (1965). In essence they describe the same effects as those predicted by Condon (1932) for homonuclear molecules and observed in H_2 molecules by Crawford and Dagg (1953) and Crawford and MacDonald (1958).

The reduction of either eq. (5.1b) or eq. (5.1c) indicates that the appropriate mode coefficient $-(\partial^3\Phi/\partial\mathcal{E}\,\partial E\,\partial u_j)$ with $\Gamma(j)=F_{2g}$ has only one independent component, say a. The matrix form for $\Delta\boldsymbol{e}_j$ is determined through the procedure developed in §5.3. An equation analogous to eq. (5.12) takes the form

$$\delta\Phi = -a\left[(\mathcal{E}_y E_z + \mathcal{E}_z E_y)u_x + (\text{cyclic permutation})\right]. \tag{6.1}$$

By inspection of the right-hand side of eq. (5.12) one finds the following matrix for $\Delta\boldsymbol{e}_j$

$$\Delta\boldsymbol{e}_j = a\begin{bmatrix} \cdot & E_z & E_y \\ E_z & \cdot & E_x \\ E_y & E_x & \cdot \end{bmatrix}. \tag{6..2}$$

In writing $\delta\Phi$ in the form of eq. (6.1) we have taken into consideration that both sets of functions (u_x, u_y, u_z) and $[(\mathcal{E}_y E_z + \mathcal{E}_z E_y)$, c.p.] transform like the basis functions (yz, zx, xy) of the F_{2g} mode. The row index indicates the component (μ) of the associated induced electric moment and also the component of the incident infrared electromagnetic radiation which will be absorbed. The columns index (σ) refers to the polarization of the phonon involved. This helps in interpreting and applying the polarization selection rules. Recently Kanamori et al. (1976b) and Matsumoto and Kanamori (1977) reported the tabulation for all 32 point groups of the electric-field-induced dipole moments for all Raman-active modes (i.e. matrices like the one of eq. (6.2) multiplied by the appropriate column $u_{j\sigma}$).

When comparing the mode coefficient $-(\partial^3\Phi/\partial\mathcal{E}\,\partial E\,\partial u_j)$ with the Raman tensor defined in eq. (4.5b) we may assume that these two quantities are identical not only in symmetry form but quantitatively too. The reason for this is that, to a very good approximation, the electric polarizability exhibits no frequency dispersion in the region of the spectrum between low and optical frequencies. Thus, an experimental determination of the mode coefficient $|-(\partial^3\Phi/\partial\mathcal{E}\,\partial E\,\partial u_j)|$ is important in that it very closely represents an absolute value for the Raman tensor of the mode F_{2g}. Furthermore, the identical symmetry properties of these two coefficients result in the same polarization selection rules. In this connection, one has to remember that the polarizations of the incident and scattered radiation in a forward scattering Raman experiment correspond to the polarizations of the incident infrared radiation and the applied electric field respectively in the field-induced absorption experiment. The same remarks of course apply for any mode of any other structure, provided again that $\boldsymbol{\alpha}$ exhibits no significant frequency dispersion in the spectral region of interest. In §6.3 we will discuss an equivalent experimental technique which may in fact lead to a measurement of the exact absolute value of the Raman tensor, that is, the mode property $|-(\partial^3\Phi/\partial\mathcal{E}\,\partial\mathcal{E}'\partial u_j)|$.

6.2. Experimental

The first observation of field-induced infrared absorption in diamond was based on a comparison of the recordings of two transmission spectra taken with and without a d.c. electric field E_{D} applied across the diamond plate (Anastassakis et al. 1966). Similar measurements were reported later, taken under improved experimental conditions (Anastassakis and Burstein 1970). The light beam was first chopped and then focused on the Type IIa diamond sample. The transmitted signal was spectroscopically analyzed, synchronously detected and recorded. A typical recording in the neighborhood of ω_j is shown in fig. 1.

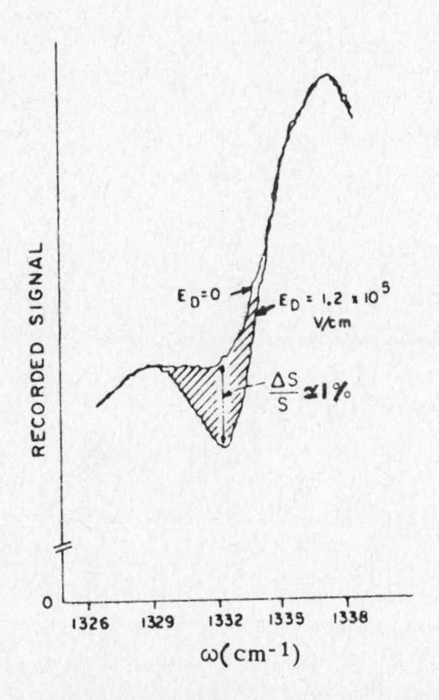

Fig. 1. Recorded transmission signal of diamond with zero and non-zero field. The relative change of the transmittance is about 1% for $E_{\mathrm{D}} = 1.2\times10^5\,\mathrm{V/cm}$ (Anastassakis and Burstein 1970).

Although simple in principle, the d.c. technique for such experiments is neither convenient nor sensitive enough to allow a detailed study of the absorption band. In addition, the band may be superimposed on a zero-field spectrum, owing to the presence of structure of other origin, principally atmospheric absorption. Instead, a differential or a.c. technique can be used to discriminate against any signal not related to the field-induced effect, thus resulting in greatly increased sensitivity. In this method an a.c. electric field (amplitude E_A) with a superimposed d.c. bias field (E_D) was applied across the crystal, to induce an a.c. change in the transmitted intensity. The latter was then synchronously detected and recorded. As expected, the induced change was found to vary linearly with either E_A or E_D. Some results of differential measurements are shown in fig. 2. From the analysis of the experimental data the following results were obtained:

(i) The frequency of ω_j at 300 K was placed at 1332 cm^{-1}, in close agreement with the value obtained from Raman scattering work. The full width at half intensity at 300 K was found to be equal to 1.75–2.0 cm^{-1}, in satisfactory agreement with the result of 1.65 cm^{-1} of Solin and Ramdas (1970). No shift or splitting of the frequency was observed or broadening of the band, even at the highest fields applied ($\sim 2\times10^5$ V/cm). This is consistent with the theoretical results of Ganesan et al. (1970) who predict that for these fields the broadening and the shifts and splittings should be

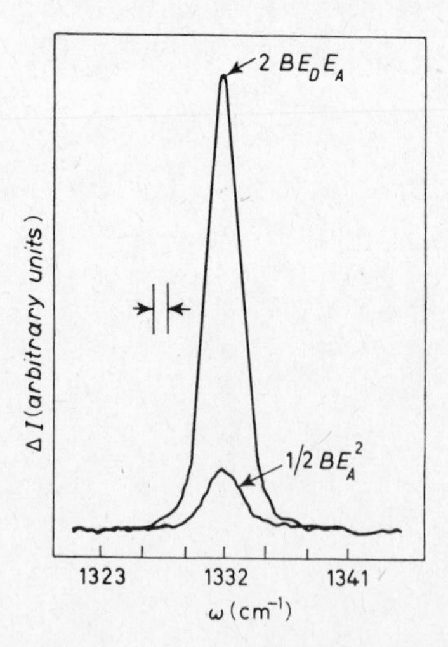

Fig. 2. Recorded transmission spectrum of diamond using modulation techniques. A d.c. field E_D together with an a.c. field E_A (frequency f=600 Hz) are simultaneously applied. $E_D = 10\times10^4$ V/cm, $E_A = 5\times10^4$ V/cm. The small band was recorded at the overtone $2f$ with $E_D = 0$. The factor B stands for a proportionality constant (Anastassakis and Burstein 1970).

of the order of $10^{-4}\,cm^{-1}$. This is too small to be detected through this experimental approach.

(ii) The absolute value of a, the single independent component of the mode coefficient $-(\partial\Phi^3/\partial\mathfrak{E}\,\partial E\,\partial u_j)$, was established as 4 ± 0.5 in Å^2 units. As stated earlier this is the first experimental determination of the magnitude of the Raman tensor of optical phonons in a solid.

(iii) The field-induced effective charge $|\Delta \boldsymbol{e}_j|$ for the mode F_{2g}, was determined as $0.8\times10^{-6}\,E_D$ in units of electron charge, with the field E_D expressed in esu.

(iv) The field-induced absorption constant $|\Delta A|$ was found to be $1.5\times 10^{-6}E_D^2$ in cm^{-1} (again with E_D in esu).

(v) It can be shown that the Raman tensor itself exhibits a quadratic dependence on the applied field. With a special experimental arrangement it was estimated that the ratio of this quadratic term to the intrinsic constant a is roughly $-1.0\times10^{-8}E_D^2$.

A useful verification of some of these results was provided by similar experiments in diamond performed by Angress et al. (1968). Their d.c. technique clearly exhibited the transmittance minimum at $1332\,cm^{-1}$ as expected. The absorption band is shown in fig. 3 on top of a structure-free background. This was accomplished by evacuating the entire spectrometer system.

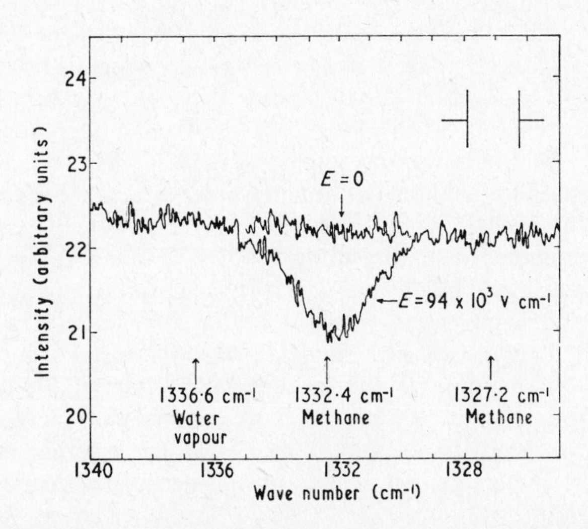

Fig. 3. Recorded traces of transmitted intensity of diamond in the presence and absence of a static electric field $\boldsymbol{E}$. The polarization of the incident field $\mathfrak{E}$ is along [110] (Angress et al. 1968).

The absolute value 4 Å^2 for the Raman tensor of diamond has since been confirmed by several independent workers, using a variety of experimental techniques. MacQuillan et al. (1970) found a value of 4.6 Å^2 from measurements of the absolute Raman scattering intensity. Grimsditch and Ramdas (1975) obtained a value of 4.4 Å^2 by measuring the ratio of integrated absolute intensities of Brillouin and Raman scattering. Unfortunately none of these experiments yields a sign for a. From a theoretical calculation Maradudin and Burstein (1967) found the values −3.8, −7.0 and +12.0 Å^2 for diamond, Si and Ge respectively. In their bond polarizability model, the experimental values for the photoelastic constants of Ramachandran (1950) were used for diamond. These constants have been checked and corrected by a number of workers (Schneider 1970 unpublished, Grimsditch and Ramdas 1975, Grimsditch et al. 1979). When the correct values for the photoelastic constants are used the model of Maradudin and Burstein yields $a = +0.53$ Å^2, which disagrees with the experimental results (Leigh and Szigeti 1970). Better agreement was obtained by Swanson and Maradudin (1970) who find −1.82, +6.73 and +40.53 Å^2 respectively, from a pseudopotential calculation. Recent progress in band structure calculations has made it possible to determine the sign of the Raman tensor for diamond (−), Si(+) and Ge(+) (Cardona et al. 1974 and references therein). The sign for Si has been verified indirectly from independent experimental sources (Cardona et al. 1974). From a calculation based on a non-linear extension of the bond charge model, Go et al. (1975) obtain the values +3.50, +13.95 and +43.10 Å^2 for diamond, Si and Ge respectively. These are in reasonable agreement with those of Swanson and Maradudin (1970), except that the sign for diamond is opposite. More recently, Hellwarth (1977) reached an approximate expression for the field-induced absorption constant in diamond, following a Born–Oppenheimer analysis of the third-order optical susceptibility. The value of $|a|$ implied from this calculation is 7.2 Å^2. Finally, Calleja et al. (1978) have succeeded in measuring the first- and second-order Raman scattering in diamond below, but near, the fundamental gap at 6 eV. From their data and from pseudopotential theory calculations they obtain a value for the Raman tensor of diamond, which is placed at −2.5 Å^2. It is also argued that the positive sign for diamond found by Go et al. (1975) is due to the inclusion of a weak $E_g(\Gamma_{12})$ component of the second-order Raman tensor, which is not fully justified from an experimental point of view. Thus considering all the available facts, the sign for diamond appears to be negative.

Besides diamond no other material seems to have been investigated in experiments of a field-induced absorption. It is not surprising that the results for diamond were successful in view of its high dielectric breakdown. Similar experimental efforts for Si were unsuccessful (Anastassakis et al. 1966). Russel (1965) measured the absolute ratio of Raman tensors per unit volume as $|\beta'_{Si}/\beta'_{d}|\simeq 2.7$ and Parker et al. (1967) established the ratio of Raman tensors for Si and Ge as $|a_{Si}/a_{Ge}|\simeq 0.5$–$0.8$. These ratios are in reasonable agreement with those based on the experimental value of $|a_d|$ and the theoretical results for $|a_{Si}|$ and $|a_{Ge}|$ quoted above.

6.3. A method for the direct measurement of the Raman tensor in crystals

As already pointed out, the field-induced infrared absorption is proportional to the square of the mode coefficient $-(\partial^3\Phi/\partial\mathcal{E}\,\partial E\,\partial u_j)$ which under certain conditions can be quantitatively identified with the Raman tensor $a \equiv -(\partial^3\Phi/\partial\mathcal{E}\,\partial\mathcal{E}'\partial u_j)$. In the field-induced infrared absorption one uses essentially a d.c. field $\boldsymbol{E}$ to induce a dipole moment which oscillates at the frequency ω_j and results in the absorption of the infrared radiation $\mathcal{E}$ at the same frequency ω_j.

We now discuss the possibility of substituting the d.c. field $\boldsymbol{E}$ by the electric field $\mathcal{E}'(\omega_0)$ of a monochromatic c.w. laser in the visible region of the spectrum (frequency ω_0). This will induce dipole moments which oscillate at the Stokes and anti-Stokes frequencies, $\omega_0-\omega_j$ and $\omega_0+\omega_j$ respectively. We restrict the following discussion to the Stokes process only. Detecting the scattered radiation (ω_S) is the subject of conventional Raman scattering spectroscopy. The intensity of the scattering is proportional to $|a|^2=|-(\partial^3\Phi/\partial\mathcal{E}\,\partial\mathcal{E}'\partial u_j)|^2$. But any dipole oscillating and emitting radiation at ω_S can also absorb at the same frequency ω_S. Therefore, if the crystal is illuminated with the continuous spectrum of a dye laser in the visible region, its transmission spectrum should exhibit a minimum at the Stokes frequency ω_S. From the relative change of the transmittance, a value for the absolute value of the appropriate mode coefficient $|-\partial^3\Phi/\partial\mathcal{E}\,\partial\mathcal{E}'\partial u_j)|$ can be obtained. This of course coincides exactly with the Raman tensor $|a|$.

The effect described above is conceptually similar to the d.c.-field-induced infrared absorption. The "applied" field $\mathcal{E}'(\omega_0)$ of the pumping laser will generate the dipole moment at ω_S. A tunable dye laser field $\mathcal{E}(\omega_S)$ can then be used to probe the induced moment (Anastassakis and Argyres 1974).

Besides providing a value for the exact Raman tensor, this approach presents the following advantages compared to the d.c. field experiments:

(i) Since the necessary field is now provided by a laser the experimentation is substantially simplified.

(ii) The transmission measurements are made in the visible rather than in the infrared. This bears all the advantages of spectroscopy in the visible, i.e. more intense light sources, use of lenses, better dispersion and detection means, absence of unrelated absorption bands, etc. A choice of ω_0 far from any of the characteristic resonances of the medium is appropriate; if either ω_0 or ω_S is close to such a resonance, a strong direct absorption of $\mathscr{E}'$ will probably mask the $\mathscr{E}'$-induced absorption.

(iii) The present approach can be applied to any material which is transparent to both the pumping and the probing laser radiation. Thus, semiconductors may be investigated by use of lasers in the appropriate region of the spectrum (semiconductors cannot be studied under a high d.c. field).

While fine focusing of the pumping laser is required to produce a high field $\mathscr{E}'$, this reduces proportionally the crystal region where absorption takes place. From the results of Anastassakis and Burstein (1970) it can be estimated that in diamond a transmission minimum of $\Delta T/T \lesssim 10^{-5}$ will be induced per watt of pumping power focused down to $10\,\mu$m. This corresponds to a measurable signal provided the dye laser noise is carefully eliminated. Different geometries may also be employed and the selection rules to be considered are the same as those for first-order Raman scattering in the analogous geometry. Phase sensitive detection techniques combined with wavelength modulation of the probing radiation may be necessary to monitor the expected weak absorption band. Modulation of the probing radiation consists in modulating the transmitted dye laser wavelength with an amplitude of at most 1 Å. Similar modulation techniques are commonly used today as an alternative method to differential Raman spectroscopy (Galeener 1976, Abstreiter and Anastassakis 1977).

It is emphasized that the experiment suggested here involves continuous radiation at low power levels. Thus the effective fields are low and only linear effects need be considered and treated semi-classically. This is to be contrasted with the situation in the work of Jones and Stoicheff (1964) on *inverse Raman scattering* (Koningstein 1972) where pulsed radiation of the order of MW power created strong absorption band in liquids owing to the substantially higher fields. The direct method proposed here has not yet been tested in solids.

6.4. Electric-field-induced second-order infrared absorption

Evidence for field-induced second-order infrared absorption has been reported by Lüth (1969a, b) in ZnO (wurtzite structure). The intrinsic second-order spectrum of ZnO is rather complicated and no effort has been made to interpret the observed field-induced changes. The spectrum was obtained by use of an a.c. technique and seems to consist of field-induced absorption, and field-induced frequency shifts of the intrinsic bands. Recently Angress and Maiden (1971) and Angress et al. (1971) have reported results of field-induced second-order infrared absorption in diamond in the region 2400–2700 cm^{-1} using an a.c. field technique. Most of the second-order activity of diamond in this region consists of Raman scattering by overtones of phonons from the critical points Γ, X, L (Solin and Ramdas 1970). As indicated in table A3, all these overtones are infrared inactive but they can become infrared active when an electric field is applied. As in the first-order case the mode coefficients entering the intrinsic second-order Raman scattering are, as far as symmetry is concerned, identical to those of field-induced second-order infrared absorption (ignoring dispersion effects). Thus, the field-induced spectrum of second-order absorption is expected to resemble very closely the spectrum of intrinsic second-order Raman scattering. This is in agreement with the experimental results of Angress et al. which are shown in fig. 4. Such a resemblance allows one to identify the major peaks of the induced

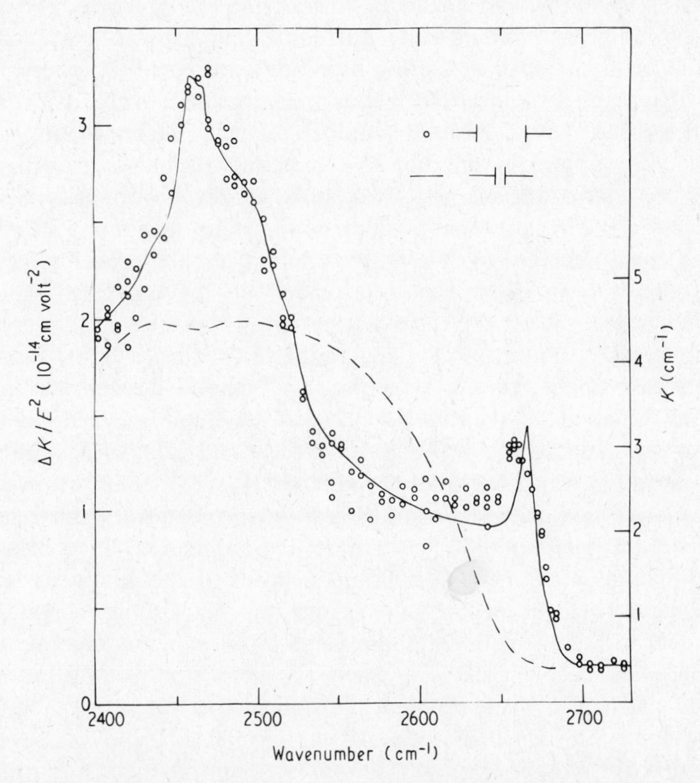

Fig. 4. Electric-field-induced changes in the two-phonon infrared absorption spectrum of diamond (Angress and Maiden 1970). Left-hand scale, open circles, $\Delta K/E^2$, where ΔK is the field-induced absorption constant. Full line, second-order Raman spectrum for a similar experimental configuration (Solin and Ramdas 1970), arbitrary units scaled to our results at the main peak. Right-hand scale, broken line, intrinsic infrared absorption spectrum.

spectrum since the complete assignment of the Raman spectrum is known (Solin and Ramdas 1970). As expected the second-order field-induced spectrum is much weaker than the first order. This imposed rather unfavorable resolution conditions ($28\,cm^{-1}$) which in turn did not allow the detailed resolution of double peaks, or their exact positions. The same measurements have been repeated by Onyango (1976) at 300 K under conditions of improved resolution ($5\,cm^{-1}$). According to Onyango the sharp peak around $2665\,cm^{-1}$ occurs at twice the frequency ω_j of the first-order Raman peak, that is at $2\times 1332\,cm^{-1} = 2664\,cm^{-1}$. The position and assignment of this peak have been and still are the subject of considerable controversy. Solin and Ramdas (1970) place it at $2667\,cm^{-1}$ (resolution $8\,cm^{-1}$), Angress et al. at $2660\,cm^{-1}$ (resolution $28\,cm^{-1}$), while Cohen and Ruvalds' (1969) theoretical model on the two-phonon bound state requires that the position of the peak depend on the amount of anharmonicity. Independent theories based on a bond polarizability model (Go et al. 1975, Tubino and Piseri 1975) attribute the peak (placed slightly above $2\omega_j$) to overtone scattering associated with the F_{2g} phonon. The same conclusion is reached by Tubino and Birman's (1976, 1977) theory which is based on a harmonic model for the potential and a bond polarizability approximation for the scattering Hamiltonian. This latter theory too leaves unexplained the systematic shift of about $3\,cm^{-1}$, between $2\omega_j$ and the calculated position of the sharp maximum. The recent Raman measurements of Washington and Cummings (1977) leave no doubt that the sharp maximum observed at $2669\,cm^{-1}$ (at 300 K) is definitely shifted by $3\,cm^{-1}$ relative to their value of the simple overtone at $2\omega_j = 2666\,cm^{-1}$, contrary to the conclusion of Onyango (1976). This raises the question about the exact relationship between the tensors entering the two experiments (i.e. the second-order Raman scattering and the electric-field-induced second-order infrared absorption). In a more recent communication, Hastings and Ruvalds (1977) show that the results of calculations based on harmonic models (referenced above) cannot account for the exact lineshape of the sharp second-order peak. In contrast, the bound-phonon model with appropriate inclusion of anharmonicity leads to a lineshape which according to Hastings and Ruvalds is in remarkable agreement with the experimental lineshape of Washington and Cummings. The observed shift of $3\,cm^{-1}$ is easily accounted for, within the bound-phonon model. The whole problem still seems to be open.

Onyango (1976) has also examined the region $2300–2400\,cm^{-1}$ to conclude that the broad band at $2333\,cm^{-1}$ of the Raman spectrum also appears in the induced spectrum at $2335\,cm^{-1}$. The nature of this band is not well understood, since no Raman-active overtone or combination is expected to occur at $2333\,cm^{-1}$. There is some speculation that it may be due to the combination $\Sigma^{(1)}(O)+\Sigma^{(2)}(O)$ which occurs at $2339\,cm^{-1}$ (Solin and Ramdas 1970).

7. *Electric-field-induced Raman scattering*

7.1. Field-induced Raman scattering – hyper-Raman effect

Eqs. (3.3c) and (4.5b) indicate that an applied electric field can induce Raman scattering provided the mode coefficient $(\partial^2\alpha/\partial E\,\partial u_j)=(\partial Z/\partial u_j)=-(\partial^4\Phi/\partial\mathfrak{E}\,\partial\mathfrak{E}\,\partial E\,\partial u_j)$ is non-zero. $\boldsymbol{Z}$ is an appropriate first-order electro-optical tensor (Nye 1964). Its reducible representation $\Gamma(\boldsymbol{Z})$ is identical to $\Gamma(\boldsymbol{d})$. Therefore, the effect can occur if $\Gamma(j)$ is included in the reduction of $\Gamma(\boldsymbol{d})$ at least once (again, we only consider symmetric Raman scattering).

For phonons of non-centrosymmetric crystals which are both Raman and infrared active it is well known that the Raman tensor of the LO component differs from that of the TO component by the so-called electro-optic contribution. The origin of this contribution is the macroscopic field which accompanies the LO phonon. In other words, the electro-optic contribution to the Raman tensor arises from the modulation of the electronic polarizability (or susceptibility) by the macroscopic electric field of the LO phonon. The symmetry properties therefore of the electro-optic Raman tensor are identical to those for the atomic displacement Raman tensor discussed so far (recall that the normal coordinate of an infrared-active phonon transforms in the same way as an electric field). Thus, a distinction between the two types of contributions to the Raman tensor is immaterial as far as the symmetry of morphic effects is concerned. For this reason we will restrict the following discussion to the atomic displacement contributions only (Pinczuk and Burstein 1975, Richter 1976).

Next, we will apply the general method of §5.3 to establish the matrix form for the field-induced Raman tensor $\Delta\boldsymbol{a}_j$ in a few cases of interest.

(i) $\Gamma(j)=F_{2u}$ *for the* O_h *structure.* This mode is the so-called *silent mode* of cubic perovskites and is infrared and Raman inactive in first order. According to eq. (5.1d), an electric field can induce Raman scattering by the F_{2u} mode with two independent components, *a* and *b*, for the field-induced Raman tensor $\Delta\boldsymbol{a}_j$. This means that there are two independent terms in eq. (5.11). The quantity $\mathfrak{E}_\mu\mathfrak{E}_\lambda E_\nu u_{j\sigma}$ transforms like a symmetric function. The two sets for $\mathfrak{E}_\mu\mathfrak{E}_\lambda E_\nu$ with F_{2u} symmetry are easily found from eq. (5.5e), i.e.

$$\left[(\mathfrak{E}_y^2-\mathfrak{E}_z^2)E_x,\text{c.p.}\right],\qquad\left[\mathfrak{E}_x(\mathfrak{E}_yE_y-\mathfrak{E}_zE_z),\text{c.p.}\right].$$

Thus the equation analogous to eq. (5.11) takes the form

$$\delta\Phi = -\tfrac{1}{2}a\left[u_x\left(\mathcal{E}_y^2 - \mathcal{E}_z^2\right)E_x + \text{c.p.}\right]$$
$$-b\left[u_x\mathcal{E}_x\left(\mathcal{E}_y E_y - \mathcal{E}_z E_z\right) + \text{c.p.}\right]. \tag{7.1}$$

The factor $\frac{1}{2}$ in front of b is dropped, since $\mathcal{E}_\mu \mathcal{E}_\lambda = \mathcal{E}_\lambda \mathcal{E}_\mu$ and this results in taking $\mathcal{E}_\mu \mathcal{E}_\lambda$ twice in the sum of eq. (5.11) when $\mu \neq \lambda$. The final matrix $\Delta\boldsymbol{a}_j$ becomes

$$\Delta\boldsymbol{a}_j = \begin{bmatrix} \cdot & bE_y & -bE_z \\ bE_y & aE_x & \cdot \\ -bE_z & \cdot & -aE_x \end{bmatrix}, \begin{bmatrix} -aE_y & -bE_x & \cdot \\ -bE_x & \cdot & bE_z \\ \cdot & bE_z & aE_y \end{bmatrix}, \begin{bmatrix} aE_z & \cdot & bE_x \\ \cdot & -aE_z & -bE_y \\ bE_x & -bE_y & \cdot \end{bmatrix}. \tag{7.2}$$

$u_{j\sigma}$: $\quad x(y^2 - z^2) \quad\quad y(z^2 - x^2) \quad\quad z(x^2 - y^2)$

These matrices give the field-induced Raman tensor for the silent mode F_{2u}. Each component is linear in the field for relatively small fields (i.e. as long as the approximation of eq. (4.4) is valid). For higher fields one would expect higher-order terms in odd powers of the field.

(ii) $\Gamma(j) = F_{1u}$ *for the* O_h *structure.* There are three independent components a, b, c for the field-induced Raman tensor $\Delta\boldsymbol{a}_j$. By inspection of eq. (5.5a) we can write

$$\delta\Phi = -\tfrac{1}{2}a\left[\left(\mathcal{E}_x^2 E_y u_y + \text{c.p.}\right) + \left(\mathcal{E}_y^2 E_x u_x + \text{c.p.}\right)\right]$$
$$-b\left[\left(\mathcal{E}_x \mathcal{E}_y E_x u_y + \text{c.p.}\right) + \left(\mathcal{E}_x \mathcal{E}_y E_y u_x + \text{c.p.}\right)\right]$$
$$-\tfrac{1}{2}c\left[\mathcal{E}_x^2 E_x u_x + \text{c.p.}\right]. \tag{7.3}$$

The field-induced Raman matrices become

$$\Delta\boldsymbol{a}_j = \begin{bmatrix} cE_x & bE_y & bE_z \\ bE_y & aE_x & \cdot \\ bE_z & \cdot & aE_x \end{bmatrix}, \begin{bmatrix} aE_y & bE_x & \cdot \\ bE_x & cE_y & bE_z \\ \cdot & bE_z & aE_y \end{bmatrix}, \begin{bmatrix} aE_z & \cdot & bE_x \\ \cdot & aE_z & bE_y \\ bE_x & bE_y & cE_z \end{bmatrix}. \tag{7.4}$$

$u_{j\sigma}$: $\quad x \quad\quad y \quad\quad z$

(iii) $\Gamma(j)=A_2$ for the D_3 structure. According to eqs. (5.2d) and (5.2e) both linear and quadratic terms exist. The quadratic effect will be examined later in terms of the equivalent stress effect. The coefficient $-(\partial^4\Phi/\partial\mathcal{E}\,\partial\mathcal{E}\,\partial E\,\partial u_j)$ has four independent components. Since $u_{j\sigma}$ transforms like z, four symmetric functions of order four are needed for $\mathcal{E}_\mu\mathcal{E}_\lambda E_\nu u_{j\sigma}$. By inspecting the basis of D_3 (eq. (5.6a, b, c)) we can write

$$\begin{aligned}\delta\Phi = &-\tfrac{1}{2}au_z\mathcal{E}_z^2E_z-\tfrac{1}{2}bu_z\left(\mathcal{E}_x^2+\mathcal{E}_y^2\right)E_z\\ &-c\left[(E_xu_z)(\mathcal{E}_x\mathcal{E}_z)+(E_yu_z)(\mathcal{E}_y\mathcal{E}_z)\right]\\ &-\tfrac{1}{2}f\left[(E_xu_z)(2\mathcal{E}_x\mathcal{E}_y)+(E_yu_z)\left(\mathcal{E}_x^2-\mathcal{E}_y^2\right)\right].\end{aligned} \tag{7.5}$$

The first two terms of the above form for $\delta\Phi$ are based on $A_2\otimes A_1\otimes A_2=A_1$. The last two terms are based on $E\otimes E=A_1\oplus\ldots$. From eqs. (5.11) and (7.2) we find

$$\Delta\boldsymbol{a}_j=\begin{bmatrix} bE_z+fE_y & fE_x & cE_x\\ fE_x & bE_z-fE_y & cE_y\\ cE_x & cE_y & aE_z\end{bmatrix}. \tag{7.6}$$

$$u_{j\sigma}:\qquad z$$

Similar matrices for all appropriate modes of all 32 point groups have been tabulated by Kanamori et al. (1976b), and by Matsumoto and Kanamori (1977).

When the incident electromagnetic field is sufficiently strong, one may expect field-induced Raman scattering due to the incident field itself. The mode coefficient of eq. (5.11) in this case takes a slightly different form, that is, $-(\partial^4\Phi/\partial\mathcal{E}_\mu\,\partial\mathcal{E}'_\lambda\,\partial\mathcal{E}_\nu\,\partial u_{j\sigma})$, where $\mathcal{E}_\mu$ and $\mathcal{E}_\nu$ are at the frequency ω_0 of the incident radiation, and $\mathcal{E}'_\lambda$ is at the Stokes (anti-Stokes) frequency $\omega_S(\omega_A)$. The process involves absorption of two photons ω_0 creation (destruction) of one phonon ω_j and scattering of a third photon $\omega_S(\omega_A)$. Conservation of energy and momentum takes the form $\omega_S=2\omega_0\mp\omega_j$ and $\boldsymbol{k}_S=2\boldsymbol{k}_0\mp\boldsymbol{q}_j$ respectively (Burstein et al. 1968). This type of "built-in" morphic effect is nowadays known as *hyper-Raman effect*. Normally it should occur for only those crystals and phonons for which the electric-field-induced Raman scattering is allowed by symmetry. Thus, it cannot occur with even-parity phonons in centrosymmetric crystals, unless one considers deviations from symmetry (due to lattice defects) or higher-order effects such as spatial dispersion effects (gradients) of the electromagnetic fields (Yu and Alfano 1975). In spite of its small strength, hyper-Raman scattering has been observed in a number of materials by the use of very

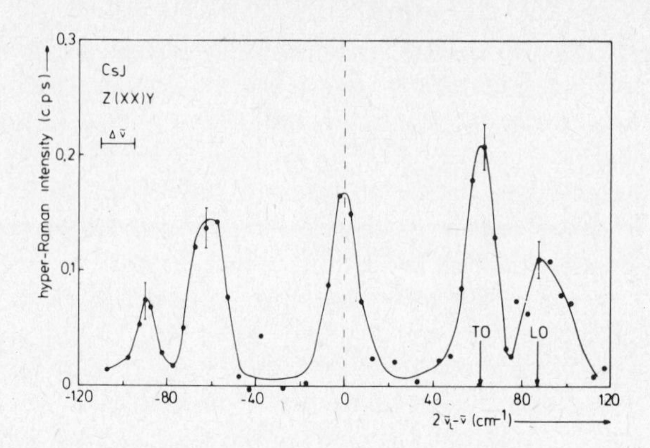

Fig. 5. Stokes and anti-Stokes hyper-Raman spectrum of CsI. c.p.s.: mean number of phonon counts per second; $\tilde{\nu}_L = 9396\,\mathrm{cm}^{-1}$: wavenumber of the Nd^{3+}YAG laser. Central peak corresponds to the second harmonic generation of $\tilde{\nu}_L$ (Vogt and Neumann 1976).

powerful pulsed lasers (Dines et al. 1976, Bancewitz et al. 1976 and references therein, Vogt and Neumann 1976, 1978 and references therein). In all these cases the incident radiation is at frequency ω_0 and the scattered radiation is detected at $\omega_S(\omega_A) = 2\omega_0 \mp \omega_j$ respectively (fig. 5).

In principle, the opposite effect should also be possible, i.e. illuminating the crystal with ω_0 and detecting two frequencies ω_{S_1} and ω_{S_2}, such that $\omega_{S_1} + \omega_{S_2} = \omega_0 - \omega_j$, where $\omega_{S_1} \gtreqless \omega_{S_2}$. The corresponding mode coefficient for this type of *hypo-Raman scattering* takes the form $-(\partial^4\Phi/\partial\mathcal{E}_\mu\partial\mathcal{E}'_\lambda\partial\mathcal{E}''_\nu\partial u_{j\sigma})$ and has the same symmetry (ignoring dispersion) as the one for hyper-Raman scattering. No such effect has been reported so far.

7.2. Experimental

The earliest observation of field-induced Raman scattering was made in α-quartz (Zubov et al. 1961). With $\boldsymbol{E}\|\hat{x}$ and the incident and scattered radiation along $\hat{y}$ and $\hat{z}$ respectively, an intensity decrease of 8–10% was observed for the line at 1159 cm^{-1} for $E = 2$ kV/cm. This line corresponds to E(TO + LO)-type phonons (Tekkipe et al. 1973) and should exhibit a change in the Raman tensor which is linear in the field, according to the criteria established in §5.2. No other substantial changes of the spectrum were reported.

In recent years electric-field-induced Raman scattering has been observed in a number of materials of O_h structure, in zincblende-structure III–V semiconductors and in wurtzite-structure II–VI semiconductors.

Since different physical mechanisms have been invoked to explain the effects of the field in these classes of materials, we review them separately. Observations of effects quadratic in the field (wherever available) are also included.

7.2.1. *Centrosymmetric structures (O_h type)*

The first observation of field-induced Raman scattering by an infrared-active Raman-inactive F_{1u}-type phonon was made in the paraelectric perovskite $KTaO_3$ (Fleury and Worlock 1967). Using an a.c. technique to discriminate against second-order Raman scattering Fleury and Worlock were able to observe scattering in the cubic phase of $KTaO_3$ by the so-called *soft* or ferroelectric mode with F_{1u}(TO) symmetry. This is shown in fig. 6. At 80 K the mode appeared at 43 cm^{-1} with an intensity proportional to E^2. The polarization selection rules were consistent with those derived from the matrices of eq. (7.4) for $\boldsymbol{E} \| \hat{z}$. From such observations, the temperature dependence of the soft mode was determined in detail in the range 8–300 K. Thus it became possible to examine the behavior of the mode close to the phase transition. The remaining two F_{1u}(TO)-type modes of $KTaO_3$ were also observed with the same a.c. technique but no F_{1u}(LO)-type mode was observed (Fleury and Worlock 1967, 1968). No scattering was induced for the silent mode F_{2u} of $KTaO_3$ although such scattering is expected, linearly in the field.

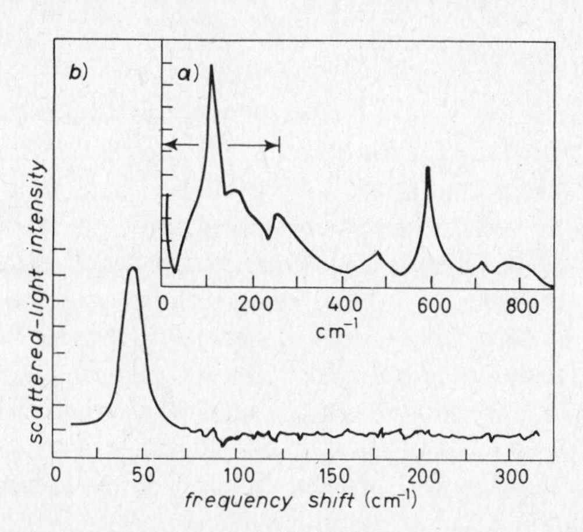

Fig. 6. Raman scattering spectrum of $KTaO_3$ at 77 K: a) intrinsic second-order spectrum taken with no applied electric field; b) electric-field-induced scattering from the low frequency normally Raman-inactive TO "ferroelectric" mode. Modulation techniques are applied with an electric field $E_A = 10$ kV/cm at 210 Hz. Arrows indicate region of a) reproduced in b) (Fleury and Worlock 1967).

Similar measurements have been made for $SrTiO_3$ which is isomorphic to $KTaO_3$ and known to undergo a structural phase transition at 105 K. Field-induced scattering by the ferroelectric F_{1u}(TO)-type mode was observed at temperatures above and below T_0 (Worlock and Fleury 1967, Fleury et al. 1968, Fleury and Worlock 1968). Induced Raman scattering was also observed by the other two F_{1u}(TO) modes and also by the silent mode (268 cm^{-1} at 10 K, Fleury and Worlock 1968). Below 105 K the intrinsic first-order Raman spectrum appears more complicated because of doubling of the unit cell (Fleury et al. 1968, Scott 1974). This results in splitting of the F_{1u}- and F_{2u}-type modes into components which are intrinsically Raman active. In addition to enhancing their Raman intensity, the applied field also causes substantial shifts and anti-crossing of their frequencies (Fleury et al. 1968). Similar a.c. techniques were used by Davis (1970) in his study of the dielectric and structural properties of the mixed crystal systems $K_xNa_{1-x}TaO_3$ and $KTa_xNb_{1-x}O_3$. These materials undergo ferroelectric transitions from non-polar to polar phases at temperatures which depend on their composition. The ferroelectric or "soft" modes were observed by Davis in both phases through electric-field-induced Raman scattering.

In all the cases discussed above, the electric field was applied *externally* and the induced scattering involved almost exclusively TO phonons of F_{1u} symmetry. A plausible physical mechanism for this is that the field induces a relative displacement of the crystal ions (Dvorak 1967) which affects the periodic potential of the crystal and thereby the orbital electronic wavefunctions of the electron–hole-pair states. As the frequencies of the F_{1u}(TO) phonons are rather low in paraelectric perovskites, the field-induced displacement in these crystals is large and this justifies the observed strength of the field-induced scattering by the F_{1u}(TO) modes.

Field-induced Raman scattering by normally Raman-inactive F_{1u}(LO)-type phonons has been observed in some families of semiconductors. In these cases the electric field is an *internal* one and a different mechanism is believed to be responsible for the scattering. In addition to the orbital electronic part of the wavefunctions of the electron–hole states, the electric field also modifies the envelope part of the wavefunction. Thus it induces a Franz–Keldysh effect which in turn is believed to cause forbidden scattering by LO phonons via a field-induced Fröhlich interaction (Pinczuk and Burstein 1975, Richter and Zeyher 1976, Richter 1976 and references therein); Brillson and Burstein (1971) have observed field-induced scattering by the Raman-inactive phonon F_{1u}(LO) in the IV–VI p-type semiconducting compounds PbTe and SnTe with the rocksalt structure (O_h). The surface space-charge electric field was applied indirectly through energy-band bending which was induced by semi-transparent Pb films. The backward scattering geometry revealed diagonal Raman scattering only in the presence of such films and under resonance conditions at the E_2 gap. The Franz–Keldysh mechanism has been invoked to explain such forbidden scattering, although the atomic displacements mechanism may also be of some importance because of the paraelectric nature of these compounds.

Scattering by the forbidden F_{1u}(LO)-type phonons has also been reported in the four members Mg_2X (X=Si, Ge, Sn, Pb) of the anti-fluorite-type II–IV semiconductor family (Anastassakis and Burstein 1971c and references therein, Onari et al. 1976, Onari and Cardona 1976). While spatial dispersion is believed to be the dominant mechanism, a Franz–Keldysh contribution is also considered to be partially responsible for the diagonal scattering observed. The electric field was not applied externally. Instead, it was assumed to exist as a built-in surface field. The forbidden backward scattering was observed under resonance conditions mainly at the E_1 gap.

7.2.2. *Non-centrosymmetric structures*

In the O_h-type structures examined in the previous section, no Raman scattering exists when no field is applied. This is because the phonon involved is one of odd parity (F_{1u}). Observations of symmetry-forbidden scattering without the need of an external field have been made in a number of semiconducting materials of zincblende and wurtzite structure. These compounds are non-centrosymmetric. This type of scattering can be observed only under resonance conditions in backward geometry. It is due to LO-type Raman and infrared-active phonons and can be best seen in those configurations for which no scattering should occur according to the intrinsic polarization selection rules. The scattering is partly due to surface built-in electric field. Spatial dispersion effects (to be discussed in §18) are known to play an important role too (Pinczuk and Burstein 1973, 1975 and references therein). Of the III–V semiconductors investigated one can mention characteristically the work on InSb (Pinczuk and Burstein 1969, Dreybrodt et al. 1973, Kiefer et al. 1975, Yu 1976 and references therein), InAs and GaSb (Corden et al. 1970, Corden 1971, Rubloff et al. 1973, Dreybrodt et al. 1973, Buchner et al. 1976a, b) and GaAs (Trommer et al. 1976, 1978). In the case of InSb an additional field was applied externally (Pinczuk and Burstein 1970) by means of well-known electrolytic techniques (Cardona et al. 1967). Metal films on InSb, InAs and GaAs have also been used to vary the surface field (Corden et al. 1970, Trommer et al. 1978). Variable carrier concentration has proved itself a useful means for monitoring the surface space-charge field in InSb and InAs and thereby changing the induced scattering intensity (Corden et al. 1970, Yu and Shen 1974, Buchner et al. 1976a, b). Extensive work has been done in recent years in the investigation of first- and second-order Raman scattering in several other III–V semiconducting compounds, mainly under resonance conditions. In all these cases, it is generally accepted that built-in surface fields play an important role in producing Raman scattering in forbidden configurations (Richter and Zeyher 1976, Richter 1976 and references therein).

The easiest way to observe and compare induced and intrinsic effects in III–V semiconductors is to work with faces which are perpendicular to cubic axes. The point group is T_d and the phonon is triply-degenerate with symmetry described by the irreducible representation F_2. The mode coefficient $-(\partial^4\Phi/\partial\mathcal{E}\,\partial\mathcal{E}\,\partial E\,\partial u_j)$ has three independent components a, b, and c. The combined field-induced and intrinsic Raman tensor (only one independent component d) takes the form

$$\begin{array}{rccc} \boldsymbol{a}_j+\Delta\boldsymbol{a}_j= & \begin{bmatrix} cE_x & bE_y & bE_z \\ bE_y & aE_x & d \\ bE_z & d & aE_x \end{bmatrix}, & \begin{bmatrix} aE_y & bE_x & d \\ bE_x & cE_y & bE_z \\ d & bE_z & aE_y \end{bmatrix}, & \begin{bmatrix} aE_z & d & bE_x \\ d & aE_z & bE_y \\ bE_x & bE_y & cE_z \end{bmatrix}. \\ u_{j\sigma}: & x & y & z \end{array} \tag{7.7}$$

The field-induced part $\Delta\boldsymbol{a}_j$ is the same as that of eq. (7.4) because the potential energy $\delta\Phi$ of eq. (7.3) turns out to be the same as for the mode F_2 of the T_d structure. In a backward scattering configuration along the $\hat{z}\|[001]$ axis, the surface field is $\boldsymbol{E}=\hat{z}E$. In the scattering configuration $z(xx \text{ or } yy)\bar{z}$ the scattering intensity observed is proportional to $(aE_z)^2$ and the phonon involved is an LO phonon. On the other hand, in the configuration $z(xy \text{ or } yx)\bar{z}$ only the intrinsic LO phonon participates in the process. Scattering from surfaces with different orientations may include interference contributions from both intrinsic and induced components of the Raman tensor. Often this procedure is used intentionally to enhance and study forbidden effects. Thus, in p-type InAs Buchner et al. (1976a, b) have observed significant differences in the LO scattering intensity from the (111) and $(\bar{1}\bar{1}\bar{1})$ faces. These surfaces are known to carry surface fields of opposite directions (Huff et al. 1968). The differences in the scattering intensity are therefore attributed to constructive and destructive interference respectively between intrinsic and surface field-induced components in the Raman tensor. The resonance curves from the two surfaces are shown in fig. 7.

Induced scattering by an E_1(LO) phonon has been observed in CdS, a wurzite-type crystal with point group symmetry C_{6v} (Shand et al. 1972). The field was applied externally by use of a high voltage pulse generator. With $\boldsymbol{E}\|\hat{x}$ and a backward scattering geometry $x(yy$ or $zz)\hat{x}$, an induced intensity was observed which increased as E^2, then leveled off and finally decreased beyond 5 kV/cm. The induced scattering was superimposed on the $\boldsymbol{q}$-dependent scattering (Martin and Damen 1971). The experiment was performed under resonance conditions. A theoretical account for this type of scattering has been given by Gay et al. (1971) who considered the ground-(1S) state exciton contribution to the field-induced scattering intensity. The experimental data are in qualitative agreement with this theory. With an appropriate choice of field $\boldsymbol{E}\|\hat{z}$ and forward scattering configuration Shand and Burstein (1973) have succeeded in observing field-induced scattering by the A_1(LO) phonon of CdS, with an intensity which varied linearly with the field. This is due to interference between allowed and forbidden field-induced terms in the Raman tensor. The same authors also report field-induced scattering by 2LO phonons.

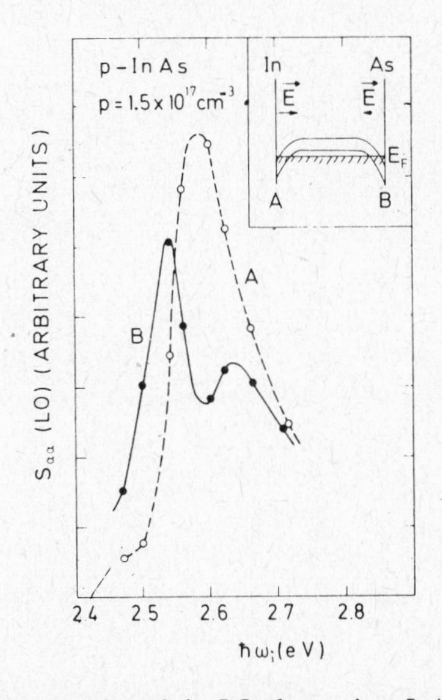

Fig. 7. Resonant Raman scattering of the LO phonon in p-InAs at 300 K from A (⟨111⟩) and B (⟨$\bar{1}\bar{1}\bar{1}$⟩) surfaces. The insert shows the surface electric fields for A and B (Buchner et al. 1976a, b).

7.2.3. Electric-field-induced Raman scattering, quadratic in the field

When the symmetry of a particular mode is such that a linear effect of an electric field on Raman scattering is forbidden, an effect quadratic in the field may still be possible. Since a quadratic factor $E_\lambda E_\nu$ transforms like a symmetric second-rank tensor we may set in eq. (5.9) $\Gamma(\boldsymbol{F})=\Gamma(\boldsymbol{EE})=\Gamma(\boldsymbol{\alpha})$. In other words the quadratic effect is, symmetry-wise, identical to the linear effect of a symmetric strain $\eta_{\lambda\nu}$, where $\Gamma(\boldsymbol{\eta})=\Gamma(\boldsymbol{\alpha})$. A situation like this arises in diamond. On top of the intrinsic Raman scattering by the optical phonon F_{2g} one can induce additional scattering. The induced Raman tensor $\Delta\boldsymbol{a}_j$ will be quadratic in the applied field. The polarization selection rules are different from those for intrinsic scattering. Following the procedure of §5.3 (see also §10) one can reach the following expressions of $\boldsymbol{a}_j+\Delta\boldsymbol{a}_j$ for diamond:

$$\boldsymbol{a}_j+\Delta\boldsymbol{a}_j=$$

$$\begin{bmatrix} cE_yE_z & fE_zE_x & fE_xE_y \\ fE_zE_x & eE_yE_z & a+\frac{1}{2}\{bE_x^2+g(E_y^2+E_z^2)\} \\ fE_xE_y & a+\frac{1}{2}\{bE_x^2+g(E_y^2+E_z^2)\} & eE_yE_z \end{bmatrix}, u_{jx}$$

$$\begin{bmatrix} eE_zE_x & fE_yE_z & a+\frac{1}{2}\{bE_y^2+g(E_z^2+E_x^2)\} \\ fE_yE_z & cE_zE_x & fE_xE_y \\ a+\frac{1}{2}\{bE_y^2+g(E_z^2+E_x^2)\} & fE_xE_y & eE_zE_x \end{bmatrix}, u_{jy}$$

$$\begin{bmatrix} eE_xE_y & a+\frac{1}{2}\{bE_z^2+g(E_x^2+E_y^2)\} & fE_yE_z \\ a+\frac{1}{2}\{bE_z^2+g(E_x^2+E_y^2)\} & eE_xE_y & fE_zE_x \\ fE_yE_z & fE_zE_x & cE_xE_y \end{bmatrix}, u_{jz}$$

(7.8)

where a and b,c,e,f,g are respectively the independent components for the intrinsic Raman tensor $\boldsymbol{a}_j$ and the mode coefficient $-(\partial^5\Phi/\partial\mathcal{E}\,\partial\mathcal{E}\,\partial E\,\partial E\,\partial u_j)$ of eq. (5.11). Notice that eq. (5.11) in this case would take the form

$$\delta\Phi=\frac{1}{4}\left(\frac{\partial^5\Phi}{\partial\mathcal{E}_\mu\partial\mathcal{E}_\lambda\partial E_\nu\partial E_\rho\partial u_{j\sigma}}\right)\mathcal{E}_\mu\mathcal{E}_\lambda E_\nu E_\rho u_{j\sigma}\equiv-\frac{1}{4}\Delta a_{j,\mu\lambda\sigma}\mathcal{E}_\mu\mathcal{E}_\lambda u_{j\sigma}.$$

(7.9)

With the electric field applied along $\hat{z}_1$ and with a $\bar{x}_1(y_1 z_1)x_1$ scattering configuration, Anastassakis et al. (1969) were able to observe a measureable change in the Raman scattering intensity, which was attributed to the quadratic effect of the electric field. The axes stand for $\hat{x}_1 // [1\bar{1}0]$, $\hat{y}_1 // [110]$ and $\hat{z}_1 // [001]$. The induced effect was observed through its interference with the intrinsic Raman scattering. The total Raman tensor which is appropriate to the scattering configuration used is $a + gE^2$. The change in the intensity then goes as $2agE^2$, and the ratio of this change to the intrinsic intensity as $2gE^2/a$. From the measured values for this ratio and the value $a = -4\,\text{Å}^2$ (see §6.2) it was possible to infer a value for g. It turned out from these experiments that g/a is of the order of -1×10^{-8} in esu units. This is in satisfactory agreement with the value obtained from field-induced infrared absorption experiments (see §6.2). Although these measurements provide only crude numerical results for the coefficients in question, they nevertheless show that it is in fact possible to observe even a weak higher-order effect when it is properly combined with a strong intrinsic effect. The polarization selection rules are of course particularly useful in situations like this.

Quadratic effects of an applied electric field on the Raman tensor have also been seen for the A_1(TO) phonon of CdS with sign opposite to that of the intrinsic tensor (Shand and Burstein 1973). The measurements were taken under strong resonance conditions at 35 K. Again, the observation of the effect was possible through the interference of the field-induced quadratic Raman tensor with the intrinsic Raman tensor for the A_1(TO) phonon.

7.3. Electric-field-induced second-order Raman scattering

The general phenomenological and group-theoretical aspects of a field-induced second-order Raman scattering are contained in the brief discussion given in §5.4. For instance, eq. (5.19) implies that no structure should be expected in the second-order Raman spectrum of a centrosymmetric crystal if no even-parity irreducible representations are included in the reduction of $D(j) \otimes D(j')$. On the other hand, eq. (5.20) states that in such cases an electric field may in fact induce scattering. Evidence for field-induced changes in the structure of the second-order Raman scattering in NaCl has been reported by Kanamori et al. (1976a). Further characteristic examples of field-induced second-order Raman activity occur in the Mg_2X ($X = $ Si, Ge, Sn)-type II–IV semiconductors (Anastassakis and Burstein 1971c, Onari and Cardona 1976). The combination $F_{2g} + F_{1u}$(LO) is normally Raman inactive. Nevertheless, it has been observed to produce very strong scattering. In Mg_2Si the combination $F_{2g}+F_{1u}(LO)$ can be observed only under resonance conditions in the region of E_0 and E_1 gaps (as in the case of forbidden scattering by the $F_1(LO)$ mode and its overtone). Similar behavior is exhibited by the $X=$ Ge, Sn members near resonance at the E_1 and $E_1+\Delta_1$, and E_1 and X energy gaps respectively (Onari and Cardona 1976). A combined mechanism has been suggested for this type of scattering, which includes deformation potential, and also Fröhlich interaction intraband transition matrix elements. The major effect of the built-in electric field is to produce a spatial separation between the electron and the hole. This results in unequal contributions from the electron and the hole to the Fröhlich interaction matrix elements (Richter and Zeyher 1976).

Resonance techniques have proved themselves extremely convenient for the study of forbidden effects in several groups of semiconductors, particularly III–V, IV–VI and II–VI types. These effects appear in the form of forbidden scattering by combinations and overtones of both optical and acoustical phonons. Built-in electric fields appear to play an important, although not exclusive, role in these interactions via the field-induced Fröhlich interaction processes mentioned earlier (Zeyher 1975, Cardona 1975, Richter 1976).

8. *Electric-field-induced changes in the phonon frequencies*

8.1. General

The linear effect of an electric field on the frequency of an optical mode of symmetry $\Gamma(j)$ is described by eq. (5.10). For a centrosymmetric structure such a linear effect is not allowed. This is because $[\Gamma(j)\otimes\Gamma(j)]_S$ and $\Gamma(\boldsymbol{E})$ reduce respectively to even- and odd-parity representations. Since the totally symmetric irreducible representation of all centrosymmetric point groups is of even parity, we conclude that $n_A=0$ in eq. (5.10), that is, no linear effect is allowed. Kanamori et al. (1976b) and Matsumoto and Kanamori (1977) recently worked out the secular equations for all those cases where an effect linear in the field is allowed to occur. Second-order effects in the field may also very well occur. This is the case in the diamond-type structures. Upon treating the second-order term $\boldsymbol{EE}$ in the same way as a symmetric strain, one finds that $n_A=3$. This results from the reduction of eq. (5.10) which becomes

$$[\Gamma(j)\otimes\Gamma(j)]_S\otimes\Gamma(\boldsymbol{EE})=[A_{1g}\oplus E_g\oplus F_{2g}]\otimes[A_{1g}\oplus E_g\oplus F_{2g}]$$
$$=3A_{1g}\oplus\ldots$$

We need three symmetric functions of order four to describe $u_\alpha u_\beta E_\mu E_\nu$. From eq. (5.5d) these are easily found to be

$$x_1^2x_2^2+y_1^2y_2^2+z_1^2z_2^2,$$
$$(x_1^2y_2^2+x_2^2y_1^2)+(y_1^2z_2^2+y_2^2z_1^2)+(z_1^2x_2^2+z_2^2x_1^2),$$
$$x_1y_1x_2y_2+y_1z_1y_2z_2+z_1x_1z_2x_2.$$

If we associate (x_1,y_1,z_1) with $u_\alpha(u_\beta)$ and $(x_2^2,\ldots,y_2z_2)$ with $E_\mu E_\nu$ and follow eq. (5.13) we find for $\delta\Phi$,

$$\delta\Phi=\tfrac{1}{4}K_1(u_x^2E_x^2+\text{c.p.})+\tfrac{1}{4}K_2(u_x^2E_y^2+u_y^2E_x^2+\text{c.p.}) + K_3(u_xu_yE_xE_y+\text{c.p.}), \tag{8.1}$$

where K_1, K_2, K_3 are the three independent components of the mode coefficient $(\partial^4\Phi/\partial u_j\partial u_j\partial E\,\partial E)$ according to eq. (5.13). The secular equation takes the form

$$\begin{vmatrix} \tfrac{1}{2}\{K_1E_x^2+K_2(E_y^2+E_z^2)\}-\lambda & K_3E_xE_y & K_3E_xE_z \\ K_3E_yE_x & \tfrac{1}{2}\{K_1E_y^2+K_2(E_z^2+E_x^2)\}-\lambda & K_3E_yE_z \\ K_3E_zE_x & K_3E_zE_y & \tfrac{1}{2}\{K_1E_z^2+K_2(E_x^2+E_y^2)\}-\lambda \end{vmatrix}=0, \tag{8.2}$$

where $\lambda\simeq 2\omega_j\Delta\omega_j$ according to eq. (3.12). Ganesan et al. (1970) have reached the same expression through a detailed microscopic theory which allowed them to obtain specific numerical values for the three independent coefficients, that is, $K_1=1.54$, $K_2=-1.195$ and $K_3=0.0548$ in units of 10^{11} $\text{cm}^2/\text{V}^2\text{sec}^2$ for diamond. On the basis of these values one estimates that for a field of 10^5 V/cm the expected change $\Delta\omega_j$ is of the order of 10^{-4} cm^{-1}. This is too small to be detected by conventional spectroscopic means. Indeed no effect was observed in either field-induced infrared absorption or field-induced changes of Raman scattering in diamond for fields up to 2×10^5 V/cm (Anastassakis et al. 1969).

8.2. Experimental

The most striking effects of an electric field on phonon frequencies have been observed in paraelectric perovskites (Scott 1974 and references therein). Fatuzzo (1964) examined these effects theoretically for $BaTiO_3$, by considering anharmonic terms of short-range forces for temperatures around the Curie point. Fleury and Worlock (1968) have done significant experimental work in connection with field-induced Raman scattering experiments. Shifts and splitting of the soft F_{1u}-type phonon were observed in $KTaO_3$ and $SrTiO_3$ at various temperatures. At 8 K the frequency of the soft mode of $SrTiO_3$ is found to shift and split from 10 cm^{-1} at zero field, to two components at 26 cm^{-1} and 44.5 cm^{-1} at 12 kV/cm. Similar behavior was observed in $KTaO_3$. From their data and the use of thermodynamical arguments Fleury and Worlock were able to infer values for the non-linear dielectric response coefficients of these crystals. As mentioned earlier, in $SrTiO_3$ the applied field produces anti-crossing between the two components of the soft mode and those of the silent F_{2u}-type mode. This is a unique example of the multiple role that the electric field can play in paraelectric perovskites. Under appropriate conditions, it can produce Raman scattering by Raman-inactive modes, it can split these modes and shift their frequencies and it can induce coupling and anti-crossing between their components.

Shifts of frequencies in non-centrosymmetric ZnO have been regarded as partially responsible for the rather rich field-induced changes of its second-order infrared absorption spectrum (Lüth 1969a, b). A detailed analysis of the spectrum observed was not possible, owing to lack of sufficient knowledge about this material and because of the complexity of its intrinsic and induced second-order infrared absorption spectrum.

9. *Stress-induced infrared absorption*

9.1. General

The use of a uniaxial stress as a controlled perturbation in the study of optical and kinetic properties associated with the electronic states of solids has been recognized as a powerful and informative experimental procedure for quite some time. However, only recently has work been done in connection with lattice-dynamical morphic effects, in particular shifts of mode frequencies, lifting of their degeneracy and inducing of phase transitions. These and other morphic effects will be reviewed in this and the following three sections. Various types of stress delivering devices have been developed through the years, falling basically into the following five categories (Pollak 1973 and references therein):

(i) the lever-arm type,
(ii) the vice-type,
(iii) the push-rod type,
(iv) the bending type,
(v) the rotating shaft and cam type.

The lever-arm type has been used in most of the studies relevant to the present chapter. Modifications of the various types have been used to allow measurements under *tensile* uniaxial stress and variable temperatures (Anastassakis 1974, Gorman et al. 1974), or configurations whereupon the stress is applied along the direction of propagation of light (Vedam and Davis 1968).

It is well known from Crystal Physics that an externally applied homogeneous stress $\boldsymbol{X}$ will result in an elastic deformation of the crystal. This deformation is homogeneous too, and is best described by the strain $\boldsymbol{\eta}$ which, in the elastic region, is proportional to the stress. Both $\boldsymbol{X}$ and $\boldsymbol{\eta}$ are treated as second-rank symmetric tensors, provided no rotation of the microscopic volume elements is caused by the strain (Nelson and Lax 1970, Anastassakis and Burstein 1974, Grimsditch and Ramdas 1977). The coefficients of proportionality are the components of a fourth-rank tensor, the well-known elastic compliance (Nye 1964). In an actual experiment, an applied force produces a macroscopic stress which in turn results in a microscopic strain. Usually, the phenomenological treatment of morphic effects is carried out in terms of the stress-induced strain; this procedure will be followed here too. In view of the proportionality between stress and strain, it makes no difference whether we consider stress- or strain-induced morphic effects.

Since an applied strain $\eta_{\mu\lambda}$ and a second-order field perturbation of the form $E_\lambda E_\mu$ are both second-rank symmetric tensors, we examine them together. Evidently $\Gamma(\boldsymbol{\eta})=\Gamma(\boldsymbol{EE})=\Gamma(\boldsymbol{\alpha})$. A strain-induced infrared absorption and a field-induced infrared absorption in second order can occur when $\Delta e_j \propto (\partial M^2/\partial\eta\partial u_j)=(\partial d/\partial u_j)$ is non-zero, i.e. when $\Gamma(\boldsymbol{d})$ includes $\Gamma(j)$. $\boldsymbol{d}$ is a third-rank tensor such as the piezoelectric tensor and reduces to identically as $\Gamma(\boldsymbol{\alpha})\otimes\Gamma(\boldsymbol{M})$. As $\boldsymbol{d}$ is a third-rank tensor, it reduces to irreducible representations with odd parity only, in centrosymmetric crystals. As a result, only odd-parity modes can exhibit the effects in these crystals. The Raman-active modes in centrosymmetric crystals have even parity and therefore are excluded (e.g. and F_{2g} mode of diamond). The effects are possible for the infrared-active mode F_{1u} and for the silent mode F_{2u} of cubic $BaTiO_3$, since $\Gamma(\boldsymbol{d})$ includes these irreducible representations three and two times respectively according to eq. (5.1d). The effects are also possible for the symmetric mode A_1 of Te (eq. (5.2d)). In this connection we notice the symmetry identity between strain-induced infrared absorption and electric field-induced Raman scattering which we examined earlier.

We consider two examples of the determination of the strain-induced effective charge. For the silent mode F_{2u} of $SrTiO_3$ there are two independent components a, b for the mode coefficient $-(\partial^3\Phi/\partial\mathcal{E}\,\partial\eta\partial u_j)$ of eq. (5.12). By inspection of the basis functions of eq. (5.5e) we can write for $\delta\Phi$:

$$\delta\Phi = -a\left[u_x\mathcal{E}_x(\eta_{yy}-\eta_{zz})+\text{c.p.}\right]-b\left[u_x(\mathcal{E}_y\eta_{xy}-\mathcal{E}_z\eta_{xz})+\text{c.p.}\right]. \tag{9.1}$$

The strain-induced effective charge is then

$$\Delta \boldsymbol{e}_j = \begin{bmatrix} a(\eta_{yy}-\eta_{zz}) & -b\eta_{xy} & b\eta_{xz} \\ b\eta_{xy} & a(\eta_{zz}-\eta_{xx}) & -b\eta_{yz} \\ -b\eta_{xz} & b\eta_{yz} & a(\eta_{xx}-\eta_{yy}) \end{bmatrix}. \tag{9.2}$$

Suppose for instance that a uniaxial strain η_{zz} is present; the only non-zero components of $\Delta\boldsymbol{e}_j$ are $\Delta e_{yy} = -\Delta e_{xx} = a\eta_{zz}$. According to the selection rules of §3.1, only the y and x components of the electromagnetic radiation are (equally) absorbed and the F_{2u}-type phonons created are polarized along y and x respectively.

Actually the above results for $\Delta\boldsymbol{e}_j$ can be obtained by inspection from the function $\delta\Phi$ which we worked out in connection with the electric-field-induced Raman scattering. In §7.1 the field-induced Raman tensor was derived for the silent mode F_{2u} of cubic $SrTiO_3$ and the matrix form was established for the three components of $u_{j\sigma}$. The tensor for $\Delta\boldsymbol{e}_j$ can be obtained by substituting $\eta_{\mu\lambda}$ for $\mathcal{E}_\mu\mathcal{E}_\lambda$ and $\mathcal{E}_\nu$ for E_ν. This is a consequence of the fact that the functional form of $\delta\Phi$, when expanded in terms of $\mathcal{E}_\mu\mathcal{E}_\lambda E_\nu u_{j\sigma}$ or in terms $\mathcal{E}_\nu\eta_{\mu\lambda}u_{j\sigma}$, is the same within a factor of 1/2 (i.e. compare eqs. (5.11) and (5.12)). Application of these to the matrices of eqs. (7.3) and (7.5) leads to the following results for the F_{1u} and A_2 modes of the O_h and D_3 structures respectively:

$$\Delta \boldsymbol{e}_j = \begin{bmatrix} a'(\eta_{yy}+\eta_{zz})+c'\eta_{xx} & b'\eta_{xy} & c'\eta_{xz} \\ b'\eta_{xy} & a'(\eta_{zz}+\eta_{xx})+c'\eta_{yy} & b'\eta_{yz} \\ c'\eta_{xz} & b'\eta_{yz} & a'(\eta_{xx}+\eta_{yy})+c'\eta_{zz} \end{bmatrix}, \tag{9.3}$$

$$\Delta \boldsymbol{e}_j = \begin{bmatrix} 2f''\eta_{xy}+c''\eta_{xz} \\ f''(\eta_{xx}-\eta_{yy})+c''\eta_{yz} \\ b''(\eta_{xx}+\eta_{yy})+a''\eta_{zz} \end{bmatrix}, \tag{9.4}$$

where the sets of primed constants a',b',c' and a'',b'',c'',f'' designate respectively the three and four independent components of the mode property $-(\partial^3\Phi/\partial\mathfrak{E}\,\partial\eta\partial u_j)$ for $\Gamma(j)=F_{1u}$ and A_2.

9.2. Experimental

Evidence for stress-induced infrared absorption by normally infrared-inactive phonons has been reported by Barker (1963, 1964). Working at 300 K on reflectivity experiments of corundum (α-Al_2O_3, point group D_{3d}), Barker observed the intrinsic spectrum of six infrared-active modes, plus a number of forbidden modes. Two of these modes showed strengths comparable to those of the infrared-active modes. Symmetry arguments, based on a phenomenological perturbation theory and the experimentally applied polarization selection rules, led to the conclusion that the forbidden activity was due to the silent mode of symmetry A_{1u}. This activity was attributed to residual surface shear strains of E_g symmetry. According to the present treatment the relevant reductions are

$$\Gamma(\boldsymbol{M})=A_{2u}\oplus E_u \tag{9.5}$$

$$\Gamma(\boldsymbol{\eta})=2A_{1g}\oplus 2E_g \tag{9.6}$$

$$\Gamma(\boldsymbol{M})\otimes\Gamma(\boldsymbol{\eta})=2A_{1u}\oplus 4A_{2u}\oplus 6E_u. \tag{9.7}$$

Thus, all the odd-type modes will be affected by a symmetric strain. In particular, the silent mode A_{1u} will become infrared active with two independent components for the mode coefficient $-(\partial^3\Phi/\partial\mathfrak{E}\,\partial\eta\partial u_j)$. This change of symmetry will manifest itself in a reflectivity spectrum through a strain-induced restrahlen band. Application of the general method developed in §5.3 leads to the following matrix form for $\Delta\boldsymbol{e}_j$, where $\Gamma(j)=A_{1u}$:

$$\Delta\boldsymbol{e}_j=\begin{bmatrix} 2a\eta_{xy}+b\eta_{xz} \\ a(\eta_{xx}-\eta_{yy})+b\eta_{yz} \\ 0 \end{bmatrix}. \tag{9.8}$$

This result indicates that a shear strain (η_{xy} or $\eta_{xx}-\eta_{yy}$) in the xy plane can induce infrared absorption by the silent mode A_{1u}. It can be observed only when the incident radiation is polarized perpendicular to the c-axis. Barker's measurements are in agreement with this. Barker also showed that the strains were indeed surface residual strains by studying surfaces of different treatment. An etched surface did not produce any activity, apparently owing to relaxation of the residual strains. By contrast, surfaces which were polished in various ways produced activity of different strengths. Furthermore, an externally applied stress of 10^3 lb/cm^2 did not produce any change in the spectrum for a polarization parallel to the c-axis. This too is in agreement with eq. (9.8).

The results of similar reflectivity measurements in $CaWO_4$ and $CaMoO_4$ have been reported by Barker (1964). These materials crystallize with C_{4h} symmetry and exhibit eight infrared-active modes ($4A_u+4E_u$), three silent modes ($3B_u$), and thirteen Raman-active modes ($3A_g+5B_g+5E_g$). Barker observed eight infrared-active modes and also a number of modes which were identified as silent modes. Including some of these forbidden modes in the classical oscillator model substantially improved the fitting to the reflectivity curves. A symmetry analysis and identification of all those modes observed in reflectivity and Raman experiments (Russel and Loudon 1965, Porto and Scott 1967) has been carried out by Scott (1968a). Using a perturbation approach to a mode-mixing model, Scott shows that a residual surface strain pattern similar to that of corundum is not sufficient to explain the experimental features. Inhomogeneous surface strains of odd-parity are presumably present and responsible for the observed breakdown of the selection rules.

With the exception of such incidental observations and interpretations of residual strain-induced first-order infrared absorption, no references appear to exist of a systematic experimental study of this effect. Some group-theoretical results have been given by Balslev (1974) while Kanamori et al. (1976b) and Matsumoto and Kanamori (1977) have completed tables, for all 32 point groups, of strain-induced dipole moments for all appropriate modes (i.e. matrices like those of eqs. (9.2)–(9.4) and (9.8) multiplied by u_j). Second-order infrared absorption induced or modified by an applied stress does not seem to have been observed at all, with the exception of some exploratory work by Hobson and Paige (1966, see §11.4). This is in contrast to the large number of experiments reported in the literature, in which an applied stress is used to induce and monitor a wide variety of optical, dynamical, kinetic and other phenomena. On the other hand, some work has been done on the shifts and splittings of second-order infrared absorption bands in Si, when subject to uniaxial stresses along [001] and [111] (Lüth 1970, Angress et al. 1975). We will discuss these in §11.4.

10. Streuss-induced Raman scatterings

We now consider the effect of an applied stress on Raman scattering by optical phonons. The relatively few experimental observations of such morphic effects have been made mostly under conditions of resonance. In the next two sections we discuss the general symmetry aspects of the problem (regardless of resonance), and the existing experimental and theoretical results obtained under resonance conditions.

10.1. Symmetry considerations

By definition, stress-induced morphic effects in Raman scattering may arise because of the modification of polarization selection rules, due to the stress-induced symmetry reduction. These effects are possible only for those modes j for which the mode coefficient $-(\partial^4\Phi/\partial\mathcal{E}\,\partial\mathcal{E}\,\partial\eta\partial u_j)=(\partial^2\alpha/\partial\eta\partial u_j)=(\partial\pi/\partial u_j)$ is non-zero. $\boldsymbol{\pi}$ is an appropriate piezo-optical coefficient, a fourth-rank tensor which reduces as $\Gamma(\boldsymbol{\alpha})\otimes\Gamma(\boldsymbol{\alpha})$. Thus, *strain-induced Raman scattering by a mode j is possible provided that* $\Gamma(j)$ *is included in the reduction of* $\Gamma(\boldsymbol{\pi})$. In centrosymmetric crystals $\Gamma(\boldsymbol{\pi})$ reduces to even-parity irreducible representations only. Infrared-active modes in centrosymmetric crystals have odd parity and are therefore excluded from exhibiting such morphic effects. The symmetry of the strain-induced Raman scattering is the same as that of quadratic field-induced Raman scattering. We consider the following two examples of determining the exact matrix form for the strain-induced Raman tensor in several modes of the O_h and D_3 structures.

(i) *For the Raman-active* F_{2g} *mode of diamond* the strain-induced Raman tensor $\Delta\boldsymbol{a}_j$ exhibits five independent components b,c,g,e,f according to eq. (5.1e). A symmetric function for $\mathcal{E}_\mu\mathcal{E}_\lambda\eta_{\nu\kappa}u_{j\sigma}$ can be found by applying Unsöld's theorem between $[u_x,u_y,u_z]$ and a second set of F_{2g}-type fourth-order basis functions which transform like $\mathcal{E}_\mu\mathcal{E}_\lambda\eta_{\nu\kappa}$. In fact, one should look for five such sets. With an appropriate choice of indices, these five sets of functions can be generated directly from those of eq. (5.5d), i.e.

$$\begin{aligned}
&[\mathcal{E}_y\mathcal{E}_z\eta_{xx},\mathcal{E}_z\mathcal{E}_x\eta_{yy},\mathcal{E}_x\mathcal{E}_y\eta_{zz}],\\
&[\mathcal{E}_x^2\eta_{yz},\mathcal{E}_y^2\eta_{zx},\mathcal{E}_z^2\eta_{xy}],\\
&[\mathcal{E}_y\mathcal{E}_z(\eta_{yy}+\eta_{zz}),\text{c.p.}],\\
&[(\mathcal{E}_y^2+\mathcal{E}_z^2)\eta_{yz},\text{c.p.}],\\
&[\mathcal{E}_x\mathcal{E}_y\eta_{zx}+\mathcal{E}_z\mathcal{E}_x\eta_{xy},\text{c.p.}].
\end{aligned}$$

From these and eq. (5.11) we can write for $\delta\Phi$

$$\begin{aligned}
\delta\Phi=&-b[u_x\mathcal{E}_y\mathcal{E}_z\eta_{xx}+\text{c.p.}]-\tfrac{1}{2}c[u_z\mathcal{E}_x^2\eta_{yz}+\text{c.p.}]\\
&-g[u_x\mathcal{E}_y\mathcal{E}_z(\eta_{yy}+\eta_{zz})+\text{c.p.}]-\tfrac{1}{2}e[u_x(\mathcal{E}_y^2+\mathcal{E}_z^2)\eta_{yz}+\text{c.p.}]\\
&-f[u_x(\mathcal{E}_x\mathcal{E}_y\eta_{zx}+\mathcal{E}_z\mathcal{E}_x\eta_{xy})+\text{c.p.}].
\end{aligned}\tag{10.1}$$

Each term in eq. (10.1) will give rise to one component $\Delta a_{j,\mu\lambda\sigma}$, which one obtains by successive differentiations according to eq. (5.11). Finally we have

$$
\Delta \boldsymbol{a}_j = \begin{bmatrix} c\eta_{yz} & f\eta_{zx} & f\eta_{xy} \\ f\eta_{zx} & e\eta_{yz} & b\eta_{xx}+g(\eta_{yy}+\eta_{zz}) \\ f\eta_{xy} & b\eta_{xx}+g(\eta_{yy}+\eta_{zz}) & e\eta_{yz} \end{bmatrix}, u_{jx}
$$

$$
\begin{bmatrix} e\eta_{zx} & f\eta_{yz} & b\eta_{yy}+g(\eta_{zz}+\eta_{xx}) \\ f\eta_{yz} & c\eta_{zx} & f\eta_{xy} \\ b\eta_{yy}+g(\eta_{zz}+\eta_{xx}) & f\eta_{xy} & e\eta_{zx} \end{bmatrix}, u_{jy} \tag{10.2}
$$

$$
\begin{bmatrix} e\eta_{xy} & b\eta_{zz}+g(\eta_{xx}+\eta_{yy}) & f\eta_{yz} \\ b\eta_{zz}+g(\eta_{xx}+\eta_{yy}) & e\eta_{xy} & f\eta_{zx} \\ f\eta_{yz} & f\eta_{zx} & c\eta_{xy} \end{bmatrix}, u_{jz}.
$$

Apart from factors of $1/2$, eqs. (10.2) and (7.6) coincide in the form of $\Delta \boldsymbol{a}_j$.

(ii) *The A_2 mode of the D_3 structure* exhibits four independent components a, b, c, e, for the strain-induced Raman tensor, according to eq. (5.2e). In order to describe $\mathcal{E}_\mu \mathcal{E}_\lambda \eta_{\nu\kappa}$ one needs four functions of order four which transform according to A_2. When multiplied by u_z (of symmetry A_2) these will give the symmetric functions for $\mathcal{E}_\mu \mathcal{E}_\lambda \eta_{\nu\kappa} u_{j\sigma}$ as needed. The four functions of symmetry A_2 are

(1) $\bar{z}z^2$ or $(x_2y_3 - x_3y_2)z_2z_3$ or $\mathcal{E}_x\mathcal{E}_z\eta_{yz} - \mathcal{E}_y\mathcal{E}_z\eta_{xz}$,

(2) $\bar{z}(x^2+y^2)$ or $(x_2y_3 - x_3y_2)(x_2x_3 + y_2y_3)$ or
$\mathcal{E}_x^2\eta_{xy} + \mathcal{E}_x\mathcal{E}_y\eta_{yy} - \mathcal{E}_x\mathcal{E}_y\eta_{xx} - \mathcal{E}_y^2\eta_{xy}$,

(3) $(2x_2y_2)(y_3z_3) - (x_2^2 - y_2^2)(x_3z_3)$ or
$(2\mathcal{E}_x\mathcal{E}_y)\eta_{yz} - (\mathcal{E}_x^2 - \mathcal{E}_y^2)\eta_{xz}$,

(4) $2\eta_{xy}(\mathcal{E}_y\mathcal{E}_z) - (\eta_{xx} - \eta_{yy})(\mathcal{E}_x\mathcal{E}_z)$.

Functions (1) and (2) are based on $A_2 \otimes A_1 = A_2$. Functions (3) and (4) are based on $E \otimes E = A_2 \oplus \ldots$. Thus $\delta\Phi$ takes the form

$$
\begin{aligned}
\delta\Phi = & - au_z(\mathcal{E}_x\mathcal{E}_z\eta_{yz} - \mathcal{E}_y\mathcal{E}_z\eta_{xz}) \\
& - \tfrac{1}{2} bu_z\left(\mathcal{E}_x^2\eta_{xy} + \mathcal{E}_x\mathcal{E}_y\eta_{yy} - \mathcal{E}_x\mathcal{E}_y\eta_{xx} - \mathcal{E}_y^2\eta_{xy}\right. \\
& - \tfrac{1}{2} cu_z\left[(2\mathcal{E}_x\mathcal{E}_y)\eta_{yz} - (\mathcal{E}_x^2 - \mathcal{E}_y^2)\eta_{xz}\right] \\
& - eu_z\left[(2\mathcal{E}_y\mathcal{E}_z)\eta_{xy} - (\mathcal{E}_x\mathcal{E}_z)(\eta_{xx} - \eta_{yy})\right].
\end{aligned} \tag{10.3}
$$

The result for $\Delta \boldsymbol{a}_j$ is

$$\Delta \boldsymbol{a}_j = \begin{bmatrix} b\eta_{xy} - c\eta_{xz} & \frac{1}{2}b(\eta_{yy} - \eta_{xx}) + c\eta_{yz} & a\eta_{yz} + e(\eta_{yy} - \eta_{xx}) \\ \frac{1}{2}b(\eta_{yy} - \eta_{xx}) + c\eta_{yz} & -b\eta_{xy} + c\eta_{xz} & -a\eta_{xz} + 2e\eta_{xy} \\ a\eta_{yz} + e(\eta_{yy} - \eta_{xx}) & -a\eta_{xz} + 2e\eta_{xy} & 0 \end{bmatrix}. \tag{10.4}$$

Similar matrices for all appropriate modes of all 32 point groups have been tabulated by Kanamori et al. (1976b), and by Matsumoto and Kanamori (1977).

10.2. Stress-induced resonant Raman scattering

Raman scattering measurements in the presence of an applied stress have been performed by a number of workers. The main objectives of such experiments are to observe and study the shift of phonon frequencies and removal of their degeneracies, and/or to induce resonant Raman scattering. Stress-induced Raman scattering far from resonance does not appear to have been observed and investigated in the literature, except in some incidental cases. For instance, Russel and Loudon (1965) report Raman scattering by the A_u mode in $CaWO_4$ at 180 cm^{-1}. This mode is infrared active and can be made Raman active only in the presence of an inhomogeneous strain of odd symmetry. Presumably this is a residual surface strain which depends on the history of the sample. Also, Uwe and Sakudo (1976) attribute to defect-induced internal (inhomogeneous) strains the weak forbidden scattering observed at 172 cm^{-1} in $SrTiO_3$ at 2 K. Under an external stress the intensity appears to increase. As already mentioned, extensive studies of the strain-induced shifts and splittings of mode frequencies have been carried out in a number of materials under nonresonant conditions (to be discussed shortly). In none of these experiments was any intensity change observed attributable to changes $\Delta \boldsymbol{a}_j$, with the exception (perhaps) of a single observation in $Bi_{12}GeO_{20}$ regarding the A_1-type phonon at 89.4 cm^{-1} (Venugopalan and Ramdas 1973). The same remark holds for second-order Raman scattering in the presence of a stress.

An applied stress, on the other hand, may very drastically affect the strength of Raman scattering by TO and LO phonons, when its application is combined with conditions suitable for resonant Raman scattering. The physical basis for this is the symmetry modification of the electronic band structure of the material. Investigations of resonant Raman scattering in the past have been limited by the small number of laser lines, although in recent years extensive work has been done using tunable dye lasers.

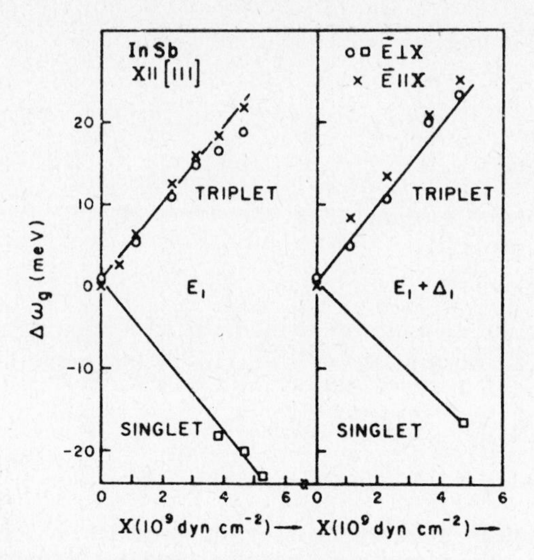

Fig. 8. Stress dependence of the E_1 and $E_1+\Delta_1$ energy gaps of InSb at 77K for $X//[111]$ according to Tuomi et al. (1970). The zero stress values of the E_1 and $E_1+\Delta_1$ gaps are 1.98 eV and 2.48 eV respectively.

Alternatively, one can vary the electronic energy gaps by means of temperature (Pinczuk and Burstein 1969, Nill and Mooradian 1969, Renucci et al. 1971a), alloying (Renucci et al. 1971b, Renucci et al. 1972) and uniaxial stress (Anastassakis et al. 1972c, 1974, Richter et al. 1976, 1978a, b). The latter technique has proved itself extremely informative, since it allows one to test experimentally the applicability of certain theoretical models of resonant Raman scattering. Essentially three semiconducting materials have been studied in stress-induced resonant Raman scattering, that is, InSb, InAs and Ge. To these, we may add the hydrostatic-pressure-induced resonant Raman scattering in GaAs (Trommer et al. 1976, to be discussed in §13.2) and the stress-induced change in the scattering intensity of CdS (Briggs and Ramdas 1976). These results are reviewed next.

(i) InSb. Fig. 8 shows the linear dependence of the E_1 and $E_1+\Delta_1$ energy gaps of InSb at 77K, on a uniaxial stress along [111] (Tuomi et al. 1970). The singlet and triplet components of these gaps correspond to electron wavevectors along $[111]//X$ and the three equivalent directions $[\bar{1}\bar{1}1]$, $[1\bar{1}\bar{1}]$ and $[\bar{1}1\bar{1}]$ respectively. It is clear that with a stress-free E_1 gap at 1.98 eV, one may drive its singlet component into the He–Ne laser excitation frequency at 1.96 eV, thus achieving resonance. The first results of such stress-induced resonant Raman scattering in InSb at 77K by TO and LO phonons in allowed configurations are shown in fig. 9 (Anastassakis et al. 1972c). Since only one excitation frequency was used a

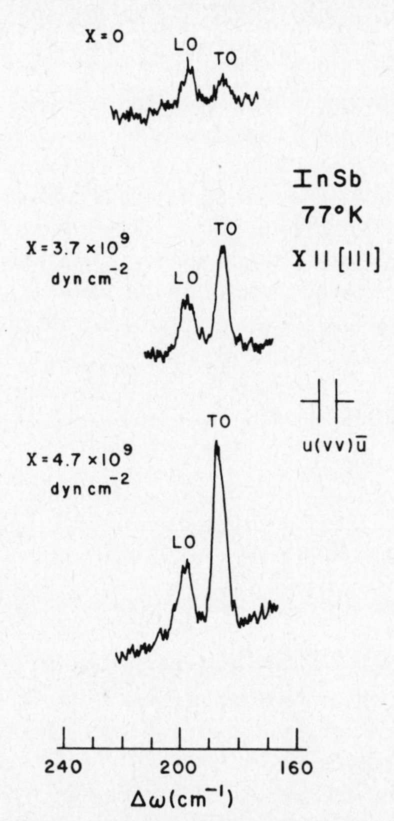

Fig. 9. Typical first-order Stokes Raman scattering spectra of InSb at 77 K using a 35 mW He–Ne laser (1.96 eV) in the configuration $u(vv)\bar{u}$ with $X//[111]$ for $X=0$, 3.7×10^9 and $4.7\times10^9\,\mathrm{dyn\cdot cm^{-2}}$. $\hat{u}//[11\bar{2}]$, $\hat{v}//[\bar{1}10]$ (Anastassakis et al. 1972c).

satisfactory description of the stress-induced scattering intensity was based on Loudon's expression of the Raman tensor for scattering by TO phonons (Loudon 1963). When integrated over all $\boldsymbol{k}$-space using a two-dimensional model for the E_1 gap, the double-pole term of that expression yields

$$I(X) \propto \left| \frac{\ln[\omega_g(X)+\omega_j-\omega_0]}{[\omega_g(X)-\omega_0]} \right|^2, \tag{10.5}$$

where $\omega_g(X)$ corresponds to the stress-dependent energy gap and ω_0 is the incident laser frequency. When written and plotted for the singlet component only, eq. (10.5) gives good agreement with the experimentally observed intensity as a function of X.

Recently, Richter et al. (1976) have investigated InSb more thoroughly using uniaxial stresses and a tunable dye laser. Their results for the TO phonon at 77 K are shown in fig. 10 together with theoretical curves based on the model for resonant Raman scattering discussed by Cardona (1973 and references therein). According to this model, at the E_1 gap the Raman tensor for deformation potential scattering by TO phonons consists of two-band and three-band terms. The former correspond to a phonon modulated energy gap, while the latter arise from wavefunction mixing of the spin–orbit split valence bands. In the presence of a uniaxial stress along [111] this expression for the Raman tensor becomes (Anastassakis et al. 1974, Richter et al. 1976, 1978b)

$$R = -\left\{ \left[\frac{2\sqrt{2}\,(\chi^+ + \chi^-)^{\mathrm{T}}}{3\Delta_1} \right] d^5_{3,0} - \frac{3}{8}\left[\left(\frac{\mathrm{d}\chi}{\mathrm{d}\omega}\right)^{\mathrm{S}} - \frac{5}{9}\left(\frac{\mathrm{d}\chi}{\mathrm{d}\omega}\right)^{\mathrm{T}} \right] d^5_{1,0} \right\} \frac{\langle \xi^2 \rangle^{1/2}}{a} + C. \tag{10.6}$$

The two terms in the curly brackets correspond respectively to the three-band and two-band processes with deformation potentials $d^5_{3,0}$ and $d^5_{1,0}$. The spin–orbit splitting is denoted by Δ_1, while χ^+ and χ^- are the contributions to the electronic susceptibility from the E_1 and $E_1 + \Delta_1$ gaps respectively. They are obtained from experimental optical constants and the values for the stress-dependent gaps of fig. 8. The indices S and T refer to contributions from the singlet and triplet gaps. $\langle \xi^2 \rangle^{1/2}$ is the thermal average of the atomic displacement ξ. The lattice constant is a and all the non-resonant terms are included in the constant C. The Raman scattering intensity is, of course, proportional to $|R|^2$. It is clear from eq. (10.6), that in materials with large spin–orbit splitting Δ_1, the prominent term in the Raman tensor is the two-band term, provided that the cancellation of the two-band contributions from S and T is reduced in the presence of the stress. This is the case of InSb, which has a relatively large value for Δ_1 ($\simeq$0.6 eV). In fitting eq. (10.6) to the experimental points, Richter et al. (1976) treat the two deformation potentials of eq. (10.6) as adjustable parameters. In this way they obtain the value of -7 for the ratio $d^5_{3,0}/d^5_{1,0}$. This compares favorably with the value of -4 obtained from pseudo-potential calculations (Richter 1976, Richter et al. 1978b).

Although all the experimental work so far has been done in allowed configurations, it is reasonable to expect stress-induced resonant Raman scattering even in configurations in which intrinsic scattering is forbidden. Such possibilities do exist as long as the symmetry restrictions of the previous section are satisfied. The problem of course becomes more

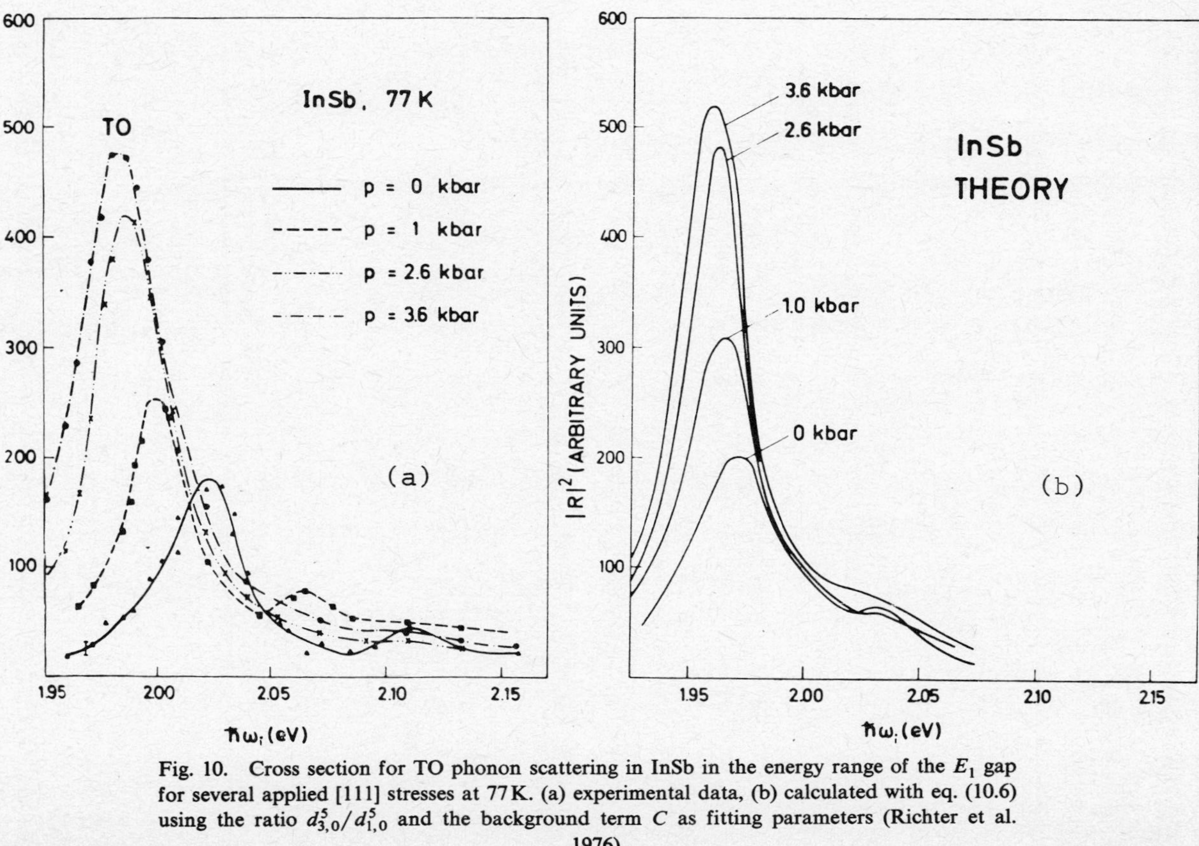

Fig. 10. Cross section for TO phonon scattering in InSb in the energy range of the E_1 gap for several applied [111] stresses at 77 K. (a) experimental data, (b) calculated with eq. (10.6) using the ratio $d^5_{3,0}/d^5_{1,0}$ and the background term C as fitting parameters (Richter et al. 1976).

complicated if forbidden scattering exists already, before even any stress is applied, owing to surface fields or spatial dispersion effects. In any case, eq. (10.6) and similar ones for forbidden configurations are the starting point for linking the phenomenological mode coefficients of the previous section with the deformation potentials and other microscopic parameters appearing in eq. (10.6). Some work in this direction has already been done by Richter (1976) and Richter et al. (1978b).

Besides scattering by the TO phonons, the applied stress also affects the scattering by the LO phonons in InSb. The data of Richter et al. show a clear splitting of the structure in the resonance curve of the LO scattering intensity, equal to the singlet–triplet splitting for the same stress. Such splitting is not observed for TO-phonon scattering, where the major peak in the spectral dependence seems to follow the singlet component of the gap. The data for the LO-phonon scattering are well fitted by theoretical models based on either $\boldsymbol{q}$-dependent intraband Fröhlich interaction between electron–hole pairs or surface electric-field-induced scattering.

(ii) InAs. The spin–orbit splitting of InAs is smaller (224 meV) and the three-band terms are expected to be important in the resonance scattering by the TO phonons. Experiments at 77 K have been performed by Anastassakis et al. (1974). The data show a clear splitting of the resonance structure comparable to that of the singlet–doublet components of the E_1 (fig. 11). The experimental results are fitted by a theoretical curve based on eq. (10.6) to give the value of -2.0 for the ratio of the deformation potentials $d^5_{3,0}/d^5_{1,0}$. No theoretical value of this ratio is available for InAs. The LO-phonon scattering does not show any drastic dependence on the applied stress.

(iii) Ge. Germanium has been studied by Richter et al. (1976, 1978b). As in InAs, the prominent contributions to the Raman tensor are found to arise from three-band terms although the two-band term $(\mathrm{d}\chi/\mathrm{d}\omega)^{\mathrm{S}}$ of eq. (10.6) also appears to participate strongly with increasing stress (spin–orbit splitting $\simeq$0.3 eV). No splitting of the resonance is observed. The ratio of deformation potential is found from the fitting to be -1.5 as compared to the value -5 obtained from pseudopotential calculations (Kane 1969).

(iv) CdS. In a recent piezospectroscopic study of CdS, Briggs and Ramdas (1976) observed a significant decrease with an applied uniaxial stress along $\hat{x}//a$ of the Raman scattering intensity at 77 K of the E_2 phonon at 42 cm^{-1} of CdS (point group C_{6v}). It is known that free and bound excitons play an important part in determining the Raman scattering intensity of CdS under resonance conditions. Since the energy levels of these excitons shift when a uniaxial stress is applied (Rowe et al. 1967), it is expected that the scattering intensity can be changed by tuning the exciton energies to the incident photon energies, i.e. a situation analogous to InSb,

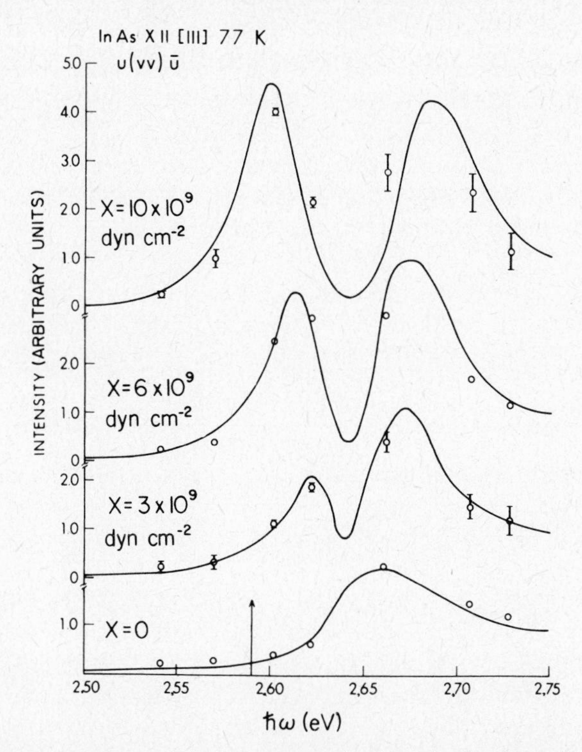

Fig. 11. Spectral dependence of the Raman scattering intensity for allowed TO phonons in InAs at 77 K at various stresses. The experimental values are indicated by circles while the solid lines are the theoretical curves (eq. (10.6)). The energy of E_1 optical gap (2.59 eV) is indicated by an arrow (Welkowski and Braunstein 1972). Representative error bars are shown (Anastassakis et al. 1974).

InAs and Ge discussed above. Such preliminary qualitative arguments have been invoked by Briggs and Ramdas (1976) to explain their observations with the 4880 and 4965 Å incident laser lines.

11. Stress-induced changes in the phonon frequencies

11.1. General

The effect of a uniaxial stress on the mode frequencies and their degeneracies has attracted the interest of many investigators in recent years. First- and second-order spectra have been studied under stress by means of Raman scattering and infrared absorption techniques respectively. A number of theoretical models also appear in the literature, concerning the microscopic treatment of the problem, while Kanamori et al. (1976b) and Matsumoto and Kanamori (1977) have worked out the secular equations for all those cases where an effect linear in the stress is allowed to occur. Considerable work has been done on crystals of the O_h, T_d and D_3 structure. We proceed by first deriving the secular equation associated with the stress-induced changes $\Delta \boldsymbol{K}_j$ of the effective force constants for the triply- and doubly-degenerate modes for O_h and D_3 respectively.

(i) For all triply-degenerate modes $\Gamma(j) = F_{1g}, F_{2g}, F_{1u}, F_{2u}$ of the O_h structure, the mode coefficient $(\partial^3\Phi/\partial\eta\partial u_j\partial u_j)$ exhibits three independent components K_1, K_2 and K_3. By analogy to eqs. (8.1) and (8.2) we find

$$\begin{aligned}\delta\Phi = {} & \tfrac{1}{2}K_1\left(u_x^2\eta_{xx} + \text{c.p.}\right)\\ & + \tfrac{1}{2}K_2\left(u_x^2\eta_{yy} + u_y^2\eta_{xx} + \text{c.p.}\right)\\ & + K_3(u_x u_y \eta_{xy} + u_y u_x \eta_{yx} + \text{c.p.}),\end{aligned} \tag{11.1}$$

and

$$\begin{vmatrix} K_1\eta_{xx} + K_2(\eta_{yy} + \eta_{zz}) - \lambda & 2K_3\eta_{xy} & 2K_3\eta_{xz} \\ 2K_3\eta_{xy} & K_1\eta_{yy} + K_2(\eta_{zz} + \eta_{xx}) - \lambda & 2K_3\eta_{yz} \\ 2K_3\eta_{xz} & 2K_3\eta_{yz} & K_1\eta_{zz} + K_2(\eta_{xx} + \eta_{yy}) - \lambda \end{vmatrix} = 0, \tag{11.2}$$

where $\lambda \simeq 2\omega_j\Delta\omega_j$. The indices x,y,z indicate the three crystallographic axes [100], [010], [001] respectively. Also, eq. (3.12) gives for the shifted frequency

$$\Omega = \omega_0 + \Delta\omega_0 \simeq \omega_0 + \lambda/2\omega_0, \tag{11.3}$$

where we have used the usual notation from the literature, i.e. $\omega_0 \equiv \omega_j$. Depending on the direction of the applied stress, different components $\eta_{\mu\nu}$ enter eq. (11.2). The number of independent values of λ which one finds by solving eq. (11.2) will determine the new frequencies Ω from eq. (11.3), and thereby the extent of degeneracy removal. The new mode eigenvectors are also found from eq. (11.2).

(ii) The effect of a homogeneous strain on the doubly-degenerate mode E of the D_3 structure requires four independent components K_1, K_2, K_3, K_4 for the mode coefficient $(\partial^3\Phi/\partial u_i \partial u_j \partial\eta)$ as seen from the reduction

$$[E\otimes E]_S \otimes \Gamma(\boldsymbol{\eta}) = (A_1 \oplus E)\otimes(2A_1 \oplus 2E) = 4A_1 \oplus \ldots. \tag{11.4}$$

The appropriate term in the potential energy can be written as

$$\begin{aligned}\delta\Phi = {} & \tfrac{1}{2}K_1(u_x^2+u_y^2)(\eta_{xx}+\eta_{yy}) \\ & +\tfrac{1}{2}K_2(u_x^2+u_y^2)\eta_{zz} \\ & +\tfrac{1}{2}K_3\left[(u_x^2-u_y^2)(\eta_{xx}-\eta_{yy})+(2u_xu_y)(2\eta_{xy})\right] \\ & +\tfrac{1}{2}K_4\left[(u_x^2-u_y^2)\eta_{yz}+(2u_xu_y)\eta_{xz}\right].\end{aligned} \tag{11.5}$$

The first two terms are based on $A_1\otimes A_1 = A_1$, the third and fourth on $E\otimes E = A_1 \oplus \ldots$. The secular equation becomes

$$\begin{vmatrix} (K_1+K_3)\eta_{xx}+(K_1-K_3)\eta_{yy}+K_2\eta_{zz}+K_4\eta_{yz}-\lambda & 2K_3\eta_{xy}+K_4\eta_{xz} \\ 2K_3\eta_{xy}+K_4\eta_{xz} & (K_1-K_3)\eta_{xx}+(K_1+K_3)\eta_{yy}+K_2\eta_{zz}-K_4\eta_{yz}-\lambda \end{vmatrix} = 0. \tag{11.6}$$

Specific applications of eqs. (11.2) and (11.6) will be discussed next in connection with experimental work.

11.2. Experimental

The first example of Raman experiments under uniaxial stress goes back to the pre-laser period of Raman spectroscopy and refers to work on α-quartz by Mariée and Mathieu (1946). In recent years a number of materials have been studied under uniaxial stress, in particular Si (Anastassakis et al. 1970), Ge, III–V and II–VI semiconductors (Cerdeira et al. 1972), α-quartz (Harker et al. 1970, Tekippe and Ramdas 1971, Tekippe et al. 1973, Briggs and Ramdas 1977), CaF_2, BaF_2 and $Bi_{12}GeO_{20}$ (Venugopalan and Ramdas 1973), TiO_2 (Peercy 1973), NbO_2 (Pollak 1973), Ti_2O_3 and Al_2O_3 (Shin et al. 1976), gadolinium molybdate (Ullman et al. 1976), CdS (Briggs and Ramdas 1976), Mg_2Si and Mg_2Sn (Onari et al. 1977), TeO_2 (Lemos et al. 1977), and diamond (Grimsditch et al. 1978). The second-order infrared absorption spectrum of Si under stress has also been reported (Lüth 1970, Angress et al. 1975). We review these materials in groups, according to their crystallographic classes.

11.2.1. Diamond-type crystals (O_h symmetry)

For a uniaxial stress $\boldsymbol{X}$ applied along either the [001] or the [111] direction one finds, by diagonalizing the secular equation (11.2), that the three-fold degeneracy of the $\boldsymbol{q}\simeq 0$ F_{2g}-type optical phonon is split into a singlet ($\Omega_3=\Omega_s$) with eigenvector parallel to the stress and a doublet ($\Omega_2=\Omega_1=\Omega_d$) with eigenvectors perpendicular to the stress. There is also a shift in the frequency of the optical phonons to $\Omega_0=\omega_0+\Delta\omega_h$ due to the hydrostatic component of the applied stress. One finds

$$\Omega_s=(\omega_0+\Delta\omega_h)+\frac{2}{3}\Delta\Omega=\Omega_0+\frac{2}{3}\Delta\Omega, \tag{11.7}$$

$$\Omega_d=(\omega_0+\Delta\omega_h)-\frac{1}{3}\Delta\Omega=\Omega_0-\frac{1}{3}\Delta\Omega, \tag{11.8}$$

$$\begin{aligned}\Delta\omega_h&=\Omega_0-\omega_0=(\Omega_s-\omega_0)-\frac{2}{3}\Delta\Omega\\ &=\frac{X}{6\omega_0}(K_1+2K_2)(s_{11}+2s_{12}),\end{aligned} \tag{11.9}$$

$$\Delta\Omega=\Omega_s-\Omega_d=\begin{cases}\dfrac{X}{2\omega_0}(K_1-K_2)(s_{11}-s_{12})\equiv\Delta\Omega^{(001)}; \quad \mathbf{X}/\!/[001] & (11.10)\\[2ex] \dfrac{X}{2\omega_0}K_3 s_{44}\equiv\Delta\Omega^{(111)}; \quad \mathbf{X}/\!/[111], & (11.11)\end{cases}$$

where X is the (negative) compressive uniaxial stress; s_{11}, s_{12} and s_{44} are the elastic compliance constants referred to the cubic axes. The suppressed index notation has been used for the stress–strain relationship (Nye 1964). It follows that an experimental determination of Ω_s (or Ω_d) and $\Delta\Omega$ for the two directions of stress allows one to determine K_1, K_2, K_3 and $\Delta\omega_h$. The consistency of the results can be checked by comparing the two values of (K_1+2K_2) which are obtained from eq. (11.9) for the two directions of stress.

Complete removal of degeneracy is achieved when the stress is applied along [110]. From eq. (11.2) one finds

$$\Omega_1=\omega_0+\Delta\omega_h-\tfrac{1}{3}\Delta\Omega^{(001)}, \tag{11.12}$$

$$\Omega_2=\omega_0+\Delta\omega_h+\tfrac{1}{6}\Delta\Omega^{(001)}+\tfrac{1}{2}\Delta\Omega^{(111)}, \tag{11.13}$$

$$\Omega_3=\omega_0+\Delta\omega_h+\tfrac{1}{6}\Delta\Omega^{(001)}-\tfrac{1}{2}\Delta\Omega^{(111)}. \tag{11.14}$$

The samples of Si used by Anastassakis et al. (1970) were oriented along $\hat{x}_1 /\!/ [110]$, $\hat{y}_1 /\!/ [\bar{1}10]$, $\hat{z}_1$ (long axis)$/\!/[001]$ and $\hat{x}_2 /\!/ [11\bar{2}]$, $\hat{y}_2 /\!/ [\bar{1}10]$ and $\hat{z}_2$ (long axis)$/\!/[111]$. Similar orientations were used for Mg_2X (X=Si, Sn) by Onari et al. (1977), for diamond (Grimsditch et al. 1978), and for Ge by Cerdeira et al. (1972) plus a third one oriented along $\hat{x}_3 /\!/ [1\bar{1}0]$, $\hat{y}_3 /\!/ [00\bar{1}]$ and $\hat{z}_3$ (long axis)$/\!/[110]$. In all cases the uniaxial stress was applied along the $\hat{z}_i$ axis at 300 K while the incident and scattered radiation were along $\hat{x}_i$ whenever Raman scattering techniques were used.

Raman scattering is a particularly useful technique for observing the splitting of the frequencies of the $\boldsymbol{q} \simeq 0$ optical phonons since one can make use of the polarization selection rules to selectively observe scattering by any one of the split components. Since the strain-induced changes in the Raman tensor are relatively small the scattering is to a good approximation determined by the stress-free first-order Raman scattering tensor. On transforming the Raman scattering tensor to the $x_1 y_1 z_1$ axes one finds that for the $x_1 y_1 z_1$ samples (uniaxial stress along $\hat{z}_1$) the singlet mode (eigenvector along $\hat{z}_1$) is observed in the configuration $\bar{x}_1(y_1 y_1)x_1$, whereas one of the doublet modes (eigenvector along $\hat{y}_1$) is observed with the same intensity in the configuration $\bar{x}_1(z_1 y_1$ or $y_1 z_1)x_1$. Similarly, one finds that for $x_2 y_2 z_2$ samples (with uniaxial stress along $\hat{z}_2$) the singlet mode (eigenvector along $\hat{z}_2$) is observed in the configuration $\bar{x}_2(z_2 z_2)x_2$, whereas one of the doublet modes (eigenvector along $\bar{y}_2$) is observed, with one-fourth of the intensity of the singlet mode, in the configuration $\bar{x}_2(z_2 y_2$ or $y_2 z_2)x_2$. Thus, switching the polarization of the incident radiation between $\hat{z}_1$ and $\hat{y}_1$ (or between $\hat{z}_2$ and $\hat{y}_2$) while leaving the polarizer for the scattered radiation fixed along $\hat{y}_1$ (or $\hat{z}_2$) allows one to selectively observe either the singlet or the doublet component.

(i) Silicon. In Si the singlet and doublet bands were recorded during the same run by operating the Raman spectrometer in a step-scanning mode and using photon counting (Anastassakis et al. 1970). The polarization of the incident laser beam was switched back and forth between directions parallel and perpendicular to the direction of the stress. In this way it became possible to successively measure the scattering intensity from the singlet and doublet component for each position of the grating. This reduced considerably the frequency reproducibility errors. The switching of the polarization was accomplished by inserting a $\lambda/2$ plate into the path of the incident beam by means of a mechanical device electronically coupled to the grating drive system. Typical spectra for the $x_2 y_2 z_2$ sample are shown in fig. 12. The Raman peak in the absence of strain occurs at $\omega_0 = 523.0 \pm 0.2$ cm^{-1} and its half width is 3.7 cm^{-1}. No change in peak width with applied stress was observed. The data of fig. 12 show that for a

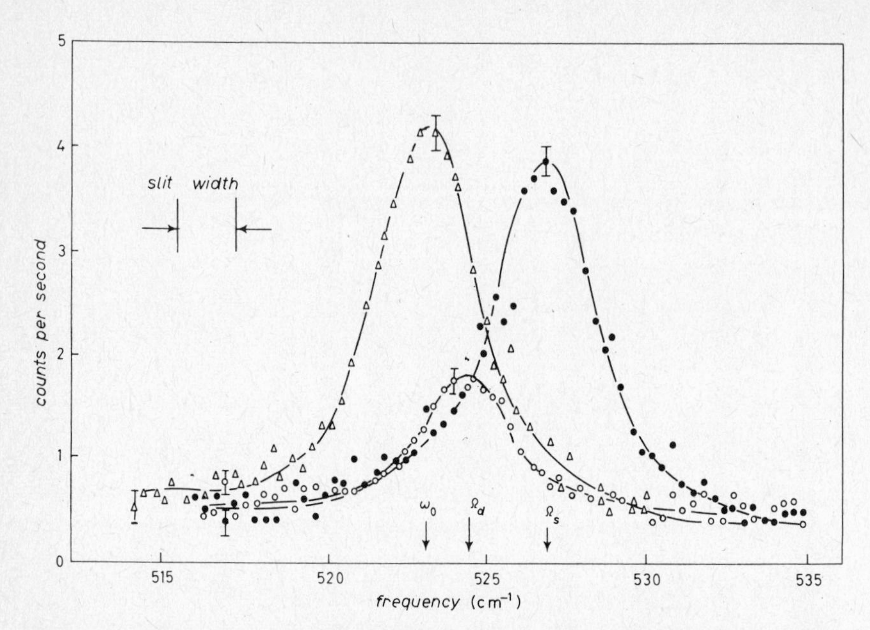

Fig. 12. The first-order Raman spectrum of Si at 300 K with a uniaxial stress of -11.5×10^9 dyn/cm^2 applied along $\hat{z}_2//[111]$. A He–Ne excitation laser line was used: $\triangle$: $x_2(z_2z_2)\bar{x}_2$, $X=0$; $\circ$: $x_2(y_2z_2)\bar{x}_2, X\neq 0$, singlet; $\bullet$: $x_2(z_2z_2)\bar{x}_2$, $X\neq 0$, doublet (Anastassakis et al. 1970).

negative uniaxial stress along [111] the splitting $\Delta\Omega = \Omega_s - \Omega_d$ is positive. In the case of the $x_1y_1z_1$ orientation (negative uniaxial stress along [001]) the observed sign of $\Delta\Omega$ is negative. After the necessary corrections were made, the ratio of the singlet to doublet peak intensity was found to be four, in agreement with the first-order selection rules. In the case of the $x_1y_1z_1$ orientation, the singlet and doublet peaks were found to have the same intensities as expected. From the slopes of the curves of Ω_s, Ω_d and $\Omega_0 = \omega_0 + \Delta\omega_h$ vs. X for the two orientations, and the values of the elastic compliance constants the following values (within 20%) were obtained for the relevant mode coefficients (in units of 10^{28} sec^{-2}): $K_1 = -1.2$, $K_2 = -1.8$ and $K_3 = -0.63$ (the values for K_1 and K_2 are obtained directly from the data for $X//[001]$). The data for the two orientations yield values of -4.7 and -5.9 for the quantity $(K_1 + 2K_2)$ which determines the shift in frequency with hydrostatic pressure, indicating a self-consistency of about 25%. These values for the coefficients agree in sign with the theoretical values $K_1 = -0.4$, $K_2 = -1.0$ and $K_3 = -1.5$ which Ganesan et al. (1970) obtained. In their quasi-harmonic approximation these authors consider only bond-stretching forces between nearest neighbors and a Morse potential for their interactions. The experimental data also provide a value of the mode Grüneisen parameter at constant temperature for the $\boldsymbol{q}\simeq 0$ optical

Table 1

Experimental results from stress-induced shifts and splittings of the $q \simeq 0$ optical phonons in Si, Ge, diamond and anti-fluorite structures at 300 K. Listed in parentheses are the values of γ determined from hydrostatic pressure measurements.

Material	ω_0^2 (10^{28} sec^{-2})	$\frac{K_1 - K_2}{2\omega_0^2}$	$\frac{K_3}{\omega_0^2}$	γ
Si[a]	0.970	0.31±0.06 0.59[p]	−0.65±0.13 −0.58[p]	0.90±0.18 (1.02±0.02)[c] (0.98±0.06)[d] (1.06),[e] 1.20[p]
Ge[b]	0.319	0.23±0.02 0.30[p]	−0.87±0.09 −0.63[p]	0.89±0.09 (1.12±0.02)[c] (1.29),[f] (1.23)[g] (1.13),[e] 1.26[p], 0.88±0.08[s]
Diamond[h]	6.323	−0.51 1.19[p]	−1.20 −1.16[p]	1.12±0.20, 1.26[p] (0.94±0.10)[i] (1.19±0.09)[j] (1.20±0.10)[k] (0.96±0.10)[q] (0.98±0.04)[r]
Mg_2Si[l](F_{2g})	0.237	–	−0.83±0.07	1.26±0.04 (1.36±0.02)[c]
Mg_2Si[l][F_{1u}(LO)]	0.418	–	−0.57±0.07[m] −0.50±0.06[n]	(1.06±0.06)[o]
Mg_2Sn[l](F_{2g})	0.174	0.265±0.03	−0.78±0.07	1.08±0.07 (1.34±0.08)[c]

[a]Anastassakis et al. (1970).
[b]Cerdeira et al. (1972).
[c]Buchenauer et al. (1971).
[d]Weinstein and Piermarini (1975).
[e]Jex (1971).
[f]Bienenstock (1964).
[g]Dolling and Cowley (1966).
[h]Grimsditch et al. (1978), best fit.
[i]Mitra et al. (1969).
[j]Parsons and Clark (1976), and Parsons (1977).
[k]Drickamer et al. (1966).
[l]Onari et al. (1977).
[m]From measurements of the 2LO peak.
[n]From measurements of the LO peak.
[o]From hydrostatic measurements of the 2LO peak.
[p]Theoretical results (corrected) of Bell (1972).
[q]Holzapfel and Anastassakis (1974).
[r]Whalley et al. (1976).
[s]Asaumi and Minomura (1978).

phonon. This is given by

$$\gamma = -\left(\frac{\partial \ln \omega_0}{\partial \ln V}\right)_T = -\left(\frac{K_1 + 2K_2}{6\omega_0^2}\right), \tag{11.15}$$

where V is the volume of the crystal. Using the average of the two values for $(K_1 + 2K_2)$ for the two orientations one obtains $\gamma = +0.9$. This value of γ for Si is essentially the same as the value 1.02 which Buchenauer et al. (1971) obtained from measurements with hydrostatic pressure. All these experimental results are summarized in table 1.

(ii) Germanium. In Ge the recording of each split band was taken independently using photon counting (Cerdeira et al. 1972). The lines of a low-pressure neon lamp were used for calibration. The same configurations were employed as for Si. In addition, the split bands Ω_1 and Ω_2 were observed with $x_3y_3z_3$ samples in the configurations $\bar{x}_3(z_3z_3)x_3$ and $\bar{x}_3(z_3y_3$ or $y_3z_3)x_3$ respectively. None of the scattering geometries which are possible with the $x_3y_3z_3$ samples allows the Ω_3 component to be observed. From the slopes of the $\Omega_{1,2}$ vs. X curves and the values for the elastic compliances the following values were obtained for the relevant mode coefficients (in units of 10^{+28} sec^{-2}) and the mode Grüneisen parameter: $K_1 = -0.47, K_2 = -0.62, K_3 = -3.47, \gamma = 0.89$ (see also table 1). The theoretical values of Ganesan et al. (1970) are $K_1 = -0.12$, $K_2 = -0.33$ and $K_3 = -0.48$. The experimental value of γ obtained from hydrostatic pressure measurements (Buchenauer et al. 1971) is 1.12. The values for γ obtained in Si and Ge from stress measurements are systematically lower than those obtained from hydrostatic pressure measurements. According to Cerdeira et al. (1972) this is due to a slight relaxation of the applied uniaxial stress in the surface region within the penetration depth. On the other hand, the discrepancy between the experimental and theoretical values for the mode coefficients $K_1 K_2, K_3$ is attributed to inadequacy of the theoretical model in which, as stated earlier, only bond-stretching forces are considered between neighboring atoms. Inclusion of third-nearest-neighbor interactions as suggested by Cerdeira et al. (1972) improves the agreement between experimental and theoretical values for K_1 and K_2, but leaves K_3 unaffected. Martin (1970) has proposed an independent model whereby two-body stretching and an average of three-body bending interactions are considered. Even so, no satisfactory agreement with the experiment is achieved. A generalized force constant model has also been used which includes two-, three- and four-body interactions. Experimental values for these constants are available. Their inclusion results in an improvement of the theoretical values for the K_1 and K_2 coefficients but does not yield any information for K_3 (Cerdeira et al. 1972). Finally, a model used by Keating

(1966) in the calculation of the second- and third-order elastic constants has been shown by Bell (1972) to be appropriate for the description of stress-induced frequency shifts and splittings of $\boldsymbol{q}\simeq 0$ optical phonons. The results of a calculation for Si and Ge are in satisfactory agreement with the experiment (table 1).*

(iii) Diamond. The shift and splitting of the F_{2g} phonon of diamond was measured recently under [111] and [001] stresses by Grimsditch et al. (1978) at 300 K. The results under a [111] stress indicate a positive shift at the rate of 0.29 cm^{-1}/kbar for the singlet and 0.05 for the doublet. The corresponding rates for a [001] stress are 0.16 and 0.09 respectively. The values of K_1, K_2, K_3 and γ can be obtained from table 1. The corresponding theoretical values of Ganesan et al. are $K_1 = -7.71$, $K_2 = -5.26$ and $K_3 = -1.07$. The experiment was performed in the $x_2(z_2z_2)y_2$ and $x_2(y_2z_2)y_2$ configurations for the singlet and doublet components respectively under [111] stress, and in the $x(zz)y$ and $x(yz)y$ configurations for the singlet and doublet components under [001] stress (x,y,z correspond to the three cubic axes). The results were analyzed by use of the three-parameter valence-field-force model of Bell (1972). From this analysis, a complete set of values was obtained for the third-order elastic constants, for which only incomplete data are available (McSkimin and Andreatch 1972).

11.2.2. *Fluorite structures (O_h symmetry)*

Venugopalan and Ramdas (1973) have carried out a complete investigation of the effect of a uniaxial stress on the frequency and degeneracy of the $\boldsymbol{q}\simeq 0$ Raman-active phonon of CaF_2 and BaF_2. The point group of these materials is O_h and the symmetry of the phonon is F_{2g}. Venugopalan and Ramdas analyze the symmetry aspects of the problem using a phenomenological Hamiltonian linear in the strain (Kaplyanskii 1964, Rodriguez et al. 1972). Three deformation potential constants a,b,c are needed to express the effect of strain on the phonon frequency, by analogy

*Several computational errors have been discovered in this work (Bell, private communications with the author). Eq. (11) and (13) should be changed as follows:

$$\gamma = -\frac{5\alpha - \beta + a_0(3\bar{\gamma} - \bar{\delta} + \bar{\varepsilon})}{6(\alpha+\beta)}, \tag{11}$$

$$\frac{r}{\omega_0^2} = \frac{\alpha + a_0\bar{\gamma}(1-\zeta)}{\alpha+\beta}. \tag{13}$$

The results of the correct calculation (shown in table 1) are still in agreement with the experimental data.

to our mode coefficients K_1, K_2, K_3 introduced earlier. The secular equation is identical to that for diamond-type crystals given by eq. (11.2). To convert the parameters $\lambda, a, b, c, \ldots$ used by Venugopalan and Ramdas (1973 and references therein) to the corresponding parameters $\lambda, K_1, K_2, K_3, \ldots$ used here, one has to multiply the former by $2\omega_0$. This conversion applies to all values of $a, b, c, \ldots$ quoted in this chapter. Again, it was assumed that the effect of the strain on the Raman tensor was negligible and thus the intrinsic Raman tensors were used to apply the polarization selection rules. The choice of intrinsic Raman tensors was dictated by the gross symmetry reduction of each mode within the new symmetry point group that describes the crystal in the presence of a stress (see relevant discussion in §5.3). The same three sample orientations $x_i y_i z_i$ $(i = 1, 2, 3)$ were used for the measurements, as those defined in §11.2.1. Since the materials are transparent the experiments were not limited to back scattering. This allowed the bulk rather than the surface effects of the applied stress to be observed. Direct recordings were sufficiently accurate to monitor the induced shifts and splittings. For the calibration of the spectra a potassium discharge lamp and a neodymium hollow-cathode lamp were used for CaF_2 and BaF_2 respectively. The measurements were taken at 15 K to obtain an overall better resolution due to the sharpening of the lines at lower temperatures. The intensity ratios were found to be in qualitative agreement with the theoretical ones. From their results, Venugopalan and Ramdas were able to calculte the rate of hydrostatic shift at 15 K as $-0.64\ \mathrm{cm}^{-1}/\mathrm{kbar}$ for CaF_2 and $-0.72\ \mathrm{cm}^{-1}/\mathrm{kbar}$ for BaF_2. These values are in close agreement with the room-temperature values of Mitra, as quoted by Ferraro et al. (1971). In table 2 we give the values of deformation potential coefficients a, b, c for CaF_2, BaF_2 and for $Bi_{12}GeO_{20}$ (to be discussed shortly). No theoretical values for a, b, c have appeared in the literature thus far, although extensive work has been reported concerning various aspects of lattice dynamics in these materials, in particular second-order Raman activity (Krishnamoorthy and Soots 1972), inelastic neutron scattering (Elkombe and Pryor 1970, Hurrell and Minkiewicz 1970), and mode force constants (Srinivasan 1958, Axe 1965).

The anti-fluorite structures Mg_2Si and Mg_2Sn were investigated recently by Onari et al. (1977) under uniaxial stress along [001] and [111] at 300 K. It was not possible to apply sufficiently high stress along [001] on Mg_2Si because of its low yield strength, and thus values for K_1 and K_2 were not obtained for this material. Neither of the other two II–IV semiconductors (Mg_2Ge, Mg_2Pb) was investigated under stress, although extensive work has been done on their Raman scattering properties (Anastassakis and Burstein 1971c, Onari and Cardona 1976). The stress-induced shifts and splittings of the F_{2g} Raman-active phonons of Mg_2Si and Mg_2Sn were

Table 2

Deformation potential constants in cm^{-1} per unit strain of Raman lines of CaF_2, BaF_2 and $Bi_{12}GeO_{20}$ at 15 K (Venugopalan and Ramdas 1973), of CdS at 77 K (Briggs and Ramdas 1976) and of α-quartz at liquid helium temperature (Briggs and Ramdas 1977).

Material	Line positions (cm^{-1})	Phonon symmetry	a	b	c	d	e	$\lvert f \rvert$	g
CaF_2	327.3	F_{2g}	−1287	−271	−103				
BaF_2	246.6	F_{2g}	−785	−281					
$Bi_{12}GeO_{20}$	124.0		−467	−165	−106	∼0			
	57.5	F	−295	−251	24	∼0			
	207.2, 209.0		−251	−454	84	∼0			
	87.2						−139	70	
	67.7	E					−250	87	
	130.2						−226	47	
	89.4	A							−210
CdS	41.8	E_2	∼96	∼104	16^b				
	255.7		−404	−483	107				
	242.6	E_1(TO)	−235	−330	57				
	306.9	E_1(LO)							
	234.7	A_1(TO)	−526	−328					
	303.6	A_1(LO)							
α-quartz[a]	205.		−800	−900					
	354.	A_1	55	−100					
	464.		−255	−470					
	1081.		0	−220					
	128.		−255	−150	175	−205			
	263.		−195	−160	−75	245			
	695.	E	−435	−200	−210	325			
	1160.		60	205	−140	435			
	393–403		−70	55	220	−500			
	450–509		−325	−300	−140	365			
	796–809	E	−370	−575	245	−485			
	1064–1231		−5	−275					

[a]Typical uncertainties, ± 20 cm^{-1} per unit strain
[b]Upper limit for $|c|$.

found to be similar to those of Si and Ge. In addition, the Raman-inactive F_{1u}(LO) phonon of Mg_2Si was observed to shift linearly with stress at nearly the same rate as the stress-split doublet of the F_{2g} phonon. Recall that forbidden scattering due to the F_{1u}(LO) phonon and its overtone has been observed and attributed to surface electric field effects and/or spatial dispersion effects (see §§7.2.1 and 18.2). By definition this LO phonon is non-degenerate. Furthermore, because of the scattering configuration it propagates perpendicularly to the stress. Therefore it is a singlet component as far as the macroscopic field-induced LO–TO splitting is concerned, and it is a doublet component as far as the stress-induced splitting is concerned (see also next section). The rate of shift of its overtone frequency was also measured and found to be twice that of the fundamental frequency. The experimental results are included in table 1. A conclusive theoretical interpretation of these results was not possible since insufficient information was available on various microscopic lattice-dynamical parameters of these materials.

11.2.3. Non-centrosymmetric cubic structures

A number of non-centrosymmetric cubic materials have been investigated following procedures similar to those discussed above. The main difference from the centrosymmetric cubic materials is that the triply-degenerate Raman-active modes are also infrared active. This results in a TO–LO splitting which is due to the long-range Coulomb interactions. The stress dependence of the LO and TO frequencies in this case can be simply obtained after the LO and TO components have been identified for a given experimental configuration. Thus, for a stress applied along $\hat{z}_3 /\!/ [111]$ and with a $x_3 y_3$ scattering plane the LO phonon will lie on the $x_3 y_3$ plane. This means that the LO phonon can be only one of the two components of the doublet of eq. (11.8). In other words, the doublet of eq. (11.8) is now intrinsically split by the long-range Coulomb forces. The applied stress will shift both of them uniformly, assuming their LO–TO splitting is small (Cerdeira et al. 1972). On the other hand, the singlet component of eq. (11.7) being parallel to $\boldsymbol{X}$ can only be a transverse component. Thus, we can rewrite eqs. (11.7) and (11.8) as follows:

$$\Omega_{\mathrm{LO}} = \Omega_{\mathrm{d}/\!/} = (\omega_{\mathrm{LO}} + \Delta\omega_{\mathrm{h}}) - \tfrac{1}{3}\Delta\Omega, \tag{11.16}$$

$$\Omega_{\mathrm{TO}\perp} = \Omega_{\mathrm{d}\perp} = (\omega_{\mathrm{TO}} + \Delta\omega_{\mathrm{h}}) - \tfrac{1}{3}\Delta\Omega, \tag{11.17}$$

$$\Omega_{\mathrm{TO}/\!/} = \Omega_{\mathrm{s}/\!/} = (\omega_{\mathrm{TO}} + \Delta\omega_{\mathrm{h}}) + \tfrac{2}{3}\Delta\Omega, \tag{11.18}$$

where $\omega_{\mathrm{LO}}, \omega_{\mathrm{TO}}$ are the stress-free LO and TO frequencies. The subscripts $/\!/, \perp$ designate phonon polarizations parallel and perpendicular to the applied stress respectively. $\Delta\omega_{\mathrm{h}}$ and $\Delta\Omega$ are given by their expressions in eqs. (11.9) and (11.10), (11.11) respectively. If the stress dependence of the LO–TO splitting becomes significant the same expressions for $\Delta\Omega$ and $\Delta\omega_{\mathrm{h}}$ are applicable but with a different set of force constant coefficients K_1, K_2, K_3 for the TO and LO components. Expressions analogous to those of eqs. (11.16)–(11.18) can be worked out for $\boldsymbol{X} /\!/ [001]$ and $\boldsymbol{X} /\!/ [110]$.

(i) *III–V and II–VI semiconductors.* Cerdeira et al. (1972) have investigated GaAs, GaSb and InAs of the III–V family and also ZnSe of the II–VI family. The point group is $\mathrm{T_d}$ and the symmetry of the Raman-

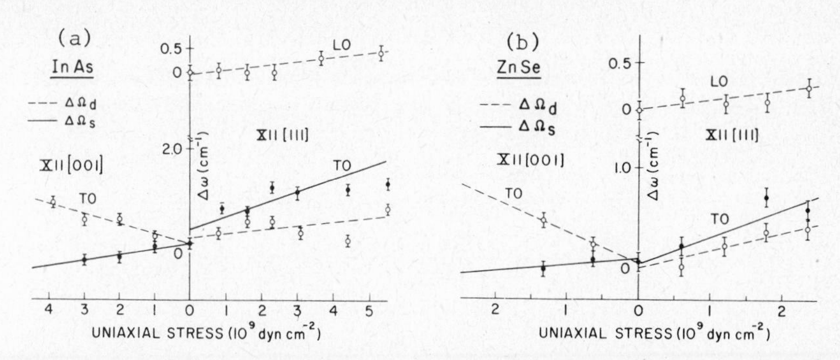

Fig. 13. Stress dependence of the TO and LO phonon frequencies in InAs(a) and ZnSe(b) for stress along the [111] and [001] directions. The solid and dashed lines represent a linear least-square fit. For $X//[001]$ with the sample configuration used in this experiment ([1$\bar{1}$0] face) scattering from the LO phonon is forbidden. In contrast to the zincblende-type materials investigated the stress dependence of the LO phonon of ZnSe is not equal to that of the TO phonon (Cerdeira and Cardona 1972).

active infrared-active phonon is F_2. It can be shown that the same secular equation (11.2) applies here as well. The sample orientations and experimental techniques are the same as those for Ge described earlier. It was found that eqs. (11.16)–(11.18) were applicable to all the compounds with the exception of ZnSe. Fig. 13 shows the observed linear dependence of the three components $\Omega_{LO}, \Omega_{TO\perp}, \Omega_{TO\parallel}$ for $X\parallel[111]$ and $X\parallel[001]$ for InAs and ZnSe. One can clearly see the parallel shift of the Ω_{LO} and $\Omega_{TO\perp}$ components in the case of InAs, and their independent shift in the case of ZnSe. The latter is believed to arise from a pronounced dependence on the stress of the effective charge associated with the mode. This suggests that the infrared spectra of ZnSe would probably show a dependence on an applied stress.

From their data, Cerdeira et al. estimate the mode Grüneisen parameters and the values for K_1, K_2 and K_3 for the materials investigated. This information is contained in table 3 together with the results for InSb. The latter were obtained from the study of stress-induced resonant Raman scattering (Anastassakis et al. 1972c). Theoretical values for GaAs and InSb are also included in table 3 (Bell 1972).

In an independent effort to study the bulk effect of a uniaxial stress on the TO, LO components of GaAs, Weinstein and Cardona (1972) measured its restrahlen spectrum under uniaxial stresses along [001] and [111]. The symmetry analysis of the shifts and splittings of the F_2-type phonon is the same as before. Because of the longer wavelengths used in these infrared reflectance measurements the corresponding absorption length (2200 Å at ω_{TO}) is considerably larger than that in the Raman experiments (900 Å).

Table 3

Experimental results from stress-induced shifts and splittings of the $q \simeq 0$ transverse optical phonons in non-centrosymmetric cubic semiconductors at 300 K (Cerdeira et al. 1972). Listed in parentheses are the values of γ from hydrostatic pressure measurements.

Material	ω_0^2 (10^{28}sec^{-2})	$\frac{K_1-K_2}{2\omega_0^2}$	$\frac{K_3}{\omega_0^2}$	γ
GaAs	0.256	0.1 ± 0.1	-0.2 ± 0.2	0.90 ± 0.30
		0.35[a]	−0.46[a]	1.23[a]
				(1.34 ± 0.08)[b]
GaAs[c]		0.30 ± 0.15[d]	-0.8 ± 0.4[e]	0.7 ± 0.3[d]
				1.00 ± 0.3[e]
		0.40 ± 0.25[f,d]	-0.8 ± 0.3[f,e]	0.9 ± 0.5[f,d]
				0.90 ± 0.25[f,e]
GaAs[c,g]	0.301	0.35 ± 0.08[d]	-0.60 ± 0.15[e]	0.80 ± 0.15[d]
				1.00 ± 0.15[e]
GaSb	0.184	0.22 ± 0.04	-1.08 ± 0.2	1.10 ± 0.22
				(1.23 ± 0.02)[b]
InAs	0.169	0.57 ± 0.12	-0.76 ± 0.15	0.85 ± 0.13
InSb[h]	0.119	0.42 ± 0.15	-0.60 ± 0.25	1.50 ± 0.20
		0.84[a]	−0.43[a]	1.05[a]
ZnSe	0.148	0.62 ± 0.19	-0.43 ± 0.12	1.80 ± 0.36
				(1.7)[i]

[a]Theoretical results (corrected) of Bell (1972).
[b]Buchenauer et al. (1971).
[c]Weinstein and Cardona (1972).
[d]From $X /\!/ [001]$.
[e]From $X /\!/ [111]$.
[f]Doped samples.
[g]Refers to measurements of the LO phonon.
[h]Anastassakis et al. (1972c) at 80 K.
[i]Mitra et al. (1969).

This means that the surface stress relaxation effects should be less important. Thus, the values of the coefficients K_1, K_2, K_3 and γ should be larger and, generally speaking, more representative of the bulk response of the material. Because of the broad profile of the reststrahlen spectrum in the region of ω_{TO} it became necessary to use doped materials ($N = 5 \times 10^{18}$ electrons/cm^3). Under these circumstances the reflectivity band exhibits a much sharper dip in the region of ω_{TO} and thus it becomes possible to follow the frequency ω_{TO} more accurately as a function of the stress. In the same work the LO component was also studied, in undoped samples only. This possibility does not exist in Raman scattering because of the more restrictive selection rules. Thus a set of values K_1', K_2', K_3' and γ' was

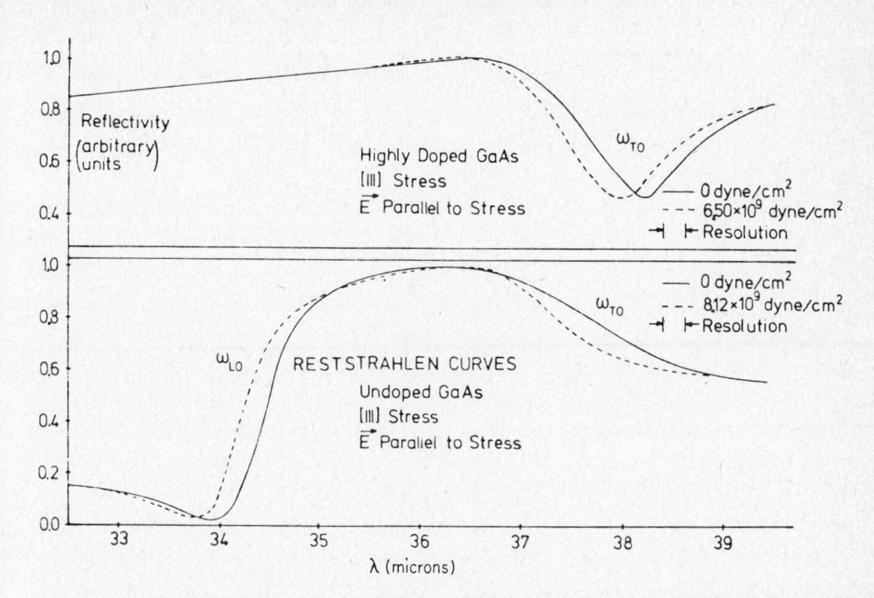

Fig. 14. Reflection spectra of an undoped and a heavily doped ($N = 5 \times 10^{18}$ electrons/cm^3) GaAs sample in the reststrahlen region at room temperature for zero stress and for a high stress along [111] (Weinstein and Cardona 1972).

obtained for the LO phonon. No meaningful comparison can be made, however, between these values and the ones for the TO phonon because of the large experimental errors. The numerical results are incorporated in table 3, while fig. 14 shows some typical reststrahlen spectra with and without a uniaxial stress.

(ii) *Bismuth–germanium-oxide ($Bi_{12}GeO_{20}$).* The effects of a uniaxial stress on the Raman spectrum of $Bi_{12}GeO_{20}$ have been studied by Venugopalan and Ramdas (1973). This material crystallizes with the point group T and is well known for its unusual optical and electro-optical properties (Greenwald and Anastassakis 1975 and references therein, Lenzo et al. 1966, Onoe et al. 1967). Its intrinsic first-order Raman spectrum is expected to consist of eight $\boldsymbol{q} \simeq 0$ optical phonons of symmetry A, eight doubly-degenerate (E) and twenty-four triply-degenerate (F). The F-type modes are also infrared active and thus an LO–TO splitting may appear in the Raman spectrum depending on scattering configuration. A total of thirty-six lines in the region 40–720 cm^{-1} was observed and identified by Venugopalan and Ramdas (1972) at 15 K.

Following the same perturbation approach as in the case of CaF_2 and BaF_2, Venugopalan and Ramdas (1973) obtained the secular equations for all three types of phonons of symmetry F, E, A. For each F-type phonon it is shown that one needs four independent deformation potentials a, b, c, d for the description of its shifts and splittings. Similarly, three independent deformation potentials e, u, v (or $e, f \equiv u + iv$) are needed for the description of the shifts and splittings of each E-type mode, and only one (g) for the shift of each non-degenerate totally symmetric mode A.

The experimental procedure for observing and identifying the split bands was similar to that followed for the CaF_2 and III–V materials. The LO–TO splitting for the F-type modes was found to be independent of the stress, thus revealing a rather weak stress dependence of the associated effective charge. It is interesting that application of the far more complex and sensitive polarization selection rules in the presence of a stress enabled these workers to identify independently and/or confirm the symmetry assignments of the phonons observed. From their measurements and the room-temperature values for the compliance coefficients (Onoe et al. 1967) numerical values were obtained for the sets (a, b, c, d), $(e, |f|)$ and g, for some prominent phonons of symmetry F, E, and A respectively. These values are listed in table 2. It was not possible to estimate independent values for u and v.

11.2.4. Uniaxial structures

Only a limited number of uniaxial structures have been studied in the presence of a stress by means of Raman scattering techniques. Here the maximum degeneracy of phonons is two, which makes this type of morphic effect useful but not as interesting as for the triply-degenerate phonons in cubic materials. The materials studied are α-quartz, Al_2O_3, Ti_2O_3, $Gd(MoO_4)_3$, NbO_2, CdS and TeO_2.

(i) *α-quartz.* Tekippe and Ramdas (1971) and Tekippe et al. (1973) have studied thoroughly the effect of a uniaxial stress on the frequencies of the $\boldsymbol{q} \simeq 0$ optical phonons of α-quartz. Prior to them, Mariée and Mathieu (1946) and Harker et al. (1969, 1970) reported some results of similar studies. The point group of α-quartz is D_3. It exhibits four symmetric A_1-type Raman-active infrared-inactive phonons, four A_2-type infrared-active Raman-inactive phonons and eight doubly-degenerate E-type Raman- and infrared-active phonons which exhibit an LO–TO splitting. Actually four of the E modes (those at 128, 263, 695 and 1160 cm^{-1}) exhibit a negligible splitting, which is unresolvable under usual infrared absorption or Raman scattering experimental procedures. The intrinsic Raman spectrum of α-quartz has been the subject of extensive work by a number of researchers in the past (Tekippe et al. 1973 and references therein). A uniaxial stress can only shift the A_1 modes, and shift and split the E modes according to the secular equation (11.6). Following the perturbation approach of Rodriguez et al. (1972), Tekippe and Ramdas (1973) reach the same form of secular equation for these two types of modes in the presence of a stress. Two deformation potentials (a,b) are employed for the A_1 mode, and four (a,b,c,d) for the E mode. Again, multiplying these deformation potentials by $2\omega_0$ gives the corresponding coefficients $K_1, K_2, \ldots$ used here (see §11.2.2). The measurements of Tekippe and Ramdas were carried out at liquid helium temperature and under similar experimental conditions to those of Venugopalan and Ramdas (1973). Only the E (128 cm^{-1}) and A_1 (205 cm^{-1}) modes were studied quantitatively, although definite shifts and splittings were observed for the unresolved E(LO+TO) modes at 263, 695 and 1160 cm^{-1}. The deformation potentials for the E(128) and A_1(205) modes were evaluated by use of the elastic compliance data at 196 K of MacSkimin et al. (1965).

A complete study at liquid helium temperature of all four A_1 modes and eight E modes of α-quartz under uniaxial stress was reported more recently by Briggs and Ramdas (1977). With appropriate "45°-cut" samples and various directions of $\boldsymbol{X}$ they were able to observe and measure shifts for all the A_1 modes and splittings for all the E modes. In fact, from the extrapolation of the stress-dependent LO and TO components, it became possible to deduce the values of the stress-free TO–LO splitting for the "unresolved" doublets. The deformation potentials a,b and a,b,c,d are included in table 2. It is interesting that with appropriate directions of $\boldsymbol{q}$, the LO–TO splitting exhibits a dependence on the applied stress similar to that in ZnSe (Cerdeira et al. 1972). All observations are consistent with the modified polarization selection rules in the presence of stress. The hydrostatic shifts for the E and A_1 modes determined from these measurements are found to be in good agreement with the values that Asell and Nicol (1968) and Mitra et al. (1969) obtained at room temperature from hydrostatic pressure experiments.

(ii) *Al_2O_3 and Ti_2O_3*. These materials crystallize according to the centrosymmetric point group D_{3d}. They both exhibit two Raman-active A_{1g}-type phonons and five Raman-active E_g-type phonons. The main difference in their electronic structure is that Al_2O_3 has no d-electrons while there is one d-electron per Ti^{3+} ion in Ti_2O_3. Shin et al. (1976) have studied the effects of a uniaxial stress on the Raman-active phonons for these two materials, presumably at 300 K. It is shown that one needs two deformation potentials (e,f) for the A_{1g} mode and four deformation potentials (a,b,c,d) for the E_g modes. We notice that the point group D_{3d} actually arises from that of α-quartz (D_3) as its direct product with the inversion operator. This explains the fact that all three materials, i.e. α-quartz, Al_2O_3 and Ti_2O_3, behave alike in the presence of stress as far as symmetry is concerned. The experimental observations of Shin et al. are consistent with the polarization selection rules. From these data and the elastic compliance values of Chi and Sladek (1973) for Ti_2O_3, and Wachtman et al. (1960) for Al_2O_3, numerical values were obtained for the six deformation potentials a, b, c, d, e, f. These are listed in table 4 for the two materials. Definite splittings of the E_g phonons were observed in Al_2O_3 but none in Ti_2O_3. Since it is the movement of oxygen atoms that is involved in the E_g-type phonons, Shin et al. conclude that there must be a strong interaction between the oxygen atoms and the d-electrons of Ti^{3+}. Similar conclusions have been reached from independent X-ray studies (Simonyi and Raccah 1973).

(iii) *CdS.* A complete study of the $\boldsymbol{q}\simeq 0$ phonons of CdS (point group C_{6v}) under uniaxial stress at 77 K has been reported by Briggs and Ramdas (1976). The stress was applied along the c-axis or the a-axis. The doubly-degenerate E_2-type phonon at 256 cm^{-1} splits into two components with a splitting linear in the stress along $\hat{x} \| a$. No splitting was observed for the low-frequency E_2 phonon at 42 cm^{-1}. In addition to the LO–TO splitting of the E_1 mode the stress splits the TO and shifts the LO component when $\boldsymbol{X} \| a$, while the A_1(TO) and A_1(LO) lines simply shift. Shifts, but no splittings, were observed for all phonons with $\boldsymbol{X} \| c$, as expected. It turns out that one needs 3, 3, 3, 2 and 2 deformation potential coefficients to describe the shifts and splittings of the E_2, E_1(TO), E_1(LO), A_1(LO) and A_1(TO) phonons respectively. It is assumed that the stress has no effect on the LO–TO splitting itself, so that the same set of deformation potential coefficients describes both the LO and the TO components of the infrared-active E_1 and A_1 phonons. The values for these coefficients as derived from the measurements of Briggs and Ramdas are included in table 2. Applied stress along c was found to decrease the frequencies of the overtones of the E_1(LO) phonons at the rate of 0.229 cm^{-1} per kbar per overtone order. From this, it is concluded that the overtones are predominantly of E_1 nature.

(iv) *Other materials.* Stress-induced shifts have been observed in NbO_2 (Pollak 1973). This material undergoes a second-order semiconductor-to-metal transition at about 1100 K. The point group in the semiconducting phase is C_{4h}. Twenty-three $\boldsymbol{q}\simeq 0$ optical phonons have been confirmed by Raman scattering measurements, that is, seven of A_g-type, six of B_g-type and ten of E_g-type (doubly degenerate). Their frequencies lie between 150 and 827 cm^{-1}. Stress parallel to both the c- and the a-axes have produced shifts for some of these phonon frequencies, with rates which vary from about 0.6 to -0.3 cm^{-1}/kbar. No splittings of the E_g modes are reported.

Gadolinium molybdate (GMO) is known to undergo a first-order ferroelectric transition from a tetragonal (point group D_{2d}) to an orthorhombic

Table 4

Deformation potential constants in cm^{-1} per unit strain for the A_{1g} and E_g modes of Ti_2O_3 and Al_2O_3 (Shin et al. 1976), and for representative modes of TeO_2 (Lemos et al. 1977).

Material	Mode (cm^{-1})	*e*	*f*	*a*	*b*	*c*	*d*
	$A_{1g}(235)$	588 ± 72	-1114 ± 14				
	$A_{1g}(512)$	-766 ± 187	-559 ± 137				
	$E_g(277)$			562 ± 133	-35 ± 71	0	0
Ti_2O_3	$E_g(304)$			-224 ± 26	-392 ± 44	0	0
	$E_g(346)$			68 ± 6	-384 ± 57	0	0
	$E_g(457)$			-467 ± 53	-818 ± 93	0	0
	$E_g(560)$			-1638 ± 319	-1648 ± 290	0	0
	$A_{1g}(417)$	-188 ± 16	-844 ± 71				
	$A_{1g}(644)$	-1637 ± 168	-549 ± 56				
	$E_g(378)$			-725 ± 75	-291 ± 3	-183 ± 11	-163 ± 8
Al_2O_3	$E_g(430)$			-601 ± 50	-161 ± 17	196 ± 18	-133 ± 26
	$E_g(448)$			-458 ± 94	145 ± 15	16 ± 20	17 ± 8
	$E_g(576)$			-885 ± 122	-384 ± 106	132 ± 17	10 ± 16
	$E_g(750)$			-817 ± 94	-320 ± 75	-35 ± 9	99 ± 19
	$A_1(148)$	187	−2				
	$B_1(61)$	−7	442				
TeO_2	$B_2(154)$	147	−67				
	E(121, 123)			19	482	1000[a]	
	$E_{TO}(173)$			163	114		
	$E_{LO}(196)$			102	138		

[a]Absolute value.

phase on cooling below 432 K (Ganguly et al. 1975, Ullman et al. 1976 and references therein). It is believed that this transition is initiated by the softening of two degenerate zone-boundary modes which become two A_1-type zone-center modes below the transition temperature. At 300 K these two lines appear as one broad line at about 47 cm^{-1}. Both the frequency and damping constant of this line are sensitive to temperature below 432 K and also to stress, since the transition temperature is known to shift by 29.5 grad/kbar under a hydrostatic pressure. Such effects with variable uniaxial stress and temperature have been observed by Ganguly et al. (1975) and by Ullman et al. (1976). The results show a significant non-linear frequency increase and a narrowing of the band with the applied stress. By fitting to a Lorentzian line shape it turns out that the force constant parameter is rather insensitive to both stress and temperature, whereas the damping constant decreases non-linearly with stress. It is concluded that anomalous damping phase transition mechanisms are not operative in GMO (also Kim et al. 1978 and references therein).

Representative $\boldsymbol{q}\simeq 0$ Raman-active phonons of paratellurite were studied recently under uniaxial stress by Lemos et al. (1977) at 300 K. Paratellurite (TeO_2) belongs to the point group D_4 and exhibits twenty-two Raman-active phonons, i.e. four A_1-type, five B_1-type, four B_2-type and nine doubly-degenerate E-type (infrared active). To describe the shift and splitting of the non-degenerate (A_1, B_1, B_2) and doubly-degenerate (E) modes one needs two (e,f) and four (a,b,c,d) deformation potentials respectively, one set for each mode. These values were determined for the modes $A_1(148\ \mathrm{cm}^{-1})$, $B_1(61)$, $B_2(154)$, $E_{TO,LO}(121, 123)$ and $E_{TO,LO}(173, 196)$ and are included in table 4. The mode Grüneisen parameters obtained are in good agreement with those calculated from hydrostatic measurements (Peercy et al. 1975). A non-linear shift was observed for the A_1 and B_1 modes with $\mathbf{X}//[100]$. This is tentatively associated with the transition $D_4 \rightarrow D_2$ which occurs under an orthorhombic distortion of the same type as the one produced by the [100] stress.

11.3. Other possibilities

In addition to the study of shifts and splittings of normal modes, application of a uniaxial stress constitutes a useful experimental technique for a variety of lattice-dynamical problems. Some typical examples of such "indirect" morphic effects are examined next.

11.3.1. Anharmonic contributions to the phonon self-energies.

As shown by Peercy (1973), measurements of the stress dependence of the Raman-active phonon frequencies can be helpful in evaluating precisely the anharmonic contributions to the phonon self-energies. Anharmonic interactions are observed in the temperature dependence of the phonon frequencies. Experimentally one measures $(\partial \ln \omega_j / \partial T)_P$ which can easily be written as

$$\left(\frac{\partial \ln \omega_j}{\partial T}\right)_P = \alpha - \frac{\beta_v}{\kappa_v}\left(\frac{\partial \ln \omega_j}{\partial P}\right)_T + \left(\frac{\partial \ln \omega_j}{\partial T}\right)_V$$
$$\equiv \delta(V) + \delta(T), \tag{11.19}$$

where β_v and κ_v are respectively the volume thermal expansion coefficient and the volume compressibility. The parameter α is identically zero for cubic and isotropic crystals. For crystals of lower symmetry (i.e. uniaxial crystals) α depends on the change of the ratio c/a with either pressure or temperature (Samara and Peercy 1973, Fritz 1973). In a first approximation α may be ignored (cubic approximation) but in general one has to consider corrections due to α. As shown by Peercy (1973) an expression can be reached for α in terms of such experimental parameters as $(\partial \ln \omega_j / \partial X_1)_{T,c}$, and $(\partial \ln \omega_j / \partial X_3)_{T,a}$, where X_1 and X_3 are stresses along a and c respectively. Furthermore one easily shows that

$$\left(\frac{\partial \ln \omega_j}{\partial P}\right)_T = 2\left(\frac{\partial \ln \omega_j}{\partial X_1}\right)_{T,c} + \left(\frac{\partial \ln \omega_j}{\partial X_3}\right)_{T,a}. \tag{11.20}$$

The sum of the first two terms ($\equiv \delta(V)$) in the right-hand side of eq. (11.19) represents the pure volume contribution due to thermal expansion of the crystal. The second term $\delta(T)$ in eq. (11.19) corresponds to the pure temperature contribution and is due to higher-order anharmonic terms. It is this term that one needs in order to analyze the phonon anharmonicities. According to eq. (11.19), a value for $\delta(T)$ can be found provided $\delta(V)$ is known. For cubic and isotropic materials $\delta(V)$ can be easily measured from hydrostatic pressure experiments. For crystals of lower symmetry one has to calculate α and $(\partial \ln \omega_j / \partial P)_T$ from stress experiments. Thus, measurements of $(\partial \ln \omega_j / \partial X_i)_T$ from Raman scattering experiments lead to an exact value for $\delta(V)$ and thereby for $\delta(T)$. Peercy applied this approach to the modes A_{1g}, B_{1g} and E_g of rutile (TiO_2) which has a tetragonal symmetry (point group D_{4h}). The shifts of their frequencies were small and no splitting was observed for the E_g mode. The results of his measurements and calculations are included in table 5 together with the values which are obtained assuming a harmonic approximation (in parentheses) for comparison. It is concluded that in this particular material there is only about a 10% improvement in the calculation of the intrinsic coefficient $(\partial \ln \omega_j / \partial T)_V$ over the one based on the cubic approximation (i.e. ignoring α). This provides a useful feature for deciding how accurate the cubic approximation is in the calculation of Raman-active phonon anharmonicities and/or elastic constants.

11.3.2. Phonon line broadening

Systematic investigations of the effect of an external stress on the line width of normal modes appear to have been carried out for only a few materials. From a lattice-dynamical point of view this represents an interesting but rather complicated problem because of the multiplicity of primary and secondary effects that may arise in the presence of a stress, uniaxial or otherwise.

The full width at half intensity of the most prominent Raman peaks of α-quartz has been studied by Harker et al. (1969, 1970) for stresses up to 3 kbar. The 128 cm^{-1} line shows a monotonic (non-linear) decrease in width

Table 5

Uniaxial-stress and hydrostatic-pressure dependences for the Raman-active modes in TiO_2. Also included are the corrections α for these modes. Separation of the isobaric frequency shifts into their pure-volume and pure-temperature dependences using the results of the uniaxial stress measurements, and comparison with the corresponding analysis for hydrostatic-pressure results (numbers in parenthesis). $\delta(V)$ is the pure-volume contribution (Peercy 1973).

Mode	ω (cm^{-1})	$(\partial \ln\omega/\partial X_3)_{\alpha,T}$ (10^{-3}/kbar)	$(\partial \ln\omega/\partial X_1)_{c,T}$ (10^{-3}/kbar)	$(\partial \ln\omega/\partial P)_T$ (10^{-3}/kbar)	α (10^{-5}/K)	$(\partial \ln\omega/\partial T)_P$ (10^{-5}/K) =	$\delta(V)$ (10^{-5}/K) +	$(\partial \ln\omega/\partial T)_V$ (10^{-5}/K)
B_{1g}	143	-0.2 ± 0.1	-1.2 ± 0.2	-2.6 ± 0.5[a]	0.6	0.6 ± 0.4	12.4	-11.8
				-2.4 ± 0.2[b]			(11.8)[d]	(-11.2)[d]
E_g	449	0.4 ± 0.1	0.4 ± 0.1	1.2 ± 0.3[a]	0.4	-6.3 ± 0.4	-6.1	-0.2
				1.15 ± 0.2[b]			(-5.7)[d]	(-0.6)[d]
A_{1g}	612	not measured	0.4 ± 0.1	...	0.05[c]	0.6 ± 0.4	-3.7	4.3
				0.75 ± 0.2[b]			(-3.7)[d]	(4.3)[d]

[a]Based on eq. (11.20).
[b]Samara and Peercy (1973).
[c]Calculated assuming $(\partial \ln\omega/\partial X_3)_{\alpha,T} = 0.15\times10^{-3}$/kbar.
[d]Values assuming a cubic approximation (Samara and Peercy 1973).

by almost 1 cm^{-1}. A qualitative argument is given according to which this decrease is consistent with the analogous effect of a decreasing temperature. Furthermore, it can be associated with stress-dependent contributions of the imaginary part of second-order self-energy graphs (Maradudin and Fein 1962, Cowley 1966). The same arguments cannot explain the behavior of the other peaks (e.g. at 207, 352, 398, 466 cm^{-1}) the width of which appears first to decrease and then to increase at higher stresses.

Surface strains are capable of producing broadening of the phonon lines in Raman scattering spectra. Evans and Ushioda (1974) have reached this conclusion after a detailed study of the Raman line shape for a number of III–V semiconductors as a function of sample surface treatment. The surfaces studied were variously polished, etched, annealed or cleaved. Depending on the preparation of the surface, different microscopic damage is generated on the surface which results in a broadening of the line by at least two independent mechanisms. First, owing to imposed imperfections (dislocations, point defects) the lattice symmetry at the surface is lowered. This opens more channels of anharmonic decay and scattering of the phonons and this in turn causes additional broadening of the line. Second, the strains within the scattering volume are quite inhomogeneous in magnitude and sign. Thus the observed phonon band is a superposition of various Lorentzian bands shifted in both directions and by different amounts relative to the position of the strain-free phonon line. In the case of GaAs, Evans and Ushioda compare the line of polished and annealed samples as a function of temperature and conclude that the broadening in the polished samples is mainly due to the second mechanism, that is, to surface inhomogeneous strains. Similar observations have been made in GaAs in connection with absorption and infrared reflectance studies (Jones and Hilton 1965, Lisitsa et al. 1968). On the basis of their Raman measurements of phonon line broadening as a function of such variable parameters as temperature, exciting laser frequency (skin depth) and surface preparation, Evans and Ushioda suggest a new approach to exploring the status of surface strains at various depths below the surface of the crystal.

Although not explicitly recognized, a broadening in the first- and second-order Raman bands of CdS seems to be contained in the spectra reported by Konstantinov et al. (1976). This is in addition to the clearly visible shifts of the bands, which are induced when the sample is illuminated. Konstantinov et al. attribute this behavior to the breaking of valence bonds which occurs when the electrons are raised from the valence to the conduction band as a result of external illumination of the sample. It is believed that in this way the elastic and dynamical characteristics of the material (including the frequencies of the normal modes) are subject to changes. The qualitative theoretical explanation given for the observed changes seems reasonable although other mechanisms (such as free-carrier effects) should not be excluded.

A physical situation opposite to those discussed above occurs in heavily doped semiconductors. The application of a uniaxial stress removes the broadening of the Raman band which was caused by heavy doping (Cerdeira et al. 1973 and references therein). It has been observed by Cerdeira and Cardona (1972) that heavy doping in p-type Si and Ge produces broadening and a downward shift of the frequency of the $\boldsymbol{q}\simeq 0$ Raman-active phonon. In the case of Si Cerdeira et al. (1973) have shown that these effects can be described by a Fano-type interference between discrete one-phonon scattering and continuum one-electron excitations from filled to empty valence states. Furthermore, it is suggested that these effects and a similar one concerning the elastic constant C_{44} (Keyes 1976) have an electronic character of common origin, that is the splitting of degeneracy of the valence bands, by the phonon distortion. Cerdeira et al. (1973) have further confirmed this suggestion by studying the behavior of the Raman band of p-type Si under a uniaxial stress. It is known that a uniaxial stress along [100] splits the degeneracy of the Γ_8 states at the top of the valence band, and in fact for high enough stress the two bands are sufficiently separated so that all carriers are left in one band. During this process the extent of Fano discrete-continuum interaction decreases gradually, and ultimately disappears at excessively high stresses. It is expected therefore that the width and the frequencies of the singlet and doublet components will tend to their intrinsic values. Similar arguments were presented for the effect of doping on the elastic constant C_{44} of p-type Si. Experimenting with boron-doped (p-type) Si at 77 and 300 K, Cerdeira et al. (1973) did indeed observe changes in the width and frequency of the Raman bands (singlet and doublet) and their tendency with increasing stress towards the intrinsic values. Using ultrasonic pulse-echo techniques Fjeldly et al. (1973) confirmed a similar behavior for the elastic constant C_{44}, thus substantiating the theoretical prediction that the effects of doping on the Raman frequency and on the elastic constants of p-type Si have a common origin. Chandrasekhar et al. (1978), have recently undertaken a theoretical and experimental study of the same problem in heavily doped n-type Si. Here the situation is a little different. The three-fold degenerate $\langle 100\rangle$ conduction band valleys are split by a [100] stress into a doublet and a singlet. The phonon doublet is coupled to the carrier depleted doublet valleys and therefore behaves as in p-type Si, i.e. its linewidth and frequency tend to their intrinsic values as the stress is increased. In contrast, the phonon singlet couples to the carrier enriched singlet valley and this results in an enhanced Fano interaction as the stress is increased.

More recently, a uniaxial stress along [010] and [110] has been used to induce a tetragonal to triclinic phase transition in $SrTiO_3$ (Uwe and Sakudo 1976) and $KTaO_3$ (Uwe and Sakudo 1977). Among other changes in the phonon characteristics, a definite broadening was observed in the Raman width of the A_1 component of the F_{1u} mode. This broadening is found to increase hyperbolically as the applied stress approaches the critical value X_c at which the phase transition occurs. The broadening is attributed to the stress dependence of the mode damping constant. No detailed theoretical interpretation is given for this effect, although it is suggested that it may require a more-than-four phonon process to account for the observed behavior. Similar in character are the stress-induced changes of linewidth of the soft mode at 47 cm^{-1} that Ganguly et al (1975) have observed in gadolinium molybdate at 300 K. A related discussion has already been presented in §11.2.

11.3.3 Localized modes

Vibrational modes other than lattice normal modes are also capable of exhibiting morphic effects in the presence of a force. Maradudin et al. (1967) have examined from a microscopic point of view the consequences of applying an electric field or uniaxial stress to a crystal containing substitutional impurities which allow localized vibrational modes to occur. They consider only impurity atoms occupying sites of T_d or O_h symmetry in non-ionic host crystals, and derive explicit expressions for the shifts and splittings of the triply-degenerate localized modes in terms of such microscopic parameters as cubic anharmonic force constants and first- and second-order dipole moment coefficients. Although no calculations based on any particular model are undertaken, the qualitative conclusions concerning the physical mechanisms and the symmetry aspects of the problem are quite general and applicable to a variety of different situations (e.g. gap modes). An independent microscopic theory has been developed by Ganesan (1971) to analyze field-induced Raman scattering by localized modes. The intensity of such scattering in a system like, for instance, NaCl : H is shown to be much weaker than that of the intrinsic second-order Raman scattering with no field present.

Significant experimental work has also been done on this line. Hayes et al. (1965) and Hayes and MacDonald (1967) have used infrared absorption techniques to study the fundamental and second harmonic of the triply-degenerate localized modes associated with U centers at the F^- sites in CaF_2. By applying a uniaxial stress or an electric field along [111] at 20 K they succeeded in splitting the fundamental vibrational line (965.5 cm^{-1}) of H^- in CaF_2 into a singlet and a doublet, in a pattern similar to that of fig. 13a for $X//[001]$. The shift and splitting were found to be linear in the applied stress or electric field. Similar observations were also made with D^- ions. As expected, the frequency splittings obtained were smaller than those for H^- by a factor of $\sqrt{2}$. Relevant calculations have been published by Clayman et al. (1971).

The effect of an electric field on the absorption and line shape of OH^- in KBr has been studied by Handler and Aspnes (1966). For light polarized parallel to the electric field the absorption band becomes narrower and stronger, while it becomes weaker when the light is polarized perpendicular to the electric field. In both cases the changes observed were found to be quadratic in the field. From the analysis of their data Handler and Aspnes conclude that the Devonshire model does not fully explain the properties of OH^- in an alkali halide lattice at 300 K (Devonshire 1936).

Nolt and Sievers (1966, 1968) have studied stress-induced shifts and splittings of the low-frequency single-ion local modes of Li^+ and Ag^+ in KBr and KI respectively. Using far-infrared interferometric techniques they were able to observe substantial shifts and splittings linear in the applied stress (figs. 15, 16). From their measurements they conclude that both defect ions occupy normal sites of O_h symmetry. They also calculate the local mode coefficients A, B, C (analogous to K_1, K_2, K_3 used here) using host lattice compliances (rather than the more appropriate local compli-

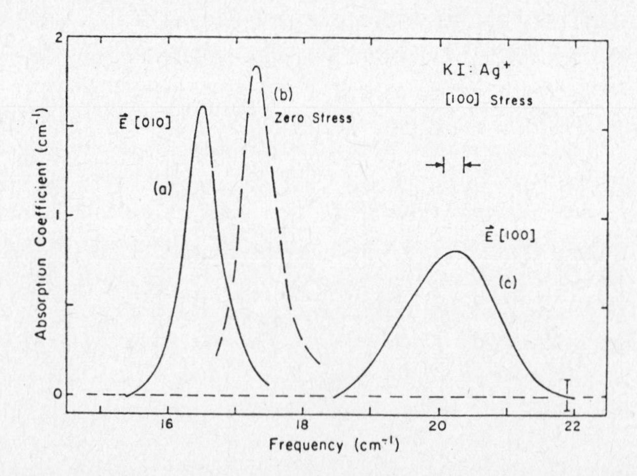

Fig. 15. KI:Ag^+ absorption coefficient for a [100] stress of 3.3 kg/mm^2. This absorption coefficient is calibrated against that of a crystalline quartz sample in the same system and shows the largest frequency splitting observed in these studies. The instrumental resolution is 0.3 cm^{-1} and is shown by the arrows above curve (c). The integrated absorption strength was measured to be 1.03, 1.23 and 1.1 ± 0.1 cm^{-2} for curves (a), (b) and (c) respectively (Nolt and Sievers 1968).

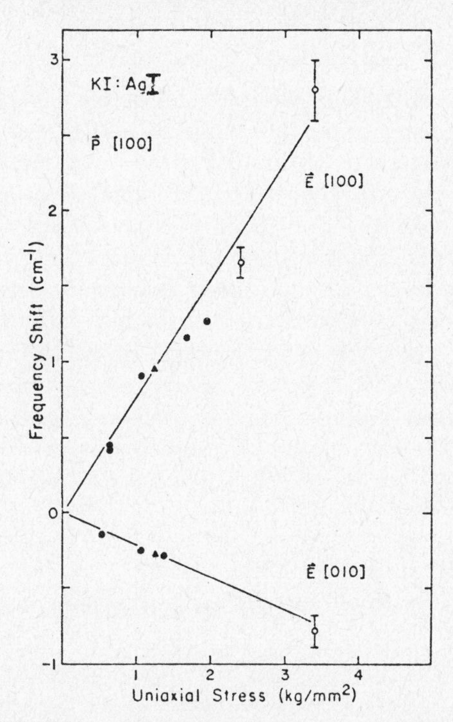

Fig. 16. KI:Ag$^+$ frequency shift induced by [100] stress. The results for two different samples are shown. Both samples contained Ag ions having the natural isotopic abundance for which the zero-stress frequency is 17.32 cm^{-1}. Because of the generally weaker absorption level in the KI:AgI samples the precision of the frequency determination is only ± 0.1 cm^{-1} (Nolt and Sievers 1968).

ances). The results of a similar investigation on these systems and also on NaCl : Cu$^+$ and CsI : Tl$^+$ were reported recently by Kahan et al. (1976).

Localized impurity pair-modes in several alkali-halide systems have also been studied, in addition to the single-ion modes discussed above. To mention a few examples, a Na$^+$–Na$^+$ infrared-active pair-mode at 44 cm^{-1} in KCl : NaCl (Templeton and Clayman 1971), an Ag$^+$–Ag$^+$ Raman-active pair-mode at 47 cm^{-1} in NaCl : AgCl (Moeller et al. 1970), and five F$^-$–F$^-$ infrared-active pair-modes at 32.7, 38.0, 40.2, 44.7 and 48.4 cm^{-1} in NaCl : NaF (Becker and Martin 1972). In cubic crystals there are seven possible symmetries for a pair-mode. Kaplyanskii (1964) has tabulated the splitting patterns and intensity ratios of the pair-mode components which are generated when the crystal is subject to an externally applied uniaxial stress along [100], [110] and [111]. Templeton and Clayman (1972) use the results of this symmetry analysis and their data to conclude that the symmetry of the Na^+–Na^+ pair-mode in KCl:NaCl is tetragonal. Their far-infrared absorption spectra of the pair-mode exhibit a frequency shift linear in the applied stress along the directions given above. No evidence of splitting of the pair-mode is reported, nor any dependence of the absorption strength on the polarization of the beam. Once the symmetry of the pair-mode is determined through its behavior under a uniaxial stress, model calculations from first principles are possible in general (Jaswal 1965, Templeton and Clayman 1972).

The few characteristic examples given above demonstrate the wide-range applicability of a perturbing force as a means of studying perfect and distorted lattices. For a thorough treatment of the subject, the reader is referred to the work of Barker and Sievers (1975). Among other aspects, the effects of external perturbations are reviewed, in this article, on the optical and vibrational properties of disordered solids due to such distortions as impurity modes, pair modes, colour centers, etc.

11.4. Second-order effects linear in the strain

Silicon is perhaps the only material whose second-order infrared absorption spectrum has been studied under stress (Lüth 1970, Angress et al. 1975). Because of its center of symmetry no first-order infrared absorption is allowed. A homogeneous strain does not remove the center of symmetry and thus one may expect a uniaxial stress to modify only the second-order infrared absorption spectrum.

The first exploratory results of a theoretical and experimental study of the effect of a uniaxial stress on the second-order infrared absorption of Si were reported by Hobson and Paige (1966). Their data indicate changes in the absorption constant from 5 to 20% for stresses up to 60 kg/mm^2, and also a strong dependence on the polarization of the incident light. Lüth (1970) investigated the region 500–1000 cm^{-1} in the presence of pulsed stresses along [100]. The data clearly indicate shifts of the rather complicated second-order spectrum for both polarizations of the light beam, parallel and perpendicular to the stress axis. The observed asymmetries also indicate a change in the absorption constant; although not shown in the spectra stress-induced splittings are not excluded. No detailed symmetry analysis is given. A topological analysis is presented for the critical points where the stress effects are expected to manifest themselves more clearly. By combining the data obtained in the two polarizations of the incident light beam, Lüth calculates the hydrostatic pressure component of the spectrum in the entire region studied. He also deduces the rate of shift of the frequencies of the two major combination bands as a function of hydrostatic pressure. These are the TO–TA combination at the L-point

(610 cm^{-1}) and the LO–LA combination at the K-point (740 cm^{-1}). The obtained values of 0.7×10^{-3} and 1.0×10^{-3} (in units of cm^{-1} per kp·cm^{-2}) compare favorably with those calculated by Jex (1971) from a shell model, i.e. 0.57×10^{-3} and 0.9×10^{-3} respectively.

It was mentioned in §5.4 that Balslev (1974) has determined by inspection methods the number of split levels of two-phonon states, and the remaining degeneracies of highly symmetric modes under $\boldsymbol{X}/\!/[001]$ and $\boldsymbol{X}/\!/[111]$, for the two-phonon processes in diamond- and zincblende-type crystals. He also calculates the relative intensities of these sublevels for the three Raman components Γ_1, Γ_{12} and Γ_{15} and for the single infrared absorption component Γ_{15}. These results are included in the Appendix as tables A4, A5 and A6.

Through a direct group-theoretical analysis, Angress et al. (1975) have determined the behavior of the degeneracies and polarization selection rules of Si at the X and L points under [001] and [111] stresses. According to this analysis only a few changes in the polarization selection rules are to be expected. For instance, combinations of such split bands as $TA_{/\!/}(X)\pm TA_{\perp}(X)$ under [111] stress, and $LO(L)\pm TA(L)$ and $LA(L)\pm TO(L)$ under [001] stress become infrared active. $/\!/$ and $\perp$ indicate phonon polarizations which are respectively parallel and perpendicular to the plane of $\boldsymbol{q}$ and $\boldsymbol{X}$. Their experimental study was confined to the combinations $TO(X)\pm TA(X)$, $TO(L)\pm TA(L)$ and $LO(L)\pm LA(L)$. The symmetry behavior of these combinations under stress and relevant numerical values are included in tables 6 and 7 respectively. It is shown that the stress-induced frequency shifts of the intrinsic band combinations and the split band combinations are proportional to an appropriate anharmonic parameter G. This parameter (included in table 7) is defined as

$$G(\boldsymbol{q},j\pm j')=\omega_{\boldsymbol{q}j}\gamma_{\boldsymbol{q}j}\pm\omega_{\boldsymbol{q}j'}\gamma_{\boldsymbol{q}j'}. \tag{11.21}$$

ω and γ correspond to the frequency and Grüneisen coefficient of the mode $\boldsymbol{q}j$. G is the only parameter that can be determined from the stress measurements. The results of table 6 are in agreement with the results of Balslev (1974), except for the strengths of the various $TO(L)\pm TA(L)$ combinations under [001] stress (the combinations $TO(X)-TA(X)$ appear in the work of Angress et al. for the first time). Related theoretical values by Dolling and Cowley (1966), by Johnson and Loudon (1964) and by Jex (1971) are also included in table 7 for comparison. The discrepancy in the values of $G(L, TO\pm TA)$ is attributed by Angress et al. to the fact that the Grüneisen values for the TA modes at the X and L points are not taken negative enough. Further support for this conjecture is offered by the work of Richter et al. (1975) who have measured $\gamma_{X,TA}=-1.4\pm0.2$ for Si, from

Table 6

Branch splittings and second-order infrared activities of Si under uniaxial stress. The last two columns give the intensities relative to the unstressed case. Starred values are those strictly obtainable from symmetry alone while unstarred values are those obtainable in the limit of zero stress (Angress et al. 1975).

Stress direction	Combination (unstressed)	Wavevector equivalence	Branch splitting		Relative strengths $E\parallel X$	Relative strengths $E\perp X$
[001]	TO±TA(L)	all L points equivalent	$TO_{\perp}$	$\pm TA_{\perp}$	0.5	0.5
			$TO_{\parallel}$	$\pm TA_{\parallel}$	0.5	0.5
			$TO_{\perp}$	$\pm TA_{\parallel}$	0*	0
			$TO_{\parallel}$	$\pm TA_{\perp}$	0*	0
[001]	LO±LA(L)	all L points equivalent	LO	±LA	0.5	0.5
[001]	TO±TA(X)	X_{001}	TO	±TA	1	0*
		X_{100}, X_{010}	TO	±TA	0*	1
[111]	TO±TA(X)	all X points equivalent	$TO_{\perp}$	$\pm TA_{\perp}$	0.5	0.5
			$TO_{\parallel}$	$\pm TA_{\parallel}$	0.5	0.5
			$TO_{\perp}$	$\pm TA_{\parallel}$	0*	0*
			$TO_{\parallel}$	$\pm TA_{\perp}$	0*	0*

Table 7

Column 1: combination bands of Si studied under uniaxial stress (Angress et al. 1975); columns 2 and 3: frequencies of infrared features; columns 4 and 5: experimental values of G obtained with [001] and [111] uniaxial stresses; columns 6 and 7: theoretical values of G.

Combination	Combination frequency (cm^{-1}) [a]	[b]	$G(\boldsymbol{q}, j\pm j')$ in cm^{-1} experimental [001] stress	experimental [111] stress	theoretical [b,c]	theoretical [b,d]
TO+TA(X)	611.5	612			558	513
TO−TA(X)	313.5	314	1066±200	1009±150	803	691
TO+TA(L)	603.6	605	334±50		541	551
LO+LA(L)	802.2	800	757±80		763	688

[a] Angress et al. (1975).
[b] Johnson and Loudon (1964).
[c] Dolling and Cowley (1966).
[d] Jex (1971).

Raman scattering under hydrostatic pressure. The corresponding values of Dolling and Cowley and of Jex are -0.82 and -0.6 respectively.

Recent studies under uniaxial stress of the second-order Raman spectrum of $KTaO_3$ at 2K have shown appreciable changes in the frequencies of the second-order structure (Uwe and Sakudo 1977). The peaks and bands observed and their stress dependence are associated with the effect of stress on the critical points of the two-phonon energy density of states. A brief relevant discussion will be postponed until the end of the next section.

12. Stress-induced phase transitions

Through a series of well-established theoretical and experimental results it has become clear that a uniaxial stress (as well as a hydrostatic pressure, to be discussed shortly) constitutes a useful variable for the study of phase transitions due to lattice instabilities.

A stress-induced ferroelectric transition has been extensively studied in $SrTiO_3$. With no stress present, this material exhibits a displacive phase transition at $T_0 = 105$ K from O_h to D_{4h} symmetry, owing to condensation (at 105 K) of the F_{2u} zone-boundary phonon mode (Unoki and Sakudo 1967, Fleury et al. 1968, Thomas and Müller 1968, Shirane and Yamada 1969, Uwe and Sakudo 1976). The ferroelectric transformation of this material is due to the F_{1u} zone-center phonon mode which never condenses at any temperature, as becomes evident from inelastic neutron scattering experiments (Yamada and Shirane 1969) and electric-field-induced Raman scattering (Fleury and Worlock 1968). The changes in the mode symmetries below T_0 are shown in table 8.

In the presence of a uniaxial stress along [111] at temperatures below 105 K, $SrTiO_3$ is known to undergo a tetragonal-to-trigonal structural phase transition. Experimental evidence for this is provided by the EPR studies under a [111] uniaxial stress of Müller et al. (1970), the piezo-fluorescence measurements of Burke and Pressley (1969) in Cr^{3+}-doped crystals, and by the Raman work of Chang et al. (1970), Rokni and Wall (1971) and Burke et al. (1971). The Raman investigation of Burke et al. (1971) below 105 K concentrates on the A_{1g} and E_g modes whose progenitor above 105 K is the silent mode F_{2u} according to table 8. The effect consists in a small frequency increase when the stress passes through a critical value X_{c1} at which the first-order phase transition is believed to occur. The values of X_{c1} observed were 12 kp/mm^2 and 25 kp/mm^2 at temperatures of 77 K and 4.2 K, respectively.

A similar transition from the cubic directly to the trigonal phase has been induced in $SrTiO_3$ by a [111] stress at 115 K (Wall et al. 1971). With no stress applied the F_{2u} mode is Raman inactive at 115 K. When the stress

Table 8

Correlation table for the F_{1u} (ferroelectric) and F_{2u} (silent) modes of $SrTiO_3$ as a function of temperature $T \gtrless T_0 = 105\,K$ and uniaxial stress $X \gtrless X_{c1}//[010]$, $X \gtrless X_{c2}//[110]$. Basis functions x, y, z and X, Y, Z correspond to [100], [010], [001] and [110], [110], and [001] respectively (Uwe and Sakudo 1976).

$T > T_0$	$T < T_0$					
stress free		[010] stress		[110] stress		
O_h^1	D_{4h}^{18}	$D_{2h}(X < X_{c1})$	$C_{2v}(X > X_{c1})$	D_{4h}^{18}	$D_{2h}(X < X_{c2})$	$C_{2v}(X > X_{c2})$
$F_{1u}(\Gamma_{15})$	$A_{2u}(z)$	$B_{1u}(z)$	$B_1(z)$	$A_{2u}(Z)$	$B_{1u}(Z)$	0 — $A_1(Z)$
(ferroelectric)	$E_u(x,y)$	$B_{2u}(y)$	$B_2(y)$	$E_u(X,Y)$	$B_{2u}(Y)$	$B_2(Y)$
		$B_{3u}(x)$	0 — $A_1(x)$		$B_{3u}(X)$	$B_1(X)$
$F_{2u}(\Gamma_{25})$	0 — A_{1g}	A_g	$A_1(x)$	A_{1g}	A_g	$A_1(Z)$
(structural)	E_g	B_{2g}	$B_1(z)$	E_g	B_{2g}	$B_1(X)$
		B_{3g}	A_2		B_{3g}	$B_2(Y)$

reaches the critical value X_{c2} the cubic-to-tetragonal phase transition occurs and the Raman-inactive mode F_{2u} reduces to the Raman-active modes A_{1g} and E_g. Their frequencies ω_A and ω_E ($>\omega_A$) increase at different rates as the stress increases beyond X_{c2}, and in fact at a second critical value X_{c3} ($>X_{c2}$) they become equal. Beyond X_{c3} the Raman measurements show that $\omega_A > \omega_E$. From such observations one can obtain numerical values for constants which appear in related theoretical models (Slonczewski 1970).

While a [111] stress can only induce a structural phase transition as discussed above, a stress along either [010] or [110] has been shown to induce a second-order ferroelectric transition. From the dielectric measurements of Burke and Pressley (1971) in $SrTiO_3$ under [010] and [110] uniaxial stresses at 4.2 K, it is concluded that under these conditions a remanent polarization is induced. The critical values of the stresses along [010] and [110] for the ferroelectric transition are placed at 10.1 kg/mm^2 and 53 kg/mm^2 respectively, or 0.99×10^9 dyn/cm^2 and 5.2×10^9 dyn/cm^2 respectively. These values have been questioned by Uwe and Sakudo (1976) as will be discussed shortly. Additional experimental evidence for this transition is offered by the optical second-harmonic generation experiments of Fujii et al. (1970). A phenomenological theory employing a thermodynamic free energy and an electrostrictive mechanism has been proposed to account for the experimental findings (Thomas and Müller 1968, Slonszewski 1970, Slonczewski and Thomas 1970, Uwe and Sakudo 1975).

Recently, Uwe and Sakudo (1976) studied in great detail the ferroelectric transition discussed above at liquid-helium temperature using their results of dielectric measurements and Raman scattering experiments. In table 8 the whole sequence of the symmetry evolution is included for the two modes of interest, that is, the silent (F_{2u}) and the ferroelectric mode (F_{1u}). The two critical stresses X_{c1} and X_{c2} ($>X_{c1}$) designate the values at which the ferroelectric transition occurs for the two stress directions [010] and [110] respectively. From their combined dielectric and Raman scattering measurements, Uwe and Sakudo obtain the values $X_{c1}=(1.6\pm0.2)\times 10^9$ dyn/cm^2 and $X_{c2}=(5.8\pm0.2)\times10^9$ dyn/cm^2 which are somewhat different from those of Burke and Presley (1971). They also conclude that the stress-induced polarization $\boldsymbol{P}_s$ is perpendicular to the plane of the [010] stress direction and the structural tetrad axis. Its magnitude is roughly 2 to 4 μC/cm^2 at a stress 2 kbar higher than X_{c1}. Typical spectra for the totally symmetric component $A_1(x)$ of the ferroelectric mode F_{1u} are shown in fig. 17 for $\boldsymbol{X}$//[010] approaching X_{c1} from above. The tendency of the frequency towards zero is clearly seen as X approaches the critical value. Both ferroelectric components observed in these configurations ($A_1(x)$ and $B_2(y)$) are found to shift in accordance with the behavior of the dielectric

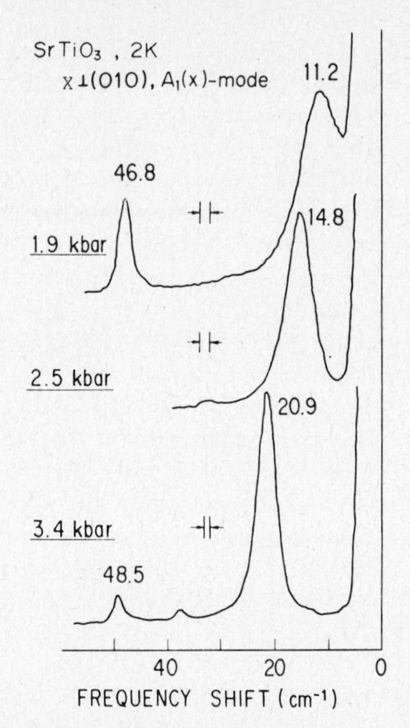

Fig. 17. Variation of the Raman spectra of the total symmetric ferroelectric mode $A_1(x)$ in $SrTiO_3$ as the applied [010] stress approaches the critical value $X_{c1} = 1.6$ kbar at 2 K (Uwe and Sakudo 1976).

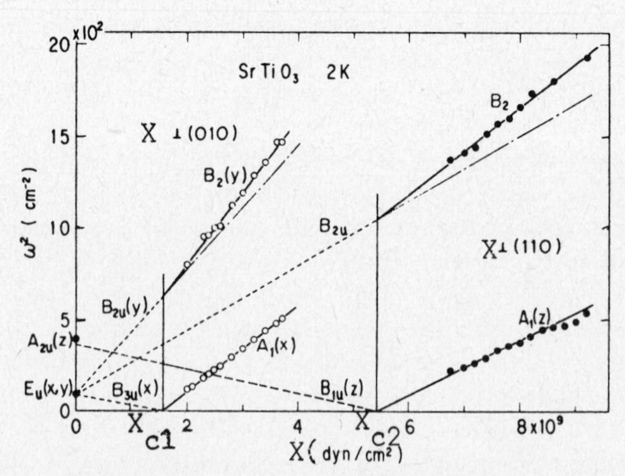

Fig. 18. Squared frequencies of the ferroelectric modes as a function of the [010] or [110] stress. Broken lines below X_{c1} and X_{c2} are calculated ones. Points at $X = 0$ are also theoretical ones (Uwe and Sakudo 1976).

constants and the LST relation (fig. 18). The Raman-active components of the F_{2u} mode are also observed to depend very critically on the applied stress and also to couple with those of the F_{1u} mode as they anti-cross. On the basis of a Landau–Devonshire phenomenological analysis, expressions are obtained for the stress-dependent dielectric constants and mode frequencies. Furthermore, comparison between experimental and theoretical results yields consistent numerical values for those parameters which enter the expression for the free energy. The dependence on the linewidth of the ferroelectric components has already been discussed in §11.3.2.

Similar studies of [010] stress-induced ferroelectricity have been made for $KTaO_3$ (Uwe and Sakudo 1977) based on Raman scattering measurements at 2 K. The transition is found to take place at $X_{c1} = 5.25$ kbar, in agreement with previous dielectric measurements (Uwe and Sakudo 1975). The analysis proceeds as in the case of $SrTiO_3$. Extended first-order spectra (up to 1200 cm^{-1}) were recorded with and without a uniaxial stress present. Observation and identification of various zone-center modes in the presence of stress gave results comparable to those obtained by independent workers using different techniques (Fleury and Worlock 1968, Perry and Tornberg 1969, Axe et al. 1970, Yacoby and Linz 1974). A summary of the first-order Raman results is given in table 9. The phonons at 184, 198, 544, 829 cm^{-1} are observed even in the paraelectric phase ($X < X_{c1}$) although according to table 8 they should be Raman inactive. The observed scattering is attributed to localized dislocations due to impurities (Yacoby and Linz 1974). The line at 274 cm^{-1} is associated with the F_{2u} silent mode. This is also supported by the results of measurements in mixed-crystal systems (Perry and Tornberg 1969, Manlief and Fan 1972, Yacoby and Linz 1974).

The studies in $KTaO_3$ under uniaxial stress were extended to include second-order Raman activity (Uwe and Sakudo 1977). The stress-free second-order spectrum has been associated with Van Hove critical points of the two-phonon energy density of states. The frequencies of these structures are listed in table 10 and they are regarded as arising from pairs of non-zone-center phonons (Yacoby and Linz 1974). Under a 9 kbar [010] stress all the critical point frequencies are found to increase as seen from the values in parentheses of table 10. In fact they appear to be far more sensitive to the uniaxial stress than the zone-center phonons. The most interesting effect of the stress seems to be that on the 2TA overtone at the stress-free frequency of 89 cm^{-1}. Increasing the stress seems to split the overtone into two components, one which slightly decreases in frequency with the stress and one which increases with the stress. According to Uwe and Sakudo, the two branches correspond to two different ways that the Ta-ions can vibrate while participating in the TA vibrational mode, that is, perpendicular and parallel to the stress axis respectively.

Table 9

Frequencies of zone-center phonons (in cm^{-1}) of $KTaO_3$ as determined by various techniques at the temperatures indicated (Uwe and Sakudo 1977).

	IR[a] (12 K)	Electric field-induced Raman[b] (10 K)	Neutron[c] (6 K)	Differential Raman[d] (80 K)	Stress-induced Raman[e] (2 K)
TO_1	25	18	25		21 ± 3[f]
LO_1	183			188	184
TO_2	196	198		202	198
LO_2/TO_3				282	274
LO_3	421			422	
TO_4	547	556		554	544
LO_4	837			830	829

[a]Perry and Tornberg (1969).
[b]Fleury and Worlock (1968).
[c]Axe et al. (1970).
[d]Yacoby and Linz (1974).
[e]Values normalized to zero stress as shown by Uwe and Sakudo (1977).

Table 10

Interpretation of the second-order Raman structures for free or stressed (9 kbar, values in parentheses) states in terms of phonons at two critical points. After Yacoby and Linz (1974), phonon frequencies (in cm^{-1}) are separated in two sets (Uwe and Sakudo 1977).

Yacoby and Linz (77 K)	(2 K)	45 (52)	172 (172)	191 (203)	517 (528)
55	45 (52)	89 (104)	217 (224)	236 (255)	562 (528)
173	172 (172)			364 (370)	687 (695)
195	191 (203)				
520	517 (528)				1033 (1048)
		294 (302)	441 (441)		
296	294 (302)		735 (743)		
444	441 (441)		881 (881)		

13. Hydrostatic-pressure-induced morphic effects

Hydrostatic pressure, or shortly pressure, is a scalar force, i.e. it transforms according to the A_1 irreducible representation of the point group. Any mode coefficients like those entering eqs. (5.11), (5.12) or (5.13) will have exactly the same tensorial character as the corresponding intrinsic mode property. This means that pressure can change only the magnitude but not the component array of the mode effective charge $\boldsymbol{e}_j$, the Raman tensor $\boldsymbol{a}_j$ and the force constant $\boldsymbol{K}_j$. A pressure will not modify the polarization selection rules for absorption and scattering and will not lift the degeneracy of vibrational modes. It appears therefore that no morphic effects should be expected, in the sense that a pressure will only affect the strength of absorption and scattering and shift the mode frequencies. This is true, provided that the structural phase of the material is not affected by the applied pressure. As we shall see, a hydrostatic pressure can, in some cases, induce a phase transition which is accompanied by an overall change in the symmetry of the lattice. Thus it is possible for morphic effects to occur. Furthermore, the pressure can shift the electronic energy bands which, as was already discussed in the case of uniaxial stress, can induce resonant Raman scattering.

It becomes clear that pressure-induced morphic effects are perfectly legitimate as such, and that the present discussion should be extended to include some of these possibilities. Of course, experimenting in the presence of pressures has been a very popular direction of research owing to the ample information one can obtain through such studies. Similarly, in light-scattering problems a significant amount of work has been reported, especially following the recent development of very high (by laboratory standards) pressure technology. The fact that extremely small samples (submillimeter size) can be probed under such pressures in Raman spectroscopy constitutes an additional advantage of this experimental procedure. In the next two sections we hope only briefly to touch on some characteristic possibilities from the abundance of experiments which have appeared in the literature.

13.1. Pressure-induced changes in the phonon frequencies

Early work on infrared and far-infrared absorption includes a number of investigations of microquantities of materials, using a high-pressure diamond cell (Lippincott et al. 1966, Brasch and Jakobsen 1965, Postmus et al. 1968a–c and references therein, Cundill and Sherman 1968). Similar systematic studies in the field of Raman spectroscopy start with the work of Postmus et al. (1968a–c) in powdered HgI_2, and Assell and Nicol (1968) in α-quartz at pressures up to 40 kbar. No theoretical analysis of the results was undertaken in these early studies. At the same time, Mitra et al. (1969) reported an extensive study of pressure-induced shifts of long-wavelength TO and LO phonons in diamond-, zincblende- and wurtzite-type crystals for pressures up to 10 kbar. From these studies experimental values were obtained for the mode Grüneisen parameters for the TO and LO components of the $\boldsymbol{q} \simeq 0$ phonons of the zincblende-type materials ZnSe, ZnTe, GaP and SiC and for the LO phonon of ZnS. The LO component of the E_2 phonons of the wurtzite-type compounds CdS and ZnO was also studied. No change of the phonon half width with the pressure was observed and the shift of the frequency was in most cases linear with the pressure. The values of γ_{LO} were found to be nearly the same for all the compounds investigated and consistently smaller than γ_{TO}. The ratio γ_{TO}/γ_{LO} increased with ionicity, while the values of γ_{E_2} for the low-frequency E_2-type modes of CdS and ZnO were found to be negative. This is connected with the close relation between the wurtzite and zincblende structures and also to the anomalous negative volume coefficient of thermal expansion which characterizes diamond-like materials.

In fig. 19 the γ_{TO} values of several materials are plotted as a function of the effective ionic charge per electron. The nearly linear increase of γ_{TO} reflects clearly its dependence on ionicity (Brafman and Mitra 1971). Experimental information about the mode Grüneisen parameters along

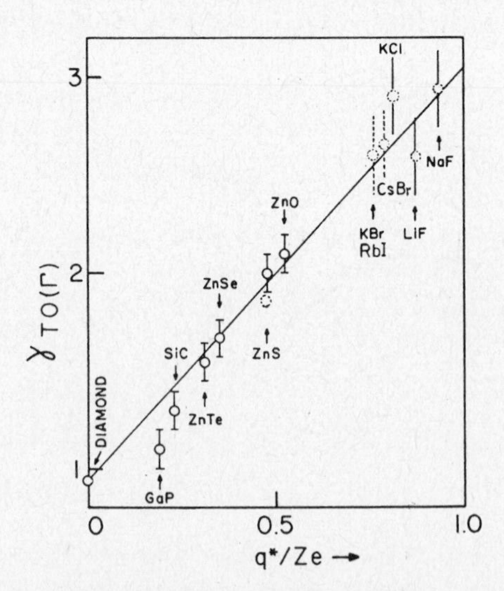

Fig. 19. The long-wavelength TO mode Grüneisen parameter versus the effective ionic charge per electron for several materials. Solid circles, Raman data. Dotted circles, infrared data (Brafman and Mitra 1971).

with the pressure derivatives of the elastic and dielectric constants are particularly useful in lattice-dynamical calculations (Namjoshi et al. 1971 and references therein). As an extension of the above experimental work, Brafman and Mitra (1971) also studied the pressure-induced frequency shifts of second-order Raman bands in CdS, ZnO, ZnS, ZnTe and ZnSe. It is shown that through such measurements a reliable identification of second-order structures may be reached as well as numerical values for non-zone-center phonon Grüneisen parameters. Knowledge of such parameters is critical in understanding various mechanisms in related topics, such as lattice dynamics, phase transitions, thermal expansion, etc. Recently, the first- and second-order Raman spectra of Mn-doped CdS and Mn- and Co-doped ZnS were investigated at pressures up to 40 kbar (Zigone et al. 1976). Symmetry assignments for the impurity modes became possible from these measurements.

Raman experiments with pressures up to 8 kbar have been performed by Buchenauer et al. (1971) for Ge, Si, GaAs, AlSb, Mg_2Si, Mg_2Ge and Mg_2Sn. The results on these materials and those of other workers for the relevant pressure coefficients and Grüneisen parameters are accumulated in table 11. The Grüneisen constants $\gamma_{\pm}$ are defined as

$$\gamma_+ = -\frac{d\ln(\omega_{LO}+\omega_{TO})}{d\ln V}, \tag{13.1}$$

$$\gamma_- = -\frac{d\ln\Delta}{d\ln V} = -\frac{d\ln(\omega_{LO}-\omega_{TO})}{d\ln V}. \tag{13.2}$$

As becomes clear from table 11, the stress-deduced values for γ are systematically lower than those from pressure experiments. It has already been commented in §11.2.1 that this is probably due to a strain relaxation at the surface where the scattering process occurs (Cerdeira et al. 1972).

Germanium–silicon alloys have been studied under pressures up to 10 kbar (Renucci et al. 1971b). The relevant Grüneisen constants of the Raman frequencies for the Ge–Ge and Si–Si pairs are found to increase from the value for the pure element when approaching the 50% concentration, reaching the values 1.3 ± 0.1 and 1.24 ± 0.02 respectively. The value for the Ge–Si pair mode also increases, with a maximum value of 1.20 ± 0.04. These characteristics are associated with the dominant role of vibrational eigenstates from the zone boundary as the disorder increases.

Negative Grüneisen constants have been measured for all three Raman-active modes ($A_1 + 2E$) of Se and Te (Richter et al. 1973, Fjeldly and Richter 1975). This has been related to the resemblance between these structures and a distorted simple cubic lattice with one atom per unit cell (Grosse 1969). Calculations based on a valence force model by Nakayama

Table 11

Pressure coefficients and Grüneisen parameters for several materials investigated by different workers using various techniques.

Material	$\frac{d\omega_0}{dP}$ or $\frac{d\omega_{TO}}{dP}$ (cm^{-1}/kbar)[b]	$\frac{d\Delta}{dP}$ (cm^{-1}/kbar)[b]	γ b	γ c (uniaxial work)	γ_+ b	γ_- b
diamond	0.28 ± 0.02[a]	—	0.94 ± 0.10[a]	1.12[o]	—	—
	0.36 ± 0.03[j]		1.19 ± 0.09[j]			
	0.29 ± 0.02[p]		1.20 ± 0.10[d]			
	0.32 ± 0.02[o]		0.96 ± 0.10[p]			
			0.98 ± 0.04[t]			
Si	0.54 ± 0.01	—	1.02 ± 0.02	0.90 ± 0.18[e]	—	—
			0.98 ± 0.06[i]			
			1.06[k]			
Ge	0.46 ± 0.01	—	1.12 ± 0.02	0.89 ± 0.1	—	—
			1.29,[f] 1.23[g]			
			1.13,[k] 0.88[u]			
Mg_2Si	0.64 ± 0.01	—	1.36 ± 0.02	—	—	—
Mg_2Ge	0.68 ± 0.02	—	1.45 ± 0.05	—	—	—
Mg_2Sn	0.72 ± 0.05	—	1.34 ± 0.08	—	—	—
GaAs	0.49 ± 0.03	−0.0075 ± 0.014	1.34 ± 0.08	0.90 ± 0.30	1.23 ± 0.08	−0.25 ± 0.50
GaSb	0.50 ± 0.005	0.012 ± 0.024	1.23 ± 0.02	1.10 ± 0.22	1.20 ± 0.02	0.70 ± 1.40
AlSb	0.67 ± 0.03	0.026 ± 0.020	1.23 ± 0.05	—	1.15 ± 0.05	0.70 ± 0.60
			1.09 ± 0.03[l]			
GaP[a,h,r]	0.43	0.05	1.07 ± 0.10	—	1.05	1.2
			1.19 ± 0.04[m]			
ZnO[a,h]	0.50[s]	—	2.10[r]	—	—	—
	−0.11[q]					
ZnS[a,h]	0.74	−0.18	1.85	—	1.45	−1.8
ZnSe[a,h]	0.57	−0.09	1.7	1.8 ± 0.36	1.35	−1.9
ZnTe[a,h]	0.54	−0.15	1.6	—	1.28	−2.9
			1.5[n]			
CdS[a,h]	−0.20[q]					
	0.45[r]	—	1.1[r]	—	—	—

[a]Mitra et al. (1969).
[b]Buchenauer et al. (1971).
[c]Cerdeira et al. (1972).
[d]Drickamer et al. (1966).
[e]Anastassakis et al. (1970).
[f]Bienenstock (1964).
[g]Dolling and Cowley (1966).
[h]Brafman and Mitra (1971).
[i]Weinstein and Piermarini (1975).
[j]Parsons and Clark (1976), and Parsons (1977).
[k]Jex (1971).
[l]Weinstein and Piermarini (1974).
[m]Weinstein et al. (1973).
[n]Vetelino et al. (1970).
[o]Grimsditch et al. (1978), best fit.
[p]Holzapfel and Anastassakis (1974).
[q]Low-frequency E_2 mode at 99 cm^{-1} (ZnO), 42 cm^{-1} (CdS).
[r]TO at 365 cm^{-1} (GaP); TO at 407 cm^{-1} (ZnO); LO at 305 cm^{-1} (CdS).
[s]High-frequency E_2 mode at 438 cm^{-1}.
[t]Whalley et al. (1976).
[u]Asaumi and Minomura (1978).

and Odajima (1972) yield values for the pressure coefficients (in cm^{-1}/kbar) which are in reasonable agreement with the ones measured. Negative Grüneisen coefficients have also been observed in As and Sb for the Raman-active modes E_g and A_{1g} and in Bi for the A_{1g} modes (Renucci 1974, Richter et al. 1978a).

The effect of pressures up to 7 kbar on the Raman-active A_{1g} and B_{3g} modes of the IV–VI orthorhombic layer compounds GeS and GeSe has been reported by Chandrasekhar et al. (1977). The pressure coefficients and mode Grüneisen parameters of the rigid layer modes are found to be up to one order of magnitude larger than those of the rocksalt-type modes. Such behavior is regarded as consistent with the weaker interlayer forces of these materials as compared to intralayer forces.

Grüneisen parameters for some non-zone-center modes have been obtained by Weinstein et al. (1973) and by Weistein and Piermarini (1974) from second-order Raman measurements under pressure in GaP. The Grüneisen parameter for the TA phonons along the line from X to the K point of the zone is found to be negative. Such negative values for zone-edge TA phonons have been previously observed in Ge by Payne (1964). They are also anticipated in the calculations of Dolling and Cowley (1966), Bienenstock (1964) and Jex (1971), and seem to be related to the fact that long-range forces become more important as the volume decreases. Richter et al. (1975) have studied the 2TA peak in the second-order Raman spectrum of Si under pressures up to 6 kbar. This peak has been assigned to an overtone of phonons near the X point in the Brillouin zone. The Grüneisen parameter of this overtone was found to be -1.4, in disagreement with the results (-0.6) of theoretical models where up to second-nearest neighbors are included. It was concluded that anharmonic forces between further neighbors should be considered.

The use of Raman spectroscopy under pressure has also proved itself very informative in the study of more complex molecular crystals which are capable of exhibiting internal and external modes. In the case of the ring-molecule elemental crystal orthorhombic sulfur, and the layer-structure chalcogenide crystal As_2S_3, Zallen (1974) has determined through such studies satisfactory relations between mode frequencies and mode Grüneisen parameters. The gross connection seems to follow the relation $\gamma_j \sim \omega_j^{-2}$, while the large range of γ-values appears to be related to the range of force constants of the material.

Parsons and Clark (1976) and Parsons (1977) have studied IIb-type diamond with pressures up to 24 kbar. Linear pressure coefficients and Grüneisen parameters have been determined for the first-order and most of the major second-order Raman scattering features (table 11). It is interesting that the rate of frequency shift of the sharp second-order peak at 2667 cm^{-1} is twice that of the first-order peak at 1332.4 cm^{-1}. Similar measurements on the first-order peaks by Holzapfel and Anastassakis (1974) for pressures up to 80 kbar have not revealed any quadratic pressure dependence of the $\boldsymbol{q} \simeq 0$ optical phonon frequency. The rate of shift observed by Holzapfel and Anastassakis agrees with the one reported by Mitra et al. (1969) (see table 11).

Members of the ferroelectric arsenates family have been investigated under pressures up to 10 kbar by Leung et al. (1976) using light-scattering techniques. The pressure coefficients $(\partial\omega_j/\partial P)$ and $(\partial\Gamma_j/\partial P)$ have been measured, where Γ_j is the damping constant of the ferroelectric mode. These values were then used to test various aspects of the pseudospin model of hydrogen-bonded ferroelectrics (de Gennes 1963, Kobayashi 1968, Young and Elliot 1974).

A number of copper halides, such as CuI, CuBr and CuCl, have been studied by Hochheimer et al. (1976) at variable temperatures below 300 K and under pressures up to 7 kbar. By combining pressure and temperature coefficients these authors succeeded in separating anharmonic contributions into volume effects and multiphonon effects. In addition, Shand et al. (1976) have looked into the anomalous TO-phonon region of CuCl in more detail, at 40 K and pressures up to 7 kbar. Particular attention is given to the so-called β and γ peaks which are explained as both being associated with the TO phonon. According to theory, this arises from coupling by cubic anharmonicity of the harmonic TO phonon to a two-phonon density of states characterized by a Van Hove P_3 singularity at its high-frequency end. These predictions have been further substantiated by the recent measurements of Shand and Hanson (1977) in CuCl under pressures up to 31 kbar.

In all the studies mentioned above the range of pressures was rather low, rarely exceeding 40 kbar. In recent years, however, with the rapid development of high-pressure technology, extremely high (by optical measurements standards) pressures beyond 100 kbar have been achieved (Piermarini and Block 1975, Block and Piermarini 1976). This has opened a practically new field of lattice-dynamical investigations, with particular emphasis on higher-order anharmonic interactions between near neighbors, negative thermal-expansion coefficients, etc. Thus, for instance, Sherman et al. (1971) have used pressures up to 300 kbar in conjunction with spectroscopic methods to study internal modes of vibration of polyatomic impurity ions isolated in an alkali halide. A number of cubic materials have also been studied under very high pressures. The first-order Raman spectrum of $PbTiO_3$ consisting of eight phonons was investigated by Cerdeira et al. (1975) at 300 K, at pressures up to 80 kbar (fig. 20). A

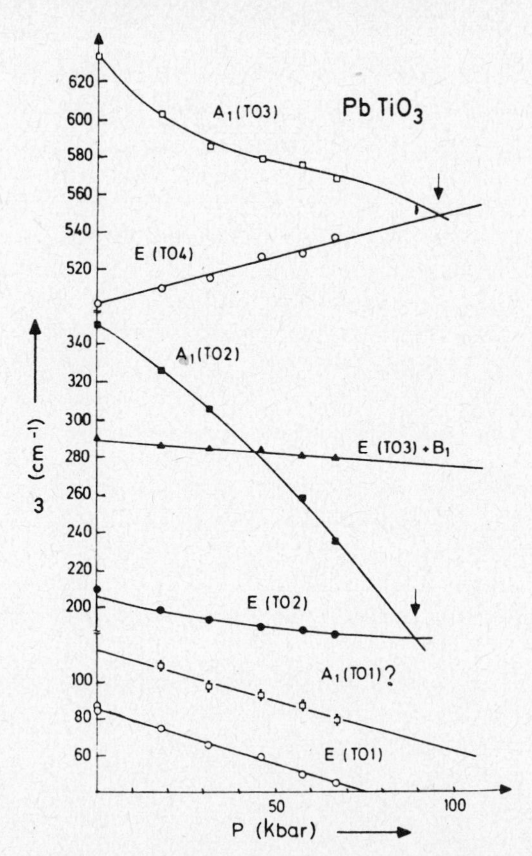

Fig. 20. Pressure dependence of the frequencies of the different modes of $PbTiO_3$ at 300 K. The points correspond to measured values while the solid lines represent least-square fits to the data using polynomials with up to cubic terms in P. The arrows indicate extrapolated crossing points (Cerdeira et al. 1975).

two-anvil sapphire cell was used. Linear pressure coefficients and Grüneisen parameters were obtained for all but one phonon. Furthermore, three of the modes showed a quadratic pressure dependence and one mode even showed a term cubic in pressure. These results are summarized in table 12. Included in table 12 are also the results of Weinstein and Piermarini (1975), who have examined the first- and second-order Raman spectrum of GaP and Si at 300 K under pressures up to 135 kbar using a diamond anvil cell. In this way the general effects of pressure on the phonon dispersion curves were studied (fig. 21). Quadratic pressure coefficients and mode Grüneisen parameters were also obtained for critical points at the zone

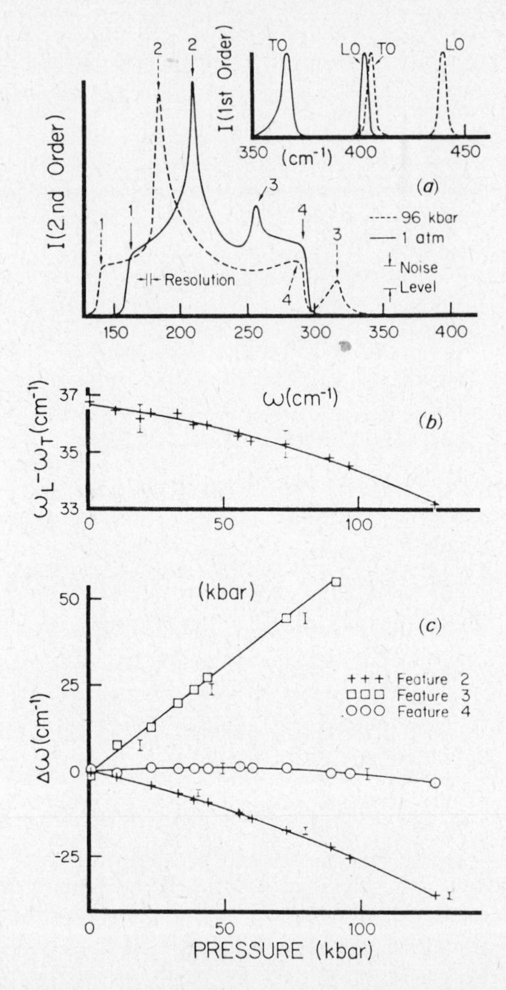

Fig. 21. (a) Measured first- and second-order room temperature Raman spectra of GaP at 1 atm and 96kbar. Intensity scales are arbitrary. (b) Energy splitting of the one-phonon peaks as a function of pressure. (c) Frequency shift of features 2–4 as a function of pressure. Features 1 (omitted for clarity) and 2 have a similar pressure dependence. Note the large positive shift for the difference mode, feature 3. For clarity, error flags are displaced. Features 2 and 4 correspond to the overtones 2TA (X and Σ) and 2A (W and Σ) (Weinstein and Piermarini 1975).

Table 12

First- and higher-order pressure coefficients of $PbTiO_3$ (Cerdeira et al. 1975), Si and GaP (Weinstein and Piermarini 1975).

Mode	$\omega_0(P=0)$ (cm^{-1})	$\frac{d\omega_0}{dP}$ (cm^{-1}/kbar)	$10^2 \frac{d^2\omega_0}{dP^2}$ ($cm^{-1}/kbar^2$)	$10^3 \frac{d^3\omega_0}{dP^3}$ ($cm^{-1}/kbar^3$)	γ
E(TO1)	85.0 ± 0.3	-0.60 ± 0.20	—	—	-6.0 ± 0.2
A_1(TO1)?	117 ± 3	-0.55 ± 0.05	—	—	-4.0 ± 0.5
E(TO2)	205 ± 3	-0.43 ± 0.10	-0.4 ± 0.2	—	-1.8 ± 0.4
E(TO3) + B_1	289 ± 0.5	-0.15 ± 0.03	—	—	-0.44 ± 0.09
A_1(TO2)	350 ± 1	-1.15 ± 0.08	-1.7 ± 0.3	—	-2.8 ± 0.2
E(TO4)	501.5 ± 0.6	0.49 ± 0.05	—	—	0.84 ± 0.08
A_1(TO3)	634 ± 1.5	-2.4 ± 0.2	7 ± 2	1.2 ± 0.6	-3.2 ± 0.3
Si	519.5 ± 0.8	0.52 ± 0.03	-0.14 ± 0.04	—	0.98 ± 0.06
GaP	365.5 ± 1[a]	0.49 ± 0.02[c]	-0.07 ± 0.01	—	1.09 ± 0.03
	402.5 ± 1[b]	0.53 ± 0.02[c]	-0.08 ± 0.02	—	0.95 ± 0.02

[a]Refers to the $\boldsymbol{q} \simeq 0$ TO phonon.
[b]Refers to the $\boldsymbol{q} \simeq 0$ LO phonon.
[c]Weinstein et al. (1973).

boundary as well as at the zone center (table 11). Thus, for instance, the frequency of the $\boldsymbol{q} \simeq 0$ phonon of Si is found to follow an expression like

$$\omega_j(P) = \omega_{j0} + aP + bP^2, \tag{13.3}$$

where a and b are the first- and second-order pressure coefficients with values (0.52 ± 0.03) $cm^{-1} \cdot kbar^{-1}$ and (-0.0007 ± 0.0002) $cm^{-1} \cdot kbar^{-2}$ respectively, and $\omega_{j0} = (519.5 \pm 0.8)$ cm^{-1}. The thermal expansion coefficient of Si as a function of temperature was calculated and compared with the experimental values. Similar work has been reported by Trommer et al. (1976) for GaAs. The $\boldsymbol{q} \simeq 0$ TO and LO phonon frequencies show a non-linear shift with pressures up to 72 kbar. The evaluated strong dependence (decrease) of the effective charge on the pressure indicates a pressure-dependent ionicity.

13.2. Pressure-induced phase transitions and resonant Raman scattering

Another aspect of the effects a hydrostatic pressure can have on a material is the possibility of pressure-induced phase transitions with consequent changes in the symmetry of the lattice. Again, Raman spectroscopy is a unique tool for probing such morphic effects.

Thallium iodide is one of the first materials that has been investigated in this connection (Brafman et al. 1969). At 300 K and 1 atm, TlI exhibits an orthorhombic structure (space group D_{2h}^{17} which changes to that of CsCl at 716 K (1 atm) or 5 kbar (300 K) (Samara et al. 1967). Five bands corresponding to first-order Raman scattering by Raman-active phonons in the orthorhombic phase were observed to decrease and diminish as the pressure approached the critical value of 5 kbar. The frequency shifts of these five bands with pressures below 5 kbar gave an average value of 3 for the mode Grüneisen parameter. A similar study has been reported for SbSI by Balkanski et al. (1971). This material exhibits a paraelectric orthorhombic phase (space group D_{2h}^{16}) above $T_0 \simeq 293$ K and a ferroelectric orthorhombic phase (space group C_{2v}^{9}) below T_0. A hydrostatic pressure is known to cause a linear decrease of T_0 and thus it can be used to monitor the phase transition at constant temperature. The Raman spectra for the three lowest frequency Γ_1 modes show a definite and rather abrupt frequency decrease as the Curie temperature is adjusted to coincide with the temperature of the sample. Brafman and Cardona (1977) investigated the two phase transitions which occur in CuI under hydrostatic pressure. These are a cubic-to-rhombohedral (at about 14.5 kbar) and a rhombohedral-to-tetragonal (at about 40 kbar). The results are analyzed by using (i) the superstructure relationship of this phase to the zincblende structure, and (ii) the strong polarizability exhibited by the I^- ions.

Cerdeira et al. (1975) have studied the tetragonal-to-cubic phase transition of $PdTiO_3$ at 300 K with pressures up to 80 kbar. From the observed behavior of the scattering strength and frequency shifts, they deduce a value of (90 ± 4) kbar for the necessary critical pressure at which the tetragonal (C_{4v}) ferroelectric phase can change continuously into the cubic (O_h) paraelectric phase at 300 K.

The semiconductor-to-metal transition of Si has been induced by pressures of (125 ± 5) kbar (Weinstein and Piermarini 1975). The metallic phase to which Si transforms is that of α-Sn. Some common behavior is recognized for the zone-boundary configuration of the flat TA branch, between α-Sn and metallic Si. Zigone et al. (1976) have reported a phase transition of CdS from the wurtzite to the NaCl structure at (26 ± 2) kbar. The transition manifests itself from the disappearance of all Raman activity as the NaCl structure is approached. Peercy (1976) has studied the behavior of the soft mode and the paraelectric-to-ferroelectric phase transition as a function of pressure and temperature of KH_2PO_4 (or KDP) and SbSI. It is established from these measurements that the soft mode of KDP may be regarded as a propagating excitation in the paraelectric phase and that the proton motion remains coupled to an optical phonon in the ferroelectric phase. In SbSI it is shown that there is a Curie critical point at $T_0 \simeq 235$ K, $P_0 \simeq 1.40$ kbar at which the usual first-order phase transition changes to a second-order one. Paratellurite (TeO_2) has been shown from X-ray measurements to exhibit a pressure-induced phase transition at 8 kbar, from a tetragonal (space group D_4^4) to an orthorhombic (space group D_2^4) structure (Skelton et al. 1976 and references therein).

Pressure measurements can be particularly useful in elucidating different aspects of intermolecular forces and phase transitions in molecular crystals. Thus, pressure effects on the phase transitions of hexamethyl benzene (HMB), potassium nitrate (KNO_3) and ice have been studied by Bodenheimer et al. (1976) and by Kennedy et al. (1976). Also Iqbal and Christoe (1976) have reported on, and proposed a mechanism for, the pressure-induced phase transition of thiourea. Furthermore, Iqbal et al. (1977) identify a second-order phase transition in crystalline S_4N_4 at 295 K and 7 kbar pressure from the discontinuity that the pressure coefficients of the Raman-active external modes and the S–S stretching modes exhibit near 7 kbar.

One more aspect of pressure-related phenomena in Lattice Dynamics is the pressure-induced resonant Raman scattering. A unique example of such a possibility is provided by GaAs. Conventional resonant Raman scattering is rather difficult in this material, the reason for this being that

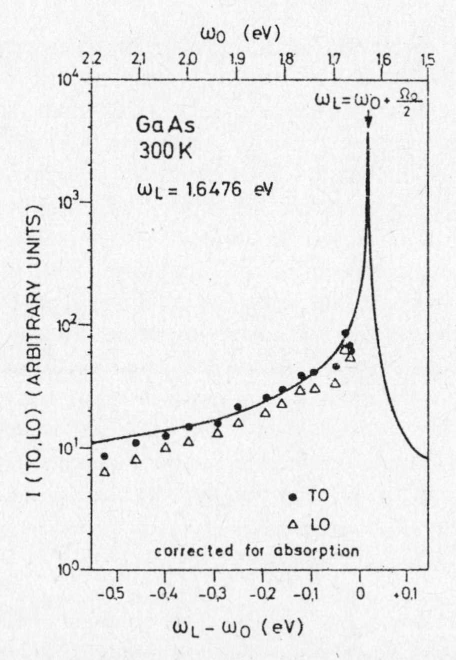

Fig. 22. Pressure-induced resonant Raman scattering at 300 K by the TO and LO long-wavelength optical phonons of GaAs at the E_0 gap. ω_0 corresponds to the pressure-dependent optical gap. Solid line, theoretical fit (Trommer et al. 1976).

the E_0 gap (1.43 eV) is beyond the range of those laser frequencies which can be used for Raman scattering measurements. However, an increasing pressure shifts the E_0 gap towards the 1.65 eV line of Kr^+ laser (Welber et al. 1976). Using this procedure Trommer et al. (1976) have succeeded in observing a pressure-induced enhancement of the first-order Raman scattering at 300 K (fig. 22). A theoretical fit (shown by the solid curve in fig. 22) based on the dielectric theory of Raman scattering by TO phonons at the E_0 gap (Weinstein and Cardona 1973) appears to give a very good description of the experimental results. Additional measurements of pressure-induced resonant Raman scattering in GaAs have been undertaken by Yu and Welber (1977). With a fixed pressure at 62 kbar and a dye-laser variable excitation frequency, Yu and Welber have investigated the entire region of first- and second-order Raman scattering and also the effect of pressure on its photoluminescence.

14. Temperature-induced morphic effects

Like pressure, temperature is a scalar force and can neither modify the polarization selection rules nor remove the degeneracy of vibrational modes. Nevertheless it can affect the frequency and linewidth of phonons owing to volume expansion and higher-order anharmonic interactions (Hart et al. 1970, Anastassakis et al. 1971, Borer et al. 1971). The effect of temperature on the Raman scattering intensity originates from the phonon population factors (Hart et al. 1970, Anastassakis et al. 1971, Ray et al. 1971) and also from the possibility of tuning an appropriate energy gap to the excitation frequency (temperature-induced resonant Raman scattering, Rennuci et al. 1971a and references therein, Anastassakis and Perry 1971). Furthermore, temperature can induce phase transitions in solids, and this results in gross changes in the symmetry of the system. Such changes can manifest themselves through Raman scattering by phonons which are Raman inactive in one phase but become Raman active in the other phase. A typical example is the second-order cubic-to-tetragonal displacive transition at 106 K of $SrTiO_3$ (Scott 1974). One could also mention the possibility of splitting of a doubly-degenerate E-type mode in a ferroelectric phase, due to a pseudospin-two-phonon coupling mechanism (Young and Elliot 1974). Such splittings have been observed in ferroelectric arsenates (Leung et al. 1976). Finally, the magnetic ordering induced at low temperatures in magnetic materials is known to cause changes in the symmetry which in turn have an effect on its lattice-dynamical behavior. This particular type of morphic effect will be examined independently in §17.

The subject of temperature effects as outlined above is far too broad to be handled within the scope of the present chapter. We can only aim at alerting the reader about these possibilities of morphic effects. A systematic study of structural phase transitions can be found in the review papers of Worlock (1971) and Scott (1974). For more specific temperature effects and types of materials, in connection always with light scattering (Raman or Brillouin), the reader is referred to the proceedings of International Conferences in *Light Scattering in Solids* (1969), ed. by G. Wright (Springer-Verlag, New York), (1971) ed. by M. Balkanski (Flammarion, Paris) and (1976) ed. by M. Balkanski et al. (Flammarion, Paris); also in *Phonons* (1971) ed. by M. A. Nucimovici (Flammarion, Paris), and in *Lattice Dynamics* (1977) ed. by M. Balkanski (Flammarion, Paris).

15. Magnetic-field-induced infrared absorption and Raman scattering

15.1. General

In all the morphic effects that we have discussed so far the forces applied were polar and time symmetric. It was stated in the introduction that axial and/or time antisymmetric forces are also capable of inducing morphic effects. The chief example of such forces is a magnetic field, which is both axial and time antisymmetric. As we shall see, the combined space–time symmetry restrictions have interesting consequences on the polarization selection rules for absorption and scattering, and also on the removal of mode degeneracies. The corresponding morphic effects are treated to first order in the magnetic field. Effects quadratic in the magnetic field are equivalent to the effects of an applied stress and therefore need not be considered. For some crystal structures and propagation configurations, the natural eigenvectors of the propagating photons and phonons are circularly polarized. In these cases we employ circular coordinates in deriving the matrix forms of the effective charge and Raman tensor. Such a procedure is very convenient, and often necessary, in interpreting circular dichroism-type effects which may arise from optical phonons in the presence of a magnetic field. We also note that the present results, which are based on the same first-order perturbation theory followed previously, apply to the case of a magnetically-ordered crystal with a non-zero average magnetic moment per unit volume (ferro-, ferrimagnets, etc.), to first order in the magnetic moment. This is because a magnetic field and a magnetic moment have the same space- and time-transformation properties. The gross reduction of symmetry due to magnetic ordering will be examined separately, in §17.

In the presence of a magnetic field $\boldsymbol{H}$ the Hamiltonian and the eigenfunctions of the system are not real in general (Laundau and Lifshitz 1958,

Messiah 1962). Instead, the Hamiltonian is known to satisfy the condition

$$\mathcal{H}(\boldsymbol{H})=\mathcal{H}^*(-\boldsymbol{H}). \tag{15.1}$$

An expression for $\mathcal{H}(\boldsymbol{H})$ linear in the field and satisfying eq. (15.1) can be written as

$$\mathcal{H}(\boldsymbol{H})\simeq\mathcal{H}^{(0)}+\mathrm{i}\mathcal{H}^{(1)}(\boldsymbol{H}), \tag{15.2}$$

where $\mathcal{H}^{(0)}$ is real and independent of $\boldsymbol{H}$ and $\mathcal{H}^{(1)}$ is real and linear in $\boldsymbol{H}$. To a higher approximation, all even and odd powers of $\boldsymbol{H}$ are included in $\mathcal{H}^{(0)}$ and $\mathcal{H}^{(1)}$ respectively. Similar expressions are valid for the crystal potential energy Φ.

15.2. Magnetic-field-induced infrared absorption

In view of eq. (15.2), the crystal electric moment per unit cell in the presence of a magnetic field can be written as

$$M_\mu(\boldsymbol{H})\simeq M_\mu^{(0)}+\mathrm{i}M_\mu^{(1)}(\boldsymbol{H})=M_\mu^{(0)}+\mathrm{i}m_{\mu\nu}H_\nu, \tag{15.3}$$

to terms linear in the field. From the condition of eq. (15.1) it follows that

$$\boldsymbol{M}(\boldsymbol{H})=\boldsymbol{M}^*(-\boldsymbol{H}), \tag{15.4}$$

which means that the parameter

$$m_{\mu\nu}\equiv\frac{\partial M_\mu^{(1)}}{\partial H_\nu} \tag{15.5}$$

is a real and, in general, non-symmetric generalized *kinetic* coefficient. It should not be identified with the well-known macroscopic magnetoelectric tensor $\lambda_{\mu\nu}=(\partial M_\mu/\partial H_\nu)\neq\lambda_{\nu\mu}$ (Bhagavantam 1966, Birss 1964), which is a *static* property of the crystal, and as such vanishes identically for all non-magnetic crystal classes (because of time reversal symmetry).

From eqs. (15.3) and (3.1) one finds

$$M_\mu(\boldsymbol{H})\simeq M_{0\mu}+(e_{j,\mu\sigma}+\mathrm{i}\Delta e_{j,\mu\sigma})u_{j\sigma}, \tag{15.3a}$$

where

$$e_{j,\mu\sigma} = \left(\frac{\partial M_\mu^{(0)}}{u_{j\sigma}} \right), \tag{15.3b}$$

$$\Delta e_{j,\mu\sigma} = \left(\frac{\partial M_\mu^{(1)}}{\partial u_{j\sigma}} \right) = \left(\frac{\partial m_{\mu\nu}}{\partial u_{j\sigma}} \right) H_\nu = \left(\frac{\partial^2 M_\mu^{(1)}}{\partial H_\nu \partial u_{j\sigma}} \right) H_\nu \equiv h_{\mu\sigma\nu} H_\nu. \tag{15.3c}$$

The mode coefficient

$$h_{\mu\sigma\nu} = \left(\frac{\partial m_{\mu\nu}}{\partial u_{j\sigma}} \right) = \left(\frac{\partial^2 M_\mu^{(1)}}{\partial H_\nu \, \partial u_{j\sigma}} \right) \tag{15.4a}$$

is real and non-symmetric. For a given crystal class and mode j, the number of its independent components is n_j, where

$$\Gamma(\boldsymbol{m}) = \Gamma(\boldsymbol{M}) \otimes \Gamma(\boldsymbol{H}) = n_j \Gamma(j) \oplus \ldots, \tag{15.5a}$$

according to the definition of $h_{\mu\sigma\nu}$ in eq. (15.4a). The last column of table A1 gives the reduction of $\Gamma(\boldsymbol{m})$ for the 32 point groups, i.e. it indicates the phonons that can exhibit this effect, together with n_j, the number of independent components of the tensor $h_{\mu\sigma\nu} \neq h_{\sigma\mu\nu}$. From eqs. (5.7) and (15.5a) we have

$$\Gamma(\boldsymbol{M}) \otimes \Gamma(\boldsymbol{M}) = n_j \Gamma(j_p) \oplus \ldots, \tag{15.5b}$$

where $\Gamma(j_p) \equiv \Gamma_p \otimes \Gamma(j)$. We conclude that $\boldsymbol{H}$-induced infrared absorption by a phonon j, whose irreducible representation is $\Gamma(j)$, is identical, from the standpoint of symmetry, to general non-symmetric Raman scattering by the phonon j_p, whose irreducible representation is $\Gamma(j_p)$. This observation enables one to write down the explicit matrix form of $\Delta \boldsymbol{e}_j(\boldsymbol{H})$ by using the non-symmetric Raman matrices for the phonon j_p. The transition from $a_{j_p,\mu\nu\sigma}$ to $\Delta e_{j,\mu\sigma}$ follows the scheme

$$a_{j_p,\mu\nu\sigma} \to \Delta e_{j,\mu\sigma} = h_{\mu\sigma\nu} H_\nu. \tag{15.6}$$

Two conclusions are obvious: (i) In centrosymmetric crystals $\Gamma(\boldsymbol{M})$ and $\Gamma(\boldsymbol{H})$ reduce only to irreducible representations of odd and even parity respectively. Thus, $\Gamma(\boldsymbol{M}) \otimes \Gamma(\boldsymbol{H})$ includes only modes of odd parity. According to eq. (15.5a) then, infrared *absorption induced by* $\boldsymbol{H}$ *in first order may occur for modes of odd parity*. *No such effect is possible for totally*

Raman-active modes (i.e. modes which can produce symmetric and/or antisymmetric Raman scattering). This again is to be contrasted with the corresponding effects induced by an applied electric field. The latter is possible only for those modes of even parity which are totally Raman active (see §6.1). (ii) For crystal classes with no improper rotations (such classes must be non-centrosymmetric), $\boldsymbol{M}$ and $\boldsymbol{H}$ transform identically, i.e. $\Gamma_p = A_1$ and $\Gamma(j_p) = \Gamma(j)$. Eq. (15.6) is then further simplified. Recently, Kanamori et al. (1976b) and Matsumoto and Kanamori (1977) completed the tabulation of the effective charges linear in the magnetic field, for all those cases where the effect is allowed. Let us take two examples.

(i) First we consider the E mode of trigonal tellurium. The point group is D_3 and according to table A1, $\Gamma_p = A_1$. The most complete (non-symmetric) Raman matrices for the Raman-active modes are (Wallis and Maradudin 1971)

$$a_{j_p,\mu\lambda\sigma} = a_{j,\mu\lambda\sigma} = \underset{A_1}{\begin{bmatrix} A & 0 & 0 \\ 0 & A & 0 \\ 0 & 0 & B \end{bmatrix}}, \quad \underset{A_2(z)}{\begin{bmatrix} 0 & F & 0 \\ -F & 0 & 0 \\ 0 & 0 & 0 \end{bmatrix}}$$

$$\underset{E(x)}{\begin{bmatrix} 0 & C & 0 \\ C & 0 & R+G \\ 0 & R-G & 0 \end{bmatrix}}, \quad \underset{E(y)}{\begin{bmatrix} C & 0 & -R-G \\ 0 & -C & 0 \\ -R+G & 0 & 0 \end{bmatrix}}, \tag{15.7a}$$

where the indices $\mu\lambda$ designate the $\mu\lambda$ component of each matrix and the index σ is determined from the type of representation. Thus, $\sigma = 1$ for A_1 and $A_2(z)$ and $\sigma = 1, 2$ for $E(x), E(y)$ respectively. The components F and G correspond to the antisymmetric part of the intrinsic tensor and are usually neglected. It is more convenient to write the last two matrices as

$$\underset{E(x)}{\begin{bmatrix} 0 & C & 0 \\ C & 0 & R' \\ 0 & G' & 0 \end{bmatrix}}, \quad \underset{E(y)}{\begin{bmatrix} C & 0 & -R' \\ 0 & -C & 0 \\ -G' & 0 & 0 \end{bmatrix}}. \tag{15.7b}$$

It is clear now that the modes A_1, A_2, E have 2, 1, 3 independent components, i.e. (A, B), F and (C, R', G') respectively, in agreement with eq. (5.2c). Each component of the matrices (15.7) leads to a corresponding component $\Delta e_{j,\mu\sigma}$ according to eq. (15.6). The results for the E mode are

$$\Delta e_j = \begin{bmatrix} cH_y & cH_x - rH_z \\ cH_x + rH_z & -cH_y \\ gH_y & -gH_x \end{bmatrix}, \tag{15.8}$$

where σ corresponds to x or y, since the mode is polarized in the xy plane. c, r and g are the three independent components of $h_{\mu\sigma\nu}$ for the mode E. In the absence of the field $\boldsymbol{H}$, $e_{j,\mu\sigma}$ is diagonal ($e_{j,xx} = e_{j,yy} \neq 0$). The field-induced components, however, are both diagonal and non-diagonal. Let us designate by $\hat{k}(\hat{\mu}\hat{\sigma})\hat{q}$ the configuration for a given experiment of first-order infrared absorption. $\hat{k}$ and $\hat{q}$ are the direction unit vectors for the incident (absorbed) photon and generated phonon respectively. They are parallel because of the principle of momentum conservation. $\hat{\mu}$ and $\hat{\sigma}$ are the unit polarization vectors associated with $\hat{k}$ and $\hat{q}$ respectively. For the configuration $z(yx)z$ and $\boldsymbol{H}=0$, no absorption should occur since $e_{j,yx}=0$. For the same configuration and $\boldsymbol{H}//\hat{z}$, however, absorption occurs with a strength proportional to

$$|\mathrm{i}\Delta e_{j,yx}|^2 = (rH_z)^2.$$

(ii) As a second example we consider the C_{3v} class for which $\Gamma_p = A_2$ and $\Gamma(\boldsymbol{M}) \otimes \Gamma(\boldsymbol{M}) = 2A_1 \oplus A_2 \oplus 3E$. If we take $\Gamma(j) = E$ then $\Gamma(j_p) = A_2 \otimes E = E$, and according to eq. (15.6) the charge $\Delta e_{j,\mu\sigma}$ can be expressed in terms of $a_{j_p,\mu\nu\sigma}$. Using the corrected matrices of Wallis and Maradudin (1971) we find

$$\Delta \boldsymbol{e}_j = \begin{bmatrix} cH_x + rH_z & -cH_y \\ -cH_y & -cH_x + rH_z \\ gH_x & gH_y \end{bmatrix}. \tag{15.9}$$

In the same way we find for the field-induced effective charge of the normally infrared-inactive mode A_2 of the C_{3v} class,

$$\Delta \boldsymbol{e}_j = \begin{bmatrix} bH_x \\ bH_y \\ cH_z \end{bmatrix}, \tag{15.10}$$

where $\hat{\sigma}$ is along $\hat{z}$, and b, c are the two independent components of $h_{\mu\sigma\nu}$ for the mode A_2. We notice that an infrared-inactive mode may become infrared active in first order with $\boldsymbol{H}$, in contrast to a Raman-inactive mode which cannot become Raman active in first order with $\boldsymbol{H}$, as will be seen shortly.

All the above analysis refers to the symmetry restrictions of the magnetic-field-induced infrared absorption. A physical mechanism which couples the $\boldsymbol{q}\simeq 0$ optical phonon to the incident photon in the presence of a magnetic field cannot be provided by group theory alone. A macroscopic theory based on an appropriate model is necessary. No such theory has been formulated so far, nor does there exist any experimental information on this particular type of morphic effect. A closely related observation of infrared absorption by the even-parity E_g phonon of CoF_2 without the need of an external magnetic field has been reported by Allen and Guggenheim (1968). These authors interpret their results by assuming an optical phonon mixing to a magnetic-dipole-active exciton level. A theoretical model for the absorption strength and other characteristics of the same effect has been proposed by Mills and Ushioda (1970), which too is based on an exciton–phonon coupling mechanism.

15.3. Magnetic-field-induced Raman scattering

When no magnetic field is present the electric polarizability per unit cell $\alpha_{\mu\lambda}$ is treated as a real and symmetric second-rank tensor. In the presence of $\boldsymbol{H}$, however, $\alpha_{\mu\lambda}(\boldsymbol{H})$ is neither symmetric nor real. To terms linear in $\boldsymbol{H}$ we can write

$$\alpha_{\mu\lambda}(\boldsymbol{H})\simeq\alpha_{\mu\lambda}^{(0)}+\mathrm{i}\alpha_{\mu\lambda}^{(1)}(\boldsymbol{H})=\alpha_{\mu\lambda}^{(0)}+\mathrm{i}f_{\mu\lambda\nu}H_{\nu}. \tag{15.11}$$

According to the properties of a generalized susceptibility (Landau and Lifshitz 1958), $\alpha_{\mu\lambda}(\boldsymbol{H})$ can be treated as a generalized kinetic coefficient satisfying the following two conditions:

$$\alpha_{\mu\lambda}(\boldsymbol{H})=\alpha_{\lambda\mu}(-\boldsymbol{H}), \tag{15.12a}$$

$$\alpha_{\mu\lambda}(\boldsymbol{H})=\alpha^{*}_{\lambda\mu}(\boldsymbol{H}). \tag{15.12b}$$

The first condition expresses Onsager's reciprocity theorem, a well-known property of any kinetic coefficient in the presence of a magnetic field. The second expresses the fact that in the absence of absorption, $\alpha_{\mu\lambda}(\boldsymbol{H})$ must be Hermitian. From eqs. (15.11) and (15.12) it follows that both $\alpha_{\mu\lambda}^{(0)}$ and $f_{\mu\lambda\nu}$ are real and that

$$\alpha_{\mu\lambda}^{(0)}=\alpha_{\lambda\mu}^{(0)}, \tag{15.13a}$$

$$f_{\mu\lambda\nu}\equiv\left(\partial\alpha_{\mu\lambda}^{(1)}/\partial H_{\nu}\right)=-f_{\lambda\mu\nu}. \tag{15.13b}$$

The antisymmetric tensor $f_{\mu\lambda\nu}$ is known as the *magneto-optical tensor*. It is a generalized *kinetic coefficient*, responsible for the Faraday effect in non-magnetic crystals (Landau and Lifshitz 1960). Its symmetry should not be identified with that of the well-known piezomagnetic tensor (Bhagavantan 1966, Birss 1964). The latter is a *static coefficient* and vanishes identically for non-magnetic crystals.

From eqs. (15.11) and (3.3a) we find

$$\alpha_{\mu\lambda}(\boldsymbol{H}) \simeq \alpha_{0\mu\lambda} + (a_{j,\mu\lambda\sigma} + \mathrm{i}\Delta a_{j,\mu\lambda\sigma})u_{j\sigma}, \tag{15.14a}$$

where

$$a_{j,\mu\lambda\sigma} = \left(\frac{\partial \alpha^{(0)}_{\mu\lambda}}{\partial u_{j\sigma}}\right), \tag{15.14b}$$

$$\Delta a_{j,\mu\lambda\sigma} = \left(\frac{\partial \alpha^{(1)}_{\mu\lambda}}{\partial u_{j\sigma}}\right) = \left(\frac{\partial f_{\mu\lambda\nu}}{\partial u_{j\sigma}}\right) H_\nu = \left(\frac{\partial^2 \alpha^{(1)}_{\mu\lambda}}{\partial H_\nu \, \partial u_{j\sigma}}\right) H_\nu \equiv b_{\mu\lambda\nu\sigma} H_\nu. \tag{15.14c}$$

The mode coefficient

$$b_{\mu\lambda\nu\sigma} \equiv \left(\frac{\partial f_{\mu\lambda\nu}}{\partial u_{j\sigma}}\right) = \left(\frac{\partial^2 \alpha^{(1)}_{\mu\lambda}}{\partial H_\nu \, \partial u_{j\sigma}}\right) H_\nu = -b_{\lambda\mu\nu\sigma} \tag{15.15}$$

is real and antisymmetric in the pair of indices μ and λ. Thus $\Delta a_{j,\mu\lambda\sigma}$ too is antisymmetric in μ and λ. The same is true for all terms of the Raman tensor which are of odd powers in $\boldsymbol{H}$. By contrast, $a_{j,\mu\lambda\sigma}$ is (for most practical purposes) symmetric in μ and λ, and so are all higher-order terms of the Raman tensor which are of even powers in $\boldsymbol{H}$. These are consequences of the non-real character of $\mathcal{H}(\boldsymbol{H})$.

The above symmetry properties of $b_{\mu\lambda\nu\sigma}$ are quite general and independent of crystal class and phonon type j. In addition to these, once the crystal class and phonon j are specified, the invariance under spatial symmetry operations will impose further restrictions on the form of $b_{\mu\lambda\nu\sigma}$. For a given crystal class and phonon j the number of independent components of the coefficient $b_{\mu\lambda\nu\sigma}$ is given according to eq. (15.15) by n_{A}, where

$$\Gamma(\mathbf{b}) = \Gamma(f) \otimes \Gamma(j) = n_{\mathrm{A}} \mathrm{A}_1 + \ldots \tag{15.16a}$$

$\Gamma(f)$ is the representation of the tensor $f_{\mu\lambda\nu}$, given by

$$\Gamma(f) = [\Gamma(\mathcal{E}) \otimes \Gamma(\mathcal{E})]_{\mathrm{A}} \otimes \Gamma(\boldsymbol{H}). \tag{15.17a}$$

$\Gamma(\boldsymbol{H})$ and $\Gamma(\mathfrak{E})$ are the representations (not necessarily irreducible) according to which $\boldsymbol{H}$ and $\mathfrak{E}$ transform respectively. $[\Gamma(\mathfrak{E})\otimes\Gamma(\mathfrak{E})]_A$ is the antisymmetric part of the double Kronecker product $\Gamma(\mathfrak{E})\otimes\Gamma(\mathfrak{E})$. Since the components of $\boldsymbol{H}$ transform as the antisymmetric combinations of the components of two polar vectors it turns out that $[\Gamma(\mathfrak{E})\otimes\Gamma(\mathfrak{E})]_A$ is identical to $\Gamma(\boldsymbol{H})$ and therefore

$$\Gamma(\boldsymbol{f})=\Gamma(\boldsymbol{H})\otimes\Gamma(\boldsymbol{H}). \tag{15.17b}$$

From eqs. (15.17b) and (5.7) we have

$$\begin{aligned}\Gamma(\boldsymbol{f})&=\Gamma_p\otimes\Gamma(\mathfrak{E})\otimes\Gamma_p\otimes\Gamma(\mathfrak{E})\\ &=\Gamma(\mathfrak{E})\otimes\Gamma(\mathfrak{E}).\end{aligned} \tag{15.17c}$$

Finally, eqs. (15.16a), (15.17c) give

$$\Gamma(\boldsymbol{b})=\Gamma(\mathfrak{E})\otimes\Gamma(\mathfrak{E})\otimes\Gamma(j)=n_A A_1\oplus\ldots \tag{15.16b}$$

or, equivalently,

$$\Gamma(\boldsymbol{f})=\Gamma(\mathfrak{E})\otimes\Gamma(\mathfrak{E})=n_j\Gamma(j)\oplus\ldots. \tag{15.16c}$$

But the integer n_j of eq. (15.16c) also gives the number of independent components of the general *non-symmetric* Raman tensor for the phonon j. Since $n_A=n_j$, we conclude that $b_{\mu\lambda\nu\sigma}$ is identically zero (i.e. $n_A\equiv 0$) for those phonons j which are *totally Raman inactive* (i.e. phonons which can exhibit no Raman activity at all, symmetric or antisymmetric). Alternatively, *a magnetic field in first order can induce Raman activity only for those phonons which normally are totally Raman active*. This is to be contrasted with the electric-field-induced Raman scattering by Raman-inactive phonons which may very well be allowed, and has in fact been observed (§7). In table A1 we give the results of the reduction of $\Gamma(\boldsymbol{f})=\Gamma(\mathfrak{E})\otimes\Gamma(\mathfrak{E})$ for all 32 crystal classes. The integer coefficient in front of each representation indicates the number of independent components of the tensor $b_{\mu\lambda\nu\sigma}=-b_{\lambda\mu\nu\sigma}$. The same column $\Gamma(\boldsymbol{f})$ of this table holds for the non-symmetric intrinsic Raman tensor $a_{j,\mu\lambda\sigma}\neq a_{j,\lambda\mu\sigma}$, i.e. these reductions indicate which phonons are generally Raman active. Furthermore, as discussed in §6.1, these are the same phonons that can exhibit infrared absorption in the presence of an applied electric field to first order in the field.

The matrix form of $\Delta a_{j,\mu\lambda\sigma}=b_{\mu\lambda\nu\sigma}H_\nu$ is easily obtained by following the general procedures developed in §5.3. It can also be obtained by inspection, if one notices that (i) $b_{\mu\lambda\nu\sigma}$ has as many independent components as the intrinsic non-symmetric Raman tensor (i.e. n_j), and (ii) by definition $b_{\mu\lambda\nu\sigma}$ has the symmetry of the product $[\mathcal{E}_\mu \mathcal{E}_\lambda]_A u_\sigma H_\nu$ (this is based on the fact that the Hamiltonian is invariant under all spatial symmetry operations of the crystal point group). The antisymmetric factor $[\mathcal{E}_\mu \mathcal{E}_\lambda]_A$ transforms as $\varepsilon_{\mu\lambda\rho} H_\rho$, where $\varepsilon_{\mu\lambda\rho}$ is the antisymmetric unit tensor. Thus, $b_{\mu\lambda\nu\sigma}$ transforms as $\varepsilon_{\mu\lambda\rho} H_\rho H_\nu u_\sigma$, which in turn has the same symmetry as the term $\varepsilon_{\mu\lambda\rho} \mathcal{E}_\rho \mathcal{E}_\nu u_\sigma$. But $\mathcal{E}_\rho \mathcal{E}_\nu u_\sigma$ transforms in the same way as the $a_{j,\rho\nu\sigma}$-component of the intrinsic non-symmetric Raman tensor. Thus, inspection of the intrinsic non-symmetric Raman tables yields immediately the matrix form of the magnetic-field-dependent contributions. The transition from the intrinsic to the induced matrices follows the scheme

$$\varepsilon_{\mu\lambda\rho} \mathcal{E}_\rho \mathcal{E}_\nu u_\sigma \rightarrow \left[\mathcal{E}_\mu \mathcal{E}_\lambda\right]_A H_\nu u_\sigma, \tag{15.17d}$$

or

$$a_{j,\rho\nu\sigma} \equiv \varepsilon_{\mu\lambda\rho} a_{j,\rho\nu\sigma} \rightarrow b_{\mu\lambda\nu\sigma} H_\nu = \Delta a_{j,\mu\lambda\rho}, \tag{15.17e}$$

where eq. (15.14c) has been used together with the conclusion above regarding the symmetry of $b_{\mu\lambda\nu\sigma}$. It is emphasized that for a given ρ the indices $\mu\lambda$ are defined so that $\varepsilon_{\mu\lambda\rho} \equiv 1$. Recently, Kanamori et al. (1976b) and Matsumoto and Kanamori (1977) worked out the magnetic-field-induced Raman matrices for all those cases where this effect is allowed.

As an example we treat the Raman-active modes of tellurium. The intrinsic Raman matrices are given by eqs. (15.7). According to the conclusions above the contributions to the Raman tensor which are linear in the applied magnetic field will be given by similar matrices with 2, 1, 3 independent components designated by $(a, b), f, (c, r, g)$ for the modes A_1, A_2 and E, respectively. The construction of these new matrices is based on eq. (15.17e). Thus, the (yzx)-component of the intrinsic Raman tensor for the mode E is R' (eq. (15.7b)). This corresponds to a component r of the induced matrix such that

$$R' = a_{j,yzx} \equiv \varepsilon_{\mu\lambda y} a_{j,yzx} \rightarrow b_{\mu\lambda zx} H_z = \Delta a_{j,\mu\lambda x},$$

where $(\mu\lambda) = (zx) = -(xz)$ and $\Gamma(j) = \mathrm{E}$. Thus,

$$R' = a_{j,yzx} \rightarrow \Delta a_{j,zxx} = b_{zxzx} H_z = -b_{xzzx} H_z \equiv r H_z.$$

In a similar manner one finds one component of the induced matrix from each independent component of the intrinsic matrix. The complete matrices for $\Delta \boldsymbol{a}_j$ are given below

$$\begin{bmatrix} 0 & bH_z & -aH_y \\ -bH_z & 0 & aH_x \\ aH_y & -aH_x & 0 \end{bmatrix}, \qquad \begin{bmatrix} 0 & 0 & fH_x \\ 0 & 0 & fH_y \\ -fH_x & -fH_y & 0 \end{bmatrix}$$
$$\mathrm{A}_1 \qquad\qquad\qquad\qquad \mathrm{A}_2(z)$$

$$\begin{bmatrix} 0 & gH_y & -cH_x-rH_z \\ -gH_y & 0 & cH_y \\ cH_x+rH_z & -cH_y & 0 \end{bmatrix}, \begin{bmatrix} 0 & -gH_x & cH_y \\ gH_x & 0 & cH_x-rH_z \\ -cH_y & -cH_x+rH_z & 0 \end{bmatrix}.$$
$$\mathrm{E}(x) \qquad\qquad\qquad\qquad \mathrm{E}(y)$$

$$(15.18)$$

The total Raman matrices in the presence of $\boldsymbol{H}$ are obviously given by the sum of the corresponding matrices from eqs. (15.7), (15.18). Suppose $\boldsymbol{H}$ is along $\hat{z}$. For a Raman scattering experiment performed in any one of the configurations $z(xx)\bar{z}, z(yy)\bar{z}, z(xy)\bar{z}, z(yz)\bar{z}$ it is clear from the matrices (15.18) that the magnetic field will not induce any changes in the Raman intensity for the modes A_2 and E in first order. The symmetric mode A_1, however, will scatter in the cross configuration $z(xy)\bar{z}$ or $z(yx)\bar{z}$ with an intensity proportional to $[\mathrm{i}bH_z]^2$ (modification of the polarization selection rules). Other directions of $\boldsymbol{H}$ will yield different results. For instance, when $\boldsymbol{H}//\hat{y}$ and the configuration $z(xy)\bar{z}$ is used, the only expected effect is an increase in the scattering intensity of the transverse mode E(x) from $[+C]^2$, when $H_y=0$, to $|C+\mathrm{i}gH_y|^2 = C^2+g^2H_y^2$ when $H_y \neq 0$.

In summary, we have discussed in this and the preceding section the effects in first order of an applied magnetic field on the Raman and infrared activity of $\boldsymbol{q}\simeq 0$ optical phonons in crystalline materials. It is shown that the corrections to the Raman tensor which are linear in the field are antisymmetric with respect to the polarizations of the incident and scattered radiation. Only phonons which are already totally Raman active can be affected by the magnetic field in Raman scattering. On the other hand, magnetic-field-induced infrared activity does not require intrinsic infrared activity (i.e. activity in the absence of the field).

The symmetry aspects of the problem have been based on appropriate phenomenological coefficients. These are given by $b_{\mu\lambda\nu\sigma} = (\partial f_{\mu\lambda\nu}/\partial u_{j\sigma})$ and $h_{\mu\sigma\nu} = (\partial m_{\mu\nu}/\partial u_{j\sigma})$ according to eqs. (15.15) and (15.4) respectively. $f_{\mu\lambda\nu} = -f_{\lambda\mu\nu}$ and $m_{\mu\nu} \neq m_{\nu\mu}$ are magneto-optical coefficients and are treated as generalized *kinetic coefficients*. We know that the intrinsic effects are possible when the electric polarizability and dipole moment of the crystal are allowed by the crystal spatial symmetry to exhibit a linear dependence on the atomic normal coordinate $u_{j\sigma}$ associated with the phonon j. By the same token the magnetic-field-induced effects are possible when the magneto-optical coefficients $f_{\mu\lambda\nu}$ and $m_{\mu\nu}$ are allowed to exhibit such a dependence. Furthermore, Onsager's principle requires that $f_{\mu\lambda\nu}$ be antisymmetric in μ and λ. In this connection we would like to stress the fact that the Raman matrices obtained here to first order in $\boldsymbol{H}$ are in general different from those corresponding to the new crystal class which describes the spatial symmetry of the system 'crystal + $\boldsymbol{H}$'. For instance, the crystal class D_3 with $\boldsymbol{H}//\hat{z}$ has the spatial symmetry of the class C_3. The Raman matrices for the class C_3 are not the same as those of eqs. (15.18) with $H_x = H_y = 0 \neq H_z$, since no restriction for their antisymmetric character was considered in deriving the Raman matrices for the class C_3. Furthermore, the latter correspond to all orders of $\boldsymbol{H}$.

Many experiments of Raman scattering with an applied magnetic field have been performed by several investigators. In most of these cases the materials are magnetically ordered. No specific effort has been made to confirm the $\boldsymbol{H}$-induced Raman scattering in paramagnetic phases or in non-magnetic materials.

The work of Vella-Coleiro (1969) on CdS refers to a situation of non-thermal phonon equilibrium, whereby a magnetic field is used to tune the splitting between various Landau levels to the frequency of the E_1(LO)-type phonon. Damen and Shah (1970) have reported magnetic-field-induced resonance Raman scattering in CdS. With fields up to 9 T the incident or scattered radiation can be tuned to the Landau levels of the Γ_7 conduction and valence bands. Depending on the frequency of the incident radiation the resonance enhancement appears for the 1LO process or its overtones.

More recently, Schaack (1975) has reported the observation of magnetic-field-induced changes in the Raman scattering efficiency of some E_g-type phonons in magnetically ordered CeF_3. At 1.9 K and with various polarizations for the incident and scattered radiation, a clearly variable scattering intensity is observed for the two magnetic-field-split components of the E_g phonon. No symmetry analysis is given. Fig. 24 (to be discussed in detail in §16.3) shows some of these results.

Forbidden Raman scattering in the magnetic semiconductors EuSe, EuS and EuTe by the Raman-inactive F_{1u}(LO) phonons has been reported by Tsang et al. (1974b), Tekippe et al. (1974a, b) and Silberstein et al. (1976). The scattering strength depends critically on the magnitude and the direction of an applied magnetic field but is observed only in the magnetically ordered phase. The mechanism for such scattering below T_c and its resonance characteristics is not yet fully understood. It is certain, though, that the phonon symmetry in the ordered phase changes to one that is consistent with the possibility of Raman scattering in the presence of a magnetic field (Safran et al. 1976). The effect on the phonon symmetries of magnetic ordering alone will be discussed in §17.

In the paramagnetic phase of EuX(X=S, Se, Te, O) the F_{1u}(LO) phonon cannot scatter in first order because of its odd parity. According to the discussion above, a magnetic field cannot turn this phonon into a Raman-active one in the paramagnetic phase. The broad band (and its overtones) that has been observed in the paramagnetic phase by various workers (Silberstein et al. 1976 and references therein, Vitings and Wachter 1975, Güntherodt 1976, Grünberg et al. 1977) has been attributed to first-order scattering by $\boldsymbol{q}\simeq 0$ optical phonons from the entire Brillouin zone. According to Tsang et al. (1974a), this is a result of breaking of the inversion and translational symmetry due to spin disorder. On the basis of this assumption, Suzuki (1976) has developed a phenomenological theory according to which such broad scattering can be accounted for by considering only symmetric scattering based on a spin-pair mechanism. The latter involves mainly the variation with lattice vibrations of the f-electron transfer. Recent work by Merlin et al. (1977) has demonstrated that contrary to the assumption of Suzuki (1976), an appreciable F_{1g}-type antisymmetric contribution to the $\boldsymbol{q}\neq 0$ phonon Raman scattering does exist. Such antisymmetric scattering is attributed to spin disorder of the Eu^{2+} ions and the spin–orbit interaction of the 4f-electrons.

15.4. Circular coordinate formalism – circular dichroism of optical phonons

So far the whole treatment of infrared absorption and Raman scattering has been based on the use of Cartesian coordinates with unit vectors $(\hat{x},\hat{y},\hat{z})$. A better physical insight to the problem can, however, be obtained by using circular coordinates. We consider the system of unit vectors $(\hat{\varepsilon}_R,\hat{\varepsilon}_L,\hat{z})$ as defined by Koster et al. (1963), where $\hat{\varepsilon}_{R(L)}\equiv(\pm i/\sqrt{2})(\hat{x}\mp i\hat{y})$. The upper (lower) sign corresponds to a right (left) unit vector. Thus a propagating electric field $\mathcal{E}_L=\mathcal{E}_0\hat{\varepsilon}_L$ designates a left-circularly polarized field when viewed against the direction of propagation. Similarly all degenerate optical phonon displacements $u_{j\sigma}$ can be expressed in circular coordinates u_R, u_L for all crystal classes (for simplicity we omit the mode index j for the rest of this section).

In general, any mode property can be transformed from the $(\hat{x},\hat{y},\hat{z})$ to the $(\hat{\varepsilon}_R,\hat{\varepsilon}_L,\hat{z})$ system through a formal unitary transformation. One can then follow the methods presented in §5.3 for Cartesian coordinates, with one important point to be remembered, that is, the new basis is no longer real (in fact, $u_R^*=u_L$ and $u_L^*=u_R$). When the transformation is not real, as is the case when transforming to the circular coordinate system $(\hat{\varepsilon}_{\mathrm{R}}, \hat{\varepsilon}_{\mathrm{L}}, z)$, where

$$\begin{pmatrix} \varepsilon_{\mathrm{R}} \\ \varepsilon_{\mathrm{L}} \end{pmatrix} \equiv \frac{\pm \mathrm{i}}{\sqrt{2}} (x \mp \mathrm{i} y), \tag{15.19}$$

one should be careful in constructing combinations of complex quantities which transform as scalars. Let the capitals $P, Q, S, N, \dots$ designate any of the circular coordinate indices $\mathrm{R}, \mathrm{L}, z$. The crystal Hamiltonian will include terms of the form $M_P \mathfrak{E}_P^*$ to ensure invariance. The corresponding component of electric moment is therefore defined as

$$M_P = -\left(\frac{\partial \Phi}{\partial \mathfrak{E}_P^*} \right). \tag{15.20}$$

In the same spirit the definitions of polarizability, effective charge, Raman tensor, etc. can be written as follows:

$$\alpha_{PQ} \equiv \left(\frac{\partial M_P}{\partial \mathfrak{E}_Q'} \right) = -\left(\frac{\partial^2 \Phi}{\partial \mathfrak{E}_P^* \, \partial \mathfrak{E}_Q'} \right), \tag{15.21}$$

$$e_{PS} \equiv \left(\frac{\partial M_P}{\partial u_S} \right) = -\left(\frac{\partial^2 \Phi}{\partial \mathfrak{E}_P^* \, \partial u_S} \right), \tag{15.22}$$

$$h_{PSN} \equiv \left(\frac{\partial^2 M_P}{\partial u_S \, \partial H_N} \right) = -\left(\frac{\partial^3 \Phi}{\partial \mathfrak{E}_P^* \, \partial H_N \, \partial u_S} \right), \tag{15.23}$$

$$a_{PQS} \equiv \left(\frac{\partial \alpha_{PQ}}{\partial u_S} \right) = \left(\frac{\partial^2 M_P}{\partial \mathfrak{E}_Q' \, \partial u_S} \right) = -\left(\frac{\partial^3 \Phi}{\partial \mathfrak{E}_P^* \, \partial \mathfrak{E}_Q' \, \partial u_S} \right), \tag{15.24}$$

$$b_{PQNS} \equiv \left(\frac{\partial^2 \alpha_{PQ}}{\partial u_S \, \partial H_N} \right) = -\left(\frac{\partial^4 \Phi}{\partial \mathfrak{E}_P^* \, \partial \mathfrak{E}_Q' \, \partial H_N \, \partial u_S} \right), \tag{15.25}$$

where H_N is the Nth component of an applied field $\boldsymbol{H}$. It becomes clear from the above definitions and eq. (15.2) that the most appropriate expansion of $\delta \Phi$ will have the form

$$\delta \Phi = -e_{PS} \mathfrak{E}_P^* u_S - \mathrm{i} h_{PSN} \mathfrak{E}_P^* H_N u_S \\ -\frac{1}{2} (a_{PQS} \mathfrak{E}_P^* \mathfrak{E}_Q' u_S + \mathrm{i} b_{PQNS} \mathfrak{E}_P^* \mathfrak{E}_Q H_N u_S), \tag{15.26}$$

where the M_P and α_{PQ} terms are omitted. The next step is to determine group-theoretically all the non-zero combinations of $\mathfrak{E}, \mathfrak{E}', \boldsymbol{u}, \boldsymbol{H}$, etc., that may appear in eq. (15.26). Once the expression for $\delta \Phi$ is written down, the non-zero components of $e_{PS}, h_{PSN}, a_{PQS}, b_{PQNS}$ are immediately determined by comparison with eq. (15.26).

We notice that the term $\mathcal{E}_P^* \mathcal{E}_Q' u_S$ of eq. (15.26) describes a Stokes process according to which an incident photon ($\mathcal{E}_P$) is destroyed, while a phonon (u_S) and a photon ($\mathcal{E}_Q'$) are created. In an anti-Stokes process an incident photon ($\mathcal{E}_P$) and a phonon (u_S) are destroyed and a photon ($\mathcal{E}_Q'$) is created. The corresponding term would be $\mathcal{E}_P^* \mathcal{E}_Q' u_S^*$ which can be written as $(\mathcal{E}_P \mathcal{E}_Q'^* u_S)^*$. The latter is the complex conjugate of the transpose of the term $\mathcal{E}_Q \mathcal{E}_P'^* u_S$, or the adjoint of the Stokes term $\mathcal{E}_P'^* \mathcal{E}_Q u_S$. In other words, *the anti-Stokes Raman matrices are the adjoints of the Stokes Raman matrices* (the transposition always involves the two electromagnetic radiation indices). This establishes unambiguously the differences in the polarization selection rules involved in the two processes.

Next we consider specific examples.

(i) Degenerate modes of trigonal tellurium (E-type). The mode properties $e_{PS}, h_{PSN}, a_{PQS}, b_{PQNS}$ of eq. (15.26) have 1, 1, 2, 1 independent components respectively, for the mode E of trigonal tellurium and $\boldsymbol{H}//\hat{z}$. In this example both the radiation and phonon coordinates can provide a basis for the irreducible representation E of the crystal point group D_3 in either Cartesian or circular form. Therefore, writing down an expression like the one of eq. (15.26) is a matter of straightforward use of group-theoretical tables (Koster et al. 1963). The same procedure can be applied to the infrared-active E-type modes of all the uniaxial crystal classes except $D_4, C_{4v}, D_{2d}, D_{4h}$. One finds

$$\begin{aligned}\delta\Phi = & -e(\mathcal{E}_L u_R + \mathcal{E}_R u_L) - ih(i\mathcal{E}_L u_R - i\mathcal{E}_R u_L) H_z \\ & -\frac{1}{2}C(\mathcal{E}_R \mathcal{E}_R' u_R + \mathcal{E}_L \mathcal{E}_L' u_L) - \frac{1}{2}D(-i\mathcal{E}_z \mathcal{E}_L' u_R + i\mathcal{E}_z \mathcal{E}_R' u_L) \\ & -\frac{1}{2}D(-i\mathcal{E}_L \mathcal{E}_z' u_R + i\mathcal{E}_R \mathcal{E}_z' u_L),\end{aligned} \tag{15.27}$$

where $e, h, (C, D)$ are the independent components of e_{PS}, h_{PSN} and $a_{PQS} = a_{QPS}$ respectively. All numerical normalization factors, if any, have been incorporated in these components. The same code applies to all similar expansions for the rest of this section. The b_{PQNS} terms will be discussed later. Eq. (15.27) can be written in the form of eq. (15.26), i.e.

$$\begin{aligned}\delta\Phi = & -e(\mathcal{E}_R^* u_R + \mathcal{E}_L^* u_L) - h(\mathcal{E}_L^* u_L - \mathcal{E}_R^* u_R) H_z \\ & -\frac{1}{2}C(\mathcal{E}_L^* \mathcal{E}_R' u_R + \mathcal{E}_R^* \mathcal{E}_L' u_L) - \frac{1}{2}iD(\mathcal{E}_z^* \mathcal{E}_R' u_L - \mathcal{E}_z^* \mathcal{E}_L' u_R) \\ & -\frac{1}{2}iD(\mathcal{E}_L^* \mathcal{E}_z' u_L - \mathcal{E}_R^* \mathcal{E}_z' u_R).\end{aligned} \tag{15.28}$$

Comparison between eqs. (15.26) and (15.28) yields the following matrices for $e_{PS}(\boldsymbol{H}) \simeq e_{PS} + \mathrm{i}h_{PSz}H_z$ and a_{PQS} in circular coordinates (R, L, z)

$$e_{PS}(H_z) \simeq \begin{pmatrix} e - hH_z & 0 \\ 0 & e + hH_z \end{pmatrix}, \tag{15.29}$$

$$a_{PQS} = \underset{u_\mathrm{R}}{\begin{pmatrix} 0 & 0 \\ C & 0 \end{pmatrix}}, \quad \underset{u_\mathrm{L}}{\begin{pmatrix} 0 & C \\ 0 & 0 \end{pmatrix}}, \tag{15.30}$$

i.e. $e_{\mathrm{RR}}(\boldsymbol{H}) = e - hH_z$, $a_{\mathrm{LRR}} = a_{\mathrm{RLL}} = C$, etc. (note that the index S always designates the phonon polarization). We have considered propagation of all three quasi-particles (phonon and two photons) along z only, i.e. parallel to $\boldsymbol{H}$ (Faraday configuration), since it is for this direction only that the circular unit vectors $\hat{\varepsilon}_\mathrm{R}, \hat{\varepsilon}_\mathrm{L}$ are eigenvectors for the photon and E-type phonon modes (for different directions of propagation the circular coordinate formalism leads to results with no obvious physical meaning). Thus, one can ignore the third row and column of the normally (3×3) matrices $e_{PS}(H_z)$ and a_{PQS}, since for propagation along z, both photons and phonons have transverse character (i.e. they are polarized in the xy-plane). It is emphasized that the matrices of eq. (15.30) are applicable only when all three quasi-particles propagate along z. It is for this configuration only that the circular components involved in eq. (15.27) are consistent. To apply eq. (15.30) for backward scattering, one has to go one step further and conjugate the column index. This is because reversal of the direction of propagation of the scattered photon (from z to $-z$), implies conjugation of its polarization. Thus for backward scattering we have, $a_{\mathrm{LLR}} = a_{\mathrm{RRL}} = C$, etc. Furthermore, in a forward scattering configuration the phonon direction can be along z or $-z$, depending on the relative magnitude of the wavevectors of the two photons. The latter is established from the behavior of the photon dispersion near the zone center. If for some reasons the wavevector of the scattered photon is larger in magnitude than that of the incident photon, the Stokes phonon will be propagating along $-z$, i.e. opposite to the two photons. In order to use the matrices (15.30), one then has to conjugate the phonon polarizations, namely interchange u_R and u_L. Similar considerations apply to an anti-Stokes process in addition to the comments following eq. (15.26).

According to eq. (15.29) R-photons can only generate R-phonons with an absorption strength $\sim|e - hH_z|^2$, and L-photons can only generate L-phonons with an absorption strength $\sim|e + hH_z|^2$. Two conclusions become immediately obvious from the above: (i) the angular momentum is conserved, and (ii) R- and L-photons are unequally absorbed. The latter corresponds to *magnetic-field-induced circular dichroism for optical phonons.*

The full Raman matrices of eq. (15.30) involve two independent parameters C and D. These are the same as those appearing in the symmetric part of the Cartesian matrices of eq. (15.7a) within a normalization factor. This is because eq. (15.30) can be reached independently by applying the present unitary transformation to eq. (15.7a). The non-symmetric appearance of eq. (15.30) is due to the complex conjugation of the incident field $\mathcal{E}_P$ as required by eq. (15.26). Thus, in eq. (15.30) we have $C=a_{\mathrm{LRR}}$, derived from the term $\mathcal{E}_{\mathrm{L}}^* \mathcal{E}_{\mathrm{R}}' u_{\mathrm{R}} = \mathcal{E}_{\mathrm{R}} \mathcal{E}_{\mathrm{R}}' u_{\mathrm{R}}$ of eq. (15.27). Its symmetric component, according to the usual definition, is a_{RLR}, which requires a term $\mathcal{E}_{\mathrm{R}}^* \mathcal{E}_{\mathrm{L}}' u_{\mathrm{R}} = \mathcal{E}_{\mathrm{L}} \mathcal{E}_{\mathrm{L}}' u_{\mathrm{R}}$ in eq. (15.27). This term, however, is forbidden by symmetry. In summary, one can argue that the matrices of eq. (15.30) are symmetric in the sense that they derive from *symmetric Cartesian matrices*. Inclusion of the antisymmetric part of eq. (15.7a) would introduce an additional term in eq. (15.27).

The field-induced contributions $\Delta a_{PQS} = b_{PQNS} H_N$ in circular coordinates can be derived by substitution from the antisymmetric invariant $(xH_y' - yH_x')H_z$, with H_y' transforming as $\mathcal{E}_z \mathcal{E}_x' - \mathcal{E}_x \mathcal{E}_z'$, etc. When the applied field is along z this invariant (expressed in circular coordinates) takes the form

$$(\mathcal{E}_z^* \mathcal{E}_{\mathrm{L}}' - \mathcal{E}_{\mathrm{R}}^* \mathcal{E}_z') u_{\mathrm{R}} H_z + (\mathcal{E}_z^* \mathcal{E}_{\mathrm{R}}' - \mathcal{E}_{\mathrm{L}}^* \mathcal{E}_z') u_{\mathrm{R}} H_z,$$

which leads to the following general results, with all three particles propagating along z:

$$\Delta a_{PQS} = b \begin{bmatrix} 0 & 0 & -H_z \\ 0 & 0 & 0 \\ 0 & H_z & 0 \end{bmatrix}_{u_{\mathrm{R}}}, \quad b \begin{bmatrix} 0 & 0 & 0 \\ 0 & 0 & -H_z \\ H_z & 0 & 0 \end{bmatrix}_{u_{\mathrm{L}}}. \tag{15.31}$$

The factor b stands for the single independent component of b_{PQNS}. Again, the third row and column can be dropped, leading to the conclusion that H_z has no effect on the Raman tensor in first order. The case of $\boldsymbol{H}$ taken along a general direction can be best treated in terms of Cartesian coordinates.

(ii) The degenerate mode of zincblende (F_2-type). Here again the radiation and the phonon coordinates transform according to the same irreducible representation (F_2) of the crystal point group T_{d}. In this case one can write eq. (15.26) in Cartesian form as usual and then replace x,y,z by $\mathrm{i}/\sqrt{2}$ $(\varepsilon_{\mathrm{L}} - \varepsilon_{\mathrm{R}}), 1/\sqrt{2}\,(\varepsilon_{\mathrm{L}} + \varepsilon_{\mathrm{R}})$ and z with similar substitutions for $\mathcal{E}_x \ldots u_z$, etc. We assume for simplicity that the propagation is along the z-axis. Regardless of choice of the coordinate system the mode properties e_{PS}, h_{PSN} and

a_{PQS} have only one independent component each for this crystal class and type of phonon. Let us postpone the treatment of the last term eq. (15.26) for a while. The invariants representing the first three terms of eq. (15.26) are easily constructed group-theoretically and have the following Cartesian forms, respectively:

$$xx' + yy' + zz',$$
$$H_x H'_x + H_y H'_y + H_z H'_z,$$
$$xy'z'' + x''yz' + x'y''z,$$

where $\boldsymbol{H}$, and $\boldsymbol{H}'$ designate axial vectors (H_x transforms as $yz' - y'z$, etc.). In terms of these invariants we can write

$$\delta\Phi = -e(\mathcal{E}_x u_x + \mathcal{E}_y u_y + \mathcal{E}_z u_z) - \mathrm{i}h'\left[(\mathcal{E}_y u_z - \mathcal{E}_z u_y)H_x + (\mathcal{E}_z u_x - \mathcal{E}_x u_z)H_y + (\mathcal{E}_x u_y - \mathcal{E}_y u_x)H_z\right] - \frac{a'}{2}(\mathcal{E}_x \mathcal{E}_y u_z + \mathcal{E}'_x \mathcal{E}'_y u_z + \text{c.p.}). \qquad (15.32)$$

The transformation of $\mathcal{E}$ and $\boldsymbol{u}$ from Cartesian to circular coordinates yields

$$\delta\Phi = -e(\mathcal{E}_\mathrm{R} u_\mathrm{L} + \mathcal{E}_\mathrm{L} u_\mathrm{R} + \mathcal{E}_z u_z) - \frac{\mathrm{i}h'}{\sqrt{2}}\left[(\mathcal{E}_\mathrm{R} u_z - \mathcal{E}_z u_\mathrm{R})(H_x + \mathrm{i}H_y) + (\mathcal{E}_\mathrm{L} u_z - \mathcal{E}_z u_\mathrm{L})(H_x - \mathrm{i}H_y) + \frac{\mathrm{i}}{\sqrt{2}}(\mathcal{E}_\mathrm{L} u_\mathrm{R} - \mathcal{E}_\mathrm{R} u_\mathrm{L})H_z\right]$$
$$- \frac{\mathrm{i}a'}{4}\left[(\mathcal{E}_\mathrm{L} - \mathcal{E}_\mathrm{R})(\mathcal{E}'_\mathrm{L} + \mathcal{E}'_\mathrm{R})u_z + (\mathcal{E}_\mathrm{L} + \mathcal{E}_\mathrm{R})\mathcal{E}'_z(u_\mathrm{L} - u_\mathrm{R}) + \mathcal{E}_z(\mathcal{E}'_\mathrm{L} - \mathcal{E}'_\mathrm{R})(u_\mathrm{L} + u_\mathrm{R}) + (\text{same terms with } \mathcal{E}, \mathcal{E}' \text{ interchanged})\right]. \qquad (15.33)$$

Comparison between eqs. (15.26) and (15.33) yields the matrix forms for e_{PS}, h_{PSN}, a_{PQS}. Since the total effective charge to first order in $\boldsymbol{H}$ is $e_{PS}(\boldsymbol{H}) = e_{PS} + \mathrm{i}h_{PSN}H_N$, we have for $\boldsymbol{H}//\hat{z}$,

$$e_{PS}(\boldsymbol{H}) \simeq \begin{bmatrix} e - hH_z & 0 & 0 \\ 0 & e + hH_z & 0 \\ 0 & 0 & e \end{bmatrix}, \qquad (15.34)$$

i.e. $e_{\mathrm{RR}}(\boldsymbol{H}) = e - hH_z$, $e_{\mathrm{LL}}(\boldsymbol{H}) = e + hH_z$, etc. Recall that $\mathcal{E}^*_\mathrm{L} = \mathcal{E}_\mathrm{R}$, $\mathcal{E}^*_z = \mathcal{E}_z$, etc. The intrinsic Raman matrices a_{PQS} with all three quasi-particles propagating along z, are found to be

$$a_{PQS} = \mathrm{i}a\begin{bmatrix} 0 & 0 & 0 \\ 0 & 0 & -1 \\ -1 & 0 & 0 \end{bmatrix}_{u_{\mathrm{R}}}, \quad \mathrm{i}a\begin{bmatrix} 0 & 0 & 1 \\ 0 & 0 & 0 \\ 0 & 1 & 0 \end{bmatrix}_{u_{\mathrm{L}}}, \quad \mathrm{i}a\begin{bmatrix} 0 & 1 & 0 \\ -1 & 0 & 0 \\ 0 & 0 & 0 \end{bmatrix}_{u_z}. \tag{15.35}$$

The magnetic-field-induced Raman tensors $b_{\mu\lambda\nu\sigma}H_\nu$ and $b_{PQNS}H_N$ are obtained from the invariant

$$\left[(H_x' H_y + H_y' H_x)u_z + \text{c.p.}\right].$$

Since H_x' transforms as $\mathcal{E}_y\mathcal{E}_z' - \mathcal{E}_z\mathcal{E}_y'$, etc., one finds by substitution the form of this invariant in circular coordinates,

$$\begin{aligned}&\left[(\mathcal{E}_{\mathrm{L}}\mathcal{E}_{\mathrm{R}}' - \mathcal{E}_{\mathrm{R}}\mathcal{E}_{\mathrm{L}}')H_{\mathrm{R}} + (\mathcal{E}_{\mathrm{R}}\mathcal{E}_z' - \mathcal{E}_z\mathcal{E}_{\mathrm{R}}')H_z\right]u_{\mathrm{R}} + \left[(\mathcal{E}_{\mathrm{R}}\mathcal{E}_{\mathrm{L}}' - \mathcal{E}_{\mathrm{L}}\mathcal{E}_{\mathrm{R}}')H_{\mathrm{L}}\right.\\ &\left.+ (\mathcal{E}_{\mathrm{L}}\mathcal{E}_z' - \mathcal{E}_z\mathcal{E}_{\mathrm{L}}')H_z\right]u_{\mathrm{L}} + \left[(\mathcal{E}_{\mathrm{R}}\mathcal{E}_z' - \mathcal{E}_z\mathcal{E}_{\mathrm{R}}')H_{\mathrm{R}} + (\mathcal{E}_{\mathrm{L}}\mathcal{E}_z' - \mathcal{E}_z\mathcal{E}_{\mathrm{L}}')H_{\mathrm{L}}\right]u_z.\end{aligned}$$

The Cartesian and circular forms of the Raman tensor $\Delta \boldsymbol{a}$ for $\boldsymbol{H}//\hat{z}$ are obtained from the above two invariants respectively, i.e.

$$\Delta a_{\mu\lambda\sigma} = b'\begin{bmatrix} 0 & 0 & -H_z \\ 0 & 0 & 0 \\ H_z & 0 & 0 \end{bmatrix}_{u_x}, \quad b'\begin{bmatrix} 0 & 0 & 0 \\ 0 & 0 & H_z \\ 0 & -H_z & 0 \end{bmatrix}_{u_y}, \quad b'\begin{bmatrix} 0 & 0 & 0 \\ 0 & 0 & 0 \\ 0 & 0 & 0 \end{bmatrix}_{u_z}, \tag{15.36}$$

$$\Delta a_{PQS} = b\begin{bmatrix} 0 & 0 & 0 \\ 0 & 0 & H_z \\ -H_z & 0 & 0 \end{bmatrix}_{u_{\mathrm{R}}}, \quad b\begin{bmatrix} 0 & 0 & H_z \\ 0 & 0 & 0 \\ 0 & -H_z & 0 \end{bmatrix}_{u_{\mathrm{L}}}, \quad b\begin{bmatrix} 0 & 0 & 0 \\ 0 & 0 & 0 \\ 0 & 0 & 0 \end{bmatrix}_{u_z}, \tag{15.37}$$

where b' and b are the single independent components of $b_{\mu\lambda\nu\sigma}$ and b_{PQNS} respectively. Notice that the matrices of (15.37) are not antisymmetric, whereas those of (15.36) have been constructed to be so, according to eq. (15.15).

The present procedure also applies to all degenerate infrared-active phonons transforming according to those representations for which the set of circular coordinates $(\varepsilon_{\mathrm{R}}, \varepsilon_{\mathrm{L}}, z)$ do not form a basis (Koster et al. 1963).

These are the degenerate infrared-active phonons from all the cubic crystal classes and the four uniaxial classes D_4, C_{4v}, D_{2d} and D_{4h}.

For the propagation along z the general results of eqs. (15.34), (15.35) and (15.37) reduce to

$$e_{PS} \simeq \begin{pmatrix} e-hH_z & 0 & 0 \\ 0 & e+hH_z & 0 \end{pmatrix}, \tag{15.34a}$$

$$a_{PQS} \simeq \underset{u_R}{\begin{pmatrix} 0 & 0 \\ 0 & 0 \end{pmatrix}}, \quad \underset{u_L}{\begin{pmatrix} 0 & 0 \\ 0 & 0 \end{pmatrix}}, \quad \underset{u_z}{\mathrm{i}a\begin{pmatrix} 0 & 1 \\ -1 & 0 \end{pmatrix}}, \tag{15.35a}$$

$$\Delta a_{PQS} = 0. \tag{15.37a}$$

The third row in the matrix (15.34a) is dropped because it corresponds to the z-component of the electromagnetic radiation, which for propagation along z is zero. The third column, on the other hand, corresponds to the z-component of the generated phonon, i.e. the longitudinal optical (LO) phonon, which unlike case (i) may very well occur. However, the corresponding components of the $\boldsymbol{H}$-induced effective charge are zero in the Faraday configuration. Again, we have the same circular dichroism interpretation for an infrared absorption process as in example (i). The Raman matrices (15.35a) and (15.37a) indicate that for this configuration ($\boldsymbol{H}//\hat{z}$, all particles along $\hat{z}$) one can only expect the intrinsic Raman scattering by the F_2(LO) phonon component.

To summarize, for propagation along z, an applied magnetic field in the Faraday configuration ($\boldsymbol{H}//\hat{z}$) can produce only circular dichroism, with no effect on the Raman tensor to first order in H_z.

(iii) The infrared-inactive degenerate mode of diamond (F_{2g}-type). This phonon and all the degenerate infrared-inactive ones from all uniaxial and cubic crystal classes *do not* belong to the same representation as the radiation field i.e. the one with basis (x,y,z) or $(\varepsilon_R, \varepsilon_L, z)$ as in case (i), or (x,y,z) only as in case (ii). Instead, the three phonon normal coordinates (u_x, u_y, u_z) of the F_{2g} mode transform as (yz, zx, xy) respectively. Circularly polarized phonons of this type are constructed again, by taking (u_R, u_L, u_z) for propagation along the axis z, where

$$\begin{bmatrix} u_R \\ u_L \end{bmatrix} \equiv \frac{\pm \mathrm{i}}{\sqrt{2}} (u_x \mp \mathrm{i} u_y).$$

Otherwise the procedure is the same as in case (ii).

An expression similar to eq. (15.32) can be derived from the invariant $yzy'z' + zxz'x' + xyx'y'$ of the O_h point group. The first two terms of eq.

(15.32) do not exist, since $e = h \equiv 0$ for this type of phonon. Thus, only the last term of eq. (15.32) applies to the present case, as can be seen by taking (u_x, u_y, u_z) as $(y'z', z'x', x'y')$ in the above invariant. Similarly the last term of eq. (15.33) and the subsequent results of eqs. (15.35)–(15.37) are still valid. Consequently, an applied magnetic field in the Faraday configuration will have no effect on infrared absorption and Raman scattering to first order in H_z. For a general direction of propagation the analysis is far more complicated.

Before closing this section it should be mentioned that the use of circularly-polarized coordinates has proved itself extremely useful in the study of configuration and conformation properties of complicated molecular systems of biological importance. The physical basis for this is that the scattering intensity due to molecular vibrations depends on the sense of polarization of the circularly-polarized incident radiation. A relevant technique is the *magnetic-field-induced optical activity* in molecules which lack natural optical activity. This technique has been used extensively in recent years in the study of structural and vibrational properties of the systems mentioned above (Barron 1976 and references therein).

16. *Magnetic-field-induced changes in the phonon frequencies*

16.1. General

We now use the definition of eq. (15.2), or the corresponding one for the crystal potential energy, to define the necessary correction $\Delta \boldsymbol{K}_j$ to the phonon force constant $\boldsymbol{K}_j$ due to the applied magnetic field $\boldsymbol{H}$. From eqs. (15.2) and (4.3) we have

$$K_{j,\alpha\beta}(\boldsymbol{H}) \simeq K^{(0)}_{j,\alpha\beta} + \mathrm{i}\Delta K_{j,\alpha\beta}(\boldsymbol{H}), \tag{16.1}$$

where

$$K^{(0)}_{j,\alpha\beta} = \left(\frac{\partial^2 \Phi^{(0)}}{\partial u_{j\alpha}\, \partial u_{j\beta}} \right), \tag{16.2a}$$

$$\Delta K_{j,\alpha\beta}(\boldsymbol{H}) = \left(\frac{\partial^2 \Phi^{(1)}}{\partial u_{j\alpha}\, \partial u_{j\beta}} \right) = \left(\frac{\partial^3 \Phi^{(1)}}{\partial u_{j\alpha}\, \partial u_{j\beta}\, \partial H_\nu} \right) H_\nu \equiv K_{j,\alpha\beta\nu} H_\nu. \tag{16.2b}$$

As usual, the indices α, β run from 1 to the degree of degeneracy. The fact that $\Phi^{(1)}$ is linear in $\boldsymbol{H}$ has also been taken into consideration.

On the basis of the theory of the generalized susceptibility (Landau and Lifshitz 1958) it can be shown that the effective force constant as defined above is a generalized *kinetic* coefficient satisfying the following two conditions:

$$K_{j,\alpha\beta}(\boldsymbol{H})=K_{j,\beta\alpha}(-\boldsymbol{H}), \tag{16.3a}$$

$$K_{j,\alpha\beta}(\boldsymbol{H})=K^{*}_{j,\beta\alpha}(\boldsymbol{H}). \tag{16.3b}$$

The first equation expresses Onsager's reciprocity theorem, a well-known principle of any kinetic coefficient in the presence of a magnetic field. The second equation expresses the fact that $K_{j,\alpha\beta}(\boldsymbol{H})$ is Hermitian so that attenuation of phonons does not occur. From eqs. (16.1)–(16.3) it follows that $K^{(0)}_{j,\alpha\beta}$ and $K_{j,\alpha\beta\nu}$ are both real and that

$$K^{(0)}_{\alpha\beta}=K^{(0)}_{\beta\alpha}, \tag{16.4a}$$

$$K_{j,\alpha\beta\nu}=-K_{j,\beta\alpha\nu}. \tag{16.4b}$$

The antisymmetric character of $K_{j,\alpha\beta\nu}$ is a consequence of the non-real character of the Hamiltonian in the presence of $\boldsymbol{H}$. Recall that in the presence of polar forces (electric field, strains, etc.), the Hamiltonian is real and the corresponding coefficients $K_{j,\alpha\beta\nu}$ are symmetric. To summarize, *the magnetic-field-induced force constant, linear in the field, must be antisymmetric*. In addition, the invariance of the crystal under the spatial symmetry operations will impose further restrictions on the matrix form of the real coefficient $K_{j,\alpha\beta\nu}$. It is interesting to notice the symmetry identity between $K_{j,\alpha\beta\nu}$ and two well-known kinetic macroscopic crystal properties, i.e. the magneto-optical tensor (Landau and Lifshitz 1960) $f_{\alpha\beta\nu}=(\partial\varepsilon^{(1)}_{\alpha\beta}/\partial H_{\nu})=-f_{\beta\alpha\nu}$ and the Hall tensor (Birss 1964, Bhagavantan 1966) $R_{\alpha\beta\nu}=(\partial\sigma^{(1)}_{\alpha\beta}/\partial H_{\nu})=-R_{\beta\alpha\nu}$. $\varepsilon^{(1)}_{\alpha\beta}$ and $\sigma^{(1)}_{\alpha\beta}$ are the antisymmetric parts of the dielectric constant $\varepsilon_{\alpha\beta}(\boldsymbol{H})=\varepsilon^{(0)}_{\alpha\beta}+\mathrm{i}\varepsilon^{(1)}_{\alpha\beta}(\boldsymbol{H})$ and the conductivity $\sigma_{\alpha\beta}(\boldsymbol{H})=\sigma^{(0)}_{\alpha\beta}+\mathrm{i}\sigma^{(1)}_{\alpha\beta}(\boldsymbol{H})$. In fact we can write down the matrix form of $K_{j,\alpha\beta\nu}$ by inspection, since it is identical to that of $R_{\alpha\beta\nu}$ and the latter can be found in the literature (Bhagavantan 1966). One can also proceed independently following the general methods developed in §5.

For a magnetic field, the secular equation (3.11) takes the form

$$|\mathrm{i}\Delta K_{j,\alpha\beta}-\lambda\delta_{\alpha\beta}|=0, \tag{16.5}$$

where

$$\lambda=\Omega_j^2-\omega_j^2\simeq 2\omega_j\Delta\omega_j \tag{16.6}$$

and

$$\Omega_j \simeq \omega_j + \Delta\omega_j. \tag{16.7}$$

To solve eq. (16.5) it is necessary to know the non-zero components of $\Delta K_{j,\alpha\beta}$ or, according to eq. (16.2b), the non-zero components of the *antisymmetric* axial tensor $K_{j,\alpha\beta\nu}$. It follows from eq. (16.2b) that the number of independent components of $K_{j,\alpha\beta\nu}$ for a given structure and mode j is n_A, where

$$[\Gamma(j)\otimes\Gamma(j)]_A \otimes \Gamma(\boldsymbol{H}) = n_A A_1 \oplus \ldots. \tag{16.8}$$

A number of conclusions can be derived from eq. (16.8):

(i) The product $\Gamma(j)\otimes\Gamma(j)$, for *real non-degenerate modes* in any crystal structure is always purely symmetric (A_1). *Therefore* $n_A \equiv 0$, *and* $K_{j,\alpha\beta\nu} \equiv 0$ *for all non-degenerate modes. The magnetic field in first order has no effect on* ω_j. *All biaxial crystals have exclusively non-degenerate modes.*

(ii) *In uniaxial and cubic crystals* it is possible to have several different modes $j_1, j_2, \ldots$ of the same degeneracy. Since $[\Gamma(j_1)\otimes\Gamma(j_1)]_A = [\Gamma(j_2)\otimes \Gamma(j_2)]_A = \ldots$ it is concluded that the field $\boldsymbol{H}$ will affect the frequencies $\omega_{j1}, \omega_{j2}, \ldots$ in the same way. Thus, by examining the effects of $\boldsymbol{H}$ on one of these modes (preferably the infrared-active one which transforms like x, y, z) we immediately have the results for all the other modes of the same structure with the same degeneracy. This is the same conclusion which was reached earlier for the symmetric force constants (see §5.2).

(iii) In *uniaxial crystals* the only degenerate modes are doubly degenerate (E) and they are polarized in the plane normal to the optical axis (z-axis). The antisymmetric part $[E\otimes E]_A$ for these modes is actually the representation $\Gamma(x_1y_2 - y_1x_2) = \Gamma(H_z)$, where H_z is the z component of any axial vector which transforms like $x_1y_2 - y_1x_2$. Thus the only way for n_A of eq. (16.8) to be non-zero is to take $\boldsymbol{H}//\hat{z}$, for then $\Gamma(H_z)\otimes\Gamma(H_z) = A_1$. In this case the coefficient $K_{j,\alpha\beta\nu}$ will have only one single independent component (K_E). Either by inspection of the Hall tensor (Bhagavantam 1966) or by applying the same general methods of §5.3 one finds that the only non-zero components for all uniaxial crystals are

$$K_{j,xyz} = -K_{j,yxz} \equiv K_E. \tag{16.9}$$

(iv) In *cubic crystals* there are real doubly-degenerate (E) and triply-degenerate (F) modes. The field $\boldsymbol{H}$ will have no effect on the E modes, since $\Gamma(\boldsymbol{H})$ is always an F-type representation and $[E\otimes E]_A$ includes no F-type representations. The left-hand side of eq. (16.8) can never include A_1. Thus $n_A \equiv 0$ and $K_{j,\alpha\beta\nu} \equiv 0$ for all doubly-degenerate modes of cubic crystals. On the other hand the field will affect the triply-degenerate modes. It turns out that $[F\otimes F]_A=\Gamma(\boldsymbol{H})=F_1$ for all components of $\boldsymbol{H}$. Therefore eq. (16.8) gives $n_A=1$. Again one finds easily that the non-zero components of $K_{j,\alpha\beta\nu}$ are

$$K_{j,xyz}=K_{j,yzx}=K_{j,zxy}=-K_{j,yxz}=-K_{j,zyx}=-K_{j,xzy}\equiv K_F. \tag{16.10}$$

Similar results can be reached for pairs of complex conjugate modes which can occur in a number of crystal classes, as we shall see shortly. Recently, Kanamori et al. (1976b) and Matsumoto and Kanamori (1977) worked out all the matrices $K_{j,\alpha\beta\nu}$ linear in the magnetic field, for all those cases where the effect is allowed.

After the component form of $K_{j,\alpha\beta\nu}$ has been established one can easily proceed and diagonalize the field-dependent dynamical matrix.

(i) Uniaxial crystals. From eqs. (16.2b), (16.4b) and (16.9) we find

$$\begin{vmatrix} -\lambda_E & iK_EH_z \\ -iK_EH_z & -\lambda_E \end{vmatrix}=0, \qquad \lambda_E=\pm K_EH_z, \tag{16.11}$$

$$\Omega_E\simeq\omega_E\pm\frac{K_EH_z}{2\omega_E}. \tag{16.12}$$

The new eigenvectors are $(1/\sqrt{2})(1,-i,0)$ and $(1/\sqrt{2})(1,+i,0)$ respectively. These results show that the *frequencies of all the doubly-degenerate modes of any uniaxial crystal will split linearly with the z component of an applied magnetic field and symmetrically about the degenerate frequency* ω_E. *The corresponding phonons are left- and right-circularly polarized.*

(ii) Cubic crystals. From eqs. (16.2b), (16.4b) and (16.10) we have

$$\begin{vmatrix} -\lambda_F & iK_FH_z & -iK_FH_y \\ -iK_FH_z & -\lambda_F & iK_FH_x \\ iK_FH_y & -iK_FH_x & -\lambda_F \end{vmatrix}=0, \tag{16.13}$$

$$\lambda_F=0, \qquad \pm K_FH,$$

$$\Omega_F\simeq\begin{cases}\omega_F+K_FH/2\omega_F\\ \omega_F\\ \omega_F-K_FH/2\omega_F\end{cases}, \tag{16.14}$$

where H is the magnitude of $\boldsymbol{H}$. *The eigenvectors are one parallel to* $\boldsymbol{H}$ *(frequency unshifted) and two left- and right-circularly polarized in the plane perpendicular to* $\boldsymbol{H}$ *(frequencies split symmetrically about* ω_F*). The splittings are isotropic in the field.*

So far we have considered degenerate modes with real irreducible representations; they are linearly polarized, and it is only because of the first-order effects of a static external magnetic field that they may become left- and right-circularly polarized and non-degenerate. These effects have their origin in a field-dependent antisymmetric force constant. The latter is a kinetic coefficient whose existence is consistent with spatial symmetry, Onsager's principle, and the assumption that no attenuation occurs. However, there are 10 crystal classes, 8 uniaxial ($C_4, S_4, C_{4h}, C_3, C_{3i}, C_6, C_{3h}, C_{6h}$) and 2 cubic ($T, T_h$) which exhibit non-real, one-dimensional, complex-conjugate pairs of modes (E_1, E_2). Normally, if time reversal symmetry is present these pairs are doubly degenerate and can be considered as a pair of modes with right and left senses of polarization u_+, u_-, or equivalently, with two mutually perpendicular linear polarizations u_1, u_2 (a linear vibration can always be considered as a superposition of two circular vibrations with the same frequency and amplitude and opposite senses of rotation). Since the only reason that E_1 and E_2 are degenerate is the presence of time reversal symmetry, it follows that whenever time symmetry is removed, as by the application of an external magnetic field in the present case, the two modes will become non-degenerate. The removal of the degeneracy comes about through the appearance of a field-dependent force constant $K_{12}(\boldsymbol{H})$. Inasmuch as we are interested only in effects which are linear in the strength of the magnetic field, we will assume that $K_{12}(\boldsymbol{H})$ is a linear function of the magnetic field components. If the damping of phonons is neglected as before, $K_{12}(\boldsymbol{H})$ obeys the Hermiticity relations of eq. (16.3b). At the same time the condition on the Hamiltonian expressed by eq. (15.2) has the consequence that $K_{12}(\boldsymbol{H})$ is an antisymmetric force constant, i.e.

$$K_{12}(\boldsymbol{H}) = \left(\frac{\partial^2 \Phi(\boldsymbol{H})}{\partial u_1 \, \partial u_2} \right) = -K_{21}(\boldsymbol{H}). \tag{16.15}$$

In addition to exhibiting the above properties, $K_{12}(\boldsymbol{H})$ must be compatible with the crystal's spatial symmetry as well. It turns out that $K_{12}(\boldsymbol{H})$ can only exist for the 8 uniaxial classes mentioned above. This can be easily seen as follows: $K_{12}(\boldsymbol{H})$ belongs to the $[\Gamma(u_1) \otimes \Gamma(u_2)]_A \otimes \Gamma(\boldsymbol{H})$ representation according to eq. (16.15). It can therefore be non-zero only when this representation includes or coincides with A_1. A straightforward use of group-theoretical tables leads to the conclusion that the representation $[\Gamma(u_1) \otimes \Gamma(u_2)]_A$ coincides with A_1 for all 10 classes listed at the beginning of this section. Thus $K_{12}(\boldsymbol{H})$ can be non-zero only when $\Gamma(\boldsymbol{H})$ coincides with A_1. This is the case for the 8 uniaxial classes given earlier provided the field is along z. This is not the case for the two cubic classes T and T_h for which $\Gamma(\boldsymbol{H}) = F_1$. In summary, *a magnetic field induces a linear splitting of the non-real, complex-conjugate modes only in the 8 uniaxial classes quoted above.*

An interesting case similar to those considered here arises when certain paramagnetic materials are cooled below the transition temperature T_c to become magnetically ordered. The time reversal symmetry is removed because of the ordering and therefore one may expect splittings of the degeneracies of $\boldsymbol{q} \simeq 0$ optical modes. In fact it will be shown in §17 that there are 14 magnetic crystal classes for which such splittings can occur, always involving a real doubly-degenerate E mode above T_c, which splits below T_c into two circular components with different frequencies. In the spirit of eq. (16.15) one could explain these splittings as due to the existence of an intrinsic antisymmetric force constant. The latter is zero above T_c, being forbidden by both spatial and time reversal symmetry. Below T_c on the other hand, the time reversal symmetry does not exist and the spatial symmetry is different (in general, the paramagnetic point group is different from the magnetic point group). Thus the magnetic spatial symmetry may allow for such an antisymmetric force constant, whereas the corresponding non-magnetic spatial symmetry does not.

In summary we have discussed the linear effects of a magnetic field on the degeneracy of the $\boldsymbol{q} \simeq 0$ optical phonons in crystalline materials. These effects are due to an antisymmetric force constant which the magnetic field induces in first order. Thermodynamic and spatial symmetry considerations impose restrictions on the existence and form of such a constant. It is found that real non-degenerate phonons are not affected by $\boldsymbol{H}$. In uniaxial crystals all the real doubly-degenerate modes split symmetrically and become circularly polarized about H_z (H_z is the only necessary component of $\boldsymbol{H}$). In cubic crystals the doubly-degenerate modes are not affected by $\boldsymbol{H}$ but the triply-degenerate modes are symmetrically split for any direction of $\boldsymbol{H}$. One of the modes (unshifted) is linearly polarized along $\boldsymbol{H}$ while the other two (symmetrically shifted) are circularly polarized in a plane perpendicular to $\boldsymbol{H}$. Degenerate complex-conjugate modes of uniaxial (but not cubic) crystals will also split linearly with the field.

It should be emphasized that the antisymmetric character of the magnetic-field-dependent force constant originates from arguments not related to space symmetry. Spatial symmetry considerations cannot lead to results which are consistent with eqs. (16.3a) and (16.3b). Take for instance the class D_3 with H_y present. If only spatial symmetry is considered, it can be shown that D_3 reduces to C_2 and the doubly-degenerate mode E of D_3 goes over to the non-degenerate modes A+B of C_2. It can further be shown that such a splitting can be linear in the y component of an applied electric field (i.e. E_y). As the point group D_3 has no improper rotations the forces $\boldsymbol{H}$ and $\boldsymbol{E}$ transform identically and therefore the above splitting could be linear in H_y as well. This result is obviously wrong when further non-spatial restrictions are added. In fact, as we have already seen, splittings linear in $\boldsymbol{H}$ can only be antisymmetric and require a non-zero H_z component. The H_y component will produce no linear splitting of the E mode in structures with D_3 symmetry.

Finally, we mention the interesting resemblance between the present effect and the magneto-optical effect. In the latter, the applied magnetic field induces a rotation of the plane of polarization of the electromagnetic radiation (Faraday effect or magnetic-field-induced optical activity, Landau and Lifshitz 1960). In the present case we have splitting of degenerate optical modes into two circularly polarized components which propagate independently at slightly different frequencies.

16.2. Microscopic treatment

In this section an example of a formal microscopic treatment of morphic effects is given based on group-theoretical arguments and a detailed expansion of the potential energy (Anastassakis et al. 1972a). The notation differs in places from that used in the preceding sections, but since this treatment is completely independent of the previous sections this should not matter. An independent microscopic treatment has also been given by Holz (1972) who examines the effects of a magnetic field on the dispersion curves of a non-magnetic ionic crystal. No reference is made to the symmetry aspects of the problem.

We begin by formally expanding the second-order atomic force constants of a crystal to first order in the components of an external magnetic field:

$$\Phi_{\alpha\beta}(l\kappa;l'\kappa') = \Phi^{(0)}_{\alpha\beta}(l\kappa;l'\kappa') + \mathrm{i}\sum_{\gamma} \Phi^{(1)}_{\alpha\beta\gamma}(l\kappa;l'\kappa')H_\gamma + \dots . \tag{16.16}$$

Here α, β, γ label the Cartesian axes, l and l' the primitive unit cells of the crystal, and κ and κ' the atoms within a primitive unit cell. A factor i has been inserted in the second term. By doing so the mode coefficients $\{\Phi^{(1)}_{\alpha\beta\gamma}(l\kappa;l'\kappa')\}$ are made purely real as a consequence of the transformation properties of the potential energy of the crystal under the operation expressed by eq. (15.1).

Under an operation of the space group of the crystal the force constants transform (Maradudin and Vosko 1968) according to

$$\Phi_{\alpha\beta}(LK;L'K') = \sum_{\alpha'\beta'} S_{\alpha\alpha'} S_{\beta\beta'} \Phi_{\alpha'\beta'}(l\kappa;l'\kappa'), \tag{16.17}$$

where (LK) and $(L'K')$ are lattice sites into which the sites $(l\kappa)$ and $(l'\kappa')$ respectively transform under the symmetry operation. $\boldsymbol{S}$ is the 3×3 real orthogonal matrix which represents the rotational element of the space group operation. The magnetic field transforms as an axial vector under a space group operation:

$$H'_\alpha = |\boldsymbol{S}| \sum_\beta S_{\alpha\beta} H_\beta. \tag{16.18}$$

Combining eqs. (16.16)–(16.18) yields the transformation laws for the coefficients $\{\Phi^{(0)}_{\alpha\beta}(l\kappa; l'\kappa')\}$ and $\{\Phi^{(1)}_{\alpha\beta\gamma}(l\kappa; l'\kappa')\}$,

$$\Phi^{(0)}_{\alpha\beta}(LK; L'K') = \sum_{\alpha'\beta'} S_{\alpha\alpha'} S_{\beta\beta'} \Phi^{(0)}_{\alpha'\beta'}(l\kappa; l'\kappa'), \tag{16.19}$$

$$\Phi^{(1)}_{\alpha\beta\gamma}(LK; L'K') = |\boldsymbol{S}| \sum_{\alpha'\beta'\gamma'} S_{\alpha\alpha'} S_{\beta\beta'} S_{\gamma\gamma'} \times \Phi^{(1)}_{\alpha'\beta'\gamma'}(l\kappa; l'\kappa'). \tag{16.20}$$

Applied to the potential energy of the crystal from which the force constants of eq. (16.16) are derived, the transformation law of eq. (15.1) yields the equation

$$\Phi^{(0)}_{\alpha\beta}(l\kappa; l'\kappa') + \mathrm{i} \sum_\gamma \Phi^{(1)}_{\alpha\beta\gamma}(l\kappa; l'\kappa') H_\gamma + \dots$$
$$= \Phi^{(0)}_{\alpha\beta}(l\kappa; l'\kappa')^* + \mathrm{i} \sum_\gamma \Phi^{(1)}_{\alpha\beta\gamma}(l\kappa; l'\kappa')^* H_\gamma + \dots. \tag{16.21}$$

The reality of the atomic displacements has been used in obtaining this result.

We conclude that both $\Phi^{(0)}_{\alpha\beta}(l\kappa; l'\kappa')$ and $\Phi^{(1)}_{\alpha\beta\gamma}(l\kappa; l'\kappa')$ are real, as has been noted earlier. To insure that the frequencies of the normal modes are real, i.e. that the phonons are not attenuated, we require that the force constants $\{\Phi^{(0)}_{\alpha\beta}(l\kappa; l'\kappa')\}$ be Hermitian. This requirement leads to the additional conditions on the force constants

$$\Phi^{(0)}_{\beta\alpha}(l'\kappa'; l\kappa) = \Phi^{(0)}_{\alpha\beta}(l\kappa; l'\kappa'), \tag{16.22a}$$

$$\Phi^{(1)}_{\beta\alpha\gamma}(l'\kappa'; l\kappa) = -\Phi^{(1)}_{\alpha\beta\gamma}(l\kappa; l'\kappa'). \tag{16.22b}$$

The frequencies of the $\boldsymbol{q}\simeq 0$ modes of a crystal are obtained from the solutions to the eigenvalue equation

$$\omega^2_{sa} e_\alpha(\kappa/sa\lambda) = \sum_{\kappa'\beta} d_{\alpha\beta}(\kappa\kappa') e_\beta(\kappa'/sa\lambda), \tag{16.23}$$

where the dynamical matrix $d_{\alpha\beta}(\kappa\kappa')$ is defined by

$$d_{\alpha\beta}(\kappa\kappa') = \frac{1}{(M_\kappa M_{\kappa'})^{1/2}} \sum_{l'} \left\{ \Phi^{(0)}_{\alpha\beta}(l\kappa; l'\kappa') + \mathrm{i} \sum_{\gamma} \Phi^{(1)}_{\alpha\beta\gamma}(l\kappa; l'\kappa') H_\gamma + \ldots \right\}$$

$$= d^{(0)}_{\alpha\beta}(\kappa\kappa') + \mathrm{i} \sum_{\gamma} d^{(1)}_{\alpha\beta\gamma}(\kappa\kappa') H_\gamma + \ldots \tag{16.24}$$

and M_κ is the mass of the κth kind of atom. We have indexed the $3r$ solutions of eq. (16.23), where r is the number of atoms in a primitive unit cell, by the triple index $(sa\lambda)$. Here s labels the irreducible representation of the point group of the crystal to which the mode belongs, λ distinguishes the partner functions in the case that the representation s is multi-dimensional, a is a repetition index which distinguishes modes of different frequencies which belong to the same irreducible representation.

We now formally expand $e_\alpha(\kappa/sa\lambda)$ and ω_{sa} in powers of the components of the magnetic field as

$$e_\alpha(\kappa/sa\lambda) = e^{(0)}_\alpha(\kappa/sa\lambda) + e^{(1)}_\alpha(\kappa/sa\lambda) + \ldots, \tag{16.25a}$$

$$\omega_{sa} = \omega^{(0)}_{sa} + \omega^{(1)}_{sa} + \ldots. \tag{16.25b}$$

Because the $\{e^{(0)}_\alpha(\kappa/sa\lambda)\}$ are the eigenvectors of a real, symmetric matrix $d^{(0)}_{\alpha\beta}(\kappa\kappa')$, with no loss of generality we can assume them to be real, and it is convenient to do so. Then, according to first-order degenerate perturbation theory $2\omega^{(0)}_{sa}\omega^{(1)}_{sa}$ is obtained from the roots of the determinantal equation

$$|\mathrm{i}K^{(sa)}_{\lambda\lambda'} - 2\omega^{(0)}_{sa}\omega^{(1)}_{sa}\delta_{\lambda\lambda'}| = 0, \ \lambda, \lambda' = 1, 2, \ldots, d_s, \tag{16.26}$$

where

$$K^{(sa)}_{\lambda\lambda'} = \sum_{\gamma} K_\gamma(sa\lambda; sa\lambda') H_\gamma, \tag{16.27}$$

$$K_\gamma(sa\lambda; sa\lambda') = \sum_{\kappa\alpha} \sum_{\kappa'\beta} e^{(0)}_\alpha(\kappa/sa\lambda) \times d^{(1)}_{\alpha\beta\gamma}(\kappa\kappa') \times e^{(0)}_\beta(\kappa'/sa\lambda'). \tag{16.28}$$

The matrix $\boldsymbol{K}^{(sa)}$ is a $d_s \times d_s$ matrix, where d_s is the dimensionality of the irreducible representation s, and consequently the degree of degeneracy of the mode $(sa\lambda)$. Since no irreducible representation of any of the 32 crystallographic point groups has a dimensionality greater than three, $\boldsymbol{K}^{(sa)}$ is at most a 3×3 matrix.

Combining eqs. (16.22b) and (16.24) we find that

$$d^{(1)}_{\beta\alpha\gamma}(\kappa'\kappa) = -d^{(1)}_{\alpha\beta\gamma}(\kappa\kappa'). \tag{16.29}$$

From this result and eq. (16.28) it follows that

$$K_\gamma(sa\lambda'; sa\lambda) = -K_\gamma(sa\lambda; sa\lambda'). \tag{16.30}$$

It is an immediate consequence of this result and eq. (16.26) that the frequency of a non-degenerate mode cannot be shifted by an external magnetic field to first order in the field strength.

To determine the effects of an external magnetic field on doubly- and triply-degenerate modes at $\boldsymbol{q} \simeq 0$, we need to know the transformation law for the matrix elements $\{K_\gamma(sa\lambda; sa\lambda')\}$ when the crystal is subjected to an operation of its space group.

With the aid of eqs. (16.20) and (16.24) we readily obtain the transformation law obeyed by $d^{(1)}_{\alpha\beta\gamma}(\kappa\kappa')$,

$$d^{(1)}_{\alpha\beta\gamma}(KK') = |\boldsymbol{S}| \sum_{\alpha'\beta'\gamma'} S_{\alpha\alpha'} S_{\beta\beta'} S_{\gamma\gamma'} d^{(1)}_{\alpha'\beta'\gamma'}(\kappa\kappa'). \tag{16.31}$$

Changing the sums in eq. (16.28) to sums over transformed lattice sites and then applying eq. (16.31) results in

$$K_\gamma(sa\lambda; sa\lambda') = |\boldsymbol{S}| \sum_{K\alpha} \sum_{K'\beta} \sum_{\alpha'\beta'\gamma'} e^{(0)}_\alpha(K/sa\lambda) \times e^{(0)}_\beta(K'/sa\lambda') S_{\alpha\alpha'} S_{\beta\beta'} S_{\gamma\gamma'} d^{(1)}_{\alpha'\beta'\gamma'}(\kappa\kappa'), \tag{16.32}$$

or

$$K_\gamma(sa\lambda; sa\lambda') = |\boldsymbol{S}| \sum_{\kappa\alpha} \sum_{\kappa'\beta} \sum_{\kappa_1\kappa_2} \sum_{\alpha'\beta'\lambda'} \times e^{(0)}_\alpha(\kappa_1/sa\lambda) T_{\alpha\alpha'}(\kappa_1\kappa/\boldsymbol{S}) d^{(1)}_{\alpha'\beta'\gamma'}(\kappa\kappa') \times e^{(0)}_\beta(\kappa_2/sa\lambda') T_{\beta\beta'}(\kappa_2\kappa'/\boldsymbol{S}), \tag{16.33}$$

where the matrix $T_{\alpha\beta}(\kappa\kappa'/\boldsymbol{S})$ is defined by

$$T_{\alpha\beta}(\kappa\kappa'/\boldsymbol{S}) = S_{\alpha\beta}\delta(\kappa, F_0(\kappa'; S)). \tag{16.34}$$

We have denoted by $F_0(\kappa; S) \equiv K$ the index of the atom into which the atom κ is sent by a space group operation whose rotational element is described by the matrix $\boldsymbol{S}$. The matrix $T_{\alpha\beta}(\kappa\kappa'/\boldsymbol{S})$ is a real, orthogonal matrix

$$T^{-1}_{\alpha\beta}(\kappa\kappa'/\boldsymbol{S}) = T_{\beta\alpha}(\kappa'\kappa'/\boldsymbol{S}). \tag{16.35}$$

The effect of multiplying the eigenvector $e_\alpha^{(0)}(\kappa/sa\lambda)$ by the matrix $T_{\alpha\beta}(\kappa\kappa'/\boldsymbol{S})$ has been shown to be (Maradudin and Vosko 1968)

$$\sum_{\kappa'\beta} T_{\alpha\beta}(\kappa\kappa'/\boldsymbol{S}) e_\beta^{(0)}(\kappa'/sa\lambda) = \sum_{\lambda'} \tau_{\lambda'\lambda}^{(s)}(\boldsymbol{S}) e_\alpha^{(0)}(\kappa/sa\lambda'), \tag{16.36}$$

where $\boldsymbol{\tau}^{(s)}(\boldsymbol{S})$ is a $d_s \times d_s$ unitary matrix representation of the sth irreducible representation of the point group of the crystal. Tables of the matrices $\{\boldsymbol{\tau}^{(s)}(\boldsymbol{S})\}$ for all irreducible representations of the 32 crystallographic point groups are available in the literature (Kovalev 1964).

Certain point groups ($C_4, S_4, C_{4h}, C_3, S_6, C_6, C_{3h}, C_{6h}, T, T_h$) have pairs of complex-conjugate one-dimensional irreducible representations which are degenerate by time reversal symmetry. The remaining irreducible representations of these groups, and of all other crystallographic point groups, are equivalent to real representations, and can be assumed to be real with no loss of generality. Each of the pairs of complex-conjugate one-dimensional irreducible representations may be replaced by a real, orthogonal, two-dimensional matrix representation (Maradudin et al. 1971) according to

$$\boldsymbol{\tau}^{(ss^*)}(\boldsymbol{S}) = \begin{pmatrix} \frac{1}{2}(\boldsymbol{\tau}^{(s)}(\boldsymbol{S}) + \boldsymbol{\tau}^{(s*)}(\boldsymbol{S})) & \frac{1}{2\mathrm{i}}(\boldsymbol{\tau}^{(s)}(\boldsymbol{S}) - \boldsymbol{\tau}^{(s*)}(\boldsymbol{S})) \\ -\frac{1}{2\mathrm{i}}(\boldsymbol{\tau}^{(s)}(\boldsymbol{S}) - \boldsymbol{\tau}^{(s*)}(\boldsymbol{S})) & \frac{1}{2}(\boldsymbol{\tau}^{(s)}(\boldsymbol{S}) + \boldsymbol{\tau}^{(s*)}(\boldsymbol{S})) \end{pmatrix}, \tag{16.37}$$

where

$$\boldsymbol{\tau}^{(s*)}(\boldsymbol{S}) = \boldsymbol{\tau}^{(s)}(\boldsymbol{S})^*. \tag{16.38}$$

Doing so removes the inconsistency which would otherwise arise from having the eigenvectors $\{e_\alpha^{(0)}(\kappa/sa\lambda)\}$, which we have assumed to be real, transform according to a complex irreducible representation when multiplied by the real matrix $T_{\alpha\beta}(\kappa\kappa'/\boldsymbol{S})$, according to eq. (16.36).

If we treat pairs of complex-conjugate one-dimensional irreducible representations in this fashion, all the matrices $\{\boldsymbol{\tau}^{(s)}(\boldsymbol{S})\}$ entering eq. (16.36) become real, orthogonal matrices. Consequently, eq. (16.36) is equivalent to

$$\sum_{\kappa'\beta} e_\beta^{(0)}(\kappa'/sa\lambda) T_{\beta\alpha}(\kappa'\kappa/\boldsymbol{S}) = \sum_{\lambda'} \tau_{\lambda\lambda'}^{(s)}(\boldsymbol{S}) e_\alpha^{(0)}(\kappa/sa\lambda'). \tag{16.39}$$

It follows from this result and eq. (16.33) that the transformation law for $K_\gamma(sa\lambda; sa\lambda')$ is

$$K_\gamma(sa\lambda; sa\lambda') = |\boldsymbol{S}| \sum_{\gamma'} \sum_{\lambda_1\lambda_2} S_{\gamma\gamma'} \tau^{(s)}_{\lambda\lambda_1}(\boldsymbol{S}) \tau^{(s)}_{\lambda'\lambda_2}(\boldsymbol{S}) \times K_{\gamma'}(sa\lambda_1; sa\lambda_2). \quad (16.40)$$

Including the additional symmetry condition given by eq. (16.30) we can rewrite eq. (16.40) in the form

$$K_\gamma(sa\lambda; sa\lambda') = \frac{1}{2} |\boldsymbol{S}| \sum_{\gamma'\lambda_1\lambda_2} S_{\gamma\gamma'} \big[\tau^{(s)}_{\lambda\lambda_1}(\boldsymbol{S}) \tau^{(s)}_{\lambda'\lambda_2}(\boldsymbol{S}) - \tau^{(s)}_{\lambda'\lambda_1}(\boldsymbol{S}) \tau^{(s)}_{\lambda\lambda_2}(\boldsymbol{S}) \big] K_{\gamma'}(sa\lambda_1; sa\lambda_2). \quad (16.41)$$

Knowledge of the rotation matrices $\{\boldsymbol{S}\}$ for the point group of the crystal, together with the irreducible representation matrices $\boldsymbol{\tau}^{(s)}(\boldsymbol{S})$, allows one to obtain the independent, non-zero components of the tensor $K_\gamma(sa\lambda; sa\lambda')$.

Less detailed, but still useful, information about the matrix $K_\gamma(sa\lambda; sa\lambda')$ can be obtained more easily. The number of independent, non-zero elements of $K_\gamma(sa\lambda; sa\lambda')$ may be determined from the character formula

$$N^{(s)} = \frac{1}{2g} \sum_{S \subset G} |\boldsymbol{S}| \chi(S) \big[\chi^{(s)}(S)^2 - \chi^{(s)}(S^2) \big], \quad (16.42)$$

where g is the order of the point group of the crystal, G,

$$\chi(S) = \mathrm{Tr}\, \boldsymbol{S}, \quad (16.43)$$

and

$$\chi^{(s)}(S) = \mathrm{Tr}\, \boldsymbol{\tau}^{(s)}(\boldsymbol{S}). \quad (16.44)$$

From its definition, eq. (16.43), we see that $\chi(S)$ is the character of that reducible representation of G which when decomposed yields the polar vector irreducible representations of G, i.e. the irreducible representations of the infrared-active modes.

The results obtained by the methods of this section are in complete agreement with those obtained in §16.1. We find that $K_\gamma(sa\lambda; sa\lambda')$, has the same form for all the two-dimensional irreducible representations, and this is also the case for the three-dimensional representations. These forms are displayed explicitly in eqs. (16.9) and (16.10) respectively. In the non-cubic groups the axis of highest symmetry has been chosen as the x_3-axis.

From these results the first-order shifts in the frequencies of doubly-degenerate modes are found to be given by

$$\omega_{sa}^{(1)} = \pm \frac{1}{2\omega_{sa}^{(0)}} H_3 |K_3(sa1; sa2)|. \tag{16.45}$$

Thus, the external magnetic field must have a component along the axis of highest symmetry in the crystal to induce frequency shifts linear in the field strength. The eigenvectors of the corresponding perturbation matrix, which give the correct linear combinations of the unperturbed eigenvectors $\{e^{(0)}(\kappa/sa\lambda)\}$, are found to be

$$\boldsymbol{a}_+ = \frac{1}{\sqrt{2}} \begin{pmatrix} 1 \\ -\mathrm{i}\,\mathrm{sgn}\, K_3(sa1, sa2) \end{pmatrix}, \tag{16.46a}$$

$$\boldsymbol{a}_- = \frac{1}{\sqrt{2}} \begin{pmatrix} 1 \\ \mathrm{i}\,\mathrm{sgn}\, K_3(sa1; sa2) \end{pmatrix}. \tag{16.46b}$$

The first-order shifts in the frequencies of triply-degenerate modes are given by

$$\omega_{sa}^{(1)} = \frac{1}{2\omega_{sa}^{(0)}} \begin{cases} H|K_1(sa2; sa3)| \\ -H|K_1(sa2; sa3)| \end{cases}, \tag{16.47}$$

where H is the strength of the applied magnetic field.

In the case of the cubic crystal classes, therefore, application of the magnetic field along any direction in the crystal results in a complete splitting of the three-fold degeneracy of the mode. The eigenvectors of the corresponding perturbation matrix are

$$\boldsymbol{a}_0 = \begin{bmatrix} \hat{H}_1 \\ \hat{H}_2 \\ \hat{H}_3 \end{bmatrix} \tag{16.48a}$$

which is associated with the unshifted frequency, and

$$\boldsymbol{a}_+ = \frac{1}{\sqrt{2}\,(\hat{H}_1^2 + \hat{H}_2^2)^{1/2}} \begin{bmatrix} -\hat{H}_1\hat{H}_3 - \mathrm{i}\hat{H}_2 \\ -\hat{H}_2\hat{H}_3 + \mathrm{i}\hat{H}_1 \\ \hat{H}_1^2 + \hat{H}_2^2 \end{bmatrix} = \boldsymbol{a}_-^*. \tag{16.48b}$$

$\hat{\boldsymbol{H}}$ is a unit vector in the direction of the applied magnetic field. From eqs.

(16.46) and (16.48) we see that in both the doubly- and triply-degenerate cases the modes whose frequencies are shifted linearly by the magnetic field become right- and left-circularly polarized.

16.3. Experimental

Experimental observations of magnetic-field-induced splittings have been reported only recently. Schaack has extensively studied members of the family of rare-earth trifluorides and trichlorides. The first results of paramagnetic CeF_3 (point group D_{3d}) at 1.9 K and fields up to 6 T along the *c*-axis show a definite splitting of two of the twelve doubly-degenerate modes E_g (Schaack 1975). The splitting is linear in the field for low ($\leqslant 3$ T) fields and reaches a saturation for higher fields (figs. 23, 24). The linewidth decreases with the field while the intensity shows significant field dependence. No effect is observed for the A_g modes or for any modes with $\boldsymbol{H}|c$. In addition the scattered radiation is found to be circularly polarized (Schaack 1976a). Terms in the phonon force constant which are quadratic

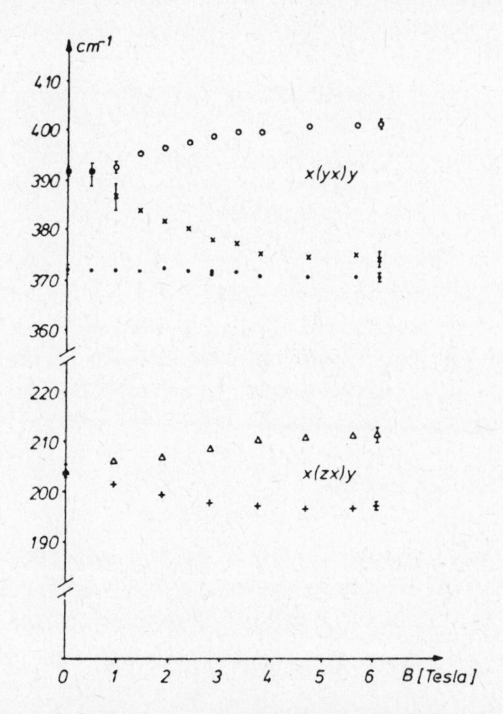

Fig. 23. Dependence of phonon frequencies (E_g-symmetry) in CeF_3 on an external magnetic field parallel to the trigonal axis at 1.9 K. The values of B have been corrected for demagnetization (Schaack 1975).

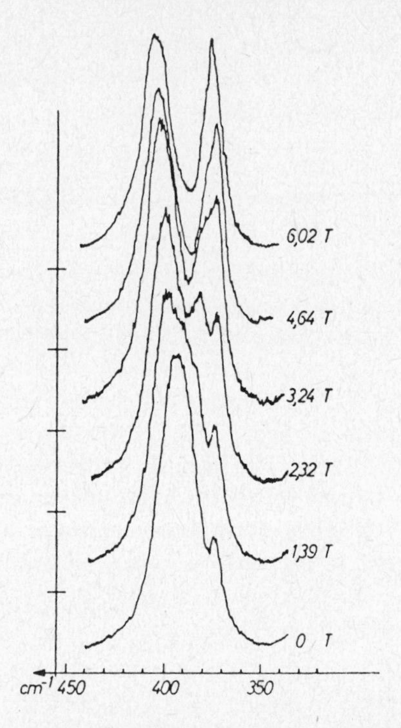

Fig. 24. Raman spectra of CeF_3 at different magnetic fields and $T = 1.9$ K. Scattering geometry $x(yx)y$. These spectra correspond to the frequencies plotted in the upper half of fig. 23 (Schaack 1975).

in the field may be partially responsible for the non-linearities and asymmetries in the phonon shifts observed at higher fields. All these observations are consistent with the symmetry analysis of the previous two sections. Assuming an effective spin $\frac{1}{2}$, Schaack has further fitted the observed splittings by the expression

$$\Delta\omega_j = \Delta\omega_j^{(s)} \tanh(\mu H / kT), \tag{16.49}$$

where $\Delta\omega_j^{(s)}$ is the saturation splitting and μ the magnetic moment of the ion $Ce^{+3}(4f^1; {}^2F_{5/2})$ in the ground state. The fit is very good, for low fields at least, and in fact a phonon g-factor is determined, equal to

$$g_{ph} = g_{el} \frac{\Delta\omega_j^{(s)}}{kT}, \tag{16.50}$$

where g_{el}, the electronic g-factor in the ground state, is typically of the order of 2. Values of $g_{ph,\parallel}=25.1$ and 11.4 and $g_{ph,\perp}=15.2$ and 10.6 are found for the 392 cm^{-1} and 204 cm^{-1} E_g-type phonons respectively. The overall behavior of the phonons in the presence of the field indicates a strong dependence of the field-induced force constants on the alignment of the angular moment and spins of the 4f electrons.

More complete measurements have been reported by Schaack (1976b) for paramagnetic CeF_3 and NdF_3. A total of five E_g-type phonons in NdF_3 show a splitting linear in $\boldsymbol{H}$ with a somewhat weaker behavior. No effect is reported for the diamagnetic member of the series PrF_3. Similar measurements in rare-earth trifluorides have also revealed an analogous behavior. In $CeCl_3$ (Schaack 1977a) two E_g-type phonons show a splitting linear in $\boldsymbol{H}$ for $\boldsymbol{H}//c$ and a quadratic splitting for $\boldsymbol{H}\perp c$. The latter is anticipated for relatively high fields and has the same symmetry characteristics as a quadratic electric field effect or a homogeneous strain. Such quadratic effects for $\boldsymbol{H}\perp c$ are shown in the spectra of fig. 25. It is also found (Schaack 1976b) that the same modes which split in $CeCl_3$ will also split in $PrCl_3$ and $NdCl_3$, the size of the splitting depending on the type of the 4f-ions. No splitting is observed in $EuCl_3$ and $GdCl_3$. This latter behavior has to be associated with the electronic structures of Eu^{+3} and Gd^{+3} ions respectively. The free ion Eu^{+3} ($^7F_0$) is diamagnetic in its ground state and the small crystal-field-induced Van Vleck paramagnetic moment is not sufficiently large to induce any phonon frequency changes. Similarly, the free ion Gd^{+3} ($^8S_{7/2}$) in its ground state exhibits only spin contributions to the magnetic moment. This leads to interactions with the crystal field, which are too weak to cause any observable splitting.

Two general conclusions are drawn from all these experiments:

(i) A permanent (and not just field-induced) magnetic moment of the ground-state 4f-ion seems to be necessary for any splitting to occur in both orientations $\boldsymbol{H}//c$ and $\boldsymbol{H}\perp c$.

(ii) Exchange or superexchange effects do not play any important role in the splitting. This becomes clear from measurements in $GdCl_3$ above and below its Curie point at 2.211 K. No change was observed in passing from the paramagnetic to the ferromagnetic phase (Schaack 1977b). On the other hand, magnetic ordering alone is not sufficient to induce splittings. One should really examine the evolution of the phonon symmetry as the crystal undergoes the ferromagnetic phase transition (Anastassakis and Burstein 1972, also see next section).

One possible basis for a microscopic theoretical model appropriate for the explanation of the magnetic-field-induced splittings is the cooperative Jahn–Teller effect (Schaack 1977a and references therein). The latter consists in an ordering of ground-state ion magnetic and electric multipoles. Essential for this ordering is a virtual phonon exchange mechanism.

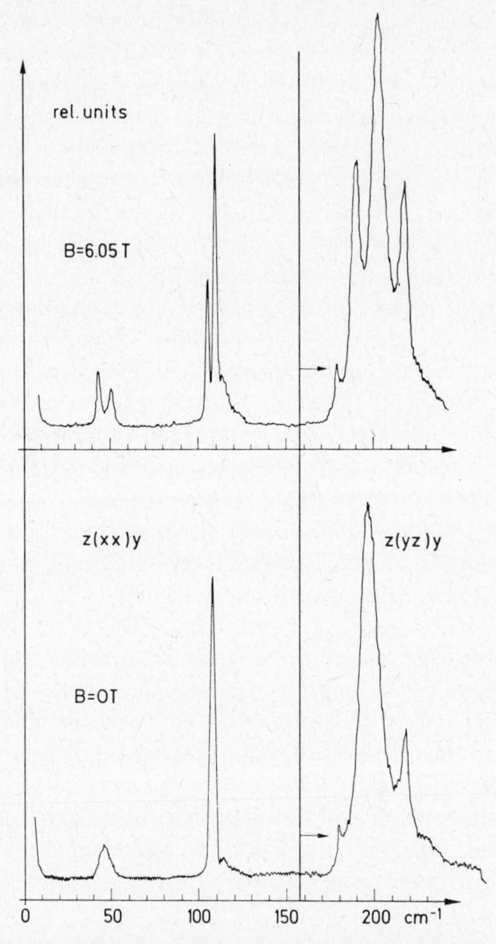

Fig. 25. Raman spectra of $CeCl_3$($T=2$ K) at zero magnetic field and at 6 T perpendicular to the optic axis. The spectra shown here are composed for clarity from two different orientations of the crystal (Schaack 1977a).

As a result, a reduction of symmetry takes place in some rare-earth compounds below a certain temperature. Application of a magnetic field along the crystal axis of some paramagnetic rare-earth compounds results in a rather complicated precessional motion of the magnetic moments. Owing to the strong spin–orbit coupling of the 4f-electrons, the multipole moments participate in the motion. It can be shown that the resulted dynamical components in the charge distribution may in fact remove the degeneracy of certain degenerate phonons in accordance with the experimental observations.

Thalmeier and Fulde (1977) have proposed an independent model for the magnetic-field-induced splittings and broadenings which are observed in $CeCl_3$, based on a magnetoelastic interaction mechanism. According to their calculation a measurable splitting should be observed with $\boldsymbol{H}//c$ provided the frequencies of the E_{1g} and E_{2g} modes are nearly equal to the separation between a pair of crystal-field-split energy levels of the Ce ion. While this is the case for the single E_{1g} mode, only one of the three different E_{2g} modes is split, as the frequencies of the other two are *off resonance*. Furthermore, the magnetoelastic coupling constant for these two E_{2g} modes may be too small to yield a splitting within experimental error. The theory of Thalmeier and Fulde predicts opposite circular polarization for the two split phonon components, and a monotonic quadratic splitting when $\boldsymbol{H}\perp c$. Finally, a second-order magnetoelastic interaction mechanism is invoked for the field-induced-broadening of the phonon bands. The experimental results on rare-earth trichlorides of Schaack (1977a, b) and on the mixed systems $Ce_xLa_{1-x}F_3$ $(1 \leqslant x \leqslant 0.2)$ of Ahrens and Schaack (1978a, b) are generally in agreement with these theoretical predictions, thus indicating that magnetoelastic interactions do in fact play a dominant role in these phenomena.

Harley et al. (1977) have reported the results of Raman scattering experiments in $DyVO_4$ in the presence of a magnetic field along the x-axis. This material is known to undergo a cooperative Jahn–Teller structural phase transition $D_{4h} \rightarrow D_{2h}$ at 14 K. The results for the magnetic-field-induced splitting of the E_g^I phonon show a non-linear dependence on the field at temperatures above and below 14 K. A saturation splitting is reached for all temperatures. The temperature and field dependence of the splittings observed are not in good agreement with the results of a calculation of the orthorhombic distortions, based on a mean-field theoretical model (Pytte 1974).

Finally, magnetic-field-induced splittings have been reported by Billardon et al. (1976) for KI doped with F centres. The results for the splittings observed and the polarization selection rules applied are in qualitative agreement with the symmetry analysis of these authors.

17. The effect of time reversal on the symmetry of the long-wavelength optical phonons

17.1. General

The purpose of this section is to examine the changes in the first-order photon–optical phonon interactions (infrared absorption and Raman scattering) which may arise when the time averaged magnetic moment density of a crystal becomes non-zero. This is the case of a paramagnetic crystal which is cooled below the transition temperature (T_n) to become a magnetically ordered crystal (ferromagnetic, antiferromagnetic, etc.). As the crystal passes from the paramagnetic ($\mathcal{P}$) state to the magnetically ordered ($\mathcal{M}$) state the time reversal symmetry is destroyed and the point group of the system changes from the grey group ($\boldsymbol{g}$) to one of the magnetic point groups ($\boldsymbol{M}$). The latter includes unitary–antiunitary point operations which leave invariant both the crystal and the ordered atomic magnetic moments. The changes from the $\mathcal{P}$ state to the $\mathcal{M}$ state, or equivalently the changes of symmetry from $\boldsymbol{g}$ to $\boldsymbol{M}$ (i.e. $\boldsymbol{g} \rightarrow \boldsymbol{M}$), result in changes in the symmetry of the $\boldsymbol{q} \simeq 0$ optical phonons. These changes are expected to manifest themselves in first-order Raman scattering in the usual three ways, that is, (i) change in the polarization selection rules corresponding to the appearance of new components in the Raman tensor, (ii) observation of Raman scattering by normally Raman-inactive modes, and (iii) splitting of the degeneracy of doubly-degenerate modes. Similar effects may be expected in connection with first-order infrared absorption (Anastassakis and Burstein 1972). It is emphasized that the transition $\boldsymbol{g} \rightarrow \boldsymbol{M}$ does not involve any structural change; it is purely a magnetic phase transition.

If one looks upon the removal of the time reversal symmetry and the consequent effects as a result of a phase transition due to cooling of the crystal below T_n, these effects constitute examples of a temperature-induced morphic effect.

A situation similar to that of a crystal in the $\mathcal{M}$ state arises when a crystal (magnetic or otherwise) is subject to an external magnetic field which, like a non-zero magnetic moment, destroys the time reversal symmetry. The symmetry of the $\boldsymbol{q} \simeq 0$ optical phonons of a crystal in the presence of an external magnetic field can therefore be described by an appropriate magnetic point group; in other words, the system "crystal+ magnetic field" can be looked upon as an ordered ferromagnet. Alternatively, the first-order perturbation approach discussed in §15 may be followed.

The possible experimental configurations for observing first-order Raman scattering by any type of Raman-active excitations are determined from the non-zero components of the Raman tensor. It is important to note that it is only the symmetry character of the excitation that determines which components of its Raman tensor are non-zero. This implies that two physically different types of excitations with identical symmetry will have *identical* Raman tensors. In the present case it will be shown that the symmetry of a $\boldsymbol{q} \simeq 0$ optical phonon in the $\mathcal{M}$ state is identical to the symmetry of a physically different excitation, namely a $\boldsymbol{q}\simeq 0$ optical magnon. Therefore we can use the Raman tensor for the magnon in order to describe Raman scattering processes involving phonons and vice versa. The complete matrix form for the Raman tensor for all the Raman-active magnons and all the 58 coloured magnetic classes has been worked out by Cracknell (1969). In the same reference it is also shown that the Raman matrices of the 32 grey groups are, within a unitary transformation, the same as the Raman matrices of the 32 ordinary point groups as derived by Ovander (1960). In what follows we will try to establish the connection between the symmetry aspects of infrared absorption and Raman scattering by magnons and phonons in the $\mathcal{M}$ state. In treating the group-theoretical analysis of the problem, Wigner's theory of corepresentations is used (Wigner 1959). Specific examples will be treated in detail.

A brief summary of some related definitions for magnetic groups is given next. There are 122 magnetic point groups classified into three types (Bhagavantan 1966, Dimmock and Wheeler 1962a).

(i) The 32 *colourless* magnetic point groups. These are identical to the ordinary 32 non-magnetic point groups ($\boldsymbol{G}$).

(ii) The 32 *grey* groups ($\boldsymbol{g}$), one for each ordinary point group.

(iii) The 58 *double coloured* or *black and white* magnetic point groups ($\boldsymbol{M}$).

Type (iii) magnetic point groups are the ones that we are primarily concerned with. Their construction and tabulation can be found in a number of places. We use here the corrected tables (I) of Cracknell (1968). Throughout this section we use the notation of Cracknell (1966, 1968). Since the order of listing of the 58 groups in the tables (I) of Cracknell (1968) is consistent, we will often identify the magnetic point groups by their order of appearance in those tables.

By definition $\boldsymbol{M}$ and $\boldsymbol{g}$ are given by

$$\boldsymbol{M}=\boldsymbol{H}+\boldsymbol{R}(\boldsymbol{G}-\boldsymbol{H})\equiv\boldsymbol{G}[\boldsymbol{H}], \tag{17.1}$$

$$\boldsymbol{g}=\boldsymbol{G}+\boldsymbol{R}\boldsymbol{G}, \tag{17.2}$$

where $\boldsymbol{G}$ is the parent unitary point group, that is, the ordinary point group describing the unitary symmetry of the state, $\boldsymbol{H}$ is one of the invariant halving unitary subgroups of $\boldsymbol{G}$ and $\boldsymbol{R}$ is the time reversal operator. For a given $\boldsymbol{G}$, one can in general find more then one invariant halving subgroup $\boldsymbol{H}$, thus making possible the construction of more than one $\boldsymbol{M}$ for the same parent point group $\boldsymbol{G}$. A helpful diagram which gives all the invariant (but not necessarily halving) subgroups for the 32 point groups ($\boldsymbol{G}$) can be found in figs. 5, 1, 12-1 of Koster et al. (1963), Dimmock and Wheeler (1962a) and Dimmock and Wheeler (1964) respectively. The number of symmetry operations of $\boldsymbol{M}$ is the same as that of $\boldsymbol{G}$. Half of them (those of $\boldsymbol{H}$) are unitary while the rest (those of $\boldsymbol{R}(\boldsymbol{G}-\boldsymbol{H})$) are antiunitary (due to the antiunitary character of $\boldsymbol{R}$). A word of caution; $\boldsymbol{M}$ and $\boldsymbol{H}$ as used here should not be confused with the same symbols used earlier to designate the electric moment and the magnetic field respectively.

The irreducible representations of $\boldsymbol{M}$ are called *irreducible corepresentations* (Wigner 1959); their construction is based on the representations of $\boldsymbol{H}$ and not of $\boldsymbol{G}$. If $\Gamma^{(i)}$ designates a given representation of $\boldsymbol{H}$, the corepresentation of $\boldsymbol{M}$ derived from $\Gamma^{(i)}$ is designated by $D\Gamma^{(i)}$. Depending on $\Gamma^{(i)}$, the corepresentation $D\Gamma^{(i)}$ can be one of the following three types (Dimmock and Wheeler 1962b).

(a) $\Gamma^{(i)}$ leads to $D\Gamma^{(i)}$ with the same dimensionality.

(b) $\Gamma^{(i)}$ leads to $D\Gamma^{(i)}$ with twice the dimensionality of $\Gamma^{(i)}$.

(c) $\Gamma^{(i)}$ and $\Gamma^{(m)}$ (two non-equivalent representations of $\boldsymbol{H}$) *stick together* to produce a degenerate corepresentation $D\Gamma^{(k)}$ of $\boldsymbol{M}$.

A complete list of the types of $D\Gamma$ for any Γ and all 32 groups ($\boldsymbol{G}$) appears in table 8 of Cracknell (1968). In table 2 of the same reference all type (c) corepresentations are listed separately.

Since the present problem refers to phonons we are concerned only with single-valued point groups and their single-valued corepresentations. It can be shown that for single-valued representations, all the representations of $\boldsymbol{G}$ with real characters lead to a type (a) corepresentation of $\boldsymbol{g}$, and all the representations of $\boldsymbol{G}$ with complex characters lead to a type (c) corepresentation of $\boldsymbol{g}$. Type (b) corepresentations do not exist for single-valued grey groups (Dimmock and Wheeler 1962a, Dimmock and Wheeler 1964). Thus, all pairs of non-degenerate complex-conjugate representations of any ordinary point group lead to a doubly-degenerate corepresentation of the corresponding grey group.

17.2. Raman scattering by $q\simeq 0$ optical magnons and $q\simeq 0$ optical phonons

Following the definitions of the previous section, we designate by $\boldsymbol{G}$ the crystal point group in the $\mathcal{P}$ state, and by $\boldsymbol{H}$ the unitary invariant halving subgroup of $\boldsymbol{G}$ on the basis of which the crystal magnetic point group $\boldsymbol{M}$ is constructed, that is, $\boldsymbol{M}=\boldsymbol{G}[\boldsymbol{H}]$. As shown in §5.2 for the general case of first-order non-symmetric Raman scattering by *optical phonons*, the requirement for Raman activity is fulfilled by those phonons j whose irreducible representation $\Gamma(j)$ can be found in the reduction of the direct Kronecker product $\Gamma(\mathfrak{E})\otimes\Gamma(\mathfrak{E})$, i.e.

$$\Gamma(\mathfrak{E})\otimes\Gamma(\mathfrak{E})=h_j\Gamma(j)\oplus\ldots. \tag{17.3a}$$

Here $\Gamma(\mathfrak{E})$ is a three-dimensional representation (not necessarily irreducible) for the electric field vector $\mathfrak{E}$ (or any polar vector) in the single non-magnetic point group ($\boldsymbol{G}$) which describes the symmetry of the crystal. Since $\Gamma(\mathfrak{E})$ can in general be reduced as $\Sigma_\nu \Gamma_\nu$, where Γ_ν is one of the irreducible representations of the point group, eq. (17.3a) can be written as

$$\Gamma(\mathfrak{E})\otimes\Gamma(\mathfrak{E})=\sum_{\mu,\lambda} C_{\mu\lambda,j}\Gamma(j)\oplus\ldots, \tag{17.3b}$$

where $C_{\mu\lambda,j}$ are the Clebsch–Gordan coefficients (Hammermesh 1964, Birman and Berenson 1974) or coupling coefficients (Koster et al. 1963), for the Kronecker product

$$\Gamma_\mu\otimes\Gamma_\lambda=\sum_\kappa C_{\mu\lambda,\kappa}\Gamma_\kappa. \tag{17.3c}$$

Γ_μ and Γ_λ are any two of the irreducible representations occurring in $\Gamma(\mathfrak{E})=\Sigma_\nu\Gamma_\nu$ and Γ_κ can be any of the irreducible representations of the point group. The values for $C_{\mu\lambda,j}$ for all combinations $(\mu\lambda j)$ are readily found in the multiplication tables of Koster et al. (1963).

For first-order Raman scattering by an *optical magnon* j the same arguments can be followed as for optical phonons except that we now have to use corepresentations (DΓ) rather than representations (Γ). Relations analogous to eqs. (17.3a, b, c) are

$$\mathrm{D}\Gamma(\mathfrak{E})\otimes\mathrm{D}\Gamma(\mathfrak{E})=\mathrm{D}n_j\,\mathrm{D}\Gamma(j)\oplus\ldots \tag{17.4a}$$

$$=\sum_{\mu,\lambda} d_{\mu\lambda,j}\mathrm{D}\Gamma(j)\oplus\ldots, \tag{17.4b}$$

$$\mathrm{D}\Gamma_\mu\otimes\mathrm{D}\Gamma_\lambda=\sum_\kappa d_{\mu\lambda,\kappa}\mathrm{D}\Gamma_\kappa, \tag{17.4c}$$

where $d_{\mu\lambda,j}$ and $d_{\mu\lambda,\kappa}$ are the corresponding Clebsch–Gordan coefficients for the reduction of corepresentation products. These coefficients have been calculated for all magnetic point groups by several workers and can be found in a number of places (Rudra and Sikdar 1976 and references therein). It should be remembered that the representations (Γ) appearing in eqs. (17.3a, b, c) belong to the point group $\boldsymbol{G}$ whereas those of eq. (17.4a, b, c) belong to the point subgroup $\boldsymbol{H}$ because, by definition, the construction of a corepresentation DΓ of the magnetic point group $\boldsymbol{M}=\boldsymbol{G}[\boldsymbol{H}]$ is based on the representations Γ of $\boldsymbol{H}$. Eqs. (17.3a, b, c) are quite general and apply for any point group. For convenience in what follows we will assume that eqs. (17.3a, b, c) refer to the subgroup $\boldsymbol{H}$.

One notes that *for each magnon of symmetry corepresentation* $D\Gamma(j)$ *there corresponds a representation* $\Gamma(j)$ *of the unitary subgroup* $\boldsymbol{H}$. It can be shown that the necessary and sufficient condition for a particular coefficient $d_{\mu\lambda,\kappa}$ of eq. (17.4c) to be non-zero is that the coefficient $C_{\mu\lambda,\kappa}$ of eq. (17.3c) be non-zero. A non-zero coefficient $C_{\mu\lambda,\kappa}$ means that the representation $\Gamma(j) = \Gamma_\kappa$ of $\boldsymbol{H}$ is included in the reduction of the representation of a non-symmetric second-rank tensor, according to eq. (17.3b). It is therefore concluded that *a magnon of symmetry* $D\Gamma(j)$ *is Raman active (i.e.* $d_{\mu\lambda,j} \neq 0$*) if and only if the representation* $\Gamma(j)$ *of the unitary subgroup* $\boldsymbol{H}$ *can describe a Raman-active phonon of any structure from the point group* $\boldsymbol{H}$. The reverse statement is also true. As shown by Bradley and Davies (1968, table V) the coefficients $d_{\mu\lambda,\kappa}$ and $C_{\mu\lambda,\kappa}$ are not in general equal but they are *simply* related. Thus, the number of independent components Dn_j of the Raman tensor for the magnon of symmetry $D\Gamma(j)$ is not necessarily equal to the number of independent components n_j of the Raman tensor for a phonon in any crystal structure with point group $\boldsymbol{H}$, whose symmetry happens to be $\Gamma(j)$. More about the symmetry properties of magnons and their interactions with light in infrared absorption and Raman scattering processes can be found in Cracknell's *Magnetism in Crystalline Materials* (Cracknell 1975).

The connection between the representation $\Gamma_G(j)$, for a phonon j in the $\mathscr{P}$ state (whose spatial symmetry is described by the point group $\boldsymbol{G}$) and the corepresentations of the corresponding magnetic point group $\boldsymbol{M}$, can easily be obtained following the same procedure that Cracknell uses to describe the behavior of atomic levels in a crystalline field when the crystal undergoes a magnetic phase transition (Cracknell 1968; for the correct form of group 33 see Bradley and Cracknell 1972, table 7.1). In the present case we simply consider the changes in the symmetry of the crystal, following the scheme

$$\boldsymbol{g} \rightarrow \boldsymbol{G} \rightarrow \boldsymbol{H} \rightarrow \boldsymbol{M}, \tag{17.5a}$$

where $\boldsymbol{g}$ is the grey group in the $\mathscr{P}$ state. A given phonon j is described by the corepresentation $D\Gamma_G(j)$ in $\boldsymbol{g}$ and by the representation $\Gamma_G(j)$ in $\boldsymbol{G}$. The trivial relationship between $D\Gamma_G(j)$ and $\Gamma_G(j)$ was discussed in §17.1. Next, the representation $\Gamma_G(j)$ reduces to at least one representation $\Gamma(j)$ of the point group $\boldsymbol{H}$. This is always possible since $\boldsymbol{H}$ is a subgroup of $\boldsymbol{G}$ by definition. For the reduction $\Gamma_G(j) \rightarrow \Sigma\Gamma(j)$ the compatibility tables of Koster et al. (1963) are particularly useful. The different notations used by different authors are reconciled in the Appendix (table A2, also Cracknell 1968, appendix). For convenience we choose the notation $\Gamma(j)$ rather than the more appropriate one $\Gamma_H(j)$. Accordingly, the representations change as follows:

$$\mathrm{D}\Gamma_G(j)\rightarrow\Gamma_G(j)\rightarrow\Sigma\Gamma(j)\rightarrow\Sigma\mathrm{D}\Gamma(j). \qquad (17.5\mathrm{b})$$

The corepresentations $\mathrm{D}\Gamma(j)$ *of* $\boldsymbol{M}$ *which occur in the above reduction describe the symmetry in the* $\mathcal{M}$ *state of the original phonon j.*

A final comment should be made in the interpretation of the point group $\boldsymbol{G}$. In the preceding discussion it has been assumed that, (i) $\boldsymbol{G}$ is the point group in the $\mathcal{P}$ state, and (ii) one of its unitary halving subgroups ($\boldsymbol{H}$) is used for the construction of the crystal magnetic point group $\boldsymbol{M}=\boldsymbol{G}[\boldsymbol{H}]$. There are crystals however for which $\boldsymbol{G}$, as defined in (i), is not the same as that in (ii), namely, the magnetic point group is constructed on the basis of another point group $\tilde{\boldsymbol{G}}$, of lower symmetry than that of $\boldsymbol{G}$. Thus $\tilde{\boldsymbol{M}}=\tilde{\boldsymbol{G}}[\tilde{\boldsymbol{H}}]$. Again $\tilde{\boldsymbol{M}}$ is one of the 58 magnetic classes, and $\tilde{\boldsymbol{H}}$ is as usual a halving subgroup of $\tilde{\boldsymbol{G}}$. This further reduction of the lattice symmetry from $\boldsymbol{G}$ to $\tilde{\boldsymbol{G}}$ can arise from the presence of additional forces due to a non-zero magnetization below T_{n}, for example magnetostriction. The transition element oxides, like NiO, FeO and MnO, are materials of this type. Above T_{n} they have the NaCl structure ($\boldsymbol{G}=\mathrm{O_h}$) whereas below T_{n} they become slightly distorted from the cubic to the rhombohedral structure i.e. $\tilde{\boldsymbol{G}}=\mathrm{D_{2h}}$ (Cracknell and Joshua 1969). The magnetic point group is $\tilde{\boldsymbol{M}}=\mathrm{D_{2h}[C_{2h}]}=\mathrm{m^1m^1m}$. Under these circumstances eq. (17.5a) is modified as follows:

$$\boldsymbol{g}\rightarrow\boldsymbol{G}\rightarrow\tilde{\boldsymbol{G}}\rightarrow\tilde{\boldsymbol{H}}\rightarrow\tilde{\boldsymbol{M}}, \qquad (17.5\mathrm{c})$$

which, from the representation point of view, is equivalent to

$$\boldsymbol{g}\rightarrow\boldsymbol{G}\rightarrow\tilde{\boldsymbol{H}}\rightarrow\tilde{\boldsymbol{M}}. \qquad (17.5\mathrm{d})$$

The reduction of the representations of $\boldsymbol{G}$ to those of $\tilde{\boldsymbol{H}}$ is obtained easily from the compatibility tables of the groups $\boldsymbol{G}$ and $\tilde{\boldsymbol{H}}$ (Koster et al. 1963). Thus the changes of the mode symmetry as described by eq. (17.5b) are still valid except that now the representations $\Gamma(j)$ are those of $\tilde{\boldsymbol{H}}$ and the corepresentations $\mathrm{D}\Gamma(j)$ are those of $\tilde{\boldsymbol{M}}$.

In the following sections we will treat in detail specific examples demonstrating the most important types of effects that may be exhibited in first-order Raman scattering and infrared absorption by the optical phonons of a magnetically ordered crystal.

17.3. Modification of the polarization selection rules of Raman-active phonons

As an application of the general procedure described above we examine crystals with the rutile structure in the $\mathcal{P}$ state, for example $\mathrm{MnF_2}$, $\mathrm{CoF_2}$,

FeF_2. They all become antiferromagnetic below T_n and have a symmetry described by the magnetic point group $4^1/mmm$ (more correctly $4^1/mmm^1$) or $D_{4h}[D_{2h}]$, i.e. $\boldsymbol{G}=D_{4h}, \boldsymbol{H}=D_{2h}$. There are five generally non-symmetric Raman-active modes with symmetries $\Gamma_G(j)=$ $A_{1g}, A_{2g}, B_{1g}, B_{2g}, E_g$. The Raman tensor matrices in the $\mathcal{P}$ state are (Ovander 1960)

$$
\underset{A_{1g}}{\begin{bmatrix} a & \cdot & \cdot \\ \cdot & a & \cdot \\ \cdot & \cdot & b \end{bmatrix}}, \quad
\underset{A_{2g}}{\begin{bmatrix} \cdot & c & \cdot \\ -c & \cdot & \cdot \\ \cdot & \cdot & \cdot \end{bmatrix}}, \quad
\underset{B_{1g}}{\begin{bmatrix} d & \cdot & \cdot \\ \cdot & -d & \cdot \\ \cdot & \cdot & \cdot \end{bmatrix}},
$$

$$
\underset{B_{2g}}{\begin{bmatrix} \cdot & e & \cdot \\ e & \cdot & \cdot \\ \cdot & \cdot & \cdot \end{bmatrix}}, \quad
\underset{E_g}{\left\langle \begin{bmatrix} \cdot & \cdot & f \\ \cdot & \cdot & \cdot \\ g & \cdot & \cdot \end{bmatrix}, \begin{bmatrix} \cdot & \cdot & \cdot \\ \cdot & \cdot & f \\ \cdot & g & \cdot \end{bmatrix} \right|}, \tag{17.6}
$$

where a, b, c, d, e, f, g are all real. The bra ($\langle|$) sign indicates that the mode E_g is doubly degenerate. The same matrices describe the Raman tensor in the grey group $\boldsymbol{g}=D_{4h}+\boldsymbol{R}D_{4h}$, since all five modes $\Gamma_G(j)$ lead to type (a) corepresentations of $\boldsymbol{g}$. In order to find the corepresentations of the five modes in the magnetic group $D_{4h}[D_{2h}]$, we consider the reduction of eq. (17.5b), with $\Gamma(j)$ standing for the representations of $\boldsymbol{H}=D_{2h}$. We have

$\boldsymbol{g}$	$\boldsymbol{G}$	$\boldsymbol{H}$	$\boldsymbol{M}$	
$DA_{1g}\to$	$A_{1g}\to$	A_g	$\to DA_g$	(a)
$DA_{2g}\to$	$A_{2g}\to$	B_{1g}	$\to DB_{1g}$	(a)
$DB_{1g}\to$	$B_{1g}\to$	A_g	$\to DA_g$	(a)
$DB_{2g}\to$	$B_{2g}\to$	B_{1g}	$\to DB_{1g}$	(a)
$DE_g\to$	$E_g\to$	$B_{2g}+B_{3g}$	$\to DB_g$	(c).

It should be remembered that the representations in the first and second columns refer to $\boldsymbol{G}=D_{4h}$. Those of the third and fourth column refer to $\boldsymbol{H}=D_{2h}$. The last column indicates the type of the corresponding corepresentation in $\boldsymbol{M}$, according to the table of Cracknell (1966). Notice the *sticking* of the two non-degenerate (real) representations B_{2g} and B_{3g} of D_{2h}, to produce a doubly-degenerate corepresentation BD_g of $\boldsymbol{M}$. The

Raman matrices for the corresponding corepresentations of $\boldsymbol{M}$ are (Cracknell 1969)

$$\underset{\mathrm{DA_g}}{\begin{bmatrix} A & \cdot & \cdot \\ \cdot & A^* & \cdot \\ \cdot & \cdot & B \end{bmatrix}}, \quad \underset{\mathrm{DB_{1g}}}{\begin{bmatrix} \cdot & C & \cdot \\ C^* & \cdot & \cdot \\ \cdot & \cdot & \cdot \end{bmatrix}}, \quad \underset{\mathrm{DA_g}}{\begin{bmatrix} D & \cdot & \cdot \\ \cdot & D^* & \cdot \\ \cdot & \cdot & Q \end{bmatrix}},$$

$$\underset{\mathrm{DB_{1g}}}{\begin{bmatrix} \cdot & E & \cdot \\ E^* & \cdot & \cdot \\ \cdot & \cdot & \cdot \end{bmatrix}}, \quad \underset{\mathrm{DB_g}}{\left\langle \begin{bmatrix} \cdot & \cdot & F \\ \cdot & \cdot & \cdot \\ G & \cdot & \cdot \end{bmatrix}, \begin{bmatrix} \cdot & \cdot & \cdot \\ \cdot & \cdot & -F^* \\ \cdot & -G^* & \cdot \end{bmatrix} \right|\right|}, \tag{17.7}$$

where B and Q are real. A direct comparison of matrices (17.6) and (17.7) reveals all the changes which occur in the Raman tensor when the crystal undergoes the transition from the $\mathscr{P}$ state to the $\mathscr{M}$ state. No splittings are observed in the present example. However, changes in the polarization selection rules do occur for one of the present modes: the zz component of the $\mathrm{B_{1g}}$ matrix is zero in the $\mathscr{P}$ state. In the $\mathscr{M}$ state the mode goes over to $\mathrm{DA_g}$ of $\boldsymbol{M}$ and the zz component is a non-zero real quantity (Q). This means that for $T<T_\mathrm{n}$ one should in principle be able to observe a band in the Raman spectrum at the frequency of the mode $\Gamma_G(j)=\mathrm{B_{1g}}$. The incident and scattered radiation must be polarized along the tetragonal axis z and the strength will be proportional to Q^2. The four-fold symmetry axis of $\mathrm{D_{4h}}$ coincides with one of the three two-fold symmetry axes of $\mathrm{D_{2h}}$. We take this to be the common z axis for both $\mathrm{D_{4h}}$ and $\mathrm{D_{2h}}$. As the crystal is heated above T_n, the band should disappear. No experimental observation of this effect has been reported. The Raman spectrum of phonons in several antiferromagnetic materials has been studied (Porto et al. 1967, Fleury 1968, Fleury et al. 1969, Macfarlane and Ushioda 1970) but only in the $\mathscr{P}$ state. In the case of FeF_2 and MnF_2 Porto et al. (1967) simply observe a shift of the frequency of the $\mathrm{B_{1g}}$ mode at 20 K relative to that at 300 K. Since the transition temperatures for these materials are 78.5 K and 67.5 K respectively (Fleury and Loudon 1968), this mode should more correctly be referred to as $\mathrm{DA_g}$ for $T<T_\mathrm{n}$ according to the above discussion. These spectra were not taken in the zz configuration.

17.4. Raman scattering by normally Raman-inactive phonons

(i) *Paramagnetic crystals with a centre of symmetry.* When $\boldsymbol{G}$, the point group in the $\mathscr{P}$ state, is a centrosymmetric point group, the alignment of spins below T_n may or may not preserve the centre of symmetry. In other words, the halving subgroup $\boldsymbol{H}$ may or may not be a centrosymmetric point group. We examine these cases separately.

(a) Assuming $\boldsymbol{G}$ is a centrosymmetric point group, we can always write it as

$$\boldsymbol{G}=\boldsymbol{H}'+\boldsymbol{I}\boldsymbol{H}', \tag{17.8}$$

where $\boldsymbol{H}'$ is a well-defined non-centrosymmetric halving point group, and $\boldsymbol{I}$ is the spatial inversion operator. In this case the Raman tensor can be found in a particularly straightforward manner, as can be seen from the following arguments. Let $\Gamma_G(j_e)$ and $\Gamma_G(j_o)$ be the representations of $\boldsymbol{G}$ for an even-parity phonon and its corresponding odd-parity phonon respectively. The compatibility relations between the representations of $\boldsymbol{G}$ and those of $\boldsymbol{H}'$ require that both $\Gamma_G(j_e)$ and $\Gamma_G(j_o)$ go over to the same representation $\Gamma(j)$ of $\boldsymbol{H}'$; the corresponding phonon j may be both Raman and infrared active (since $\boldsymbol{H}'$ lacks a centre of symmetry). In this case the Raman tensor for $\Gamma(j)$ has the same matrix form as that for $\Gamma_G(j_e)$. We now examine those crystals for which the magnetic point group is $\boldsymbol{M}'=\boldsymbol{G}[\boldsymbol{H}']$. There are 21 magnetic point groups of the type $\boldsymbol{M}'=\boldsymbol{G}[\boldsymbol{H}']$. These are listed in page 740 of Dimmock and Wheeler (1964). It follows from eqs. (17.1) and (17.8) that

$$\boldsymbol{M}'=\boldsymbol{H}'+\boldsymbol{R}\boldsymbol{I}\boldsymbol{H}'. \tag{17.9}$$

The reduction scheme of (17.5b) can now be written as

$$\left.\begin{matrix} \mathrm{D}\Gamma_G(j_e) \\ \mathrm{D}\Gamma_G(j_o) \end{matrix}\right\} \rightarrow \left.\begin{matrix} \Gamma_G(j_e) \\ \Gamma_G(j_o) \end{matrix}\right\} \rightarrow \Gamma(j) \rightarrow \mathrm{D}\Gamma(j),$$

where $\Gamma(j)$ and $\mathrm{D}\Gamma(j)$ correspond to $\boldsymbol{H}'$ and $\boldsymbol{M}'$ respectively. It turns out (Cracknell 1969) that the Raman tensor for the corepresentations of $\boldsymbol{M}'$ is the same as that for the grey group

$$\boldsymbol{g}'=\boldsymbol{H}'+\boldsymbol{R}\boldsymbol{H}'. \tag{17.10}$$

Furthermore, the Raman tensor for the phonon j is the same in $\boldsymbol{H}'$ and $\boldsymbol{g}'$, $\boldsymbol{g}'$ being the grey group of $\boldsymbol{H}'$ (Cracknell 1969). Hence the Raman tensors for $\Gamma(j)$ and $\mathrm{D}\Gamma(j)$ are the same. From the above, it follows that the Raman tensors for $\Gamma_G(j_e)$, $\Gamma_G(j_o)$, $\Gamma(j)$ and $\mathrm{D}\Gamma(j)$ are all the same in the $\mathcal{M}$ state, and are represented by the Raman tensor of the Raman-active phonon j_e in the $\mathcal{P}$ state.

As an example let us consider the antiferromagnetic material Cr_2O_3 for which $\boldsymbol{G}=\mathrm{D}_{3d}=\mathrm{D}_3+\boldsymbol{I}\mathrm{D}_3$ and $\boldsymbol{M}'=\mathrm{D}_{3d}[\mathrm{D}_3]=\mathrm{D}_3+\boldsymbol{R}\boldsymbol{I}\mathrm{D}_3$. Half of the representations of D_{3d} are of even parity and half of odd parity. Because of the complementary principle of first-order Raman scattering and infrared absorption, the odd-parity phonons may only be infrared active. It has just been shown that the Raman-inactive (and infrared-active) phonons with symmetry A_{1u}, E_u in the centrosymmetric point group D_{3d} will become Raman active at temperatures below T_n. The Raman tensor for these phonons in the $\mathfrak{M}$ state will be the same as that of the Raman-active phonons with symmetries A_{1g}, E_g in the $\mathcal{P}$ state. The Raman-inactive phonon A_{2g} remains inactive below T_n and so does the corresponding phonon of odd parity A_{2u}. The symmetric part of the Raman matrices for the Raman-active phonons with symmetries A_{1g}, E_g is (Loudon 1964),

$$\underset{A_{1g}}{\begin{bmatrix} A & \cdot & \cdot \\ \cdot & A & \cdot \\ \cdot & \cdot & B \end{bmatrix}}, \underset{E_g}{\left\langle \begin{bmatrix} C & \cdot & \cdot \\ \cdot & -C & D \\ \cdot & D & \cdot \end{bmatrix} \begin{bmatrix} \cdot & -C & -D \\ -C & \cdot & \cdot \\ -D & \cdot & \cdot \end{bmatrix} \right\rangle}.$$

According to the above discussion the same matrices will describe the matrix form of the Raman tensor in the $\mathfrak{M}$ state for the normally Raman-inactive phonons A_{1u}, E_u. One can easily show that these matrices are, within a unitary transformation, the same as the symmetric part of the matrices for the corepresentations DA_1 and DE of the magnetic point group $D_{3d}[D_3]$ given by Cracknell (1969).

It is emphasized that the preceding results refer to the magnetic point groups $\boldsymbol{M}' = \boldsymbol{G}[\boldsymbol{H}']$ with $\boldsymbol{G}$ and $\boldsymbol{H}'$ related through eq. (17.8). They were treated explicitly because the results can be obtained in a particularly straightforward manner, that is, the Raman tensor in the $\mathfrak{M}$ state of a phonon which is Raman inactive in the $\mathcal{P}$ state, is the same as the Raman tensor of the corresponding even-parity Raman-active phonon in the $\mathcal{P}$ state. We notice that $\boldsymbol{M}' = \boldsymbol{H}' + \boldsymbol{RIH}'$ and $\boldsymbol{G} = \boldsymbol{H}' + \boldsymbol{IH}'$, i.e. it is only the spin alignment that causes the removal of the centre of symmetry and consequently breaks the complementary principle for infrared absorption and Raman scattering. Such a situation may be encountered in antiferromagnetic materials, when identical atoms interchangeable through inversion carry equal and opposite spins.

(b) $\boldsymbol{H}$ is a non-centrosymmetric point group other than $\boldsymbol{H}'$ of eq. (17.8), for example $D_{4h}[D_{2d}], D_{3d}[C_{3v}]$. In this case the general procedure developed in §17.3 is followed. Again, the removal of centre of symmetry may result in breaking of the complementary principle and therefore Raman-inactive modes may become Raman active.

(c) Both $\boldsymbol{G}$ and $\boldsymbol{H}$ are centrosymmetric point groups, for example $D_{4h}[D_{2h}]$ or $D_{3d}[C_{3i}]$. The latter describes the symmetry of ferromagnetic

VBO_3. Here the centre of symmetry is not removed below T_n; thus the silent modes remain silent. The infrared-active modes remain infrared active and they cannot become Raman active because the complementary principle continues to hold. This case applies to cubic ferromagnets. As an example we consider EuO, which has the rock-salt structure in the $\mathcal{P}$ state. In the presence of an external magnetic field $\boldsymbol{B}$ partial or complete alignment of the atomic spins can be achieved. The direction of the magnetization depends on the direction of $\boldsymbol{B}$. Thus it is actually the direction of $\boldsymbol{B}$ that determines the magnetic point group $\boldsymbol{M}=\boldsymbol{G}[\boldsymbol{H}]$ of the ordered ferromagnet. Since neither a magnetic field nor a net magnetic moment can remove the centre of symmetry, $\boldsymbol{H}$ remains centrosymmetric and therefore the parity selection rules remain in effect. Thus, the triply-degenerate (F_{1u}) infrared-active, Raman-inactive phonon of EuO will preserve its odd parity below T_n, regardless of direction of the applied field. Generally speaking any structure whose atoms carry aligned spins and are located at centres of symmetry will retain the centre of symmetry below T_n. Antiferromagnetic materials may also belong to this category, for example MnF_2 and FeF_3 with symmetry $D_{4h}[D_{2h}]$ and $D_{3d}[C_{3i}]$ respectively.

(ii) *Paramagnetic crystals with no centre of symmetry.* For magnetic point groups $\boldsymbol{M}=\boldsymbol{G}[\boldsymbol{H}]$ for which $\boldsymbol{G}$ is not a centrosymmetric point group the results about the Raman activity of a normally Raman-inactive mode can be obtained by following the general procedure described in §17.3. As an example we take the silent modes B_1 and B_2 of the point group $\boldsymbol{G}=D_6$. For a crystal which in the $\mathcal{M}$ state has the symmetry of the magnetic point group $D_6[D_3]$ these antiferromagnetic modes follow the reduction scheme of eq. (17.5b), i.e.

$$DB_1 \rightarrow B_1 \rightarrow A_2 \rightarrow DA_2,$$

$$DB_2 \rightarrow B_2 \rightarrow A_1 \rightarrow DA_1.$$

The Raman matrices for B_1 and B_2 in the $\mathcal{M}$ state are the same as those for DA_2 and DA_1 respectively of the magnetic point group $D_6[D_3]$ as established by Cracknell (1969), namely,

$$\underset{DA_1}{\begin{bmatrix} A & \cdot & \cdot \\ \cdot & A & \cdot \\ \cdot & \cdot & B \end{bmatrix}} \underset{DA_2}{\begin{bmatrix} \cdot & C & \cdot \\ -C & \cdot & \cdot \\ \cdot & \cdot & \cdot \end{bmatrix}}.$$

The originally silent modes B_1 and B_2 will now exhibit diagonal and antisymmetric Raman scattering respectively.

17.5. Splitting of doubly-degenerate modes

(i) Type (a) corepresentations. The reduction $\Gamma_G(j) \to \Sigma\Gamma(j)$ of eq. (17.5b) may require that a degenerate representation $\Gamma_G(j)$ split into two type (a) representations $\Gamma^1(j)$ and $\Gamma^2(j)$. These will remain split in $\boldsymbol{M}$ as $D\Gamma^1(j)$ and $D\Gamma^2(j)$ since they are type (a) corepresentations and therefore preserve their degeneracies, as discussed in §17.1. Being a degenerate representation in $\boldsymbol{G}$, $\Gamma_G(j)$ has the same degeneracy in $\boldsymbol{g}$. The overall effect of the transition $\boldsymbol{g} \to \boldsymbol{M}$ on the degeneracy is that a doubly-degenerate mode of $\boldsymbol{g}$ (or $\boldsymbol{G}$) in the $\mathcal{P}$ state may split into two non-degenerate modes of $\boldsymbol{M}$ in the $\mathcal{M}$ state. It turns out that the splitting described above can occur in the following 14 magnetic classes: 15, 20, 22, 29, 30, 31, 33, 38, 41, 46, 52, 54, 55, 58 (Cracknell 1966). It always involves a doubly-degenerate real representation E in $\boldsymbol{G}$ (or DE in $\boldsymbol{g}$) which splits into two non-degenerate complex-conjugate representations E^1, E^2 in $\boldsymbol{H}$. The latter leads to two non-degenerate complex-conjugate corepresentations DE^1, DE^2 of $\boldsymbol{M}$. The corresponding phonons are left- and right-circularly polarized and with slightly different frequencies. As the crystal returns to the $\mathcal{P}$ state, the two phonons reform the original doubly-degenerate linearly-polarized phonon E (or DE). For instance, such splittings are expected for the E_g-type modes of antiferromagnetic FeF_3 for which $\boldsymbol{M} = D_{3d}[C_{3i}]$ with the magnetic moment aligned along the six-fold axis. These are the only cases of type (a) corepresentations in which a degeneracy may be lifted below T_n. In the remaining classes with type (a) corepresentations (25, 26, 28, 34, 35, 39, 40, 42, 47, 48, 49, 50, 51, 56, 57) the degeneracy is not removed. The whole process is shown schematically in fig. 26a.

(ii) Type (b) corepresentations. There are only 3 single magnetic classes with type (b) corepresentations (13, 14, 19). They all behave in the same way. No splitting is observed as $\boldsymbol{g} \to \boldsymbol{M}$ (fig. 26b).

(iii) Type (c) corepresentations. There are 14 magnetic classes with type (c) corepresentations (Cracknell 1968) falling into two groups: 17, 18, 32, 36, 37, 43, 44, 45, 53 and 16, 21, 23, 24, 27. Neither of these schemes (fig. 26c) represents a net splitting as $\boldsymbol{g} \to \boldsymbol{M}$. Notice the difference between fig. 26a and 26c$_2$; in the former, the two representations E^1, E^2 lead to type (a) corepresentations of $\boldsymbol{M}$ and therefore remain split as DE^1, DE^2 in $\boldsymbol{M}$. In the latter, B_1 and B_2 stick together, leading to a type (c) corepresentation DB(2) of double degeneracy.

(iv) $\boldsymbol{M}$ is a colourless magnetic point group. As discussed in §17.1, there are magnetic classes for which $\boldsymbol{M}$ coincides with one of the ordinary point groups $\boldsymbol{G}$. In this case the transition $\boldsymbol{g} \to \boldsymbol{M}$ actually becomes $\boldsymbol{g} \to \boldsymbol{G}$. Therefore all the non-degenerate complex-conjugate modes of $\boldsymbol{G}$ (if any) which in the $\mathcal{P}$ state are stuck together as DE(2) of $\boldsymbol{g}$, will now split to

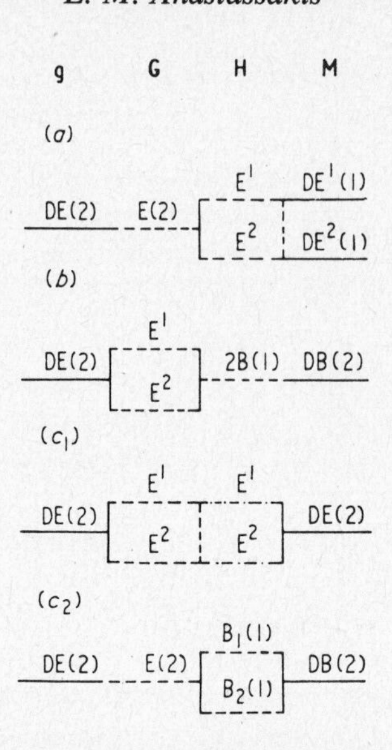

Fig. 26. Changes in the $q\simeq 0$ optical phonon degeneracy as the crystal undergoes the transition from the paramagnetic to the magnetically ordered state. E^1 and E^2 stand for complex conjugate modes. The number in parenthesis designates the degeneracy of the corresponding mode. The three types (a, b, c) correspond to the corepresentations of M. There are 14, 3, 10, 5 magnetic classes with phonons following the schemes a, b, c_1, c_2 respectively (Anastassakis and Burstein 1972).

E^1, E^2 of G and will remain so below T_n. Although the spatial symmetry does not change during the transition, the mere removal of the time reversal symmetry will result in a splitting of the normally doubly-degenerate mode DE(2) of g. For instance, the E_g-type mode of ferromagnetic $CrBr_3$ ($G = M = C_{3i}$ with the magnetic moment along the six-fold axis) is expected to show such splittings below T_n. The magnetic crystal classes which can exhibit the above splitting are $C_3, C_4, C_{4h}, S_4, C_{3i}, C_{3h}, C_6, C_{6h}$. It can be shown that these classes can support a non-zero antisymmetric term in the effective "force constant" of the mode, only as long as time reversal symmetry is absent. A phenomenological approach to the above splitting can be based entirely on the existence of such an antisymmetric term (see §16.1).

17.6. Time reversal symmetry and first-order infrared absorption

The general discussion presented in the previous sections regarding first-order Raman scattering by optical magnons and phonons in the $\mathcal{M}$ state can be readily extended to first-order infrared absorption by these excitations. The symmetry requirement for infrared absorption by optical phonons in the $\mathcal{P}$ state is that $\Gamma(j)$ be included in $\Gamma(\mathfrak{E})$, the three-dimensional, generally reducible, representation for a polar vector $\mathfrak{E}$. In principle the same requirement applies for infrared absorption by optical magnons through an electric dipole-type interaction. Thus, a magnon with symmetry $D\Gamma(j)$ is infrared-active in first order when $D\Gamma(j)$ is included in $D\Gamma(\mathfrak{E})$. Again, it can be shown that this is so if and only if the corresponding representation $\Gamma(j)$ of $\boldsymbol{H}$ is included in $\Gamma(\mathfrak{E})$ of $\boldsymbol{H}$; that is, if j could describe an infrared-active phonon in any crystal structure with point group $\boldsymbol{H}$.

Having established a method for tracing the changes of the phonon symmetry imposed by the phase transition $\boldsymbol{g} \rightarrow \boldsymbol{M}$, we can now examine the effect of these symmetry changes on infrared absorption by phonons in the $\mathcal{M}$ state. The crystal property which characterizes the process of infrared absorption is the mode effective charge $\boldsymbol{e}_j$. Modification of polarization selection rules may be expected due to new components of the effective charge in the $\mathcal{M}$ state which are normally zero in the $\mathcal{P}$ state. It is also possible that normally infrared-inactive phonons become infrared active below T_n. Consider as an example the magnetic point group $D_6[D_3]$. In the $\mathcal{P}$ state the modes B_1, B_2 and E_2 of $\boldsymbol{G}=D_6$ are infrared inactive. The reductions of eq. (17.5b) become

$$DB_1 \rightarrow B_1 \rightarrow A_2 \rightarrow DA_2 \qquad \text{(a)},$$

$$DB_2 \rightarrow B_2 \rightarrow A_1 \rightarrow DA_1 \qquad \text{(a)},$$

$$DE_2 \rightarrow E_2 \rightarrow E \rightarrow DE \qquad \text{(a)}.$$

The mode A_1 of $\boldsymbol{H}=D_3$ is infrared inactive while the modes $A_2(z)$ and $E(x,y)$ are infrared active (the component in parentheses indicates the polarization of the mode and therefore the component of the incident electromagnetic radiation which is absorbed). Accordingly, DA_1 corresponds to an infrared-inactive phonon in the $\mathcal{M}$ state with the same sense of polarization. In summary, the infrared-inactive phonon B_2 remains so in the $\mathcal{M}$ state, while the infrared-inactive phonons B_1 and E_2 become infrared active in the $\mathcal{M}$ state. No splitting of the E_2 mode is expected in the $\mathcal{M}$ state. Detailed tables for the effective charge tensor in the $\mathcal{M}$ state, similar to the Raman tensor tables, do not exist, although in principle they can be worked out in the same way as for the Raman tensor (Cracknell 1969).

17.7. General comments and experimental results

We have examined effects due to the removal of the time reversal symmetry operation as the crystal structure undergoes the transition from the $\mathcal{P}$ state to the $\mathfrak{M}$ state. Accordingly, the crystal point group changes from a grey point group to a magnetic point group; any qualitative symmetry arguments can be now expressed by the use of corepresentation theory. Criteria for Raman activity are discussed for first-order Raman scattering by $\boldsymbol{q}\simeq 0$ optical magnons and the connection of their Raman tensors to those of $\boldsymbol{q}\simeq 0$ Raman-active optical phonons in the $\mathfrak{M}$ state is established. It is shown through specific examples that below T_{n} the polarization selection rules may be modified, normally Raman-inactive phonons may become Raman active, and doubly-degenerate modes may split. Similar effects are expected for first-order infrared absorption by phonons in the $\mathfrak{M}$ state. As already mentioned in §17.1, the present approach can be applied to any crystal structure in the presence of a magnetic field $\boldsymbol{B}$, the reason being that, like a non-zero magnetic moment, $\boldsymbol{B}$ destroys the time reversal symmetry. In the presence of a magnetic field one can show that the overall new symmetry of the system 'crystal + $\boldsymbol{B}$' can be described by *one* of the 90 magnetic point groups, defined here as *the equivalent magnetic point group*. However, in comparing the present results for the equivalent magnetic group with those of §16, one should remember that the former arise from all orders of $\boldsymbol{B}$ and are not necessarily antisymmetric. Furthermore it was demonstrated in §17.4 that a phonon which in the paramagnetic state is Raman inactive may become Raman active in the magnetic state (which is described by a magnetic point group $\boldsymbol{M}$). At first sight, this seems to contradict the conclusion reached in §16.3 that only Raman-active modes may be affected by a magnetic field *in first order*. However, in the former case the effects are considered to all orders of $\boldsymbol{B}$, while in the latter case only terms which are linear in $\boldsymbol{B}$ are taken into consideration. Moreover, the equivalent magnetic point groups are not among those magnetic groups $\boldsymbol{M}$ which can turn a Raman-inactive phonon into a Raman-active one.

All the effects which arise from the change of symmetry $\boldsymbol{g}\rightarrow\boldsymbol{M}$ are due to the alignment of the spins within each crystal magnetic sublattice. Neither the magnitude (strength) of these effects nor the mechanism involved can be determined group-theoretically. Group theory can only decide whether or not a particular effect is possible. An independent theoretical approach is needed for each case to establish the associated matrix elements and thereby determine the strength of the effect.

Following procedures similar to those of Moriya (1966), Sokoloff (1972) has shown that the strength of normally forbidden Raman scattering in the $\mathfrak{M}$ state is typically two orders of magnitude smaller than that of Raman scattering by single magnons. Sokoloff also estimates the splitting of the mode frequencies as about 0.2 cm^{-1}, using values available from the literature for the derivatives of the exchange interactions with respect to the mode normal coordinates (Sinha and Upadhyaya 1962). Such splittings are rather small to be observed through conventional spectroscopic methods, although they may become observable by the use of interferometric techniques. Similar, or larger, values are found by Baltensperger and Helman (1968) for the phonon frequency shifts of ferromagnetic insulators like EuO. Their model is based on a bilinear spin Hamiltonian and an exchange constant which depends on the mode normal coordinates. Such a spin–phonon coupling calculation yields a relative change in frequency of 10^{-2} to 10^{-3}. The same model has been invoked by Steigmeier and Harbeke (1970) to interpret their results of a Raman scattering study of the ferromagnetic semiconducting spinels $CdCr_2Se_4$ and $CdCr_2S_4$ as a function of temperature, using a single He–Ne excitation frequency. These materials are cubic (space group O_h^7) above their Curie temperature of 130 K and 84 K, respectively, and exhibit a number of Raman-active phonons of symmetries A_{1g}, E_g and F_{2g}. According to the results of Steigmeier and Harbeke, the intensity and (slightly) the frequency of certain of these phonons show a definite temperature dependence as the Curie point is approached from above. In fact two phonon bands of mixed symmetry exhibit a splitting of about 4 cm^{-1} in $CdCr_2S_4$. No group-theoretical treatment is undertaken in the framework of a magnetic point group analysis. Instead, the detailed eigenvectors are calculated for all the $\boldsymbol{q}\simeq 0$ Raman-active phonons. Lattice-dynamical calculations based on a simple force model with four short-range adjustable parameters have also been carried out for the same spinels by Brüsch and D'Ambrogio (1972). Scattering measurements as a function of temperature have been reported for the calcite structure compounds VBO_3 (ferromagnetic, $T_c = 32.5$ K) and $FeBO_2$ (antiferromagnetic, $T_c = 350$ K) by Shepherd (1972), and for the spinels discussed above (Koshizuka et al. 1976, 1977). The latter authors have also established that the Raman scattering intensity of the temperature sensitive phonons depends on the excitation frequency; in other words it exhibits resonance behavior. Similar resonance characteristics in $CdCr_2Se_4$ were observed by Iliev et al. (1978a) and also in $HgCr_2Se_4$ by Iliev et al. (1978b).

With regard to the magnetic ordering effect, Suzuki and Kamimura (1973) have developed a general theory of the spin-dependent phonon Raman scattering in magnetic crystals which seems to explain adequately the results of Steigmeier and Harbeke (1970). According to this theory, the mechanism for the magnetic-ordering-dependent Raman scattering of $CdCr_2S_4$ is the variation with the lattice vibrations of the d-electron transfer. As already discussed, the results of Koshizuka et al. (1976, 1977) indicate that both magnetic ordering and resonance effects are necessary for a more complete explanation of their experimental observations.

The forbidden sharp F_{1u}(LO) spectra of NaCl-type Eu-chalcogenides and their dependence on temperature and magnetic field were already mentioned in §15.3. Recently, Safran et al. (1976) have approached the problem of symmetry changes which occur below the phase transition temperature, by considering the consequences of magnetic ordering and the removal of time reversal symmetry. Their phenomenological analysis is based on the assumption that the forbidden scattering is due to the $\boldsymbol{q}$-dependent terms in the Raman tensor. These terms are so constructed as to satisfy spatial and time reversal symmetry restrictions. It is also argued that the isotropic phonon-modulated spin–spin interaction introduced by Suzuki and Kanimura (1973) is not a sufficient mechanism to explain the spin-dependent Raman scattering in the ferromagnetic phase. Anisotropic spin coupling should be considered as well. Ousaka et al. (1977), have recently worked on the same problem. Specifically, they have analyzed the experimental results on EuSe and EuTe of Silberstein et al. (1976) by considering the cross effect of the 4f spin–orbit interaction and the exciton–phonon interaction. Their results on the phonon dispersion curves and on the symmetry and intensities of the observed Raman lines are different from those of Safran et al. (1976).

18. Spatial dispersion effects

18.1. General

Spatial dispersion effects arise when gradients of the electromagnetic fields which propagate through the crystal are taken into consideration. Since these fields are imposed externally one may regard their gradients as some kind of an external force with well-defined transformation properties. Therefore, effects resulting from spatial dispersion may be regarded as morphic effects. A related effect of Crystal Physics is the optical activity, a well-known effect which arises from field gradient contributions to the dielectric susceptibility (Landau and Lifshitz 1960). In the same spirit, gradients of the phonon normal coordinates may very well lead to spatial dispersion effects. In this latter case, however, nothing is imposed externally and the applied force should be viewed as a built-in or internal force, the effects of which are usually ignored.

As one would expect, spatial dispersion effects in connection with lattice-dynamical phenomena can manifest themselves in infrared absorption and Raman scattering experiments. By this we do not mean to disregard the unique role and usefulness of neutron scattering as a procedure for the mapping and study of phonon dispersion curves as a whole.

Since all the relevant experimental work of morphic effects linear in $\boldsymbol{k}$ or $\boldsymbol{q}$ has been done in Raman scattering we restrict the discussion of this and the following sections mainly to Raman processes. As usual, one has to consider contributions to the Raman tensor which arise from field or phonon gradients. Such contributions will in general alter the polarization selection rules and will cause *forbidden* scattering to occur. The latter is particularly important and informative in the case of intrinsically Raman-inactive phonons. Normally, dispersion effects are expected to be very weak, as higher-order effects. Near resonance though, their strength increases significantly, to the level of *allowed* scattering and sometimes even more. All the experimental studies of $\boldsymbol{q}$-dependent Raman scattering have been carried out near resonance.

Assuming a plane-wave expression for the electromagnetic field one can easily establish that the field and phonon gradients are $\nabla\mathscr{E} = \mathrm{i}\boldsymbol{k}\mathscr{E}$ and $\nabla u_j = \mathrm{i}\boldsymbol{q}u_j$ respectively. Each contribution is proportional to its own wavevector and this is why spatial dispersion effects are very often referred to as $\boldsymbol{k}$- or $\boldsymbol{q}$-dependent effects. For a Raman scattering experiment, gradient contributions from all three quantities should be considered, that is, incident and scattered field and phonon normal coordinate. The three wavevectors are always related through the principle of conservation of momentum. This simplifies significantly the symmetry analysis of the corresponding higher-order tensors. For example, in a backward scattering configuration one finds that $\boldsymbol{q}_j = 2\boldsymbol{k}_\mathrm{o}$, where $\boldsymbol{k}_0$ is the incident wavevector.

Considering for a moment only $\boldsymbol{q}$-type contributions in the presence of an external field $\boldsymbol{E}$, we can write for the Raman tensor $\boldsymbol{a}_j$ of an optical phonon j

$$\begin{aligned} a_j(\boldsymbol{q}_j, \boldsymbol{E}) &\simeq a_{j0} + (\Delta a_{jE} + \Delta a_{jq} + \Delta a_{jqE}) \\ &= (a_{j0} + a_{jE}E) + \mathrm{i}q(a_{jq} + a_{jqE}E)u_j, \end{aligned} \tag{18.1}$$

where

$$a_{jE} \equiv -\left(\frac{\partial^4\Phi}{\partial\mathscr{E}\,\partial\mathscr{E}'\partial E\,\partial u_j}\right), \tag{18.2a}$$

$$a_{jq} \equiv -\left(\frac{\partial^3\Phi}{\partial\mathscr{E}\,\partial\mathscr{E}'\partial\nabla u_j}\right), \tag{18.2b}$$

$$a_{jqE} \equiv -\left(\frac{\partial^4\Phi}{\partial\mathscr{E}\,\partial\mathscr{E}'\partial E\,\partial\nabla u_j}\right), \tag{18.2c}$$

are appropriate mode coefficients. They describe, respectively, electric-field-induced Raman scattering (see eq. (4.5b)), first-order $\boldsymbol{q}$-dependent, and field-induced $\boldsymbol{q}$-dependent Raman scattering. For simplicity we write $\boldsymbol{q}$ rather than $\boldsymbol{q}_j$.

The symmetry of the mode coefficient $\boldsymbol{a}_{jq}$ is the same as that of the quantity $-(\partial^4\Phi/\partial\mathscr{E}\,\partial\mathscr{E}'\partial q_j\partial u_j)$. This is because

$$\Gamma(\nabla u_j)=\Gamma(j)\otimes\Gamma(\boldsymbol{q}_j)=\Gamma(j)\otimes\Gamma(\boldsymbol{E}), \tag{18.3}$$

according to § 5.1. Notice that, regardless of the phonon symmetry, the wavevector $\boldsymbol{q}_j$ is a polar vector and therefore $\Gamma(\boldsymbol{q}_j)=\Gamma(\boldsymbol{E})$. The fifth-order tensor $\boldsymbol{a}_{jqE}$ is identically zero for odd-parity phonons in centrosymmetric crystals and may be non-zero in all other cases. The detailed matrix form of $\boldsymbol{a}_{jq}$ will depend on the microscopic physical mechanism which leads to $\boldsymbol{q}$-dependent Raman scattering. We examine briefly some accepted possibilities regarding the origin of $\boldsymbol{q}$-dependent contributions to the atomic displacement Raman tensor and to the electro-optic Raman tensor.

(i) In the atomic displacement Raman tensor the main contributions originate from the $\boldsymbol{q}$-dependence of the resonant denominators. The corresponding Raman tensor $\boldsymbol{a}_{jq}$ has been regarded and treated, as far as symmetry is concerned, as identical to those of eqs. (18.2a) and (18.2b). In other words the electric-field-induced Raman tensors worked out in §7 are also considered to be appropriate for $\boldsymbol{a}_{jq}$. For this reason, and because of the complementary principle in centrosymmetric crystals, such terms can only occur with odd-parity phonons (Burstein and Pinczuk 1971, Richter 1976). For non-centrosymmetric crystals parity restrictions do not exist.

Additional contributions to the atomic displacement Raman tensor may arise from the $\boldsymbol{k}$-dependence of the momentum matrix elements. The latter contributions are particularly important when the interband electronic transitions are electric-dipole forbidden, in which case one has to consider electric-quadrupole and/or magnetic-dipole interactions. The resulting $\boldsymbol{k}$-dependent Raman tensors are of the form of $a_{jk}=-(\partial^4\Phi/\partial\mathscr{E}\,\partial\mathscr{E}'\partial k\partial u_j)$. Using group-theoretical techniques similar in principle to those presented in §5, Birman (1974c) has calculated the detailed matrix form of $\boldsymbol{a}_{jk}$ based on the following reduction for the reducible representation of $\boldsymbol{a}_{jk}$:

$$\Gamma(\boldsymbol{a}_{jk})=[\Gamma(\mathscr{E}\boldsymbol{k})]_{Q+M}\otimes\Gamma(\mathscr{E}')\otimes\Gamma(j)=n_{\mathrm{A}}\mathrm{A}_1\oplus\dots \tag{18.4}$$

The Kronecker product in brackets designates the reducible representation for the quadrupole + magnetic dipole part of the first-order multipole operator. The number n_{A} of independent components of $\boldsymbol{a}_{jk}$ is the same as

n_j, where

$$[\Gamma(\mathscr{E}\boldsymbol{k})]_{Q+M} \otimes \Gamma(\mathscr{E}') = n_j \Gamma(j) \oplus \ldots . \tag{18.5}$$

For the case of cubic materials of O_h symmetry this representation reduces like a non-symmetric second-rank polar tensor (see eq (5.1c)) without the A_{1g} term, provided $\mathscr{E} \cdot \boldsymbol{k} = 0$. This leads to $[\Gamma(\mathscr{E}\boldsymbol{k})]_{Q+M} = (E_g + F_{2g})_Q \oplus (F_{1g})_M$. Eq. (18.5) together with $\Gamma(\mathscr{E}') = F_{1u}$ of eq. (5.1a) allows one to find which phonons will participate in what kind of multipole interaction and with how many independent coefficients for the corresponding tensor $\boldsymbol{a}_{jk}$ (Birman 1974c). The quadrupole–dipole part of the multipole-dipole interactions considered above has also been invoked by Safran et al. (1976, 1977) to analyze the symmetry of the various magnetic phases and the resonant enhancement of first-order Raman scattering experiments in europium chalcogenides.

(ii) The $\boldsymbol{q}$-dependence of the electro-optic Raman tensor originates from resonant two-band terms only. Again, one has the same contributions as for the atomic displacement Raman tensor, and also contributions linear in $\boldsymbol{q}$ because of the $\boldsymbol{q}$-dependent intraband matrix elements of the Fröhlich electron–phonon interaction characteristic of the infrared-active LO phonons (Hamilton 1969, Martin 1971, Martin and Falikov 1975 and references therein). The Raman tensor $\boldsymbol{a}_{jq}$ is considered to have the same form as the electric-field-induced Raman tensor for infrared-active LO phonons. Such situations can only arise in non-centrosymmetric crystals.

As stated earlier all the experimental studies of spatial dispersion effects linear in $\boldsymbol{k}$ or $\boldsymbol{q}$ have been carried out by means of scattering techniques (because of resolution restrictions, neutron scattering cannot be as useful as Raman scattering in this connection). No experimental work appears to have been done on $\boldsymbol{q}$-dependent infrared absorption. In their study of $\boldsymbol{q}$-dependent splittings of doubly-degenerate infrared-active phonons of α-quartz (to be discussed in §18.3), Pine and Dresselhaus (1969) have analyzed theoretically the equivalent effect of optical activity in the infrared. For this, it is necessary to introduce the concept of the $\boldsymbol{q}$-dependent effective charge which is needed for the calculation of the infrared rotatory power. Since $\boldsymbol{q}$ is a time antisymmetric vector one can show by the use of circular coordinates, that an expression equivalent to that of eq. (15.29) for the E mode of the D_3 structure is given by

$$e_{PS}(q_z) = \begin{pmatrix} e - \eta q_z & 0 \\ 0 & e + \eta q_z \end{pmatrix}. \tag{18.6}$$

η is the single independent component of the mode coefficient $\eta_{j,\mu\nu\sigma}=(\partial^2 M_\mu/\partial q_\nu \partial u_{j\sigma})$ by analogy to eq. (15.4). It is not difficult to reach an expression like this, if one follows the same procedure that led to the magnetic-field-induced infrared absorption as in §15.4. Recall that the indices P, S refer to circular unit vectors. The interpretation of the result of eq. (18.6) is that left- and right-circularly polarized light are absorbed unequally. This corresponds to *infrared circular dichroism*, an effect that does not seem to have been observed experimentally so far. Matrices similar to that of eq. (18.6) can be easily formed for all other modes and structures which qualify for $\boldsymbol{q}$-dependent infrared absorption.

18.2. Experimental

Although a great number of materials have been investigated near resonance in the last few years, the situation is not completely clear from the point of view of uniquely identifying the $\boldsymbol{q}$- or $\boldsymbol{k}$-dependent contributions to forbidden (or allowed) Raman scattering. The complication in the analysis of the data arises from the symmetry identity of the mode coefficients $\boldsymbol{a}_{jE}$ and $\boldsymbol{a}_{jq}$ or $\boldsymbol{a}_{jk}$ discussed in the previous section. If for some reason an electric field (internal or external) is present, the $\boldsymbol{q}$-dependent scattering will be superimposed on the $\boldsymbol{E}$-induced scattering. As the selection rules for the two contributions are identical their separation requires careful experimentation.

Let us consider the situation from the point of view of eq. (18.1). In general the scattering intensity will be proportional to

$$I_j \propto |a_{j0}+a_{jE}E|^2+|a_{jq}+a_{jqE}E|^2 q^2. \tag{18.7}$$

One notices that real and imaginary terms do not interfere. The same is true even when higher-order real and imaginary terms are included.

Next we briefly consider some experimental results which are related to the various possibilities contained in eq. (18.7).

(i) Raman-active LO phonons, no fields present. The intensity will be proportional to either $|a_{j0}|^2+|a_{jq}|^2q^2$ or $|a_{jq}|^2q^2$ depending on configuration. Measurements near resonance should reveal significant intensity enhancement due to resonant $\boldsymbol{a}_{jq}$-terms. This was first observed by Leite and Porto (1966) at the fundamental gap of CdS for the E_1(LO) phonon. A number of related investigations on this material have been reported since (Martin and Falikov 1975 and references therein). Similar work at the fundamental gap has been done for ZnS (Lewis et al. 1971), ZnSe (Leite et al. 1969, Lewis et al. 1971), GaSe (Hoff and Irwin 1974) and ZnTe (Schmidt et al. 1975).

(ii) Raman-active LO phonons, with fields present. Assuming for a moment that $\boldsymbol{a}_{jqE}$-terms are not significant, the intensity will be proportional to either $|a_{j0}+a_{jE}E|^2+|a_{jq}|^2q^2$ or $|a_{jE}|^2E^2+|a_{jq}|^2q^2$ depending on configuration. Such possibilities were explored at resonance in CdS by Shand and Burstein (1973) and by Shand et al. (1972), respectively (see §7.2.2). In both cases the external field-induced intensities were observed to vary as $a_{j0}a_{jE}E$ and $|a_{jE}|^2E^2$ respectively, independently of the $|a_{jq}|^2q^2$ term.

The situation is more complicated in III–V semiconductors where built-in surface field terms may very well be participating in addition to the $\boldsymbol{q}$-terms. In InSb, Richter et al. (1976) have shown that the $\boldsymbol{q}$-terms contribute as much as a field of 10^4 V/cm. The different methods used to vary the total field have been discussed in §7.2.2. More recently, Trommer et al. (1978) applied external voltages to a Ni-film electrode evaporated on the surface of n-type GaAs. This enabled them to vary the slope of the energy bands near the surface and thereby separate the $\boldsymbol{q}$-terms from the field terms. The situation was actually more complicated owing to the presence of impurity-induced scattering. It has been suggested by Pinczuk and Burstein (1973) that in general both $\boldsymbol{q}$-dependent and field-dependent terms contribute to the scattering process. The former is more prominent at low carrier concentrations while the latter becomes more important as the carrier concentration increases. Finally the effect of a uniaxial stress on the forbidden scattering intensity of LO phonons of InSb has been discussed by Richter et al. (1978b). It is shown that the $\boldsymbol{q}$-dependent terms arising from the singlet and doublet components of the stressed E_1 gap interfere destructively, thus causing a decrease in the LO scattering intensity with increasing stress. The same conclusions were reached for the surface field-induced terms. Such decrease in the scattering intensity has indeed been observed experimentally (Anastassakis et al. 1972c, Richter et al. 1976).

(iii) Raman-inactive LO phonons. The most striking example of forbidden scattering by Raman-inactive F_{1u}(LO) phonons in first- and higher-order is that of the Mg_2X (X=Si, Ge, Sn, Pb) II–IV semiconductors. As discussed in §7.2.1, surface field-induced scattering is not excluded, although the main contributions are believed to arise from $\boldsymbol{q}$-dependent terms (Anastassakis and Burstein 1971c, Onari and Cardona 1976). In fig. 27 the spectra of the F_{2g} (allowed) and F_{1u}(LO) phonons of Mg_2Ge are shown for different laser energies in the neighborhood of the E_1 gap.

A unique mode for testing the multipole-dipole-type $\boldsymbol{q}$-dependent scattering has been shown to be the F_{1u}(LO) phonon of Cu_2O (Compaan and Cummins 1973, Yu and Shen 1975, Genack et al. 1975, 1976 and references therein). According to Compaan and Cummins (1973) a

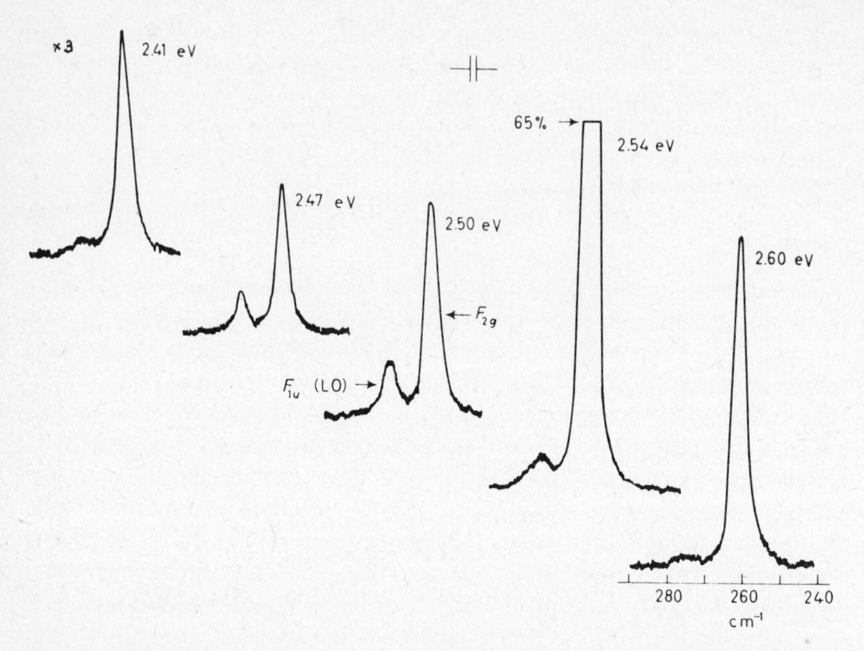

Fig. 27. First-order Raman spectrum of Mg_2Ge at 77 K using Ar^+-laser excitation lines (2.41–2.60) eV. The strong band is due to scattering by the Raman-active infrared-inactive F_{2g} phonon. The weak band is due to scattering by the normally infrared-active Raman-inactive F_{1u}(LO). The scattering is due to the finite phonon wavevector and is observed only under conditions of resonance. The estimated instrumental resolution at 2.54 eV was 1.5 cm^{-1}. The spectra have been shifted for clarity. The strong peak is at the same frequency in all cases (author's unpublished data).

quadrupole–dipole mechanism is responsible for the forbidden scattering which is observed at resonance with the dipole-forbidden electronic transitions to the 1S yellow exciton state. A magnetic dipole–dipole mechanism may also be operative under resonance with the 1S green exciton. Such mechanisms permit non-diagonal scattering (Birman 1974c). However, the results of Williams and Porto (1973) show that the scattering is diagonal, in agreement with the mechanism of intraband electron–phonon Fröhlich-type interaction. The latter involves 2S and higher-order discrete exciton states (Martin 1971, Martin and Falikov 1975) and leads to essentially diagonal scattering. It is possible that the non-diagonal components of the multipole–dipole mechanism are just too small to be detected, although allowed to exist by general symmetry considerations (Birman 1974c). More recent experimental results by Compaan et al. (1976 and references therein) on the forbidden first-order Raman scattering near the 1S yellow exciton in Cu_2O give additional support to the predictions based on the quadrupole–dipole mechanism. In fact stresses have been used by Waters and Cummins (1977) to further explore and elucidate the physical mechanism of quadrupole–dipole interactions. The corresponding symmetry analysis has been carried out (Berenson 1978).

18.3. The effect of spatial dispersion on the frequencies of the $q \simeq 0$ optical phonons

The phonon wavevector $\boldsymbol{q}$ has been assumed so far to have no effect on the phonon frequency. $\boldsymbol{q}$-dependent splittings of phonon frequencies can very easily be treated in the same spirit as a magnetic field $\boldsymbol{H}$, since both the axial vector $\boldsymbol{H}$ and the polar vector $\boldsymbol{q}$ are time antisymmetric. Thus, the contributions to the force constant linear in $\boldsymbol{q}$ must be antisymmetric, as can be seen from §16.1 on replacing $\boldsymbol{q}$ for $\boldsymbol{H}$. An analogous macroscopic crystal property is provided by the tensor $\gamma_{\alpha\beta\nu} = (\partial\varepsilon^{(1)}_{\alpha\beta}/\partial k_\nu)$ which is responsible for optical activity in crystals (Landau and Lifshitz 1960, Pine and Dresselhaus 1969). k_ν stands for the ν component of the wavevector of the electromagnetic radiation.

As far as spatial symmetry is concerned, eq. (16.8) can be replaced by

$$[\Gamma(j)\otimes\Gamma(j)]_{\mathrm{A}}\otimes\Gamma(\boldsymbol{q}) = n_{\mathrm{A}}\mathrm{A}_1 \oplus \dots . \tag{18.8}$$

Conclusions (i) and (ii) in §16.1 are still valid. Conclusion (iii) will be modified to state that only the component q_z can induce splittings, provided the crystal symmetry is such that $\Gamma(H_z)\otimes\Gamma(q_z) = \mathrm{A}_1$. This is possible in the following uniaxial crystal classes: $\mathrm{C}_3, \mathrm{D}_3, \mathrm{C}_4, \mathrm{D}_4, \mathrm{C}_6, \mathrm{D}_6$ for propagation along the optical axis (z). Finally, conclusion (iv) will state that only those crystal classes for which $\Gamma(\boldsymbol{H}) = \Gamma(\boldsymbol{q})$ will exhibit splitting of the triply-degenerate modes, which is isotropic in $\boldsymbol{q}$. There are two such classes, i.e. O and T. The doubly-degenerate modes are not affected.

We notice that centrosymmetric classes do not allow $\boldsymbol{q}$-splittings, while only a few of the non-centrosymmetric classes do. This is to be compared with the $\boldsymbol{H}$-splittings which can occur for all the doubly- and triply-degenerate modes of all uniaxial and cubic classes respectively. Otherwise, the symmetry analysis and the results of §16.1 apply to the present case as well.

The first experimental observation of splitting linear in $\boldsymbol{q}$ was reported for α-quartz by Pine and Dresselhaus (1969). The point group (D_3) of this material allows the splitting of the doubly-degenerate mode E(TO) at 128 cm^{-1}, for backward scattering along the optical axis (z). For all other directions and modes, shifts and splittings (if permissible) can only be quadratic in q. In this latter case the symmetry of the effect is identical to that of effects quadratic in an electric field or linear in a homogeneous strain.

The secular equation (16.11) written for terms linear in $\boldsymbol{q}$, gives the following phonon frequencies:

$$\omega_{\mathrm{Eq}} \simeq \omega_{\mathrm{E}} \pm \frac{K_{\mathrm{E}} q_z}{2\omega_{\mathrm{E}}}, \tag{18.9}$$

where K_{E} is the single independent component of the mode coefficient $K_{j,\alpha\beta\nu} = (\partial^3\Phi/\partial u_{j\alpha}\partial u_{j\beta}\partial q_\nu) = -K_{j,\beta\alpha\nu}$ for $\Gamma(j) = \mathrm{E}$. In addition, the eigenvectors of the two split phonon components are left- and right-circularly polarized in exactly the same way as for a magnetic field H_z. This suggests the use of intrinsic ($\boldsymbol{H}$- or $\boldsymbol{q}$-independent) Raman matrices expressed in circular coordinates. These have already been worked out for the E-mode of the D_3 structure, and are given in eq. (15.30). As commented there, the only non-zero components for *backward* scattering are $a_{LLR} = a_{RRL} = C$, where the third index corresponds to the phonon polarization. Thus, the configurations for observing the left- or right-circularly polarized split phonons are respectively $z(RR)\bar{z}$ and $z(LL)\bar{z}$. The experimental results of Pine and Dresselhaus are in full agreement with the above symmetry analysis. Typical results at 5 K with a 4880 Å laser line are shown in fig. 28. Since the splittings are well under 1 cm^{-1}, Fabry–Perot interferometric techniques are necessary. For the L-phonon the splitting measured was $0.19 \pm 0.01\ \mathrm{cm}^{-1}$ at 4880 Å. The linear dependence on $\boldsymbol{q}$ of the splitting was tested with several laser wavelengths (for backward scattering one has maximum $q = 2k_0 = 2\bar{n}/\lambda_0$, where $\bar{n}$ is an appropriate refractive index). A slope of $\mathrm{d}\omega/\mathrm{d}q = (0.86 \pm 0.05) \times 10^5$ cm/sec was deduced from these experiments. The importance of such results should be noted, inasmuch as they complement neutron scattering data in the same way that sound velocity experiments do. The overall effects of $\boldsymbol{q}$-dependent splitting of the E(TO)-mode frequency is shown to lead to *infrared optical activity* that is, a gradual rotation of the plane of polarization of the E(TO) phonon as it propagates along the optical axis. A detailed theoretical account is given by Pine and Dresselhaus for the strength and dispersion of the infrared rotatory power, in order to establish the connection between the two phenomena. For the 128 cm^{-1} phonon of α-quartz they estimate a net rotation of the plane of polarization by 0.5° along the damping length of the phonon. Such a measurable rotation remains still to be experimentally confirmed. Pine and Dresselhaus extended their study to trigonal tellurium also (D_3 symmetry). Again, splittings linear in $\boldsymbol{q}$ were observed in backward scattering for the E(TO) modes at 95.2 cm^{-1} and 142.9 cm^{-1}. With

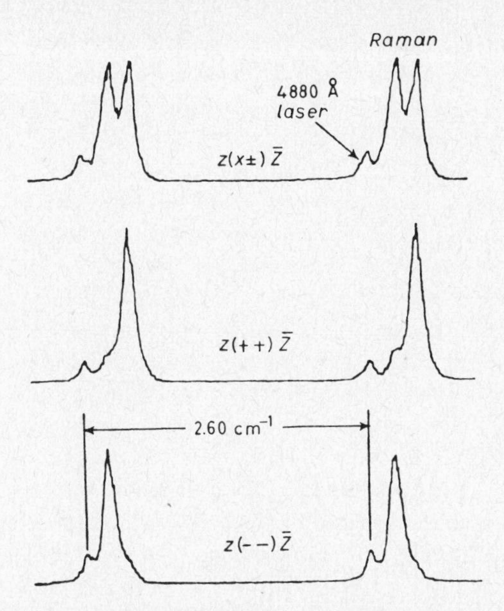

Fig. 28. High-resolution Raman scattering from the 128 cm^{-1} E-mode in α-quartz at 5 K. The fine splitting (top) is due the finite phonon wavevector. Back scattering of 4880 Å light along c-axis; Fabry–Perot analyses with free spectral range 2.60 cm^{-1}. Upper trace, linear polarization; lower traces, opposite circular polarizations (Pine and Dresselhaus 1969).

the 5145 Å laser line the splittings measured were 0.6 ± 0.1 cm^{-1} and 0.26 ± 0.1 cm^{-1} respectively. More recently, Grimsditch et al. (1977) reported the results of a similar study for the E(TO) mode of α-quartz under uniaxial stress along $\hat{x}$. Their results of 90° scattering ($\boldsymbol{q} \| \hat{y}$ and $\boldsymbol{q} \| \hat{z}$) are in agreement with the predictions of the symmetry analysis ($\boldsymbol{q}$-induced and stress-induced splittings) and also with the results of Pine and Dresselhaus (1969) and Tekippe et al. (1973).

18.4. Acoustical activity and Faraday effect of acoustic phonons

The discussion presented here on splittings linear in $\boldsymbol{q}$ and $\boldsymbol{H}$ (§§18.3 and 16.1 respectively), can very easily be extended to include $\boldsymbol{q}$- or $\boldsymbol{H}$-induced splittings of degenerate acoustic phonons propagating along acoustic axes. These splittings would manifest themselves as rotation of the phonon plane of polarization. The rotations linear in $\boldsymbol{q}$ and $\boldsymbol{H}$ are known as *acoustical activity* and the *Faraday effect of acoustic waves* respectively. Acoustical activity has been investigated by a number of workers, both theoretically

(Kluge and Scholz 1965, Portigal and Burstein 1968) and experimentally (Joffrin and Levelut 1970, Pine 1970). The Faraday effect of acoustic waves in metals has also been discussed (Quinn and Rodriguez 1964) and observed experimentally in Cu (Boyd and Gavenda 1966).

The symmetry aspects of acoustical activity are briefly outlined below. For a more detailed discussion the reader is referred to the work of Portigal and Burstein (1968). The phenomenological coefficient responsible for $\boldsymbol{q}$-splittings of acoustic waves is the derivative $d_{ijkl,\nu}=(\partial c_{ijkl}/\partial q_\nu)$, where c_{ijkl} are the compliance elastic constants. Time reversal symmetry arguments impose the restriction that $d_{ijkl,\nu}=-d_{klij,\nu}$. Otherwise $d_{ijkl,\nu}=d_{jikl,\nu}=d_{ijlk,\nu}$. From the group-theoretical point of view the tensor $\boldsymbol{d}$ belongs to the reducible representation $\Gamma(\boldsymbol{d})=[\Gamma(\boldsymbol{\alpha})\otimes\Gamma(\boldsymbol{\alpha})]_A$, were $\Gamma(\boldsymbol{\alpha})$ is the representation of a symmetric second-rank tensor. Assuming that there is no relation between the indices ν and $ijkl$, the number of independent components of $d_{ijkl,\nu}$ is determined as usual from the number of times that A_1 occurs in the reduction of $\Gamma(\boldsymbol{d})$. However, the indices ν and $ijkl$ are not completely independent, as discussed by Waterman (1959). When the direction of propagation is fixed, say $\nu=z$, it turns out from consideration of the equation of motion, that only the components $d_{zjkz,z}$ are involved, i.e. the number of independent components needed is further reduced by the requirement $\nu=i=l$. This fact simplifies the symmetry of the problem significantly for the following reason. The requirement for antisymmetry $(ijkl)=-(klij)$ now becomes $(jk)=-(kj)$ since $i=l=\nu$. Thus the symmetry of the tensor $d_{\nu jk\nu,\nu}$ is the same as that of $\gamma_{jk,\nu}=-\gamma_{kj,\nu}$ introduced in §18.3. In other words, the effect of acoustical activity occurs only in those uniaxial (propagation along z) and cubic classes which are optically active, i.e. in crystal classes C_3, C_4, C_6, D_3, D_4, D_6, O and T.

The symmetry of the Faraday effect for acoustic phonons can be viewed in the same way. The phenomenological coefficient involved is $J_{ijkl,\nu}=(\partial c_{ijkl}/\partial H_\nu)=-J_{klij,\nu}$. When the field $\boldsymbol{H}$ is applied along an acoustic axis and the wave is propagating along the same axis, only transverse acoustic modes are involved. From considerations similar to those entering the discussion of acoustical activity, it turns out that $i=l=\nu$ and therefore the symmetry of the tensor $J_{\nu jk\nu,\nu}$ is the same as that of the magneto-optical tensor $f_{jk,\nu}=-f_{kj,\nu}$ or the Hall tensor $R_{jk,\nu}=-R_{kj,\nu}$ mentioned in §15.2. Thus the Faraday effect of acoustic phonons can occur in all uniaxial and cubic crystals when the applied field $\boldsymbol{H}$ and the direction of propagation are both along an acoustic axis. When the three directions do not coincide the symmetry analysis and the whole problem become far more complicated.

18.5. Gradients of external forces

The concept of morphic effects due the spatial dispersion of static applied forces has been treated in detail by Humphreys and Maradudin (1972). It is shown that infrared absorption and Raman scattering may be modified by gradients of a strain, an electric field or a magnetic field. Strain gradients occur when the crystal is bent or twisted. Electric field gradients may exist in the depletion layers of n-type or p-type semiconductors, in which non-uniform electric fields are induced (e.g. via band bending produced by metal films deposited on their surfaces, Corden et al. 1970, Brillson and Burstein 1971, Trommer et al. 1978). Similarly, magnetic field gradients may occur in situations where an external magnetic field (or a magnetization) happens to be strongly non-uniform.

The mode coefficients responsible for such morphic effects in first order in the gradient of the force, are as follows:

(i)

$$\nabla \boldsymbol{E}\text{-induced}\begin{cases}\text{infrared absorption} & -\left(\dfrac{\partial^3\Phi}{\partial\mathscr{E}\,\partial\nabla E\,\partial u_j}\right) \qquad (18.10a)\\[2ex] \text{Raman scattering} & -\left(\dfrac{\partial^4\Phi}{\partial\mathscr{E}\,\partial\mathscr{E}\,\nabla E\,\partial u_j}\right), \qquad (18.11a)\end{cases}$$

(ii)

$$\nabla \boldsymbol{\eta}\text{-induced}\begin{cases}\text{infrared absorption} & -\left(\dfrac{\partial^3\Phi}{\partial\mathscr{E}\,\partial\nabla \eta\,\partial u_j}\right) \qquad (18.12a)\\[2ex] \text{Raman scattering} & -\left(\dfrac{\partial^4\Phi}{\partial\mathscr{E}\,\partial\mathscr{E}\,\partial\nabla \eta\,\partial u_j}\right), \qquad (18.13a)\end{cases}$$

(iii)

$$\nabla \boldsymbol{H}\text{-induced}\begin{cases}\text{infrared absorption} & -\left(\dfrac{\partial^3\Phi}{\partial\mathscr{E}\,\partial\nabla H\,\partial u_j}\right) \qquad (18.14a)\\[2ex] \text{Raman scattering} & -\left(\dfrac{\partial^4\Phi}{\partial\mathscr{E}\,\partial\mathscr{E}\,\partial\nabla H\,\partial u_j}\right). \qquad (18.15a)\end{cases}$$

For the same reasons discussed in §§15.2 and 15.3 the mode coefficients of eqs. (18.14a) and (18.15a) are purely imaginary. The latter is also antisymmetric in the indices of $\mathscr{E}\mathscr{E}$. It is a matter of straightforward application of the general procedures developed in §5.2 to show that the morphic effects above are allowed for a mode j, when $\Gamma(j)$ occurs at least once in the following reductions respectively:

$$\{\Gamma(\boldsymbol{E})\}^3 = n_j\Gamma(j)\oplus\ldots, \tag{18.10b}$$

$$\Gamma(\boldsymbol{\alpha})\otimes\Gamma(\boldsymbol{E})\otimes\Gamma(\boldsymbol{E}) = n_j\Gamma(j)\oplus\ldots, \tag{18.11b}$$

$$\Gamma(\boldsymbol{\alpha})\otimes\Gamma(\boldsymbol{E})\otimes\Gamma(\boldsymbol{E}) = n_j\Gamma(j)\oplus\ldots, \tag{18.12b}$$

$$\{\Gamma(\boldsymbol{\alpha})\}^2\otimes\Gamma(\boldsymbol{E}) = n_j\Gamma(j)\oplus\ldots, \tag{18.13b}$$

$$\{\Gamma(\boldsymbol{E})\}^3 = n_{j_p}\Gamma(j_p)\oplus\ldots, \tag{18.14b}$$

$$\{\Gamma(\boldsymbol{H})\otimes\Gamma(\boldsymbol{H})\}^2\otimes\Gamma(\boldsymbol{E}) = \{\Gamma(\boldsymbol{E})\}^3 = n_j\Gamma(j)\oplus\ldots. \tag{18.15b}$$

Notice that eqs. (18.11b) and (18.12b) are identical, as are eqs. (18.10b) and (18.15b). Also, eqs. (18.14b) and (18.15b) are identical provided that $\Gamma(j)$ and $\Gamma(j_p)\equiv\Gamma_p\otimes\Gamma(j)$ are identical, that is, $\Gamma_p = A_1$ (i.e. the crystal point group includes no improper rotations). Finally, eqs. (18.15b), and (18.2c) are identical, provided the latter describes a non-symmetric tensor in $\mathscr{E}\mathscr{E}'$. The reductions of eqs. (18.10b)–(18.15b) for all 32 point groups are given in table A7 of the Appendix. Also, table A8 shows the detailed matrices of the $\nabla\boldsymbol{\eta}$-induced Raman tensor for the odd-parity triply-degenerate mode Γ_4^- of the O_h crystal class (Humphreys and Maradudin 1972).

It should be emphasized that the present effects are due to gradients of the externally applied static forces. They should not be identified with the spatial dispersion effects which are due to gradients of the electromagnetic fields and/or the gradients of the phonon displacements. The present effects are expected to be very weak as they are essentially higher-order effects. No experimental confirmation of such effects seems to exist in the literature.

19. *Coupling of $\boldsymbol{q}\simeq 0$ optical phonons in the presence of external forces*

19.1. Dynamical vector space of $q\simeq 0$ optical phonons

A different aspect of the mode behavior in the presence of a force is now discussed, that is, the possibility of force-induced coupling (mixing) of modes with different symmetry. Group theory and perturbation theory are employed to determine the criteria and the extent of mixing respectively. We consider the force-induced changes in the force constant as a perturbation to the force-free force constant of the mode. At first, we assume a linear dependence on the force (linear perturbation). If the linear perturbation is forbidden by the symmetry the quadratic dependence is considered (quadratic perturbation). In either case perturbation theory can be employed in first- and/or second-order. The emphasis is given to non-degenerate perturbed modes where first- and/or second-order perturbation theory may lead to mixing with degenerate and/or non-degenerate perturbing modes. The present results therefore apply best to the case of biaxial crystals with polyatomic unit cells. Here only non-degenerate modes can exist. The low symmetry of the crystal and the large number of existing perturbing modes ensure many possibilities of mixing for a given non-degenerate perturbed mode and type of force. Furthermore, degenerate perturbed modes do not mix with modes of different symmetry according to first-order perturbation theory, regardless of linear or quadratic perturbation. This was essentially the basis for the procedure followed in §3.3. The general results of §3.3 implied that first-order perturbation theory might lead to

(i) frequency shift of non-degenerate modes, assuming that the force is time symmetric,

(ii) frequency splitting of degenerate modes,

(iii) mixing of the degenerate mode eigenstates *among themselves*, to produce the new eigenvectors of the perturbed and (originally) degenerate mode. As we shall see, however, mixing of degenerate perturbed modes is also possible when second-order perturbation theory is employed.

Only long-wavelength optical modes are considered. The use of phonon irreducible representations allows one to construct for each type of crystal a unitary vector space spanned by the normal coordinates of all the existing optical modes. Orthogonality within this space is guaranteed by the orthogonality of the basis functions of the corresponding irreducible representations, as required by group theory. The present lattice-dynamical problem can then be looked upon as an energy eigenvalue problem and the usual time-independent perturbation theory procedures of Quantum Mechanics can be employed in a straightforward manner (Anastassakis 1973).

We proceed to the construction of this vector space after some minor changes are introduced in the notation of normal coordinates, frequencies and force constants. Let $u_r^{(\kappa)}$ be the normal coordinate of the mode r, which transforms according to the κth row of the irreducible representation Γ_r. In general, the crystal may exhibit n_r many optical modes of degeneracy $l_r (1 \leqslant l_r \leqslant 3)$ and frequencies $\omega_r^{(i)}$, $i = 1, 2, \ldots, n_r$. We use the normal coordinate function $u_r^{(\kappa)}$ to construct a basis for a linear unitary vector space $\boldsymbol{L}$. For geometrical considerations it is convenient to treat these functions as unit vectors $\boldsymbol{u}_r^{(\kappa)}$. Owing to the great orthogonality theorem these vectors will then obey the orthonormality relation (Maradudin and Vosko 1968, Warren 1968)

$$\left(\boldsymbol{u}_r^{(\kappa)}, \boldsymbol{u}_n^{(\lambda)}\right) = \delta_{rn}\delta_{\kappa\lambda}. \tag{19.1}$$

The dimensionality of $\boldsymbol{L}$ is $3N-3$, that is, all n_r normal modes with the same symmetry and different frequencies are treated independently. A similar approach has been followed in the past for molecular vibrations, where the normal unit vectors are taken to form a $(3N-6)$-dimensional ($(3N-5)$ for linear molecules) reducible representation of the point group which describes the symmetry of the molecule (Tinkham 1964).

The dynamical eigenvalue problem can be best formulated in terms of a linear operator $\boldsymbol{K}$ which designates the generalized effective force constant of the optical mode. $\boldsymbol{K}$ is defined in $\boldsymbol{L}$ and is diagonal when referred to the basis $\boldsymbol{u}_r^{(\kappa)}$, that is,

$$K_{rn}^{(\kappa\lambda)} = K_{rr}^{(\kappa\kappa)}\delta_{rn}\delta_{\kappa\lambda}, \tag{19.2}$$

where

$$K_{rn}^{(\kappa\lambda)} \equiv \left(\frac{\partial^2 \Phi}{\partial u_r^{(\kappa)} \partial u_n^{(\lambda)}}\right) \tag{19.3}$$

by analogy to eq. (4.4c). Within the harmonic approximation the dynamical equation can be written in the form

$$\boldsymbol{K}\cdot\boldsymbol{u}_r^{(\kappa)} = \sum_{n\lambda} K_{nr}^{(\lambda\kappa)}\boldsymbol{u}_n^{(\lambda)} = \omega_r^2 \boldsymbol{u}_r^{(\kappa)}, \tag{19.4}$$

where ω_r is the frequency of the normal mode $\boldsymbol{u}_r^{(\kappa)}$. Eq. (19.4) is the exact analog of the usual decomposition of a linear operator $\boldsymbol{A}$ in terms of a complete orthonormal set $|n\rangle$ (Messiah 1962), i.e.

$$\boldsymbol{A}|m\rangle = \sum_{rn} |r\rangle\langle r|\boldsymbol{A}|n\rangle\langle n|m\rangle = \sum_{r} A_{rm}|r\rangle.$$

Eqs. (19.2) and (19.4) yield the familiar result

$$K_{rr}^{(\kappa\kappa)} = \omega_r^2, \qquad \kappa = 1, 2, \ldots, l_r. \tag{19.5}$$

The absence of the index κ from ω_r^2 indicates the l_r-fold degeneracy of the mode $\boldsymbol{u}_r^{(\kappa)}$.

In summary, it is possible to construct a linear unitary vector space $\boldsymbol{L}$ which we define here as *the dynamical vector space of the long-wavelength optical modes* of the crystal. $\boldsymbol{L}$ is spanned by the normal coordinate vectors $\boldsymbol{u}_r^{(\kappa)}$. The mode force constant operator $\boldsymbol{K}$ is diagonal when referred to the above basis and the dynamical eigenvalue problem takes the form of eqs. (19.4) and (19.5). In dealing with a more general problem of Lattice Dynamics, Waeber (1972) uses extensively the technique of spectral decomposition of the dynamical matrix. Essentially the same technique, although in a less general form, is employed here, as implied by the spectral decomposition of the dynamical vector space outlined above.

19.2. A perturbation approach

Eq. (19.4) has the form of an unperturbed Hamiltonian eigenvalue problem when written in a matrix representation, with the eigenstates of the unperturbed Hamiltonian serving as a basis for the representation. When a perturbation is present, rediagonalization of the perturbed Hamiltonian is required to obtain the new eigenstates and eigenvalues. Alternatively, perturbation theory may be employed in first and/or second order, to yield approximate solutions to the problem.

An analogous situation arises within the dynamical vector-space $\boldsymbol{L}$ of the long-wavelength optical phonons of a crystal. We may consider eq. (19.4) as the basic equation for the unperturbed dynamical eigenvalue problem and the starting point for developing a perturbation theory, when the operator $\boldsymbol{K}$ is subject to a perturbation due to an external force. The nature of the perturbation is described phenomenologically by expanding the force constant tensor in powers of the applied force $\boldsymbol{F}$. The term linear in the perturbation will be

$$
\begin{aligned}
K'^{(\kappa\lambda)}_{rn} &= \sum_{\nu} \left(\frac{\partial K^{(\kappa\lambda)}_{rn}}{\partial F_{\nu}} \right) F_{\nu} = \sum_{\nu} \left(\frac{\partial^3 \Phi}{\partial u_r^{(\kappa)} \partial u_n^{(\lambda)} \partial F_{\nu}} \right) F_{\nu} \\
&\equiv \sum_{\nu} \mathcal{K}^{(\kappa\lambda)}_{rn\nu} F_{\nu},
\end{aligned}
\tag{19.6}
$$

where $\mathcal{K}^{(\kappa\lambda)}_{rn\nu}$ is a higher-order tensor. The next higher-order term will correspond to quadratic contributions $K''^{(\kappa\lambda)}_{rn} \equiv \sum_{\nu\mu} \mathcal{K}^{(\kappa\lambda)}_{rn\nu\mu} F_{\nu} F_{\mu}$. One may consider such quadratic contributions when the linear ones vanish by symmetry, or when the applied force is sufficiently large. Whether or not the linear contribution vanishes by symmetry is determined from the reduction of the reducible representation for the mode coefficient tensor $\mathcal{K}$

i.e.

$$\Gamma(\mathcal{K})=\Gamma_r\otimes\Gamma_n\otimes\Gamma(\boldsymbol{F})=n_{\mathrm{A}}\mathrm{A}_1\oplus\ldots. \tag{19.7}$$

If $n_{\mathrm{A}}\neq 0$, then $\mathcal{K}\not\equiv 0$ for this particular choice of modes and force $\boldsymbol{F}$. The component form of $\mathcal{K}$ is then established by employing previously described procedures (§5.3). The tensors involved in the quadratic perturbation are treated analogously.

A comment seems to be in order at this point regarding the interchangeability of the indices r and n of the tensor $\mathcal{K}_{rn\nu}^{(\kappa\lambda)}$. As discussed in §3.3, $\mathcal{K}$ can be regarded as a generalized kinetic coefficient. From Onsager's reciprocity theorem and the Hermiticity of $\mathcal{K}$, it follows that $\mathcal{K}_{rn\nu}^{(\kappa\lambda)}=\mathcal{K}_{nr\nu}^{(\lambda\kappa)}$ provided that no magnetic field (or any other time antisymmetric force) is present. This imposes an additional restriction on eq. (19.7), i.e. in the reduction of $\Gamma_r\otimes\Gamma_n$ one should keep only those terms (representations) whose basis functions are obtained from a symmetric (interchangeable) product of the basis functions of Γ_r and Γ_n. This can be designated by $[\Gamma_r\otimes\Gamma_n]_{\mathrm{S}}$, that is, the "generalized" symmetrized part of the double product $\Gamma_r\otimes\Gamma_n$ (this is a generalization of the well-known group-theoretical symmetrization procedure which applies only when $\Gamma_r\equiv\Gamma_n$). As an example let us consider the $\mathrm{T_d}(\bar{4}3\mathrm{m})$ structure with $\Gamma_r=\Gamma_4$ and $\Gamma_n=\Gamma_5$ (Koster et al. 1963). We have $\Gamma_4\otimes\Gamma_5=\Gamma_2\oplus\Gamma_3\oplus\Gamma_4\oplus\Gamma_5$. From the tables of Koster et al. (1963) it follows that the basis functions of Γ_4 and Γ_5 can mix so as to give (i) diagonal (and therefore symmetrized) products which transform according to the representations Γ_2 and Γ_3 of the above reduction, (ii) non-diagonal and symmetrized products which transform according to Γ_4, and (iii) non-diagonal and *antisymmetrized* products which transform according to Γ_5. Accordingly, we can set $[\Gamma_4\otimes\Gamma_5]_{\mathrm{S}}=\Gamma_2\otimes\Gamma_3\otimes\Gamma_4$ and $[\Gamma_4\otimes\Gamma_5]_{\mathrm{A}}=\Gamma_5$. In the presence of a magnetic field (or any other time antisymmetric force) the above arguments are reversed; now the tensor $\mathcal{K}_{rn\nu}^{(\kappa\lambda)}$ is antisymmetric, i.e. $\mathcal{K}_{rn\nu}^{(\kappa\lambda)}=-\mathcal{K}_{nr\nu}^{(\lambda\kappa)}$. Therefore one should consider in eq. (19.7) only the "generalized" antisymmetrized part of $\Gamma_r\otimes\Gamma_n$ (in our example only Γ_5). It is interesting to notice that owing to the very nature of the kinetic coefficients, they appear symmetrized (antisymmetrized) according to the time symmetric (antisymmetric) character of the applied force. In different physical situations such a correlation does not exist in general.

Non-degenerate and degenerate modes are next treated separately.

19.2.1. Non-degenerate perturbed modes

Eq. (19.4) can be rewritten for a specific non-degenerate mode $\boldsymbol{u}_m$ of frequency ω_m as

$$\boldsymbol{K}\cdot\boldsymbol{u}_m=\omega_m^2\boldsymbol{u}_m, \tag{19.8}$$

where $K_{mn} = K_{mn}\delta_{mn} = K_{mm}$. Assuming that $\boldsymbol{K}'$ is small we can expand $\boldsymbol{K}$, $\boldsymbol{u}_m$ and ω_m^2 in powers of a parameter g,

$$(\boldsymbol{K} + g\boldsymbol{K}' + g^2\boldsymbol{K}'' + \dots)\cdot(\boldsymbol{u}_m + g\boldsymbol{u}'_m + g^2\boldsymbol{u}''_m + \dots)$$
$$= (\omega_m^2 + g\omega_m'^2 + g^2\omega''_m + \dots)\cdot(\boldsymbol{u}_m + g\boldsymbol{u}'_m + g^2\boldsymbol{u}''_m + \dots). \tag{19.9}$$

We assume that $\boldsymbol{u}_m(g)$ and $\omega_m^2(g)$ are analytic for $0 \leqslant g \leqslant 1$. At the end we will set $g = 1$. From eq. (19.9) we find, by equating coefficients of equal powers of g,

$$\boldsymbol{K}\cdot\boldsymbol{u}_m = \omega_m^2\boldsymbol{u}_m, \tag{19.10a}$$

$$\boldsymbol{K}\cdot\boldsymbol{u}'_m + \boldsymbol{K}'\cdot\boldsymbol{u}_m = \omega_m^2\boldsymbol{u}'_m + \omega_m'^2\boldsymbol{u}_m, \tag{19.10b}$$

$$\boldsymbol{K}\cdot\boldsymbol{u}''_m + \boldsymbol{K}'\cdot\boldsymbol{u}'_m = \omega_m^2\boldsymbol{u}''_m + \omega_m'^2\boldsymbol{u}'_m + \omega_m''^2\boldsymbol{u}_m. \tag{19.10c}$$

Eqs. (19.8) and (19.10a, b, c) correspond respectively to the unperturbed problem, to the first-order perturbation problem and to the second-order perturbation problem. Later we will examine these separately.

The first-order correction $\boldsymbol{u}'_m$ can be written as a linear combination of all the unperturbed eigenvectors $\boldsymbol{u}_r^{(\kappa)}$ regardless of degeneracy, namely,

$$\boldsymbol{u}'_m = \sum_{r\kappa} a_{mr}^{(\kappa)}\boldsymbol{u}_r^{(\kappa)}. \tag{19.11}$$

In a similar manner we can expand the second-order correction $\boldsymbol{u}''_m$,

$$\boldsymbol{u}''_m = \sum_{r\kappa} b_{mr}^{(\kappa)}\boldsymbol{u}_r^{(\kappa)}. \tag{19.12}$$

Furthermore, since $\boldsymbol{K}'$ is not diagonal when referred to the $\boldsymbol{u}_n^{(\lambda)}$ basis of L the effect of $\boldsymbol{K}'$ on $\boldsymbol{u}_n^{(\lambda)}$ is to produce a linear combination of all the basis vectors of $\boldsymbol{L}$

$$\boldsymbol{K}'\cdot\boldsymbol{u}_n^{(\lambda)} = \sum_{r\kappa} K_{rn}'^{(\kappa\lambda)}u_r^{(\kappa)}, \tag{19.13}$$

and in particular

$$\boldsymbol{K}'\cdot\boldsymbol{u}_m = \sum_{r\kappa} K_{rm}'^{(\kappa)}\boldsymbol{u}_r^{(\kappa)}. \tag{19.14}$$

The superscript $\lambda = 1$ which corresponds to the row index of the non-degenerate state $\boldsymbol{u}_m$ is omitted for simplicity.

(i) First-order perturbation

From eqs. (19.10b), (19.11) and (19.14) and the fact that $\boldsymbol{K}$ is linear we find

$$\sum_{r\kappa} a_{mr}^{(\kappa)} \boldsymbol{K}\cdot\boldsymbol{u}_r^{(\kappa)} + \sum_{r\kappa} K_{rm}'^{(\kappa)} \boldsymbol{u}_r^{(\kappa)} = \omega_m^2 \sum_{r\kappa} a_{mr}^{(\kappa)} \boldsymbol{u}_r^{(\kappa)} + \omega'^2 \boldsymbol{u}_m. \tag{19.15}$$

If we take the dot product of both sides of (19.15) with $\boldsymbol{u}_m^{(\lambda)}$ and make use of eq. (19.1) we find,

$$\omega_m'^2 = K_{mm}', \tag{19.16}$$

$$\underset{r\neq m}{a_{mr}^{(\kappa)}} = \frac{K_{rm}'^{(\kappa)}}{\omega_m^2 - \omega_r^2}. \tag{19.17}$$

(ii) Second-order perturbation

From eqs. (19.10c), (19.12) and (19.14) we obtain in a similar manner,

$$\omega_m''^2 = \sum_{\substack{r\kappa \\ r\neq m}} \frac{|K_{mr}'^{(\kappa)}|^2}{\omega_m^2 - \omega_r^2}, \tag{19.18}$$

$$\underset{r\neq m}{b_{mr}^{(\kappa)}} = \sum_{n\lambda} \frac{K_{rn}'^{(\kappa\lambda)} K_{nm}'^{(\lambda)}}{(\omega_m^2 - \omega_r^2)(\omega_m^2 - \omega_n^2)} - \frac{K_{rm}'^{(\kappa)} K_{mm}'}{(\omega_m^2 - \omega_r^2)^2} + \frac{a_{mm} K_{rm}^{(\kappa)}}{\omega_m^2 - \omega_r^2}, \tag{19.19}$$

where ω_r and ω_n correspond to the unperturbed modes $\boldsymbol{u}_r^{(\kappa)}$ and $\boldsymbol{u}_n^{(\lambda)}$ respectively.

The expressions for $a_{mr}^{(\kappa)}$ and $b_{mr}^{(\kappa)}$ obtained above are not valid for $r = m$. A normalization condition may be imposed on the perturbed states by analogy with the problem of the perturbed states of a harmonic oscillator. Such a condition can lead to the following results (Schiff 1955):

$$a_{mm} = 0, \qquad b_{mm} = -\frac{1}{2} \sum_{r\kappa} |a_{mr}^{(\kappa)}|^2. \tag{19.20}$$

The final results of first- and second-order perturbation theory for a non-degenerate perturbed mode are summarized as follows:

$$\Omega_m^2 = \omega_m^2 + \omega_m'^2 + \omega_m''^2 = \omega_m^2 + K_{mm}' + \sum_{\substack{r\kappa \\ r\neq m}} \frac{|K_{mr}'^{(\kappa)}|^2}{\omega_m^2 - \omega_r^2} \tag{19.21}$$

$$\begin{aligned} \boldsymbol{U}_m &= \boldsymbol{u}_m + \boldsymbol{u}_m' + \boldsymbol{u}_m'' \\ &= \boldsymbol{u}_m + \sum_{\substack{r\kappa \\ r\neq m}} \frac{K_{rm}'^{(\kappa)}}{\omega_m^2 - \omega_r^2} \boldsymbol{u}_r^{(\kappa)} \\ &+ \sum_{\substack{r\kappa \\ r\neq m}} \left\{ \left[\sum_{\substack{n\lambda \\ n\neq m}} \frac{K_{rn}'^{(\kappa\lambda)} K_{nm}'^{(\lambda)}}{(\omega_m^2 - \omega_r^2)(\omega_m^2 - \omega_n^2)} - \frac{K_{rm}'^{(\kappa)} K_{mm}'}{(\omega_m^2 - \omega_r^2)^2} \right] \boldsymbol{u}_r^{(\kappa)} - \frac{1}{2} \frac{|K_{rm}'^{(\kappa)}|^2}{(\omega_m^2 - \omega_r^2)^2} \boldsymbol{u}_m \right\}. \end{aligned} \tag{19.22}$$

Eq. (19.21) includes the contributions to the unperturbed frequency ω_m as established from first- and second-order perturbation theory on the basis of a linear perturbation $\boldsymbol{K}'$. The first-order term $\omega'^2 = K'_{mm}$ is allowed by symmetry, only when the applied force $\boldsymbol{F}$ exhibits totally symmetric components, i.e. components which leave the symmetry of the crystal unchanged. This follows from eq. (19.7), since $\Gamma_r \otimes \Gamma_n = \Gamma_m \otimes \Gamma_m = A_1$, where Γ_m is non-degenerate. By contrast the second-order term ω''^2_m requires components of $\boldsymbol{F}$ which do not preserve the crystal symmetry, namely $\Gamma(\boldsymbol{F}) = 0A_1 \oplus \ldots$. Again, this is because $\Gamma_r \otimes \Gamma_n = \Gamma_r \otimes \Gamma_m \neq A_1$ when $\Gamma_r \neq \Gamma_m$. From the above it follows that the same component of $\boldsymbol{F}$ cannot appear in both terms ω'^2_m and ω''^2_m simultaneously. Depending on the symmetry of the component, it will appear in either or neither of these two terms. Notice, further, that the ω''^2_m term may lead to a major change in the frequency ω_m depending on how close the perturbing frequency ω_r is to ω_m. If the force happens to be time antisymmetric, for example a magnetic field, $\boldsymbol{K}'$ is a purely imaginary and antisymmetric tensor. In this case $K'_{mm} \equiv 0$ and therefore $\omega'^2_m \equiv 0$. However, ω''^2_m can be non-zero provided that eq. (19.7) (appropriately antisymmetrized) is satisfied. If the force is time symmetric, $\boldsymbol{K}'$ is real and symmetric, i.e. $K'^{(\kappa)}_{mr} = K'^{(\kappa)}_{rm}$.

The main conclusion of eq. (19.22) is that in the presence of the force the normal modes may couple to each other, as demonstrated by the linear combinations of the vectors $\boldsymbol{u}^{(\kappa)}_r$ of eq. (19.22). The coupling coefficients between the vibrational states $\boldsymbol{u}_m$ and $\boldsymbol{u}^{(\kappa)}_r$ have an explicit dependence on the force and are defined here in terms of a phenomenological parameter, that is, the appropriate component(s) of the tensor $\mathcal{K}$. As a result of the coupling, energy may be exchanged between the two modes; this will manifest itself as an intensity transfer in an experiment such as Raman scattering. In principle one should be able to determine $\mathcal{K}$ through appropriate fitting to the experimental data. Combining eqs. (19.22) and (2.2) yields the effect of the force on the Cartesian components $\mu_\alpha(l\kappa)$ of the real displacements of the atoms in the unit cell.

The strength of the coupling will depend on the difference of the frequencies of the perturbed and perturbing modes. According to eq. (19.22), a totally symmetric force will not change the symmetry of $\boldsymbol{u}_m$ since $K'^{(\kappa)}_{rm} = K'^{(\lambda)}_{nm} \equiv 0$, as discussed earlier. This is to be expected, because such a force will not affect the crystal symmetry at all. The only effect will be a hydrostatic shift of ω_m due to K'_{mm}. A force which does not preserve the crystal symmetry, on the other hand, may result in mixing of the modes according to eqs. (19.7) and (19.22). We notice that if the mixing is forbidden by symmetry in first order ($\boldsymbol{u}'_m \equiv 0$), it will be forbidden in second order as well ($\boldsymbol{u}''_m \equiv 0$). In that case one should consider quadratic perturbations, again through first- and second-order perturbation theory. The reverse statement is also true, namely, if the mixing is allowed in first order ($\boldsymbol{u}_m \not\equiv 0$), it will be allowed in second order as well, provided that eq. (19.7) is satisfied. The last term of eq. (19.22) obviously does not result in any mixing of modes with different symmetries. As already pointed out all modes with the same symmetry and different frequency are treated independently of each other. There are n_r such modes for each representation. When the perturbed and perturbing modes $\boldsymbol{u}_m$ and $\boldsymbol{u}_r$ have the same symmetry (both are non-degenerate) the results of eqs. (19.21) and (19.22) are still valid ($\omega_m \neq \omega_r$). Since $\Gamma_m \otimes \Gamma_r = \mathrm{A}_1$, it follows that only totally symmetric forces will result in mixing. In fact mixing of such modes occurs even without the need of an external force. The force-free mode of symmetry Γ_m (or Γ_r) will simply be a linear combination of all the participating modes with identical symmetry. Furthermore, perturbing modes with the same symmetry, i.e. $\Gamma_r = \Gamma_{r'} \neq \Gamma_m$, will contribute according to eqs. (19.21) and (19.22). Clearly, the same components of the force will appear in $K'^{(\kappa)}_{rm}$ and $K'^{(\kappa')}_{r'm}$, but in general these terms will not be equal because of the different denominators.

If the crystal possesses a centre of symmetry the normal modes will be characterized by either even or odd parity. According to eq. (19.7) then, an even-parity force (like a stress) will mix modes with the same parity, and an odd-parity force (like an electric field) will mix modes of opposite parity. Otherwise the same general comments hold as before.

19.2.2. *Degenerate perturbed modes*

When the mode $\boldsymbol{u}_r^{(\kappa)}$ happens to be a degenerate one, first-order perturbation theory requires rediagonalization of the $l_r \times l_r$ dynamical matrix in order to obtain the shift and splitting of the degenerate frequency ω_r, and also to find the new eigenvectors. The whole problem of degenerate modes has been treated in detail in §§3.3 and 16, for the case of time symmetric and time antisymmetric forces respectively. It is essential to realize that first-order perturbation theory does not lead to mixing of $\boldsymbol{u}_r^{(\kappa)}$ with modes of different symmetry. The new eigenvectors are simply linear combinations of the unperturbed degenerate eigenvectors. Even inclusion of quadratic perturbations will not change the situation. Mixing will occur only when second-order perturbation theory is employed, with linear and/or quadratic perturbations. Second-order perturbation theory is also necessary when first-order theory fails to remove the degeneracy because of the symmetry of the particular modes and force under consideration, i.e.

when eq. (19.7) with $\Gamma_r = \Gamma_n$ is not satisfied. The analogy to the quantum-mechanical problem is again obvious.

Next we give a brief presentation of the procedure followed when we have, for instance, a doubly-degenerate mode $\boldsymbol{u}_m^{(1)}, \boldsymbol{u}_m^{(2)}$ in the presence of a force (Schiff 1955). Suppose that $K_{mm}^{\prime(11)} = K_{mm}^{\prime(22)} \neq 0$, but $K_{mm}^{\prime(12)} = K_{mm}^{\prime(21)} = 0$. This is the case of a uniaxial stress applied along the z axis of α-quartz. The degeneracy of the doubly-degenerate modes is not removed when first-order theory is applied. In general, since Γ_m is a degenerate representation $\Gamma_m \otimes \Gamma_m$ is not just A_1 and thus the force does not have to be totally symmetric for $K_{mm}^{\prime(11)}$ to be non-zero. We can expand to lowest order and write

$$\boldsymbol{U}_m^{(1)} \simeq a_1 \boldsymbol{u}_m^{(1)} + a_2 \boldsymbol{u}_m^{(2)} + g \sum_{l \neq m} a^{(l)} \boldsymbol{u}_l, \tag{19.23a}$$

$$\boldsymbol{U}_m^{(2)} \simeq b_1 \boldsymbol{u}_m^{(1)} + b_2 \boldsymbol{u}_m^{(2)} + g \sum_{l \neq m} b^{(l)} \boldsymbol{u}_l, \tag{19.23b}$$

$$\Omega_m^2 \simeq \omega_m^2 + g\omega_m^{\prime 2} + g^2 \omega_m^{\prime\prime 2}, \tag{19.23c}$$

where again at the end we will set $g = 1$. The procedures of §19.2.1 lead to the following results:

(i) $\omega_m^{\prime 2} = K_{mm}^{\prime(11)} = K_{mm}^{\prime(22)}$ as expected, that is, a simple hydrostatic shift without lifting of the degeneracy of ω_m.

(ii) The higher-order correction $\omega_m^{\prime\prime 2}$ is given by the eigenvalues of the matrix

$$\begin{pmatrix} A & B \\ C & D \end{pmatrix}, \tag{19.24}$$

where

$$A = \sum_{\lambda, l \neq m} \frac{K_{ml}^{\prime(1\lambda)} K_{lm}^{\prime(\lambda 1)}}{\omega_m^2 - \omega_l^2}, \qquad B = \sum_{\lambda, l \neq m} \frac{K_{ml}^{\prime(1\lambda)} K_{lm}^{\prime(\lambda 2)}}{\omega_m^2 - \omega_l^2},$$

$$C = \sum_{\lambda, l \neq m} \frac{K_{ml}^{\prime(2\lambda)} K_{lm}^{\prime(\lambda 1)}}{\omega_m^2 - \omega_l^2}, \qquad D = \sum_{\lambda, l \neq m} \frac{K_{ml}^{\prime(2\lambda)} K_{lm}^{\prime(\lambda 2)}}{\omega_m^2 - \omega_l^2}.$$

The two values of $\omega_m^{\prime\prime 2}$ will determine the splitting of ω_m.

(iii) The diagonalization procedure above will also yield the coefficients a_1, a_2, b_1, b_2 of eqs. (19.23a, b), i.e. the linear combinations $a_1 \boldsymbol{u}_m^{(1)} + a_2 \boldsymbol{u}_m^{(2)}$ and $b_1 \boldsymbol{u}_m^{(1)} + b_2 \boldsymbol{u}_m^{(2)}$ of eq. (23a, b) serve as eigenvectors for the matrix of eq.

(19.24). The mixing of $\boldsymbol{u}_m^{(1)}$ and $\boldsymbol{u}_m^{(2)}$ with modes of different symmetry is described by the summation terms of eqs. (19.23a, b). $\boldsymbol{u}_l$ stands for either degenerate ($\boldsymbol{u}_l^{(\lambda)}$) or non-degenerate ($\boldsymbol{u}_l$) perturbing modes. It can also correspond to a degenerate mode with the same symmetry as $\boldsymbol{u}_m^{(1)}$ and $\boldsymbol{u}_m^{(2)}$, but of different frequency.

(iv) The coupling coefficients $a^{(l)}$ and $b^{(l)}$ are related to a_1, a_2, b_1, b_2 through the equations

$$a^{(l)}(\omega_m^2-\omega_l^2)=a_1K_{lm}'^{(\lambda 1)}+a_2K_{lm}'^{(\lambda 2)} \tag{19.25a}$$

$$b^{(l)}(\omega_m^2-\omega_l^2)=b_1K_{lm}'^{(l1)}+b_2K_{lm}'^{(\lambda 2)}. \qquad \lambda\neq m \tag{19.25b}$$

The complexity of the problem is significantly reduced when the individual components of the tensor $\mathcal{K}$ are worked out. Furthermore it should be remembered that $\mathcal{K}$ is either symmetric or antisymmetric, depending on the behavior of the applied force with regard to the time reversal symmetry.

A similar procedure, though perhaps more complicated, can in principle be followed for triply-degenerate modes.

19.3. Application to structures of D_3 symmetry

Consider the A_2 mode of a crystal with the symmetry D_3 (i.e. $\Gamma_m = A_2$) in the presence of an electric field $\boldsymbol{E}$. For this structure we have $\Gamma(\boldsymbol{E})=A_2\oplus E$. Therefore according to eq. (19.7), $\Gamma_m \otimes A_1 \otimes \Gamma(\boldsymbol{E}) = \Gamma_m \otimes \Gamma(\boldsymbol{E}) = A_2 \otimes (A_2 \otimes E) = A_1 \oplus \ldots$ (when $\Gamma_n = A_1$), and $\Gamma_m \otimes E \otimes \Gamma(\boldsymbol{E}) = A_1 \oplus \ldots$ (when $\Gamma_n = E$). Thus, $\boldsymbol{u}_{A_2}$ can mix with both, $\boldsymbol{u}_{A_1}$ and $\boldsymbol{u}_E$. Then tensor $\mathcal{K}$ is symmetric and its non-zero components are

$$\begin{aligned}
&\mathcal{K}_{A_2A_1z}=\mathcal{K}_{A_1A_2z}\equiv\mathcal{K}_1\\
&\mathcal{K}^{(x)}_{A_2Ey}=\mathcal{K}^{(x)}_{EA_2y}=-\mathcal{K}^{(y)}_{A_2Ex}\equiv-\mathcal{K}^{(y)}_{EA_2x}\equiv\mathcal{K}_2\\
&\mathcal{K}^{(x)}_{A_1Ex}=\mathcal{K}^{(x)}_{EA_1x}=\mathcal{K}^{(y)}_{EA_1y}\equiv\mathcal{K}^{(y)}_{A_1Ey}\equiv\mathcal{K}_3\\
&\mathcal{K}^{(xx)}_{EEx}=-\mathcal{K}^{(yy)}_{EEx}=-\mathcal{K}^{(xy)}_{EEy}=-\mathcal{K}^{(yx)}_{EEy}\equiv\mathcal{K}_4.
\end{aligned}$$

The Cartesian subscript stands for the field component and the Cartesian superscript stands for the row index. The above component form of the tensor $\mathcal{K}$ can be derived following the group-theoretical procedures developed in §5.3. The results of eqs. (19.21) and (19.22) are

$$\Omega_{A_2}^2 = \omega_{A_2}^2 + \frac{\mathcal{K}_1^2 E_z^2}{(\omega_{A_2}^2 - \omega_{A_1}^2)} + \frac{\mathcal{K}_2^2(E_x^2 + E_y^2)}{(\omega_{A_2}^2 - \omega_E^2)}$$

$$U_{A_2} = u_{A_2} + \left(\frac{\mathcal{K}_1 E_z}{\omega_{A_2}^2 - \omega_{A_1}^2}\right) u_{A_1} + \frac{\mathcal{K}_2(E_y u_E^{(x)} - E_x u_E^{(y)})}{(\omega_{A_2}^2 - \omega_E^2)}$$

$$+ \left[\frac{2\mathcal{K}_2\mathcal{K}_4 E_x E_y}{(\omega_{A_2}^2 - \omega_E^2)^2} + \frac{\mathcal{K}_1\mathcal{K}_3 E_x E_z}{(\omega_{A_2}^2 - \omega_E^2)(\omega_{A_2}^2 - \omega_{A_1}^2)}\right] u_E^{(x)}$$

$$+ \left[\frac{\mathcal{K}_2\mathcal{K}_4(E_x^2 - E_y^2)}{(\omega_{A_2}^2 - \omega_E^2)^2} + \frac{\mathcal{K}_1\mathcal{K}_3 E_y E_z}{(\omega_{A_2}^2 - \omega_E^2)(\omega_{A_2}^2 - \omega_{A_1}^2)}\right] u_E^{(y)}$$

$$- \frac{1}{2}\left[\frac{\mathcal{K}_2^2(E_x^2 + E_y^2)}{(\omega_{A_2}^2 - \omega_E^2)^2} + \frac{\mathcal{K}_1^2 E_z^2}{(\omega_{A_2}^2 - \omega_{A_1}^2)^2}\right] u_{A_2}.$$

The frequency shifts with the electric field only in second order. Coupling to the u_E mode occurs in both first- and second-order in the field. Coupling to u_{A_1} occurs only in first order.

19.4. General remarks and experimental results

We have shown that it is possible to use well-known procedures of Quantum Mechanics to investigate the coupling behavior of normal vibrational modes of long wavelength under the influence of an external force. The resulting frequency shift and perturbed eigenvectors of a specific mode may be of particular interest in the case of biaxial crystals with a large number of atoms per unit cell. The coupling between modes of different symmetry turns out to depend critically on the frequency difference of the two modes. The same argument applies to modes of the same symmetry but different frequencies. Criteria for determining which modes can couple in the presence of a force are established through straightforward group-theoretical arguments. They are based on the fact that the symmetry of any mode is modified in the presence of a force. If this modified symmetry is the same as, or includes the symmetry of a given unperturbed mode, then mixing of the two is permitted by the crystal symmetry. This is described by eq. (19.7). It is also demonstrated very clearly in the application to D_3 symmetry. The functional form of the factors $E_\mu \boldsymbol{u}_r^{(\kappa)}$ and $E_\mu E_\nu \boldsymbol{u}_r^{(\kappa)}$ (or combinations of them) turns out to be consistent with the symmetry of the unperturbed mode $\boldsymbol{u}_{A_2}$. Consequently, the entire equation for the perturbed mode $\boldsymbol{U}_{A_2}$ is consistent with the representation A_2 to which $\boldsymbol{u}_{A_2}$ belongs.

It is emphasized that the whole symmetry analysis is performed on the basis of the unperturbed crystal symmetry. This enables one to determine to what order in the force the mixing, if any, occurs. Alternatively, although without this possibility, one may consider the gross change of the crystal symmetry in the presence of the force as described by the reduction of the original point group $\boldsymbol{G}$ to a subgroup $\boldsymbol{G}'$. The latter is determined by the nature and direction of the force $\boldsymbol{F}$. By definition the symmetry of $\boldsymbol{G}'$ is just high enough to preserve $\boldsymbol{F}$, i.e. $\boldsymbol{F}$ transforms in $\boldsymbol{G}'$ like a scalar. Consequently $\Gamma(\boldsymbol{F})$, the representation of $\boldsymbol{F}$ in $\boldsymbol{G}$, will reduce to a representation of $\boldsymbol{G}'$ which includes the totally symmetric representation Γ_1' of $\boldsymbol{G}'$. Furthermore, there is a relationship of *homomorphism* between the representations Γ of $\boldsymbol{G}$ and the representations Γ' of $\boldsymbol{G}'$. Thus eq. (19.7) would reduce to $\Gamma_m' \otimes \Gamma_n' \otimes \Gamma_1' = n_1' \Gamma_1' \oplus \ldots$. Here, $\Gamma_m'(\Gamma_n')$ is what $\Gamma_m(\Gamma_n)$ reduces to (when $\boldsymbol{F}$ is present) through the compatibility relations between $\boldsymbol{G}$ and $\boldsymbol{G}'$ (Koster et al. 1963). Although Γ_m is by definition an irreducible representation of $\boldsymbol{G}$, $\Gamma_m'(\Gamma_n')$ is not necessarily an irreducible representation of $\boldsymbol{G}'$. The above result $\Gamma_m' \otimes \Gamma_n' = n_1' \Gamma_1' \oplus \ldots$ indicates that only those representations of $\boldsymbol{G}$ can mix which reduce to representations of $\boldsymbol{G}'$ which are either identical (if irreducible) or have common part (if reducible).

The subject of coupled modes is one of the most appealing and extensively studied problems of Lattice Dynamics. Infrared absorption, light scattering and neutron scattering techniques have been widely utilized to observe and study the behavior of such systems as ferroelectric perovskites, improper ferroelectrics, α-quartz and SbSI-type structures, hydrogen-bonded ferroelectrics, etc. Typical findings of such investigations include strong anharmonic couplings, lineshape anomalies and interference effects. Theoretical studies on this subject have also been extensive. The reader is referred to the review papers of Worlock (1971), Fleury (1972) and Scott (1974) for an account of the work in this field. Since no external forces were involved in all (or nearly all) of these studies no connection can be made with the analysis presented in this section. It is noted that here the modes have been treated within the harmonic approximation without any damping involved. The symmetry analysis is quite general and independent of particular mechanisms involved. Thus, for instance, the comments made at the end of §19.2.1 regarding the *intrinsic* (force-free) coupling of non-degenerate modes with the same symmetry are in accord with the experimental findings on soft modes. Some of these results (including degenerate optical and transverse perturbing modes) are summarized in table 13 (Scott 1971).

Relatively few cases of force-induced mode mixing have been observed. These are mostly restricted to ferroelectric materials. Coupling of soft modes in ferroelectric $LiNbO_3$ and $LiTaO_3$ has been reported by Johnston and Kaminow (1968). It involves phonons of A_1 and E symmetry (point group C_{3v} below $T_0=1485$ K and 900 K respectively). The coupling exhibited strong temperature dependence and was attributed to temperature-dependent strain perturbations. Uwe and Sakudo (1976) have observed

Table 13
Soft modes and mode couplings (Scott 1971).

Material	Soft modes	T_0(K)	Coupled with
SiO_2	A_1, $q=0$	846	two acoustic-phonon state[a]
$AlPO_4$	A_1, $q=0$	852	A_1 optic mode and two-acoustic-phonon state[b]
$KTaO_3$	F_{1u}, $q\neq 0$	0	$q\cong 0$ acoustic phonon[c]
$SrTiO_3$ (cubic)	F_{1u}, $q=0$	0	other F_{1u} optic modes[d]
$BaTiO_3$	A_1, $q=0$	393	$q=0$ acoustic[e]
	A_1, $q=0$	393	$q=0$ optic[f]
$SrTiO_3$ (tetragonal)	A_{1g}, E_g $q=0$	106	$A_{2u}+E_u$[g,h]
$LiNbO_3$ } $LiTaO_3$ }	A_1, $q=0$	1480 900	A_1 and E, $q=0$[i]
SbSI	A, $q=0$	288	A, $q=0$ (or two acoustic?)[j]
TGS } KH_2PO_4 }	$q=0$ (tunneling)	322 122	$q=0$ acoustic[k]
CsH_2AsO_4	B_2, $q=0$ (tunneling)	143	B_2, $q_{TO}=0$[l]
KH_2AsO_4	B_2, $q=0$ (tunneling)	92	B_2, $q_{TO}=0$[l]

[a]Scott (1968b), Shapiro (1969).
[b]Scott (1968b), (1970).
[c]Axe et al. (1970).
[d]Barker and Hopfield (1964).
[e]Fleury and Lazay (1971), Fleury (1971).
[f]Rousseau and Porto (1968).
[g]Fleury et al. (1968), Scott (1969).
Scott and Remeica (1970), Worlock et al. (1969).
[h]Uwe and Sakudo (1976).
[i]Johnston and Kaminow (1968).
[j]Harbeke et al. (1970).
[k]Cummins (1967).
[l]Katiyar et al. (1971).

coupling of ferroelectric and structural soft modes of $SrTiO_3$ in the stress-induced ferroelectric phase. Experimental observation of electric-field-induced mixing of optical phonons has also been reported for $SrTiO_3$ (Worlock et al. 1969 and references therein). In the low-temperature phase ($T \leqslant T_0 = 110\,K$, point group D_{4h}) two of the normally Raman-active phonons (A_{1g}, E_g) *soften*, that is, $\omega \to 0$ as $T \to T_0$. Furthermore, the normally Raman-inactive components A_{2u}, E_u of the *ferroelectric mode* can become Raman active in first order with an applied electric field, as discussed in §7.2.1. Their frequencies depend on temperature and the electric field. They soften (decrease) as $T \to 0$ but they never reach zero. They *harden* (increase) as the field increases. The same electric field which activates A_{2u}, E_u will also cause their coupling with other modes. As the frequencies of the modes can be tuned by varying temperature and/or the electric field the strength of the coupling can be regulated experimentally. Since all four modes are Raman active in the presence of the electric field, the extent of the coupling can be studied directly from the Raman spectra. Indeed, the coupling of the pairs (A_{1g}, A_{2u}) and (E_g, E_u) was observed at fixed 8 K and variable fields in the form of *intensity exchange* between the two modes within each pair. The maximum exchange occurs for those values of the field for which the two frequencies *anti-cross* each other, that is, their difference passes through a non-zero minimum. Fig. 29 illustrates the

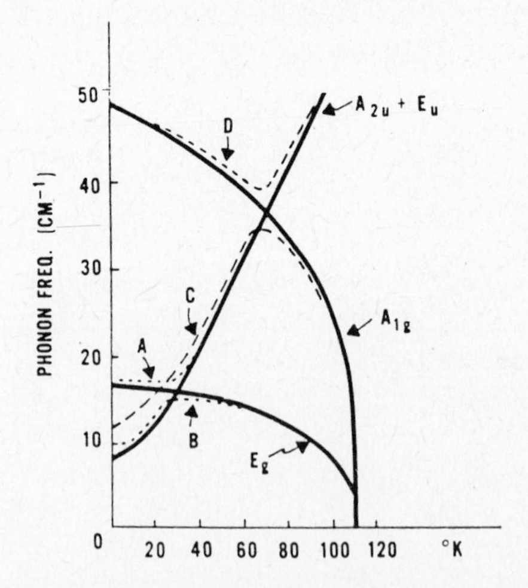

Fig. 29. Temperature dependence of the soft mode frequencies in $SrTiO_3$ (Worlock et al. 1969). In the absence of external electric fields (solid lines) the ferroelectric mode ($A_{2u} + E_u$) cannot linearly couple to the antiferroelectric soft modes (E_g, A_{1g}). An external field breaks the symmetries and permits coupling (dashed lines). Modes of A symmetry (C and D) and modes of E symmetry (A and B) interact in pairs (Fleury 1971).

temperature dependence of the four modes below T_0. Without an electric field (solid lines) their frequencies cross freely and interaction (coupling) cannot occur. As a field along [001] is applied coupling becomes possible between (A_{1g}, A_{2u}) and (E_g, E_u) but not between (A_{1g}, E_g). An appropriate extension of the procedure for a degenerate perturbed mode given in the text is needed to account for the pair (E_g, E_u). A complete theoretical account of this problem is too complicated, owing to the multiple role of the electric field. One has to take into consideration the field dependence of the frequencies $\omega_{A_{2u}}$ and ω_{E_u} (Worlock et al. 1969), the field-dependent scattering cross-section of the modes A_{2u} and E_u (i.e. the field-dependent Raman tensor), the field-dependent frequencies for the A_{1g} and E_g modes (i.e. eq. (19.21)) and finally the field-induced coupling (i.e. (19.22)).

20. *Concluding remarks*

In the course of assembling the material for this chapter, the author was surprised at the amount of work in the literature which could come under the heading of morphic effects. As a result, it has only been possible to deal briefly with each topic. The discussion has been mainly restricted to lattice-dynamical phenomena, particularly those which can be probed by absorption and scattering techniques. It is easy to visualize long review papers on morphic effects in connection with optical, electronic and lattice-dynamical phenomena probed by neutron scattering, X-ray spectroscopy, luminescence or other experimental procedures.

The attitude adopted here was to consider the effects of external perturbations on Lattice Dynamics in the widest sense possible. This allowed the inclusion of effects due to external scalar forces (e.g. pressure and temperature) and also effects due to built-in perturbations, scalar or otherwise (e.g. surface fields, residual strains, magnetic ordering, field gradients). The emphasis, however, has been given to effects which are induced by external non-scalar forces. This leads naturally to the inclusion of such indirect morphic effects as, for instance, stress-induced phase transitions or lineshape changes in heavily doped semiconductors.

Essential for understanding the kind of phenomena that may be expected in each case is the symmetry analysis. Such analysis will answer questions like *yes or no, how many or which way*. It will not answer questions like *why or how big*. These answers require an independent physical justification, always consistent with the restrictions of the symmetry analysis. It has been the principal objective of this work to present a wide variety of phenomena under a unified symmetry approach. The basis for this approach is the simple fact that the symmetry properties of a phonon can be described by some irreducible representation. Within the scope of a phenomenological theory, it is then easy to determine the appropriate mode coefficients whose existence and magnitude are responsible for the occurrence and strength of the effect.

The main use of morphic effects in Lattice Dynamics thus far has been to predict, establish and confirm symmetry assignments, to obtain numerical values for mode properties and deformation potentials and to test the validity of theoretical models proposed for the corresponding unperturbed physical problem. Most of the morphic effects that have been observed refer to first-order phenomena in the phonon normal coordinates. The few existing results relating to second-order effects appear to be very informative and promising for a more systematic investigation of second- and higher-order phenomena. In addition, it is easy to foresee the usefulness and applicability of morphic effects to such research areas as

(i) Brillouin scattering,
(ii) polaritons, plasmons, magnons,
(iii) impurity related optical and dynamical effects,
(iv) spatial dispersion and coupling of acoustical phonons with other collective excitations,
(v) multipole-dipole scattering,
(vi) surface physics,
(vii) morphic effects in liquid and molecular crystals,
(viii) non-linear effects, e.g. hyper-Raman scattering, stimulated and inverse Raman scattering, coherent anti-Stokes Raman scattering, two-photon absorption (Levenson 1977).

One can go even further into a second phase of generalization and visualize morphic effects in

(i) amorphous materials,
(ii) neutron scattering, x-ray and electron diffraction,
(iii) lattice-dynamical aspects of transport phenomena.

We believe that it will only be a matter of time before rapidly advancing technology will allow the observation and study of these and many other morphic effects, thus adding more powerful means to the methods and techniques of experimental Solid State Physics.

Acknowledgments

It is a pleasure to thank all those investigators who kindly responded to my call for reprints and other materials. Useful discussions during the writing of this chapter with M. Cardona, R. Humphreys and W. Richter are acknowledged with appreciation. Particular thanks are due to R. Humphreys and W. Richter for critical reading of the manuscript and to Ms. Yota Tsinaki for excellent typing and technical assistance.

Appendix

Table A1

List of pseudoscalar representations Γ_p for all thirty-two point groups and reduction of $\Gamma(\boldsymbol{f})=\Gamma(\boldsymbol{E})\otimes\Gamma(\boldsymbol{E})$ and $\Gamma(\boldsymbol{m})=\Gamma_p\otimes\Gamma(\boldsymbol{f})$ (Anastassakis et al. 1972b). The notation of Koster et al. (1963) is followed. (See §§5.1, 15.2, 15.3.) For conversion of notation use table A2.

Point group	$\Gamma(\boldsymbol{f})$	Γ_p	$\Gamma(\boldsymbol{m})$
C_1	$9\Gamma_1$	Γ_1	$9\Gamma_1$
$C_i(S_2)$	$9\Gamma_1^+$	Γ_1^-	$9\Gamma_1^-$
C_2	$5\Gamma_1+4\Gamma_2$	Γ_1	$5\Gamma_1+4\Gamma_2$
$C_s(C_{1h})$	$5\Gamma_1+4\Gamma_2$	Γ_2	$4\Gamma_1+5\Gamma_2$
C_{2h}	$5\Gamma_1^++4\Gamma_2^+$	Γ_1^-	$5\Gamma_1^-+4\Gamma_2^-$
$D_2(V)$	$3\Gamma_1+2\Gamma_2+2\Gamma_3+2\Gamma_4$	Γ_1	$3\Gamma_1+2\Gamma_2+2\Gamma_3+2\Gamma_4$
C_{2v}	$3\Gamma_1+2\Gamma_2+2\Gamma_3+2\Gamma_4$	Γ_3	$2\Gamma_1+2\Gamma_2+3\Gamma_3+2\Gamma_4$
$D_{2h}(V_h)$	$3\Gamma_1^++2\Gamma_2^++2\Gamma_3^++2\Gamma_4^+$	Γ_1^-	$3\Gamma_1^-+2\Gamma_2^-+2\Gamma_3^-+2\Gamma_4^-$
C_4	$3\Gamma_1+2\Gamma_2+(2\Gamma_3+2\Gamma_4)$	Γ_1	$3\Gamma_1+2\Gamma_2+(2\Gamma_3+2\Gamma_4)$
S_4	$3\Gamma_1+2\Gamma_2+(2\Gamma_3+2\Gamma_4)$	Γ_2	$2\Gamma_1+3\Gamma_2+(2\Gamma_3+2\Gamma_4)$
C_{4h}	$3\Gamma_1^++2\Gamma_2^++(2\Gamma_3^++2\Gamma_4^+)$	Γ_1^-	$3\Gamma_1^-+2\Gamma_2^-+(2\Gamma_3^-+2\Gamma_4^-)$
D_4	$2\Gamma_1+\Gamma_2+\Gamma_3+\Gamma_4+2\Gamma_5$	Γ_1	$2\Gamma_1+\Gamma_2+\Gamma_3+\Gamma_4+2\Gamma_5$
C_{4v}	$2\Gamma_1+\Gamma_2+\Gamma_3+\Gamma_4+2\Gamma_5$	Γ_2	$\Gamma_1+2\Gamma_2+\Gamma_3+\Gamma_4+2\Gamma_5$
$D_{2d}(V_d)$	$2\Gamma_1+\Gamma_2+\Gamma_3+\Gamma_4+2\Gamma_5$	Γ_3	$\Gamma_1+\Gamma_2+2\Gamma_3+\Gamma_4+2\Gamma_5$
D_{4h}	$2\Gamma_1^++\Gamma_2^++\Gamma_3^++\Gamma_4^++2\Gamma_5^+$	Γ_1^-	$2\Gamma_1^-+\Gamma_2^-+\Gamma_3^-+\Gamma_4^-+2\Gamma_5^-$
C_3	$3\Gamma_1+(3\Gamma_2+3\Gamma_3)$	Γ_1	$3\Gamma_1+(3\Gamma_2+3\Gamma_3)$
$C_{3i}(S_6)$	$3\Gamma_1^++(3\Gamma_2^++3\Gamma_3^+)$	Γ_1^-	$3\Gamma_1^-+(3\Gamma_2^-+3\Gamma_3^-)$
D_3	$2\Gamma_1+\Gamma_2+3\Gamma_3$	Γ_1	$2\Gamma_1+\Gamma_2+3\Gamma_3$
C_{3v}	$2\Gamma_1+\Gamma_2+3\Gamma_3$	Γ_2	$\Gamma_1+2\Gamma_2+3\Gamma_3$
D_{3d}	$2\Gamma_1^++\Gamma_2^++3\Gamma_3^+$	Γ_1^-	$2\Gamma_1^-+\Gamma_2^-+3\Gamma_3^-$
C_6	$3\Gamma_1+(\Gamma_2+\Gamma_3)+(2\Gamma_5+2\Gamma_6)$	Γ_1	$3\Gamma_1+(\Gamma_2+\Gamma_3)+(2\Gamma_5+2\Gamma_6)$
C_{3h}	$3\Gamma_1+(\Gamma_2+\Gamma_3)+(2\Gamma_5+2\Gamma_6)$	Γ_4	$(2\Gamma_2+2\Gamma_3)+3\Gamma_4+(\Gamma_5+\Gamma_6)$
C_{6h}	$3\Gamma_1^++(\Gamma_2^++\Gamma_3^+)$ $+(2\Gamma_5^++2\Gamma_6^+)$	Γ_1^-	$3\Gamma_1^-+(\Gamma_2^-+\Gamma_3^-)+(2\Gamma_5^-+2\Gamma_6^-)$
D_6	$2\Gamma_1+\Gamma_2+2\Gamma_5+\Gamma_6$	Γ_1	$2\Gamma_1+\Gamma_2+2\Gamma_5+\Gamma_6$
C_{6v}	$2\Gamma_1+\Gamma_2+2\Gamma_5+\Gamma_6$	Γ_2	$\Gamma_1+2\Gamma_2+2\Gamma_5+\Gamma_6$
D_{3h}	$2\Gamma_1+\Gamma_2+2\Gamma_5+\Gamma_6$	Γ_3	$2\Gamma_3+\Gamma_4+\Gamma_5+2\Gamma_6$
D_{6h}	$2\Gamma_1^++\Gamma_2^++2\Gamma_5^++\Gamma_6^+$	Γ_1^-	$2\Gamma_1^-+\Gamma_2^-+2\Gamma_5^-+\Gamma_6^-$
T	$\Gamma_1+(\Gamma_2+\Gamma_3)+2\Gamma_4$	Γ_1	$\Gamma_1+(\Gamma_2+\Gamma_3)+2\Gamma_4$
T_h	$\Gamma_1^++(\Gamma_2^++\Gamma_3^+)+2\Gamma_4^+$	Γ_1^-	$\Gamma_1^-+(\Gamma_2^-+\Gamma_3^-)+2\Gamma_4^-$
O	$\Gamma_1+\Gamma_3+\Gamma_4+\Gamma_5$	Γ_1	$\Gamma_1+\Gamma_3+\Gamma_4+\Gamma_5$
T_d	$\Gamma_1+\Gamma_3+\Gamma_4+\Gamma_5$	Γ_2	$\Gamma_2+\Gamma_3+\Gamma_4+\Gamma_5$
O_h	$\Gamma_1^++\Gamma_3^++\Gamma_4^++\Gamma_5^+$	Γ_1^-	$\Gamma_1^-+\Gamma_3^-+\Gamma_4^-+\Gamma_5^-$

Table A2

Correspondence between the Γ-notation of irreducible representations of all thirty-two point groups (Koster et al. 1963) and the standard Mulliken notation. In centrosymmetric groups use Γ^+ or Γ^- and A_g or A_u, etc., for even- and odd-parity representations respectively.

<table>
<tr><th>Point group</th><th>Γ_1</th><th>Γ_2</th><th>Γ_3</th><th>Γ_4</th><th>Γ_5</th><th>Γ_6</th></tr>
<tr><td>$C_1(1)$, $C_i(\bar{1})$</td><td>A</td><td></td><td></td><td></td><td></td><td></td></tr>
<tr><td>$C_s(m)$</td><td>A′</td><td>A″</td><td></td><td></td><td></td><td></td></tr>
<tr><td>$C_2(2)$, $C_{2h}(2/m)$</td><td>A</td><td>B</td><td></td><td></td><td></td><td></td></tr>
<tr><td>$C_{2v}(mm2)$</td><td>A_1</td><td>B_1</td><td>A_2</td><td>B_2</td><td></td><td></td></tr>
<tr><td>$D_2(222)$, $D_{2h}(mmm)$</td><td>A</td><td>B_2</td><td>B_1</td><td>B_3</td><td></td><td></td></tr>
<tr><td>$C_4(4)$, $C_{4h}(4/m)$
$S_4(\bar{4})$</td><td>A</td><td>B</td><td colspan="2">E</td><td></td><td></td></tr>
<tr><td>$D_4(422)$, $D_{4h}(4/mmm)$
$C_{4v}(4mm)$, $D_{2d}(\bar{4}2m)$</td><td>A_1</td><td>A_2</td><td>B_1</td><td>B_2</td><td>E</td><td></td></tr>
<tr><td>$C_3(3)$, $C_{3i}(\bar{3})$</td><td>A</td><td colspan="2">E</td><td></td><td></td><td></td></tr>
<tr><td>$D_3(32)$, $D_{3d}(\bar{3}m)$
$C_{3v}(3m)$</td><td>A_1</td><td>A_2</td><td>E</td><td></td><td></td><td></td></tr>
<tr><td>$C_6(6)$, $C_{6h}(6/m)$</td><td>A</td><td colspan="2">E_2</td><td>B</td><td colspan="2">E_1</td></tr>
<tr><td>$C_{3h}(\bar{6})$</td><td>A′</td><td colspan="2">E′</td><td>A″</td><td colspan="2">E″</td></tr>
<tr><td>$D_6^*(622)$, $D_{6h}^*(6/mmm)$
$C_{6v}(6mm)$</td><td>A_1</td><td>A_2</td><td>B_2</td><td>B_1</td><td>E_1</td><td>E_2</td></tr>
<tr><td>$D_{3h}(\bar{6}m2)$</td><td>A_1'</td><td>A_2'</td><td>A_1''</td><td>A_2''</td><td>E″</td><td>E′</td></tr>
<tr><td>$T(23)$, $T_h(m3)$</td><td>A</td><td colspan="2">E</td><td>F</td><td></td><td></td></tr>
<tr><td>$O(432)$, $O_h(m3m)$
$T_d(\bar{4}3m)$</td><td>A_1</td><td>A_2</td><td>E</td><td>F_1</td><td>F_2</td><td></td></tr>
</table>

*x-axis coincident with C_2' axis.

Table A3

Reducible coefficients of diamond corresponding to Raman- and infrared-active representations of overtones and combinations (Solin and Ramdas 1970). (See § 5.4.)

Combinations or overtones	Raman-active $\Gamma^{(1+)}$	Raman-active $\Gamma^{(12+)}$	Raman-active $\Gamma^{(25+)}$	Infrared-active $\Gamma^{(15-)}$
$[\Gamma^{(25+)}]_{(2)}$	1	1	1	0
$[{}^*X^{(1)}]_{(2)}$	1	1	1	0
$[{}^*X^{(2)}]_{(2)}$	1	1	1	0
$[{}^*X^{(3)}]_{(2)}$	1	1	1	0
$[{}^*X^{(4)}]_{(2)}$	1	1	1	0
${}^*X^{(1)}\otimes{}^*X^{(2)}$	0	1	0	1
${}^*X^{(1)}\otimes{}^*X^{(3)}$	0	0	1	1
${}^*X^{(1)}\otimes{}^*X^{(4)}$	0	0	1	1
${}^*X^{(2)}\otimes{}^*X^{(3)}$	0	0	1	1
${}^*X^{(2)}\otimes{}^*X^{(4)}$	0	0	1	1
${}^*X^{(3)}\otimes{}^*X^{(4)}$	0	1	0	1
$[{}^*L^{(1+)}]_{(2)}$	1	0	1	0
$[{}^*L^{(2-)}]_{(2)}$	1	0	1	0
$[{}^*L^{(3+)}]_{(2)}$	1	1	2	0
$[{}^*L^{(3-)}]_{(2)}$	1	1	2	0
${}^*L^{(1+)}\otimes{}^*L^{(2-)}$	0	0	0	1
${}^*L^{(1+)}\otimes{}^*L^{(3+)}$	0	1	1	0
${}^*L^{(1+)}\otimes{}^*L^{(3-)}$	0	0	0	1
${}^*L^{(2-)}\otimes{}^*L^{(3+)}$	0	0	0	1
${}^*L^{(2-)}\otimes{}^*L^{(3-)}$	0	1	1	0
${}^*L^{(3+)}\otimes{}^*L^{(3-)}$	0	0	0	2
$[{}^*W^{(1)}]_{(2)}$	1	1	1	0
$[{}^*W^{(2)}]_{(2)}$	1	1	1	0
${}^*W^{(1)}\otimes{}^*W^{(2)}$	0	1	2	2
${}^*W^{(2)}\otimes{}^*W^{(2)}$	1	1	1	1
$[{}^*\Sigma^{(1)}]_{(2)}$	1	1	1	0
$[{}^*\Sigma^{(2)}]_{(2)}$	1	1	1	0
$[{}^*\Sigma^{(3)}]_{(2)}$	1	1	1	0
${}^*\Sigma^{(1)}\otimes{}^*\Sigma^{(2)}$	0	0	1	0
${}^*\Sigma^{(1)}\otimes{}^*\Sigma^{(3)}$	0	0	1	1
${}^*\Sigma^{(1)}\otimes{}^*\Sigma^{(4)}$	0	1	0	1
${}^*\Sigma^{(2)}\otimes{}^*\Sigma^{(3)}$	0	1	0	1
${}^*\Sigma^{(1)}\otimes{}^*\Sigma^{(1)}$	1	1	1	1
${}^*\Sigma^{(2)}\otimes{}^*\Sigma^{(2)}$	1	1	1	1
${}^*\Sigma^{(3)}\otimes{}^*\Sigma^{(3)}$	1	1	1	1

Table A4

Vibrational direction for the strain-split one-phonon states considered. L and T denote longitudinal and transverse modes respectively, while (X) and (L) indicate points in the first Brillouin zone of the strain-free crystal. Subscripts α and β correspond to propagation parallel and non-parallel to the stress axis respectively. Subscripts 1 and 2 stand for vibrations parallel and perpendicular to the plane containing the stress axis and the wave vector. Only diamond- and zincblende-type lattices are considered (Balslev 1974). (See §§ 5.4, 11.4.)

Mode	Stress ‖ [001] Sublevel	Stress ‖ [001] Vibrational direction	Stress ‖ [111] Sublevel	Stress ‖ [111] Vibrational direction
L(X)	L_α	[001]	L	[001] and
	L_β	[010], [100]		2 equivalent
T(X)	T_α	[110], [$\bar{1}$10]	T_1	[110] and 2 eq.
	T_β	[101], [10$\bar{1}$], [011],	T_2	[1$\bar{1}$0] and 2 eq.
		[01$\bar{1}$]		
L(L)	L	[111] and	L_α	[111]
		3 equivalent	L_β	[11$\bar{1}$] and 2 eq.
T(L)	T_1	[11$\bar{2}$] and 3 eq.	T_α	[1$\bar{1}$0], [11$\bar{2}$]
	T_2	[110] and 3 eq.	$T_{\beta 1}$	[11$\bar{2}$] and 2 eq.
			$T_{\beta 2}$	[1$\bar{1}$0] and 2 eq.

Table A5

Relative Raman intensities of two-phonon spectra for [001] stress. The symbols used for the one-phonon states involved are explained in table A4. Primes and double primes are used to indicate participation of two transverse or two longitudinal phonons. Thus, the last pair of two-phonon states are degenerate for overtones. The superscripts on Γ_{12} and Γ_{15} indicate dependence on polarization configuration. With the cubic axes as basis the corresponding polarization configurations are given in the second line of the table. For a $x'y'z$ coordinate system rotated 45° about the stress axis, the $\Gamma_{12}^{(1)}$ and $\Gamma_{15}^{(1)}$ selection rules appear in the $x'y'$ and $x'x'$ (or $y'y'$) polarization respectively. For infrared absorption the relative intensities for parallel and perpendicular polarization are given by the $\Gamma_{15}^{(1)}$ and $\Gamma_{15}^{(2)}$ columns respectively. Only diamond- and zincblende-type lattices are considered (Balslev 1974). (See §§5.4, 11.4.)

Mode	Split level	Γ_1	$\Gamma_{12}^{(1)}$ xx	$\Gamma_{12}^{(2)}$ zz	$\Gamma_{15}^{(1)}$ xy	$\Gamma_{15}^{(2)}$ xz
L′(X)±L″(X)	$L'_\alpha \pm L''_\alpha$	$\frac{1}{3}$	$\frac{1}{6}$	$\frac{2}{3}$	...	...
	$L'_\beta \pm L''_\beta$	$\frac{2}{3}$	$\frac{5}{6}$	$\frac{1}{3}$	...	...
L(X)±T(X)	$L_\alpha \pm T_\alpha$	...	...	...	0	$\frac{1}{2}$
	$L_\beta \pm T_\beta$	...	...	...	1	$\frac{1}{2}$

(continued)

Mode	Split level	Γ_1	$\Gamma_{12}^{(1)}$ xx	$\Gamma_{12}^{(2)}$ zz	$\Gamma_{15}^{(1)}$ xy	$\Gamma_{15}^{(2)}$ xz
T(X) ± T″(X)	$T'_\alpha \pm T''_\alpha$	$\frac{1}{3}$	$\frac{5}{12}$	$\frac{1}{6}$	1	0
	$T'_\beta \pm T''_\beta$	$\frac{2}{3}$	$\frac{7}{12}$	$\frac{5}{6}$	0	1
L(L) ± T(L)	$L \pm T_1$	…	$\frac{1}{4}$	1	1	$\frac{1}{4}$
	$L \pm T_2$	…	$\frac{3}{4}$	0	0	$\frac{3}{4}$
T′(L) ± T″(L)	$T'_1 \pm T''_1$	$\frac{1}{2}$	$\frac{1}{8}$	$\frac{1}{2}$	$\frac{1}{10}$	$\frac{4}{10}$
	$T'_2 \pm T''_2$	$\frac{1}{2}$	$\frac{1}{8}$	$\frac{1}{2}$	$\frac{9}{10}$	0
	$T'_1 \pm T''_2$	0	$\frac{3}{8}$	0	0	$\frac{3}{10}$
	$T'_2 \pm T''_1$	0	$\frac{3}{8}$	0	0	$\frac{3}{10}$

Table A6

Relative intensities of two-phonon spectra for [111] stress. The symbols used for indication of modes are equivalent to those of tables A4 and A5. For transverse overtone states the two last pairs of the $T' \pm T''$ sublevels are degenerate. The polarization configuration (second line) refers to any coordinate system $x''y''z''$ with $z''//$stress. For infrared absorption the intensities for parallel and perpendicular polarization are given by the $\Gamma_{15}^{(2)}$ and $\Gamma_{15}^{(3)}$ columns respectively. Only diamond- and zincblende-type lattices are considered (Balslev 1974) (See §§5.4, 11.4).

Mode	Split level	Γ_1	Γ_{12}	$\Gamma_{15}^{(1)}$ $x''x''$	$\Gamma_{15}^{(2)}$ $z''z''$	$\Gamma_{15}^{(3)}$ $x''z''$, $x''y''$
L′(L) ± L″(L)	$L'_\alpha \pm L''_\alpha$	$\frac{1}{4}$	…	$\frac{1}{4}$	$\frac{3}{4}$	0
	$L'_\beta \pm L''_\beta$	$\frac{3}{4}$	…	$\frac{3}{4}$	$\frac{1}{4}$	1
L(L) + T(L)	$L_\alpha \pm T_\alpha$	…	$\frac{1}{4}$	$\frac{1}{4}$	0	$\frac{6}{16}$
	$L_\beta \pm T_{\beta_1}$	…	$\frac{3}{8}$	$\frac{3}{8}$	1	$\frac{1}{16}$
	$L_\beta \pm T_{\beta_2}$	…	$\frac{3}{8}$	$\frac{3}{8}$	0	$\frac{9}{16}$
T′(L) ± T″(L)	$T'_\alpha + T''_\alpha$	$\frac{1}{4}$	$\frac{1}{4}$	$\frac{10}{40}$	$\frac{6}{40}$	$\frac{12}{40}$
	$T'_{\beta_1} \pm T''_{\beta_1}$	$\frac{3}{8}$	$\frac{3}{16}$	$\frac{9}{40}$	$\frac{25}{40}$	$\frac{1}{40}$
	$T'_{\beta_2} \pm T''_{\beta_2}$	$\frac{3}{8}$	$\frac{3}{16}$	$\frac{9}{40}$	$\frac{9}{40}$	$\frac{9}{40}$
	$T'_{\beta_1} \pm T''_{\beta_2}$	0	$\frac{3}{16}$	$\frac{6}{40}$	0	$\frac{9}{40}$
	$T'_{\beta_2} \pm T''_{\beta_1}$	0	$\frac{3}{16}$	$\frac{6}{40}$	0	$\frac{9}{40}$
L(X) ± T(X)	$L \pm T_1$	…	…	$\frac{1}{2}$	1	$\frac{1}{4}$
	$L \pm T_2$	…	…	$\frac{1}{2}$	0	$\frac{3}{4}$
T′(X) ± T″(X)	$T'_1 \pm T''_1$	$\frac{1}{2}$	$\frac{1}{8}$	$\frac{1}{2}$	$\frac{1}{2}$	$\frac{1}{2}$
	$T'_2 \pm T''_2$	$\frac{1}{2}$	$\frac{1}{8}$	$\frac{1}{2}$	$\frac{1}{2}$	$\frac{1}{2}$
	$T'_1 \pm T''_2$	0	$\frac{3}{8}$	0	0	0
	$T'_2 \pm T''_1$	0	$\frac{3}{8}$	0	0	0

Table A7

Reduction of the reducible representations of eqs. (18.10b)–(18.15b) corresponding to the mode coefficients of eqs. (18.10a)–(18.15a) respectively (Humphreys and Maradudin 1972). (See § 18.5.)

Point group	Eqs. (18.11b), (18.12b)	Eq. (18.13b)	Eq. (18.14b)	Eqs. (18.10b), (18.15b)
C_1	$54\Gamma_1$	$108\Gamma_1$	$27\Gamma_1$	$27\Gamma_1$
C_i	$54\Gamma_1^+$	$108\Gamma_1^-$	$27\Gamma_1^+$	$27\Gamma_1^+$
C_2	$28\Gamma_1+26\Gamma_2$	$52\Gamma_1+56\Gamma_2$	$13\Gamma_1+14\Gamma_2$	$13\Gamma_1+14\Gamma_2$
C_s	$28\Gamma_1+26\Gamma_2$	$56\Gamma_1+52\Gamma_2$	$13\Gamma_1+14\Gamma_2$	$14\Gamma_1+13\Gamma_2$
C_{2h}	$28\Gamma_1^+ +26\Gamma_2^+$	$52\Gamma_1^- +56\Gamma_2^-$	$13\Gamma_1^+ +14\Gamma_2^+$	$13\Gamma_1^- +14\Gamma_2^-$
D_2	$15\Gamma_1+13\Gamma_2+13\Gamma_3+13\Gamma_4$	$24\Gamma_1+28\Gamma_2+28\Gamma_3+28\Gamma_4$	$6\Gamma_1+7\Gamma_2+7\Gamma_3+7\Gamma_4$	$6\Gamma_1+7\Gamma_2+7\Gamma_3+7\Gamma_4$
C_{2v}	$15\Gamma_1+13\Gamma_2+13\Gamma_3+13\Gamma_4$	$28\Gamma_1+28\Gamma_2+24\Gamma_3+28\Gamma_4$	$6\Gamma_1+7\Gamma_2+7\Gamma_3+7\Gamma_4$	$7\Gamma_1+7\Gamma_2+6\Gamma_3+7\Gamma_4$
D_{2h}	$15\Gamma_1^+ +13\Gamma_2^+ +13\Gamma_3^+ +13\Gamma_4^+$	$24\Gamma_1^- +28\Gamma_2^- +28\Gamma_3^- +28\Gamma_4^-$	$6\Gamma_1^+ +7\Gamma_2^+ +7\Gamma_3^+ +7\Gamma_4^+$	$6\Gamma_1^- +7\Gamma_2^- +7\Gamma_3^- +7\Gamma_4^-$
C_4	$14\Gamma_1+14\Gamma_2+13\Gamma_3+13\Gamma_4$	$26\Gamma_1+26\Gamma_2+28\Gamma_3+28\Gamma_4$	$7\Gamma_1+6\Gamma_2+7\Gamma_3+7\Gamma_4$	$7\Gamma_1+6\Gamma_2+7\Gamma_3+7\Gamma_4$
S_4	$14\Gamma_1+14\Gamma_2+13\Gamma_3+13\Gamma_4$	$26\Gamma_1+26\Gamma_2+28\Gamma_3+28\Gamma_4$	$7\Gamma_1+6\Gamma_2+7\Gamma_3+7\Gamma_4$	$6\Gamma_1+7\Gamma_2+7\Gamma_3+7\Gamma_4$
C_{4h}	$14\Gamma_1^+ +14\Gamma_2^+ +13\Gamma_3^+ +13\Gamma_4^+$	$26\Gamma_1^- +26\Gamma_2^- +28\Gamma_3^- +28\Gamma_4^-$	$7\Gamma_1^+ +6\Gamma_2^+ +7\Gamma_3^+ +7\Gamma_4^+$	$7\Gamma_1^- +6\Gamma_2^- +7\Gamma_3^- +7\Gamma_4^-$
D_4	$8\Gamma_1+6\Gamma_2+7\Gamma_3+7\Gamma_4+13\Gamma_5$	$11\Gamma_1+15\Gamma_2+13\Gamma_3+13\Gamma_4+28\Gamma_5$	$3\Gamma_1+4\Gamma_2+3\Gamma_3+3\Gamma_4+7\Gamma_5$	$3\Gamma_1+4\Gamma_2+3\Gamma_3+3\Gamma_4+7\Gamma_5$
C_{4v}	$8\Gamma_1+6\Gamma_2+7\Gamma_3+7\Gamma_4+13\Gamma_5$	$15\Gamma_1+11\Gamma_2+13\Gamma_3+13\Gamma_4+28\Gamma_5$	$3\Gamma_1+4\Gamma_2+3\Gamma_3+3\Gamma_4+7\Gamma_5$	$4\Gamma_1+3\Gamma_2+3\Gamma_3+3\Gamma_4+7\Gamma_5$
D_{2d}	$8\Gamma_1+6\Gamma_2+7\Gamma_3+7\Gamma_4+13\Gamma_5$	$13\Gamma_1+13\Gamma_2+11\Gamma_3+15\Gamma_4+28\Gamma_5$	$3\Gamma_1+4\Gamma_2+3\Gamma_3+3\Gamma_4+7\Gamma_5$	$3\Gamma_1+3\Gamma_2+3\Gamma_3+4\Gamma_4+7\Gamma_5$
D_{4h}	$8\Gamma_1^+ +6\Gamma_2^+ +7\Gamma_3^+ +7\Gamma_4^+ +13\Gamma_5^+$	$11\Gamma_1^- +15\Gamma_2^- +13\Gamma_3^- +13\Gamma_4^- +28\Gamma_5^-$	$3\Gamma_1^+ +4\Gamma_2^+ +3\Gamma_3^+ +3\Gamma_4^+ +7\Gamma_5^+$	$3\Gamma_1^- +4\Gamma_2^- +3\Gamma_3^- +3\Gamma_4^- +7\Gamma_5^-$
C_3	$18\Gamma_1+18\Gamma_2+18\Gamma_3$	$36\Gamma_1+36\Gamma_2+36\Gamma_3$	$9\Gamma_1+9\Gamma_2+9\Gamma_3$	$9\Gamma_1+9\Gamma_2+9\Gamma_3$
C_{3i}	$18\Gamma_1^+ +18\Gamma_2^+ +18\Gamma_3^+$	$36\Gamma_1^- +36\Gamma_2^- +36\Gamma_3^-$	$9\Gamma_1^+ +9\Gamma_2^+ +9\Gamma_3^+$	$9\Gamma_1^- +9\Gamma_2^- +9\Gamma_3^-$
D_3	$10\Gamma_1+8\Gamma_2+18\Gamma_3$	$16\Gamma_1+20\Gamma_2+36\Gamma_3$	$4\Gamma_1+5\Gamma_2+9\Gamma_3$	$4\Gamma_1+5\Gamma_2+9\Gamma_3$
C_{3v}	$10\Gamma_1+8\Gamma_2+18\Gamma_3$	$20\Gamma_1+16\Gamma_2+36\Gamma_3$	$4\Gamma_1+5\Gamma_2+9\Gamma_3$	$5\Gamma_1+4\Gamma_2+9\Gamma_3$
D_{3d}	$10\Gamma_1^+ +8\Gamma_2^+ +18\Gamma_3^+$	$16\Gamma_1^- +20\Gamma_2^- +36\Gamma_3^-$	$4\Gamma_1^+ +5\Gamma_2^+ +9\Gamma_3^+$	$4\Gamma_1^- +5\Gamma_2^- +9\Gamma_3^-$
C_6	$12\Gamma_1+8\Gamma_2+8\Gamma_3+6\Gamma_4+10\Gamma_5+10\Gamma_6$	$20\Gamma_1+16\Gamma_2+16\Gamma_3+16\Gamma_4+20\Gamma_5+20\Gamma_6$	$7\Gamma_1+3\Gamma_2+3\Gamma_3+2\Gamma_4+6\Gamma_5+6\Gamma_6$	$7\Gamma_1+3\Gamma_2+3\Gamma_3+2\Gamma_4+6\Gamma_5+6\Gamma_6$
C_{3h}	$12\Gamma_1+8\Gamma_2+8\Gamma_3+6\Gamma_4+10\Gamma_5+10\Gamma_6$	$16\Gamma_1+20\Gamma_2+20\Gamma_3+20\Gamma_4+16\Gamma_5+16\Gamma_6$	$7\Gamma_1+3\Gamma_2+3\Gamma_3+2\Gamma_4+6\Gamma_5+6\Gamma_6$	$2\Gamma_1+6\Gamma_2+6\Gamma_3+7\Gamma_4+3\Gamma_5+3\Gamma_6$
C_{6h}	$12\Gamma_1^+ +8\Gamma_2^+ +8\Gamma_3^+ +6\Gamma_4^+ +10\Gamma_5^+ +10\Gamma_6^+$	$20\Gamma_1^- +16\Gamma_2^- +16\Gamma_3^- +16\Gamma_4^- +20\Gamma_5^- +20\Gamma_6^-$	$7\Gamma_1^+ +3\Gamma_2^+ +3\Gamma_3^+ +2\Gamma_4^+ +6\Gamma_5^+ +6\Gamma_6^+$	$7\Gamma_1^- +3\Gamma_2^- +3\Gamma_3^- +2\Gamma_4^- +6\Gamma_5^- +6\Gamma_6^-$
D_6	$7\Gamma_1+5\Gamma_2+3\Gamma_3+3\Gamma_4+10\Gamma_5+8\Gamma_6$	$8\Gamma_1+12\Gamma_2+8\Gamma_3+8\Gamma_4+20\Gamma_5+16\Gamma_6$	$3\Gamma_1+4\Gamma_2+\Gamma_3+\Gamma_4+6\Gamma_5+3\Gamma_6$	$3\Gamma_1+4\Gamma_2+\Gamma_3+\Gamma_4+6\Gamma_5+3\Gamma_6$
C_{6v}	$7\Gamma_1+5\Gamma_2+3\Gamma_3+3\Gamma_4+10\Gamma_5+8\Gamma_6$	$12\Gamma_1+8\Gamma_2+8\Gamma_3+8\Gamma_4+20\Gamma_5+16\Gamma_6$	$3\Gamma_1+4\Gamma_2+\Gamma_3+\Gamma_4+6\Gamma_5+3\Gamma_6$	$4\Gamma_1+3\Gamma_2+\Gamma_3+\Gamma_4+6\Gamma_5+3\Gamma_6$
D_{3h}	$7\Gamma_1+5\Gamma_2+3\Gamma_3+3\Gamma_4+10\Gamma_5+8\Gamma_6$	$8\Gamma_1+8\Gamma_2+8\Gamma_3+12\Gamma_4+16\Gamma_5+20\Gamma_6$	$3\Gamma_1+4\Gamma_2+\Gamma_3+\Gamma_4+6\Gamma_5+3\Gamma_6$	$\Gamma_1+\Gamma_2+3\Gamma_3+4\Gamma_4+3\Gamma_5+6\Gamma_6$
D_{6h}	$7\Gamma_1^+ +5\Gamma_2^+ +3\Gamma_3^+ +3\Gamma_4^+ +10\Gamma_5^+ +8\Gamma_6^+$	$8\Gamma_1^- +12\Gamma_2^- +8\Gamma_3^- +8\Gamma_4^- +20\Gamma_5^- +16\Gamma_6^-$	$3\Gamma_1^+ +4\Gamma_2^+ +\Gamma_3^+ +\Gamma_4^+ +6\Gamma_5^+ +3\Gamma_6^+$	$3\Gamma_1^- +4\Gamma_2^- +\Gamma_3^- +\Gamma_4^- +6\Gamma_5^- +3\Gamma_6^-$
T	$5\Gamma_1+5\Gamma_2+5\Gamma_3+13\Gamma_4$	$8\Gamma_1+8\Gamma_2+8\Gamma_3+28\Gamma_4$	$2\Gamma_1+2\Gamma_2+2\Gamma_3+7\Gamma_4$	$2\Gamma_1+2\Gamma_2+2\Gamma_3+7\Gamma_4$
T_h	$5\Gamma_1^+ +5\Gamma_2^+ +5\Gamma_3^+ +13\Gamma_4^+$	$8\Gamma_1^- +8\Gamma_2^- +8\Gamma_3^- +28\Gamma_4^-$	$2\Gamma_1^+ +2\Gamma_2^+ +2\Gamma_3^+ +7\Gamma_4^+$	$2\Gamma_1^- +2\Gamma_2^- +2\Gamma_3^- +7\Gamma_4^-$
O	$3\Gamma_1+2\Gamma_2+5\Gamma_3+6\Gamma_4+7\Gamma_5$	$3\Gamma_1+5\Gamma_2+8\Gamma_3+15\Gamma_4+13\Gamma_5$	$\Gamma_1+\Gamma_2+2\Gamma_3+4\Gamma_4+3\Gamma_5$	$\Gamma_1+\Gamma_2+2\Gamma_3+4\Gamma_4+3\Gamma_5$
T_d	$3\Gamma_1+2\Gamma_2+5\Gamma_3+6\Gamma_4+7\Gamma_5$	$5\Gamma_1+3\Gamma_2+8\Gamma_3+13\Gamma_4+15\Gamma_5$	$\Gamma_1+\Gamma_2+2\Gamma_3+4\Gamma_4+3\Gamma_5$	$\Gamma_1+\Gamma_2+2\Gamma_3+3\Gamma_4+4\Gamma_5$
O_h	$3\Gamma_1^+ +2\Gamma_2^+ +5\Gamma_3^+ +6\Gamma_4^+ +7\Gamma_5^+$	$3\Gamma_1^- +5\Gamma_2^- +8\Gamma_3^- +15\Gamma_4^- +13\Gamma_5^-$	$\Gamma_1^+ +\Gamma_2^+ +2\Gamma_3^+ +4\Gamma_4^+ +3\Gamma_5^+$	$\Gamma_1^- +\Gamma_2^- +2\Gamma_3^- +4\Gamma_4^- +3\Gamma_5^-$

Table A8

Elements of the $\nabla\boldsymbol{\eta}$-induced Raman tensor for the modes of Γ_4^- symmetry in crystals belonging to the crystal class O_h. Coefficients $a_1 \ldots a_{15}$ describe the fifteen independent components for the mode coefficient of eq. (18.13a), for $\Gamma(j)=\Gamma_4^-$. The matrices are symmetric (Humphreys and Maradudin 1972). (See § 18.5.)

$$\Delta \boldsymbol{a}_{j\nabla\eta} = \begin{Bmatrix} a_1\frac{\partial\eta_{xx}}{\partial x}+a_2\left(\frac{\partial\eta_{yy}}{\partial x}+\frac{\partial\eta_{zz}}{\partial x}\right)+2a_3\left(\frac{\partial\eta_{xy}}{\partial y}+\frac{\partial\eta_{xz}}{\partial z}\right) & a_4\frac{\partial\eta_{xx}}{\partial y}+a_5\frac{\partial\eta_{yy}}{\partial y}+a_6\frac{\partial\eta_{zz}}{\partial y}+2a_7\frac{\partial\eta_{xy}}{\partial x}+2a_8\frac{\partial\eta_{yz}}{\partial z} & a_4\frac{\partial\eta_{xx}}{\partial z}+a_6\frac{\partial\eta_{yy}}{\partial z}+a_5\frac{\partial\eta_{zz}}{\partial z}+2a_7\frac{\partial\eta_{xz}}{\partial x}+2a_8\frac{\partial\eta_{yz}}{\partial y} \\ & a_9\frac{\partial\eta_{xx}}{\partial x}+a_{10}\frac{\partial\eta_{yy}}{\partial x}+a_{11}\frac{\partial\eta_{zz}}{\partial x}+2a_{12}\frac{\partial\eta_{xy}}{\partial y}+2a_{13}\frac{\partial\eta_{xz}}{\partial z} & 2a_{14}\frac{\partial\eta_{yz}}{\partial x}+2a_{15}\left(\frac{\partial\eta_{xz}}{\partial y}+\frac{\partial\eta_{xy}}{\partial z}\right) \\ & & a_9\frac{\partial\eta_{xx}}{\partial x}+a_{11}\frac{\partial\eta_{yy}}{\partial x}+a_{10}\frac{\partial\eta_{zz}}{\partial x}+2a_{13}\frac{\partial\eta_{xy}}{\partial y}+2a_{12}\frac{\partial\eta_{xz}}{\partial z} \end{Bmatrix}, u_{j\alpha}=x$$

$$\begin{Bmatrix} a_{10}\frac{\partial\eta_{xx}}{\partial y}+a_9\frac{\partial\eta_{yy}}{\partial y}+a_{11}\frac{\partial\eta_{zz}}{\partial y}+2a_{12}\frac{\partial\eta_{xy}}{\partial x}+2a_{13}\frac{\partial\eta_{yz}}{\partial z} & a_5\frac{\partial\eta_{xx}}{\partial x}+a_4\frac{\partial\eta_{yy}}{\partial x}+a_6\frac{\partial\eta_{zz}}{\partial x}+2a_7\frac{\partial\eta_{xy}}{\partial y}+2a_6\frac{\partial\eta_{xz}}{\partial z} & 2a_{14}\frac{\partial\eta_{xx}}{\partial y}+2a_{15}\left(\frac{\partial\eta_{yz}}{\partial x}+\frac{\partial\eta_{xy}}{\partial z}\right) \\ & a_2\frac{\partial\eta_{xx}}{\partial y}+a_1\frac{\partial\eta_{yy}}{\partial y}+a_2\frac{\partial\eta_{zz}}{\partial y}+2a_3\left(\frac{\partial\eta_{xy}}{\partial x}+\frac{\partial\eta_{yz}}{\partial z}\right) & a_6\frac{\partial\eta_{xx}}{\partial z}+a_4\frac{\partial\eta_{yy}}{\partial z}+a_5\frac{\partial\eta_{zz}}{\partial z}+2a_8\frac{\partial\eta_{xz}}{\partial x}+2a_7\frac{\partial\eta_{yz}}{\partial y} \\ & & a_{11}\frac{\partial\eta_{xx}}{\partial y}+a_9\frac{\partial\eta_{yy}}{\partial y}+a_{10}\frac{\partial\eta_{zz}}{\partial y}+2a_{13}\frac{\partial\eta_{xy}}{\partial x}+2a_{12}\frac{\partial\eta_{yz}}{\partial z} \end{Bmatrix}, y$$

$$\begin{Bmatrix} a_{10}\frac{\partial\eta_{xx}}{\partial z}+a_{11}\frac{\partial\eta_{yy}}{\partial z}+a_9\frac{\partial\eta_{zz}}{\partial z}+2a_{12}\frac{\partial\eta_{xz}}{\partial x}+2a_{13}\frac{\partial\eta_{yz}}{\partial y} & 2a_{14}\frac{\partial\eta_{xy}}{\partial z}+2a_{15}\left(\frac{\partial\eta_{yz}}{\partial x}+\frac{\partial\eta_{xz}}{\partial y}\right) & a_5\frac{\partial\eta_{xx}}{\partial x}+a_6\frac{\partial\eta_{yy}}{\partial x}+a_4\frac{\partial\eta_{zz}}{\partial x}+2a_8\frac{\partial\eta_{xy}}{\partial y}+2a_7\frac{\partial\eta_{xz}}{\partial z} \\ & a_{11}\frac{\partial\eta_{xx}}{\partial z}+a_{10}\frac{\partial\eta_{yy}}{\partial z}+a_9\frac{\partial\eta_{zz}}{\partial z}+2a_{13}\frac{\partial\eta_{xz}}{\partial x}+2a_{12}\frac{\partial\eta_{yz}}{\partial y} & a_6\frac{\partial\eta_{xx}}{\partial y}+a_5\frac{\partial\eta_{yy}}{\partial y}+a_4\frac{\partial\eta_{zz}}{\partial y}+2a_8\frac{\partial\eta_{xy}}{\partial x}+2a_7\frac{\partial\eta_{yz}}{\partial z} \\ & & a_2\left(\frac{\partial\eta_{xx}}{\partial z}+\frac{\partial\eta_{yy}}{\partial z}\right)+a_1\frac{\partial\eta_{zz}}{\partial z}+2a_3\left(\frac{\partial\eta_{xz}}{\partial x}+\frac{\partial\eta_{yz}}{\partial y}\right) \end{Bmatrix}, z$$

References

Abstreiter, G. and E. Anastassakis (1977), unpublished.

Aggarwal, R. L., (1972), in *Semiconductors and semimetals*, Ed. by R. K. Willardson and A. C. Beer (Academic Press, New York), Vol. 9, p. 151.

Ahrens, K. and G. Schaack (1978a), in *Lattice dynamics*, Ed. by M. Balkanski (Flammarion, Paris), p. 257.

Ahrens, K. and G. Schaack (1978b), Ind. J. P. Appl. Phys. **16**, 311.

Allen, S. J. and H. J. Guggenheim (1968), Phys. Rev. Lett. **21**, 1807.

Anastassakis, E. (1969), Phys. Rev. **186**, 760.

Anastassakis, E. (1972), in *Atomic structure and properties of solids, Proceedings of the international school of physics "Enrico Fermi"*, Ed. by E. Burstein (Academic Press, New York), p. 294.

Anastassakis, E. (1973), J. Phys. C: Solid State Phys. **6**, 1870.

Anastassakis, E. (1974), Appl. Optics **13**, 1971.

Anastassakis, E. and E. Burstein (1970), Phys. Rev. B **2**, 1952.

Anastassakis, E. and E. Burstein (1971a), J. Phys. Chem. Solids **32**, 313.

Anastassakis, E. and E. Burstein (1971b), J. Phys. Chem. Solids **32**, 563.

Anastassakis, E. and E. Burstein (1971c), in *Light scattering in solids*, Ed. by M. Balkanski (Flammarion, Paris), p. 52.

Anastassakis, E. and E. Burstein (1972), J. Phys. C: Solid State Phys. **5**, 2468.

Anastassakis, E. and E. Burstein (1974), J. Phys. C: Solid State Phys. **7**, 134.

Anastassakis, E. and C. H. Perry (1971), Phys. Rev. B **4**, 1251.

Anastassakis, E. and P. Argyres (1974), Phys. Lett. **49A**, 457.

Anastassakis, E., S. Iwasa and E. Burstein (1966), Phys. Rev. Lett. **17**, 1051.

Anastassakis, E., A. Filler and E. Burstein (1969), in *Light scattering in solids*, Ed. by G. Wright (Springer-Verlag, New York), p. 421.

Anastassakis, E., A. Pinczuk, E. Burstein, F. Pollak and M. Cardona, (1970), Solid State Commun. **8**, 133.

Anastassakis, E., H. C. Hwang and C. H. Perry (1971), Phys. Rev. B **4**, 2493.

Anastassakis, E., E. Burstein, A. A. Maradudin and R. Minnick (1972a), J. Phys. Chem. Solids **33**, 519.

Anastassakis, E., E. Burstein, A. A. Maradudin and R. Minnick (1972b), J. Phys. Chem. Solids **33**, 1091.

Anastassakis, E., F. H. Pollak and G. W. Rubloff (1972c), in *Proceedings of the eleventh international conference on the physics of semiconductors* (Polish Scientific Publishers, Warsaw), p. 1188.

Anastassakis, E., F. H. Pollak and G. W. Rubloff (1974), Phys. Rev. B **9**, 551.

Angress, J. F. and A. J. Maiden (1971), J. Phys. C: Solid State Phys. **4**, 235.

Angress, J. F., C. Cooke and A. J. Maiden (1968), J. Phys. C (Proc. Phys. Soc.), Ser. 2, **1**, 1769.

Angress, J. F., G. A. Gledhill and A. J. Maiden (1971), in *Phonons*, Ed. by M. A. Nusimovici (Flammarion, Paris), p. 459.

Angress, J. F., G. A. Gledhill and A. J. Maiden (1975), J. Phys. C: Solid State Phys. **8**, 1136.

Asaumi, K. F. and S. Minomura (1978), J. Phys. Soc. Japan **45**, 1061.

Asell, J. F. and M. Nicol (1968), J. Chem. Phys. **49**, 5395.

Axe, J. D. (1965), Phys. Rev. **139**, 1215.

Axe, J. D., J. Harada and G. Shirane (1970), Phys. Rev. B **1**, 1227.

Bagguley, D. M. S., G. Vela-Coleiro, S. D. Smith and C. J. Summers (1966), in *Proceedings of*

eighth international conference on the physics of semiconductors, J. Phys. Soc. Japan, Vol. 21, Supplement, Kyoto.
Balkanski, M., M. K. Teng, M. Massot and S. M. Shapiro (1971), in *Light scattering in solids*, Ed. by M. Balkanski (Flammarion, Paris), p. 392.
Balslev, I. (1972), in *Semiconductors and semimetals*, Ed. by R. K. Willardson and A. C. Beer (Academic Press, New York), Vol. 9, p. 403.
Balslev, I. (1974), Phys. Rev. B **9**, 1707.
Baltensperger, W. and J. S. Herman (1968), Helv. Phys. Acta **41**, 668.
Bancewicz, T., S. Kielich and Z. Ozgo (1976), in *Light scattering in solids*, Ed. by M. Balkanski et al. (Flammarion, Paris), p. 432.
Barker, A. S., Jr. (1963), Phys. Rev. **132**, 1474.
Barker, A. S., Jr. (1964), Phys. Rev. **135**, A742.
Barker, A. S., Jr. and J. J. Hopfield (1964), Phys. Rev. **135**, A1732.
Barker, A. S., Jr. and A. J. Sievers (1975), Rev. Mod. Phys. **47**, Suppl. no. 2, 1.
Barron, L. D. (1976), in *Proceedings of the fifth international conference on Raman spectroscopy*, Ed. by E. D. Schmid et al. (H. F. Schulz Verlag, Freiburg), p. 677.
Becker, C. R. and T. P. Martin (1972), Phys. Rev. B **5**, 1604.
Bell, M. I. (1972), Phys. Stat. Sol. (b) **53**, 675.
Berenson, R. (1978), private communications.
Berenson, R. and J. L. Birman (1976), in *Light scattering in solids*, Ed. by M. Balkanski et al. (Flammarion, Paris), p. 437.
Bhagavantam, S. (1966), in *Crystal symmetry and physical properties* (Academic Press, New York).
Bienenstock, A. (1964), Phil. Mag. **9**, 755.
Billardon, M., M. F. Russel, J. P. Buisson and S. Lefrant (1976), J. de Phys.-Lettres **37**, L251.
Bir, G. L. and G. E. Pikus (1975), in *Symmetry and strain-induced effects in semiconductors* (Wiley, New York).
Birman, J. L. (1962), Phys. Rev. **127**, 1093.
Birman, J. L. (1963), Phys. Rev. **131**, 1489.
Birman, J. L. (1974a), in *Dynamical properties of solids*, Ed. by G. K. Horton and A. A. Maradudin, (North Holland, Amsterdam), Vol. 2, p. 97.
Birman, J. L. (1974b), in *Handbuch der Physik*, Ed. by L. Genzel, (Springer-Verlag, Berlin), Vol. XXV/2b.
Birman, J. L. (1974c), Phys. Rev. B **9**, 4518.
Birman, J. L. and R. Berenson (1974), Phys. Rev. **9**, 4512.
Birss, R. R. (1964), in *Symmetry and magnetism*, (North-Holland, Amsterdam).
Block, S. and G. Piermarini (1976), Physics Today, **29**, no. 9.
Bodenheimer, J. S., W. F. Sherman, G. R. Wilkinson and H. K. Böckelmann (1976), in *Light scattering in solids*, Ed. by M. Balkanski et al. (Flammarion, Paris), p. 923.
Borer, W. J., S. S. Mitra and K. V. Namjoshi (1971), Solid State Commun. **9**, 1377.
Born, M. and K. Huang (1962), in *Dynamical theory of crystal lattices* (Clarendon Press, Oxford).
Boyd, J. R. and J. D. Gavenda (1966), Phys. Rev. **152**, 645.
Bradley, C. J. and A. P. Cracknell (1972), in *The mathematical theory of symmetry in solids*, (Oxford University Press, London).
Bradley, C. J. and B. L. Davies (1968), Rev. Mod. Phys. **40**, 359.
Brafman, O. and M. Cardona (1977), Phys. Rev. B **15**, 1081.
Brafman, O. and S. S. Mitra (1971), in *Light scattering in solids*, Ed. by M. Balkanski (Flammarion, Paris), p. 287.

Brafman, O., S. S. Mitra, R. K. Crawford, W. B. Daniels, C. Postmus and J. R. Ferraro (1969), Solid State Commun. **7**, 449.
Brasch, J. W. and R. Jakobsen (1965), Spec. Acta **21**, 1183.
Briggs, R. J. and A. K. Ramdas (1976), Phys. Rev. B **13**, 5518.
Briggs, R. J. and A. K. Ramdas (1977), Phys. Rev. B **16**, 3815.
Brillson, L. and E. Burstein (1971), Phys. Rev. Lett. **27**, 808.
Brüsch, P. and F. D'Ambrogio (1972), Phys. Status Sol. (b) **50**, 513.
Buchenauer, J., F. Cerdeira and M. Cardona (1971), in *Light scattering in solids*, Ed. by M. Balkanski (Flammarion, Paris), p. 280.
Buchner, S., E. Burstein and A. Pinczuk (1976a), in *Light scattering in solids*, Ed. by M. Balkanski et al. (Flammarion, Paris), p. 76.
Buchner, S., L. Y. Ching and E. Burstein (1976b), Phys. Rev. B **14**, 4459.
Burke, W. J. and R. J. Pressley (1969), Solid State Commun. **7**, 1187.
Burke, W. J. and R. J. Pressley (1971), Solid State Commun. **9**, 191.
Burke, W. J., R. J. Pressley and J. C. Slonczewski (1971), Solid State Commun. **9**, 121.
Burstein, E. (1964), in *Proceedings of the international conference on lattice dynamics*, Ed. by R. F. Wallis (Pergamon Press, Copenhagen), p. 315.
Burstein, E. and S. Ganesan (1965), J. Phys. (Paris) **26**, 637.
Burstein, E. and A. Pinczuk (1971), in *The physics of opto-electronic materials*, Ed. by W. A. Albers (Plenum Press, New York), p. 33.
Burstein, E., A. A. Maradudin, E. Anastassakis and A. Pinczuk (1968), Helv. Phys. Acta **41**, 730.
Callaway, J. (1965), in *Energy band theory* (Academic Press, New York), p. 15.
Calleja, J. M., J. Kuhl and M. Cardona (1978), Phys. Rev. B **17**, 876.
Callen, H. (1968), Am. J. Phys. **36**, 735.
Callender, R. H., M. Balkanski and J. L. Birman (1971), in *Light scattering in solids*, Ed. by M. Balkanski (Flammarion, Paris), p. 40.
Cardona, M. (1969), in *Modulation spectroscopy*, (Academic Press, New York).
Cardona, M. (1973), Surf. Sci. **37**, 100.
Cardona, M. (1975), in *Light scattering in solids*, Ed. by M. Cardona (Springer-Verlag, Berlin), p. 1.
Cardona, M., K. L. Shaklee and F. H. Pollak (1967), Phys. Rev. **154**, 696.
Cerdeira, F. and M. Cardona (1972), Phys. Rev. B **5**, 1140.
Cerdeira, F., C. J. Buchenauer, F. H. Pollak and M. Cardona (1972), Phys. Rev. B **5**, 580.
Cerdeira, F., T. A. Fjeldly and M. Cardona (1973), Phys. Rev. B **8**, 4734.
Cerdeira, F., W. B. Holzapfel and D. Bäuerle (1975), Phys. Rev. B **11**, 1188.
Chandrasekhar, H. R., R. G. Humphreys and M. Cardona (1977), Phys. Rev. B **16**, 2981.
Chandrasekhar, M. and F. H. Pollak (1977), Phys. Rev. B **15**, 2127.
Chandrasekhar, M., J. B. Renucci and M. Cardona (1978), Phys. Rev. B **17**, 1623.
Chang, T. S., J. F. Holzrichter, J. F. Imbush and A. L. Schawlow (1970), Solid State Commun. **8**, 1177.
Chi, T. C. and R. J. Sladek (1973), Phys. Rev. B **7**, 5080.
Childs, M. S. and H. C. Longuet-Higgins (1961), Phil. Trans. Roy. Soc. London **254**, 259.
Childs, M. S. and H. C. Longuet-Higgins (1962), Phil. Trans. Roy. Soc. London **255**, 31.
Claus, R., L. Merten and J. Brandmüller (1975), in *Springer tracts in modern physics*, Ed. by G. Höhler, (Springer-Verlag, Berlin), Vol. 75, p. 207.
Clayman, B. P., R. D. Kirby and A. J. Sievers (1971), Phys. Rev. B **3**, 1351.
Cohen, M. H. and J. Ruvalds (1969), Phys. Rev. Lett. **23**, 1378.
Compaan, A. and H. Z. Cummins (1973), Phys. Rev. Lett. **31**, 41.
Compaan, A. et al. (1976), in *Light scattering in solids*, Ed. by M. Balkanski et al. (Flammarion, Paris), p. 39.

Condon, E. V. (1932), Phys. Rev. **41**, 759.

Corden, P. (1971), Ph.D. Thesis, University of Pennsylvania, unpublished.

Corden, P., A. Pinczuk and E. Burstein (1970), in *Proceedings of the 10th international conference on the physics of semiconductors* (U.S. Atomic Energy Commission, Washington, D. C.), p. 739.

Cowley, R. A. (1966), in *Phonons in perfect lattices and lattices with point imperfections*, Ed. by R. W. H. Stewenson (Plenum Press, New York).

Cracknell, A. P. (1966), Prog. Theor. Phys. **35**, 196.

Cracknell, A. P. (1968), Advan. Phys. **17**, 367.

Cracknell, A. P. (1969), J. Phys. C: Solid State Phys. **2**, 500.

Cracknell, A. P. (1975), in *Magnetism in crystalline materials* (Pergamon Press, Oxford).

Cracknell, A. P. and S. J. Joshua (1969), Proc. Camb. Phil. Soc. **66**, 493.

Crawford, M. F. and I. R. Dagg (1953), Phys. Rev. **91**, 1569.

Crawford, M. F. and I. R. MacDonald (1958), Can. J. Phys. **36**, 1022.

Crowther, P. A., P. J. Dean and W. F. Sherman (1967), Phys. Rev. **154**, 772.

Cummins, H. Z. (1967), in *Ferroelectricity*, Ed. by E. F. Weller (Elsevier, Amsterdam), p. 212.

Cundill, M. A. and W. F. Sherman (1968), Phys. Rev. **168**, 1007.

Dabrowski, I., P. Grünberg and J. A. Koningstein (1972), J. Chem. Phys. **56**, 1264.

Damen, T. C. and J. Shah (1970), Phys. Rev. Lett. **26**, 249.

Damen, T. C., S. P. S. Porto and B. Tell (1966), Phys. Rev. **142**, 570.

Davis, T. G. (1970), J. Phys. Soc. Japan **28**, Suppl. 245.

DeGennes, P. A. (1963), Solid State Commun. **1**, 132.

Devonshire, A. F. (1936), Proc. Roy. Soc. (London) **A153**, 601.

Dimmock, J. O. and R. G. Wheeler (1962a), J. Phys. Chem. Solids **23**, 729.

Dimmock, J. O. and R. G. Wheeler (1962b), Phys. Rev. **127**, 391.

Dimmock, J. O. and R. G. Wheeler (1964), in *The mathematics of physics and chemistry*, Ed. by H. Margenau and G. M. Murphy (Van Nostrand, New York), Vol. II.

Dines, T. J., M. J. French, R. J. B. Hall, and D. A. Long (1976), in *Proceedings of the fifth international conference on Raman spectroscopy*, Ed. by E. D. Schmid et al. (H. F. Schultz Verlag, Freiburg), p. 707.

Dolling, G. and R. A. Cowley (1966), Proc. Phys. Soc. **88**, 463.

Dreybrodt, W., W. Richter, F. Cerdeira and M. Cardona (1973), Phys. Stat. Sol. (b) **60**, 145.

Drickamer, H. G., R. W. Lynch, R. L. Clendenen and E. A. Perez-Albuerne (1966), Solid State Phys. **19**, 135.

Dvorak, V. (1967), Phys. **159**, 652.

Elcombe, M. M. and A. W. Pryor (1970), J. Phys. C: Solid State Phys. **3**, 492.

Evans, D. J. and S. Ushioda (1974), Phys. Rev. B **9**, 1638.

Falicov, L. M. (1967), in *Group theory and its physical applications*, (The University of Chicago Press, Chicago).

Fatuzzo, E. (1964), Proc. Phys. Soc. **84**, 709.

Ferraro, J. R., H. Horan and A. Quattrochi (1971), J. Chem. Phys. **55**, 664.

Fjeldly, T. A. and W. Richter (1975), Phys. Stat. Sol. (b) **72**, 555.

Fjeldly, T. A., F. Cerdeira and M. Cardona (1973), Phys. Rev. B **8**, 4723.

Fleury, P. A. (1968), Bull. Am. Phys. Soc. **12**, 420.

Fleury, P. A. (1971), J. Acoust. Soc. Am. **49**, 1041.

Fleury, P. A. (1972), Comments in Solid State Phys. **IV**, 167.

Fleury, P. A. and P. Lazay (1971), Phys. Rev. Lett. **26**, 1331.

Fleury, P. A. and R. Loudon (1968), Phys. Rev. **166**, 514.

Fleury, P. A. and J. M. Worlock (1967), Phys. Rev. Lett. **18** 665.

Fleury, P. A. and J. M. Worlock (1968), Phys. Rev. **174**, 613.

Fleury, P. A., J. F. Scott and J. M. Worlock (1968), Phys. Rev. Lett. **21**, 16.

Fleury, P. A., J. M. Worlock and H. J. Guggenheim (1969), Phys. Rev. **185**, 738.
Fritz, I. J. (1973), unpublished.
Fujii, Y., H. Uwe, H. Unoki and T. Sakudo (1970), Acta Crystallogr. **A28**, S230.
Galeener, F. L. (1976), in *Proceedings of the fifth international conference on Raman spectroscopy*, Ed. by E. D. Schmid et al. (H. F. Schulz Verlag, Freiburg), p. 754.
Ganesan, S. (1971), in *Light scattering in solids*, Ed. by M. Balkanski (Flammarion, Paris), p. 314.
Ganesan, S. and R. Srinivasan (1961), Can. J. Phys. **40**, 74.
Ganesan, S., A. A. Maradudin and J. Oitmaa (1970), Ann. Phys. (N. Y.), **56**, 556.
Ganguly, B. N., F. G. Ullman, R. D. Kirby and J. R. Hardy (1975), Phys. Rev. B **12**, 3783.
Gay, J. G., J. D. Dow, E. Burstein and A. Pinczuk (1971), in *Light scattering in solids*, Ed. by M. Balkanski (Flammarion, Paris), p. 33.
Genack, A. Z., H. Z. Cummins, M. A. Washington and A. Compaan (1975), Phys. Rev. B **12**, 2478.
Genack, A. Z., H. Z. Cummins, M. A. Washington and A. Compaan (1976), in *Light scattering in solids*, Ed. by M. Balkanski et al. (Flammarion, Paris), p. 34.
Go, S., H. Bilz and M. Cardona (1975), Phys. Rev. Lett. **34**, 580.
Gorman, M., J. Dohler and S. A. Solin (1974), Rev. Sci. Instrum. **45**, 1592.
Greenwald, R. and E. Anastassakis (1975), J. Opt. Soc. Am. **65**, 446.
Grimsditch, M. H. and A. K. Ramdas (1975), Phys. Rev. **11**, 3139.
Grimsditch, M. H. and A. K. Ramdas (1977), Phys. Rev. B. **14**, 1670.
Grimsditch, M. H., A. K. Ramdas, S. Rodriguez and V. J. Tekippe (1977), Phys. Rev. B **15**, 5869.
Grimsditch, M. H., E. Anastassakis and M. Cardona (1978), Phys. Rev. B **18**, 901.
Grimsditch, M. H., E. Anastassakis and M. Cardona (1979), Phys. Rev. B **19**, 3240.
Grosse, P. (1969), in *Springer tracts in modern physics*, No. 48, Ed. by G. Höhler (Springer-Verlag, Berlin), p. 59.
Grünberg, P., G. Güntherodt, A. Frey and W. Kress (1977), Physica B **89**, 225.
Güntherodt, G. (1976), in *Proceedings of the thirteenth international conference on the physics of semiconductors*, Ed. by F. G. Fumi (Typografia Marves, Rome), p. 291.
Hamermesh, M. (1964), in *Group theory and its applications to physical problems* (Addison-Wesley, Reading, Mass).
Hamilton, D. C. (1969), Phys. Rev. **188**, 1221.
Handler, P. and D. E. Aspnes (1966), Phys. Rev. Lett. **17**, 1095.
Harbeke, G., E. F. Steigmeier and R. K. Wehner (1970), Solid State Commun. **8**, 1765.
Harker, Y. D., C. Y. She and D. F. Edwards (1969), Appl. Phys. Lett. **15**, 272.
Harker, Y. D., C. Y. She and D. F. Edwards (1970), J. Appl. Phys. **41**, 5274.
Harley, R. T., C. H. Perry and W. Richter (1977), J. Phys. C: Solid State Phys. **10**, L187.
Hart, T. R., R. L. Aggarwal and B. Lax (1970), Phys. Rev. B **2**, 638.
Hastings, R. and J. Ruvalds (1977), private communications.
Hayashi, S. and H. Kanamori (1974), J. Phys. Soc. Japan, **37**, 1399.
Hayes, W. and H. F. MacDonald (1967), Proc. Roy. Soc. **A297**, 503.
Hayes, W., H. F. MacDonald and R. J. Elliot (1965), Phys. Rev. Lett. **15**, 691.
Hellwarth, R. W. (1977), Progr. Quant. Electr. **5**, 1.
Herzberg, G. (1964), in *Infrared and Raman spectra of polyatomic molecules* (D. Van Nostrand, Princeton).
Hobson, G. S. and E. G. S. Paige (1966), Proc. Phys. Soc. **88**, 437.
Hochheimer, H. D., M. L. Shand, J. E. Potts, P. C. Hanson and T. C. Walker (1976), Phys. Rev. B **14**, 4630.
Hoff, R. M. and J. C. Irwin (1974), Phys. Rev. B **10**, 3464.
Holz, A. (1972), Nuovo Cimento **9B**, 83.

Holzapfel, W. B. and E. Anastassakis (1974), unpublished.
Huff, H. R., S. Kawaji and H. C. Gatos (1968), Surface Sci. **12**, 53.
Humphreys, L. B. and A. A. Maradudin (1972), Solid State Commun. **11**, 1003.
Hurrel, J. P. and V. J. Minkiewicz (1970), Solid State Commun. **8**, 463.
Iliev, M. N., G. Güntherodt and H. Pink (1978a), Solid State Commun. **27**, 863.
Iliev, M. N., E. Anastassakis and T. Arai (1978b), Phys. Stat. Sol. (b) **79**, 717.
Iqbal, Z. and C. W. Christoe (1976), Chem. Phys. Lett. **37**, 460.
Iqbal, Z., D. S. Downs and C. W. Christoe, (1977), J. Phys. Chem. Solids **38**, 263.
Jaswal, S. S. (1965), Phys. Rev. **140**, A687.
Jex, H. (1971), Phys. Stat. Sol. (b) **45**, 343.
Joffrin, J. and A. Levelut (1970), Solid State Commun. **8**, 1573.
Johnston, W. D. and I. P. Kaminow (1968), Phys. Rev. **168**, 1045.
Johnson, F. A. and R. Loudon (1964), Proc. Roy. Soc. **A281**, 274.
Jones, C. E. and A. R. Hilton (1965), J. Electrochem. Soc. **112**, 908.
Jones, W. J. and B. P. Stoicheff (1964), Phys. Rev. Lett. **13**, 657.
Kahan, A. M., M. Patterson and A. J. Sievers (1976), Phys. Rev. B **14**, 5422.
Kanamori, H., N. Nakamori, S. Hayashi and Y. Saito (1976a), in *Light scattering in solids*, Ed. by M. Balkanski et al. (Flammarion, Paris), p. 428.
Kanamori, H., N. Nakamori, M. Matsumoto and S. Hayashi (1976b), in *Proceedings of the fifth international conference on Raman spectroscopy*, Ed. by E. D. Schmid et al. (H. F. Schulz Verlag, Freiburg), p. 648.
Kane, E. O. (1969), Phys. Rev. **178**, 1368.
Kaplyanskii, A. A. (1964a), Opt. Spektrosk. **16**, 329.
Kaplyanskii, A. A. (1964b), Opt. Spektrosk. **16**, 557; ibid. **16**, 1031.
Katiyar, R. S., J. F. Ryan and J. F. Scott (1971), in *Light scattering in solids*, Ed. by M. Balkanski (Flammarion, Paris), p. 436.
Keating, P. N. (1966), Phys. Rev. **144**, 637; ibid. **149**, 674.
Kennedy, J., W. F. Sherman, N. Treloar and G. R. Wilkinson (1976), in *Proceedings of the fifth international conference on Raman spectroscopy*, Ed. by E. D. Schmid et al. (H. F. Schulz Verlag, Freiburg), p. 600.
Keyes, R. W. (1976), in *Solid state physics*, Ed. by F. Seitz and D. Turnbull, and references therein, (Academic Press, New York), Vol. 20, p. 37.
Kiefer, W., W. Richter and M. Cardona (1975), Phys. Rev. B. **12**, 2346.
Kiel, A., T. Damen, S. Porto, S. Singh and F. Varsanyi (1969), Phys. Rev. **178**, 1518.
Kim, Q., F. G. Ullman, R. D. Kirby and J. R. Hardy (1978), in *Lattice dynamics*, Ed. M. Balkanski (Flammarion, Paris), p. 664.
Kleiner, W. H. (1969), Phys. Rev. **182**, 705.
Kluge, G. and G. Scholz (1965), Acoustica **16**, 60.
Kobayashi, K. (1968), J. Phys. Soc. Japan **24**, 497.
Kondo, K. and A. Moritani (1976), Phys. Rev. B **14**, 1577.
Koningstein, J. A. (1972), in *Introduction to the theory of the Raman effect* (D. Reidel Publishing Co., Dordrecht-Holland).
Konstantinov, L. L., V. P. Hinkov, J. I. Burov and M. I. Borissov (1976), J. Phys. C: Solid State Phys. **9**, 3557.
Koshizuka, N., Y. Yokouama and T. Tsushima (1976), Solid State Commun. **18**, 1333.
Koshizuka, N., Y. Yokouama and T. Tsushima (1977), Physica **89B**, 214.
Koster, G. F., J. O. Dimmock, R. G. Wheeler and H. Statz (1963), in *Properties of the thirty-two point groups*, (MIT Press, Cambridge, Mass.).
Kovalev, O. V. (1964), in *Irreducible representations of the space groups* (Gordon and Breach, New York).
Krishnamoorthy, N. and V. Soots (1972), Can. J. Phys. **50**, 1350.

Landau, L. D. and E. M. Lifshitz (1958), in *Statistical physics*, (Pergamon Press, New York).
Landau, L. D. and E. M. Lifshitz (1960), in *Electrodynamics of continuum media* (Pergamon, New York).
Lax, M. and E. Burstein (1953), Phys. Rev. **91**, 39.
Leite, R. C. C. and S. P. S. Porto (1966), Phys. Rev. Lett. **17**, 10.
Leite, R. C. C., T. C. Damen and J. F. Scott (1969), in *Light scattering in solids*, Ed. by G. Wright (Springer-Verlag, New York), p. 359.
Lemos, V., F. Cerdeira, M. A. F. Scarparo and R. S. Katiyar (1977), Phys. Rev. B **16**, 5560. Also, (1978) in *Lattice dynamics*, Ed. by M. Balkanski (Flammarion, Paris), p. 707.
Lenzo, P. V., E. G. Spencer and A. A. Ballman (1966), Appl. Optics **5**, 1688.
Leung, R. C., W. B. Spilman, N. E. Tornberg and R. P. Lowndes (1976), in *Light scattering in solids*, Ed. by M. Balkanski et al. (Flammarion, Paris), p. 796.
Levenson, M. (1977), Physics Today, May.
Lewis, J. L., R. C. Wadsack and R. K. Chang (1971), in *Light scattering in solids*, Ed. by M. Balkanski (Flammarion, Paris), p. 41.
Lippincott, E. R., L. S. Whatley and H. C. Duecker (1966), in *Applied infrared spectroscopy*, Ed. by D. N. Kendall (Reinhold, New York), p. 435.
Lisitsa, M. P., V. N. Malinko, Pidlisnyi and G. G. Tsebulya (1968), Surf. Sci. **11**, 411.
Loudon, R. (1963), Proc. Roy. Soc. (London) **A275**, 218.
Loudon, R. (1964), Advan. Phys. **13**, 423.
Lüth, H. (1969a), Solid State Commun. **7**, 585.
Lüth, H. (1969b), Phys. Stat. Sol. **33**, 267.
Lüth, H. (1970), Phys. Stat. Sol. **39**, 131.
MacQuillan, A. K., W. R. L. Clements and B. P. Stoicheff (1970), Phys. Rev. A **1**, 628.
Manlief, S. K. and H. Y. Fan (1972), Phys. Rev. B **5**, 4046.
Maradudin, A. A. and A. E. Fein (1962), Phys. Rev. **128**, 2589.
Maradudin, A. A. and E. Burstein (1967), Phys. Rev. **164**, 1081.
Maradudin, A. A. and S. H. Vosko (1968), Rev. Mod Phys. **40**, 1.
Maradudin, A. A., S. Ganesan and E. Burstein (1967), Phys. Rev. **163**, 882.
Maradudin, A. A., E. W. Montroll, G. H. Weiss and I. P. Ipatova (1971), in *Theory of lattice dynamics in the harmonic approximation*, 2nd Ed. (Academic Press, New York).
Marieé, M. and J. P. Mathieu (1946), C. R. Acad. Sci. (Paris) **223**, 147.
Martin, R. M. (1970), Phys. Rev. B **1**, 4005.
Martin, R. M. (1971), Phys. Rev. B **4**, 3676.
Martin, R. M. and T. C. Damen (1971), Phys. Rev. Lett. **26**, 86.
Martin, R. M. and L. M. Falikov (1975), in *Light scattering in solids*, Ed. by M. Cardona (Springer-Verlag, Berlin), p. 80.
Matsumoto, M. and H. Kanamori (1977), in *Memoirs of the Faculty of Industrial Arts*, Kyoto Technical University, Science and Technology, **26**, 29.
McSkimin, H. J. and P. Andreatch (1972), J. Appl. Phys. **43**, 2944.
McSkimin, H. J., P. Andreatch and R. N. Thurston (1965), J. Appl. Phys. **36**, 1624.
Merlin, R., R. Zeyher and G. Güntherodt (1977), Phys. Rev. Lett. **39**, 1215.
Messiah, A. (1962), in *Quantum mechanics*, (Wiley, New York), Vol. II.
Mills, D. L. and S. Ushioda (1970), Phys. Rev. B **2**, 3805.
Mills, D. L., A. A. Maradudin and E. Burstein (1970), Ann. Phys. (New York) **56**, 504.
Mitra, S. S., O. Brafman, W. B. Daniels and R. K. Crawford (1969), Phys. Rev. **186**, 942.
Moeller, W., R. Kaiser and H. Bilz (1970), Phys. Lett. **32A**, 171.
Moryia, T. (1966), J. Phys. Soc. Japan, **21**, 926.
Muelier, H. (1935), Phys. Rev. **47**, 947.

Müller, K. A., W. Berlinger and J. C. Slonczewski (1970), Phys. Rev. Lett. **25**, 734.
Nakayama, T. and A. Odajima (1972), J. Phys. Soc. Japan **33**, 12.
Namjoshi, K. V., S. S. Mitra and J. F. Vetelino (1971), Phys. Rev. B **3**, 4398.
Nelson, D. F and M. Lax (1970), Phys. Rev. Lett. **24**, 379.
Nill, K. W. and A. Mooradian (1969), Am. Phys. Soc. **13**, 1658.
Nolt, I. G. and A. J. Sievers (1966), Phys. Rev. Lett. **16**, 1103.
Nolt, I. G. and A. J. Sievers (1968), Phys. Rev. **174**, 1004.
Nye, J. F. (1964), in *Physical properties of crystals* (Clarendon Press, Oxford).
Onari, S. and M. Cardona (1976), Phys. Rev. B **14**, 3520.
Onari, S., E. Anastassakis and M. Cardona (1976), in *Light scattering in solids*, Ed. by M. Balkanski et al. (Flammarion, Paris), p. 54.
Onari, S., M. Cardona, E. Schönherr and W. Stetter (1977), Phys. Stat. Sol. (b) **79**, 269.
Onoe, M., A. W. Warner and A. A. Ballman (1967), IEEE Trans. Sonics Ultrason. **SU-14**, 165.
Onyango, F. N. (1976), J. Phys. C: Solid State Phys. **9**, L533.
Ousaka, Y., O. Sakai and M. Tachiki (1977), Solid State Commun. **23**, 589.
Ovander, L. N. (1960), in *Optics and spectroscopy*, **9**, 302.
Parker, J. H., Jr., D. W. Feldman and M. Ashkin (1967), Phys. Rev. **155**, 712.
Parsons, B. J. (1977), Proc. Roy. Soc. London **A352**, 397.
Parsons, B. J. and C. D. Clark (1976), in *Light scattering in solids*, Ed. by M. Balkanski et al. (Flammarion, Paris), p. 414.
Payne, R. T. (1964), Phys. Rev. Lett. **13**, 53.
Peercy, P. S. (1973), Phys. Rev. B **8**, 6018.
Peercy, P. S. (1976), in *Light scattering in solids*, Ed. by M. Balkanski et al. (Flammarion, Paris), p. 782.
Peercy, P. S., I. J. Fritz and G. A. Samara (1975), J. Phys. Chem. Solids, **36**, 1105.
Perry, C. H. and N. E. Tornberg (1969), Phys. Rev. **183**, 595.
Piermarini, G. J. and S. Block (1975), Rev. Sci. Instrum. **46**, 973.
Pinczuk, A. and E. Burstein (1969), in *Light scattering in solids*, Ed. by G. Wright (Springer-Verlag, New York), p. 429.
Pinczuk, A. and E. Burstein (1973), Surface Sci. **37**, 153.
Pinczuk, A. and E. Burstein (1975), in *Light scattering in solids*, Ed. by M. Cardona, (Springer-Verlag, Berlin), p. 23.
Pine, A. S. (1970), Phys. Rev. B **2**, 2049.
Pine, A. S. and G. Dresselhaus (1969), Phys. Rev. **188**, 1489.
Pine, A. S. and G. Dresselhaus (1971), Phys. Rev. B **4**, 356.
Placzek, G. (1934), in *Handbuch der Radiologie* (Academische Verlagsgesellschaft, Leipzig), Vol. 6, Part 2, p. 205.
Pollak, F. H. (1973), in *Phase transitions–1973*, Ed. by L. E. Cross, (Pergamon Press, New York), p. 243.
Portigal, D. L. and E. Burstein (1968), Phys. Rev. **170**, 673.
Porto, S. P. S. and J. F. Scott (1967), Phys. Rev. **157**, 716.
Porto, S. P. S., P. A. Fleury and T. C. Damen (1967), Phys. Rev. **154**, 522.
Postmus, C., J. R. Ferraro and S. S. Mitra (1968a), Inorg. Nucl. Chem. Lett. **4**, 55.
Postmus, C., J. R. Ferraro and S. S. Mitra (1968b), Phys. Rev. **174**, 983.
Postmus, C., V. A. Maroni, J. R. Ferraro and S. S. Mitra (1968c), Inorg. Nucl. Chem. Lett. **4**, 269.
Pytte, E. (1974), Phys. Rev. B **9**, 932.
Quinn, J. J. and S. Rodriguez (1964), Phys. Rev. **133A**, 1589.

Ramachandran, G. N. (1950), Proc. Indian Acad. Sci. **32A**, 171.
Ray, R. K., R. L. Aggarwal and B. Lax (1971), in *Light scattering in solids*, Ed. by M. Balkanski (Flammarion, Paris), p. 288.
Renucci, J. B. (1974), Ph.D. Thesis, L'Université Paul Sabatier de Toulouse, unpublished.
Renucci, J. B., M. A. Renucci and M. Cardona (1971a), Solid State Commun. **9**, 1235.
Renucci, J. B., M. A. Renucci and M. Cardona (1971b), Solid State Commun. **9**, 1651.
Renucci, M. A., J. B. Renucci and M. Cardona (1972), Phys. Stat. Sol. (b) **49**, 625.
Richter, W. (1976), in *Springer tracts in modern physics*, Ed. by G. Höhler (Springer-Verlag, Berlin), Vol. 78, p. 121.
Richter, W. and R. Zeyher (1976), in *Festkörperprobleme XVI*, Ed. by J. Treusch, (Vieweg, Braunschweig), p. 15.
Richter, W., J. B. Renucci and M. Cardona (1973), Phys. Stat. Sol. (b) **56**, 223.
Richter, W., J. B. Renucci and M. Cardona (1975), Solid State Commun. **16**, 131.
Richter, W., R. Zeyher and M. Cardona (1976), in *Light scattering in solids*, Ed. by M. Balkanski et al. (Flammarion, Paris), p. 63.
Richter, W., T. Fjeldly, J. Renucci and M. Cardona, 1978a, in *Lattice dynamics*, Ed. by M. Balkanski (Flammarion, Paris), p. 104.
Richter, W., R. Zeyher and M. Cardona (1978b), Phys. Rev. B **18**, 4312.
Rodriguez, S., P. Fisher and F. Barra (1972), Phys. Rev. B **5**, 2219.
Rokni, M. and L. S. Wall (1971), J. Chem. Phys. **55**, 435.
Rousseau, D. L. and S. P. S. Porto (1968), Phys. Rev. Lett. **20**, 1354.
Rowe, J. E., M. Cardona and F. H. Pollak (1967), in *II–VI semiconducting compounds*, Ed. by D. G. Thomas (Benjamin, New York), p. 112.
Rubloff, G. W., E. Anastassakis and F. H. Pollack (1973), Solid State Commun. **13**, 1755.
Rudra, P. and M. K. Sikdar (1976), J. Phys. C: Solid State Phys. **9**, 1.
Russel, J. P. (1965), Appl. Phys. Lett. **6**, 223.
Russel, J. P. and R. Loudon (1965), Proc. Phys. Soc. **85**, 1029.
Safran, S. A., G. Dresselhaus and B. Lax (1976), Solid State Commun. **19**, 1217; Erratum, (1976), **20**, vii.
Safran, S. A., G. Dresselhaus, M. S. Dresselhaus and B. Lax (1977), Physica **89B**, 229.
Samara, G. A. and P. S. Peercy (1973), Phys. Rev. B **7**, 1131.
Samara, G. A., L. C. Walters and D. A. Northrop (1967), Phys. Chem. Solids **28**, 1875.
Schaack, G. (1975), Solid State Commun. **17**, 505.
Schaack, G. (1976a), J. Phys. C: Solid State Phys. **9**, L297.
Schaack, G. (1976b), in *Light scattering in solids*, Ed. by M. Balkanski et al. (Flammarion, Paris), p. 372.
Schaack, G. (1977a), Z. Phys. B **26**, 49.
Schaack, G. (1977b), Physica **89B**, 195.
Schaufele, R. F., M. J. Weber and B. D. Silverman (1967), Phys. Lett. **25A**, 47.
Schiff, L. I. (1955), in *Quantum mechanics*, (McGraw-Hill, New York), p. 154.
Schmidt, R. L., B. D. McCombe and M. Cardona (1975), Phys. Rev. B **11**, 746.
Schneider, W. C. (1970), Ph.D. Thesis, Pennsylvania State University, unpublished.
Scott, J. F. (1968a), J. Chem. Phys. **48**, 874.
Scott, J. F. (1968b), Phys. Rev. Lett. **21**, 907.
Scott, J. F. (1969), Phys. Rev. **183**, 823.
Scott, J. F. (1970), Phys. Rev. Lett. **24**, 1107.
Scott, J. F. (1971), in *Light scattering in solids*, Ed. by M. Balkanski (Flammarion, Paris), p. 387.
Scott, J. F. (1974), Rev. Mod. Phys. **46**, 83.
Scott, J. F. and J. P. Remeika (1970), Phys. Rev. B **1**, 4182.
Scott, J. F., P. A. Fleury and J. M. Worlock (1969), Phys. Rev. **177**, 1288.

Seraphin, B. O. (1972), in *Semiconductors and semimetals*, Ed. by R. K. Willardson and A. C. Beer (Academic Press, New York), Vol. 9, p. 1.
Shand, M. and E. Burstein (1973), Surface Sci. **37**, 145.
Shand, M. L. and R. C. Hanson (1977), Bull. Am. Phys. Soc. **22**, 340.
Shand, M., W. Richter, E. Burstein and J. G. Gay (1972), J. Nonmetals, **1**, 53.
Shand, M. L., H. D. Hochheimer, M. Krauzman, J. E. Potts, R. C. Hanson and C. T. Walker (1976), Phys. Rev. B **14**, 4637.
Shapiro, S. M. (1969), Ph.D. Thesis, John Hopkins University, unpublished.
Shepherd, I. W. (1972), Phys. Rev. B **5**, 4524.
Sherman, W. F., P. P. Smulovitch, G. J. Lewis and W. F. Sherman (1971), in *Phonons*, Ed. by M. A. Nusimovici (Flammarion, Paris), p. 336.
Shin, S. H., F. H. Pollak and P. M. Raccah (1976), in *Light scattering in solids*, Ed. by M. Balkanski et al. (Flammarion, Paris), p. 401.
Shirane, G. and Y. Yamada (1969), Phys. Rev. **177**, 858.
Silberstein, R. P., L. E. Schmutz, V. J. Tekippe, M. S. Dresselhaus and R. L. Aggarwal (1976), Solid State Commun. **18**, 1173.
Simonyi, E. and P. M. Raccah (1973), Bull. Am. Phys. Soc. **18**, 339.
Sinha, K. P. and U. N. Upadhyaya (1962), Phys. Rev. **127**, 432.
Skelton, E. F., J. L. Feldman, C. Y. Liu and I. L. Spain (1976), Phys. Rev. B **13**, 2605.
Slonczewski, J. C. (1970), Phys. Rev. B **2**, 4646.
Slonczewski, J. C. and H. Thomas (1970), Phys. Rev. B **1**, 3599.
Sokoloff, J. B. (1972), J. Phys. C: Solid State Phys. **5**, 2482.
Solin, S. A. and A. K. Ramdas (1970), Phys. Rev. B **1**, 1687.
Spiro, T. G. and T. C. Strekas (1972), Proc. Nat. Acad. Sci. USA. **69**, 2622.
Srinivasan, R. (1958), Proc. Phys. Soc. (London) **72**, 566.
Steigmeier, E. F. and G. Harbeke (1970), Phys. Kondens. Materie, **21**, 1.
Strauss, E. and H. J. Riederer (1977), Solid State Commun. **21**, 429.
Suzuki, N. (1976), J. Phys. Soc. Japan, **40**, 1223.
Suzuki, N. and H. Kamimura (1973), J. Phys. Soc. Japan, **35**, 985.
Swanson, L. R. and A. A. Maradudin (1970), Solid State Commun. **8**, 859.
Szigeti, B. (1965), Diamond Conference, unpublished.
Taylor, D. W. (1975), in *Dynamical properties of solids*, Ed. by G. K. Horton and A. A. Maradudin (North Holland, New York), Vol. 2, p. 285.
Tekippe, V. J. and A. K. Ramdas (1971), Phys. Lett. **35A**, 143.
Tekippe, V. J., A. K. Ramdas and S. Rodriguez (1973), Phys. Rev. B **8**, 706.
Tekippe, V. J., R. P. Silberstein, M. S. Dresselhaus and R. L. Aggarwal (1974a), Phys. Lett. **49A**, 295.
Tekippe, V. J., R. P. Silberstein, M. S. Dresselhaus and R. L. Aggarwal (1974b), in *Proceedings of the 12th international conference on the physics of semiconductors,* Ed. by M. H. Pilkuhn (B. G. Teubner, Stuttgart), p. 449.
Templeton, T. L. and B. P. Clayman (1971), Solid State Commun. **9**, 697.
Templeton, T. L. and B. P. Clayman (1972), Phys. Rev. B **6**, 4004.
Thalmeier, P. and P. Fulde (1977), Z. Phys., B **26**, 323.
Thomas, H. and K. A. Müller (1968), Phys. Rev. Lett. **21**, 1256.
Tinkham, M. (1964), in *Group theory and quantum mechanics* (McGraw-Hill, New York).
Trommer, R., E. Anastassakis and M. Cardona (1976), in *Light scattering in solids*, Ed. by M. Balkanski et al. (Flammarion, Paris), p. 396.
Trommer, R., G. Abstreiter and M. Cardona (1978), in *Lattice dynamics*, Ed. by M. Balkanski, (Flammarion, Paris), p. 189.
Tsang, J. C., M. S. Dresselhaus, R. L. Aggarwal and T. B. Reed (1974a), Phys. Rev. B **9**, 984.
Tsang, J. C., M. S. Dresselhaus, R. L. Aggarwal and T. B. Reed (1974b), Phys. Rev. B **9**, 997.

Tubino, R. and J. L. Birman (1976), in *Light scattering in solids*, Ed. by M. Balkanski et al. (Flammarion, Paris), p. 419.
Tubino, R. and J. L. Birman (1977), Phys. Rev. B **15**, 5843.
Tubino, R. and L. Piseri (1975), Phys. Rev. B **11**, 5145.
Tuomi, T., M. Cardona and F. H. Pollak (1970), Phys. Stat. Sol. **40**, 227.
Ullman, F. G., B. N. Ganguly, J. R. Hardy and R. D. Kirby (1976), in *Light scattering in solids*, Ed. by M. Balkanski et al. (Flammarion, Paris), p. 948.
Unoki, H. and T. Sakudo (1967), J. Phys. Soc. Japan, **23**, 546.
Uwe, H. and T. Sakudo (1975), J. Phys. Soc. Japan, **38**, 183.
Uwe, H. and T. Sakudo (1976), Phys. Rev. B **13**, 271.
Uwe, H. and T. Sakudo (1977), Phys. Rev. B **15**, 337.
Van der Lage, F. C. and H. Bethi (1947), Phys. Rev. **71**, 612.
Vedam, K. and T. A. Davis (1968), J. Opt. Soc. Am. **58**, 1451.
Vella-Coleiro, G. P. (1969), Phys. Rev. Lett. **23**, 697.
Venugopalan, S. and A. K. Ramdas (1972), Phys. Rev. **5**, 4065.
Venugopalan, S. and A. K. Ramdas (1973), Phys. Rev. B **8**, 717.
Vetelino, J. F., S. S. Mitra and K. V. Namjoshi (1970), Phys. Rev. B **2**, 967.
Vitings, J. and P. Wachter (1975), Solid State Commun. **17**, 911.
Vogt, H. and G. Neumann (1976), Optics Commun. **19**, 108.
Vogt, H. and G. Neumann (1978), in *Lattice dynamics*, Ed. by M. Balkanski (Flammarion, Paris), p. 780.
Wachtman, J. B., Jr., W. E. Tefft, D. G. Lam, Jr. and R. P. Stinchfield (1960), J. Natl. Bur. Std. **64A**, 213.
Waeber, W. B. (1972), J. Phys. C: Solid State Phys. **5**, 1773.
Wall, L. S., M. Rokni and A. L. Schawlow (1971), Solid State Commun. **9**, 573.
Wallis, R. F. and A. A. Maradudin (1971), Phys. Rev. B **3**, 2063.
Warren, J. L. (1968), Rev. Mod. Phys. **40**, 38.
Washington, M. A. and H. Z. Cummins (1977), Phys. Rev. B **15**, 5840.
Waterman, P. C. (1959), Phys. Rev. **113**, 1240.
Waters, R. and H. Z. Cummins (1977), private communications.
Weinstein, B. A. and M. Cardona (1972), Phys. Rev. B **5**, 3120.
Weinstein, B. A. and M. Cardona (1973), Phys. Rev. B **8**, 2795 (1973).
Weinstein, B. A. and G. J. Piermarini (1974), Phys. Lett. **48A**, 14.
Weinstein, B. A. and G. J. Piermarini (1975), Phys. Rev. B **12**, 1172.
Weinstein, B. A., J. B. Renucci and M. Cardona (1973), Solid State Commun. **12**, 473.
Welber, B., M. Cardona, C. K. Kim and S. Rodriguez (1976), Phys. Rev. B **12**, 5729.
Welkowski, M. and R. Braunstein (1972), Phys. Rev. B **5**, 497.
Whalley, E., A. Lavergne and P. Wong (1976), Rev. Sci. Instr. **47**, 845.
Wigner, E. P. (1959), in *Group theory* (Academic Press, New York), ch. 26.
Williams, P. F. and S. P. S. Porto (1973), Phys. Rev. B **8**, 1782.
Wilson, E. B., J. C. Decius and P. C. Cross (1955), in *Molecular vibrations* (McGraw-Hill, New York), p. 331.
Worlock, J. M. (1969), in *Light scattering in solids*, Ed. by G. Wright (Springer-Verlag, New York), p. 411.
Worlock, J. M. (1971), in *Structural phase transitions and soft modes*, Ed. by E. J. Samuelsen et al. (Universitetsforlaget, Oslo), p. 329.
Worlock, J. M. and P. A. Fleury (1967), Phys. Rev. Lett. **19**, 1176.
Worlock, J. M., J. F. Scott and P. A. Fleury (1969), in *Light scattering spectra of solids*, Ed. by G. Wright (Springer-Verlag, New York), p. 689.
Yacoby, Y. and A. Linz (1974), Phys. Rev. B **9**, 2723.

Yamade, Y. and G. Shirane (1969), J. Phys. Soc. Japan **26**, 396.

Young, A. P. and R. J. Elliot (1974), J. Phys. C: Solid State Phys. **7**, 2721.

Yu, P. Y. (1976), in *Light scattering in solids*, Ed. by M. Balkanski et al. (Flammarion, Paris), p. 19.

Yu, W. and R. R. Alfano (1975), Phys. Rev. A **11**, 188.

Yu, P. Y. and Y. R. Shen (1974), in *Proceedings of the 12th international conference on the physics of semiconductors*, Ed. by M. H. Pilkuhn (B. G. Teubner, Stuttgart), p. 453.

Yu, P. Y and Y. R. Shen (1975), Phys. Rev. B **12**, 1377.

Yu, P. Y. and B. Welber (1977), Bull. Am. Phys. Soc. **22**, 316; also (1978), Solid State Commun. **25**, 209.

Zallen, R. (1974), Phys. Rev. B **9**, 4485.

Zeyher, R. (1975), Solid State Commun. **16**, 49.

Zigone, M., R. Beserman and H. D. Fair, Jr. (1976), in *Light scattering in solids*, Ed. by M. Balkanski et al. (Flammarion, Paris), p. 597.

Zubov, V. G., L. P. Osipoval and E. V. Firsova (1961), Sov. Phys.–Cryst. **6**, 623.

CHAPTER 4

The Absorption of Infrared Radiation by Multiphonon Processes in Solids

D. L. MILLS

Department of Physics,
University of California
Irvine, California 92717, USA

and

C. J. DUTHLER AND M. SPARKS

Xonics, Inc., 1333 Ocean Boulevard
Santa Monica, California 90401, USA

Dynamical Properties of Solids, edited by
G. K. Horton and A. A. Maradudin

Contents

1. *Introduction*

The study of the absorption of electromagnetic radiation by solids is a topic that has occupied a central position in solid state physics for many decades. This is because the absorption spectrum contains a wealth of information about the nature of the elementary excitations of the solid and the interactions between them.

The development of high-power laser sources within the last few years has generated considerable interest in the study of very weak absorption processes difficult to detect by conventional spectroscopic methods. From the point of view of materials technology, it is crucial to understand the nature of these weak absorptions, since they may produce appreciable heating of elements in optical systems coupled to high-power lasers. While the impetus for the recent increased interest in small residual absorptions in high-purity, nominally transparent materials has its origin in the practical concern just outlined, the underlying physics is very fundamental in nature and quite fascinating in our view.

In this article, we discuss and review recent work on one important topic in this area: the absorption of infrared radiation by multiphonon processes. We begin with a brief outline of the phenomenon.

If one examines the absorption spectrum of a simple, insulating crystalline material from the infrared, through the visible and into the ultraviolet region of the spectrum, the principal features appear as sketched in fig. 1. In the infrared region, one encounters strong, narrow absorption lines from direct excitation of long wavelength, infrared active optical phonons, if crystal symmetry permits the presence of such modes. These absorption lines have widths from a few wave numbers to a few tens of wave numbers, and the absorption coefficient β assumes values the order of $10^5\ \mathrm{cm}^{-1}$ near the center of a line. The figure is a schematic illustration of the absorption spectrum for a crystal with a single infrared active transverse optical phonon of frequency ω_0.

At frequencies above those characteristic of the lattice absorption lines, one observes the onset of electronic absorption at the photon energy $\hbar\omega = E_g$, with E_g the energy gap between the valence and conduction band.

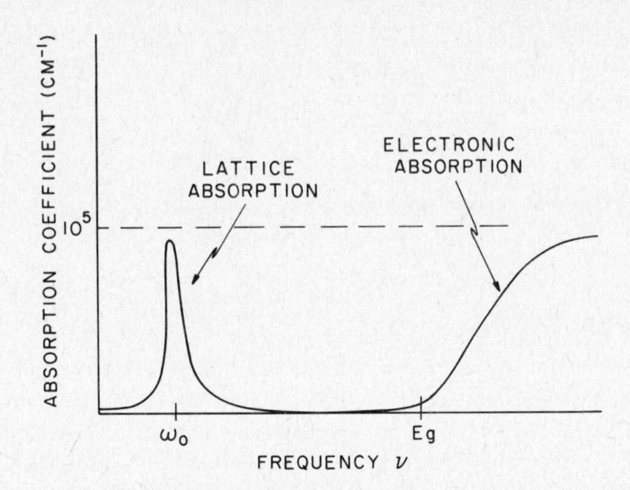

Fig. 1. A schematic illustration of the absorption spectrum of a simple insulating crystal. The figure shows a lattice absorption line (Reststrahl absorption) in the infrared, and the onset of electronic absorption by interband processes at higher frequencies.

In the region of strong electronic absorption, β again assumes values the order of 10^5 cm^{-1}, in typical cases. While the above description is schematic and over-simplified, nonetheless a large number of insulating materials are characterized by an absorption spectrum of this nature, with the electronic edge E_g in the visible region or beyond.

For frequencies above the fundamental lattice absorption peak, but below E_g, the material is nominally transparent, the value of the intrinsic absorption coefficient being well below the currently measurable values, say $\beta \ll 10^{-5}$ cm^{-1}. The value of β in this nominally transparent region is then determined by extrinsic absorption processes. Just below E_g, in the "Urbach tail" of the fundamental electronic absorption edge*, the absorption constant varies with frequency like $\exp[-A(E_g - \hbar\omega)]$, with A strongly dependent on temperature and sample purity.

Also, there is a high-frequency "wing" to the lattice absorption line at ω_0. As we shall see below, one can measure the absorption coefficient in this wing out to several times the fundamental lattice absorption frequency ω_0. The focus of the present article is this high-frequency wing of the fundamental lattice absorption line. The dominant intrinsic absorption mechanisms in the wing are the multiphonon processes, in which the incident photon of energy $\hbar\omega \gg \hbar\omega_0$ is absorbed with the creation of n

*For a discussion of mechanisms that contribute to the Urbach tail, see Dow and Redfield (1972).

phonons, with $n \approx \omega/\omega_0$. It has been a challenge to the theorist to provide a simple and useful description of such absorption processes, particularly since the experiments show that for a large number of materials the shape of the high-frequency wing is remarkably simple and universal, as we shall see.

The above remarks presume that the high-frequency wing of the lattice absorption and the Urbach tail arise from physically distinct processes. In fact, as one moves from $\hbar\omega_0$ to E_g, the intrinsic absorption coefficient $\beta(\omega)$ is a continuous, smooth function of frequency; the high-frequency wing of the lattice absorption line produced by multiphonon processes overlaps the Urbach tail, and there is clearly a frequency region where the electronic processes responsible for the Urbach tail cannot be separated from the high-frequency intrinsic lattice absorption. In wide-band-gap materials such as the alkali halides, for which the electronic edge lies well into or beyond the visible and the lattice absorptions lie in the range of a few hundred wave numbers, a clean separation between the two kinds of processes can be made for all frequencies for which the intrinsic absorption coefficient may be measured in practice. In narrow-gap materials, however, the situation can be much less clear. In the present article, we shall be concerned only with wide-band-gap materials for which the high-frequency wing of the intrinsic lattice absorption is clearly controlled by multiphonon processes.

It is useful to examine some simple numbers characteristic of the phenomenon. In the recent literature, the alkali halide KCl has been extensively studied, since it is highly transparent in the infrared region and it may prove a useful window material for infrared-laser systems. For KCl, the transverse optical phonon frequency is 143 cm^{-1}. The wavelength of the principal CO_2 laser line is 10.6 μm (943 cm^{-1}) so we have here $\omega/\omega_{TO} \approx 6.6$. The absorption of a CO_2 laser photon therefore requires creation of some six or seven phonons in the material. The Urbach tail is of little concern here since the band gap of KCl lies well beyond the visible region.

The absorption coefficient $\beta(\omega)$ quite generally may be related to $\varepsilon_2(\omega)$, the imaginary part of the frequency-dependent dielectric constant. One has, if the absorption is weak so that $\varepsilon_2(\omega) \ll \varepsilon_1(\omega)$ is satisfied, where $\varepsilon_1(\omega)$ the real part of the dielectric constant,

$$\beta(\omega) = \frac{\omega}{c} \frac{\varepsilon_2(\omega)}{\sqrt{\varepsilon_1(\omega)}} \cong \frac{\omega}{c\sqrt{\varepsilon_\infty}} \varepsilon_2(\omega). \tag{1.1}$$

In the last step, $\varepsilon_1(\omega)$ has been approximated by ε_∞, the high-frequency dielectric constant.

The simplest model of the lattice absorption in a crystal with a single infrared active tranverse-optical phonon is the damped simple harmonic-oscillator model, for which the normal coordinate $\boldsymbol{u}$ of the long wavelength optical phonon obeys the equation of motion of a damped harmonic oscillator of frequency ω_{TO} and damping constant γ. If M is the reduced mass of the unit cell, e^* the transverse effective charge, and n_c the number of unit cells per unit volume, this model gives

$$\beta(\omega)=\frac{4\pi n_c e^{*2}\omega^2}{Mc\sqrt{\varepsilon_\infty}}\frac{\gamma}{(\omega^2-\omega_{TO}^2)^2+\omega^2\gamma^2}. \tag{1.2}$$

For $\omega \gg \omega_{TO}$, eq. (1.2) reduces to

$$\beta(\omega)=\frac{4\pi n_c e^{*2}}{Mc\sqrt{\varepsilon_\infty}}\frac{\gamma}{\omega^2}, \tag{1.3}$$

an absorption coefficient that falls off as ω^{-2}, if γ is presumed independent of frequency as in many analyses. This result may be rewritten

$$\beta(\omega)=\frac{(\varepsilon_s-\varepsilon_\infty)}{c\sqrt{\varepsilon_\infty}}\frac{\gamma\omega_{TO}^2}{\omega^2}, \tag{1.4}$$

and for KCl we estimate† $\beta \approx 2\ \mathrm{cm}^{-1}$ at the 10.6 μm CO_2 laser line.

The estimate above is in gross disagreement with the room-temperature experimental value $\beta_v \approx 5\times10^{-5}\ \mathrm{cm}^{-1}$, as discussed below. Thus, the simple theory with γ independent of frequency fails in a most dramatic fashion. We should not be surprised at the inadequacy of the simple theory, although in this case its failure is striking. One knows that the simple oscillator model with frequency-independent damping constant γ fails to account for the lattice response in a number of materials, even for frequencies in the near vicinity of ω_{TO} (Barker and Hopfield 1966, Barker 1968, Rockni et al. 1972, Ushioda and McMullen 1972).

The experiments also show that the simple model gives a frequency variation that differs qualitatively from that in eq. (1.3) with γ constant. Rather that falling off like ω^{-2}, the data show that for many materials, at frequencies beyond the regime of two-phonon absorption, $\beta(\omega)\sim \exp(-A\omega)$. This strikingly simple empirical relation represents a challenge

†For this estimate, we use numbers from the table on page 190 of Kittel (1971). In addition, we suppose $\gamma=0.05\omega_{TO}$.

to the theorist. The frequency dependence of γ is key to the exponential frequency dependence of β.

The temperature variation of $\beta(\omega)$ also differs from one's simplest expectation. If we abandon the damped oscillator model with constant γ for a microscopic lattice dynamical theory of the absorption process, then it is quickly established that the integrated strength of the contribution to $\beta(\omega)$ from n-phonon processes should vary as T^{n-1} for temperatures $k_B T > \hbar\omega_M$, with ω_M the maximum vibrational frequency of the crystal. This result follows simply by counting the number of Bose–Einstein factors that enter the expression for the transition rate, then taking the high-temperature limit.* For KCl, we predict $\beta_v \sim T^5$ or T^6 for the 10.6 μm absorption constant at high temperatures, while the data shows a much slower variation. Similar deviations from the T^{n-1} law are found in many materials, as we shall see.

The experimental investigations that have provided the information on $\beta(\omega)$ described above produced a situation ideal for stimulating theoretical studies. For $\omega/\omega_0 \gg 1$, we seem to have simple empirical rules that describe both the frequency and temperature variation of $\beta(\omega)$ in a large class of insulating materials. In response to this, the theoretical studies have led to a rather clear understanding of the processes which control multiphonon absorption in insulators, although the issue of the role of higher-order dipole moments remains unresolved (§3.3 below).

In this article, §2 reviews the experimental studies, with emphasis on the past five years. §3 presents a summary of the theoretical studies. The spirit of our discussion is to outline to the general reader the principal experimental facts, as we understand them, and the theoretical concepts that form the basis of recent analyses. We do not attempt to cite each of the many papers that have appeared recently which cover detailed aspects of the subject, but we refer only to those which in our view illustrate or examine points of general interest. We call the reader's attention to a recent review by Bendow (1978), which cites the existing literature in a detailed and complete fashion.

*A simpler argument is as follows. In the harmonic approximation, the integrated strength of the fundamental absorption line at ω_{TO} is temperature independent, as one sees from eq. (1.2). The integrated strength of the n-phonon band is smaller by the factor λ^{n-1}, where $\lambda = \langle u^2 \rangle_T / a_0^2$ is the fundamental expansion parameter of lattice dynamical perturbation theory for the anharmonic lattice. Everytime we add a new phonon, a factor of λ is introduced into the expression that controls the overall strength of the absorption. Here a_0 is the lattice constant, and $\langle u^2 \rangle_T$ the mean square displacement at temperature T. Since $\langle u^2 \rangle_T \sim T$ for $k_B T \gg \hbar\omega_M$, the T^{n-1} law follows at once.

2. *Experimental studies of infrared absorption at frequencies above the characteristic lattice vibration frequencies*

2.1. Introductory remarks

In this section, we describe the principal results of experimental studies of the multiphonon processes. We focus on frequencies ω greater than twice the maximum vibration frequency ω_M of the crystal. Thus, we do not examine the extensive literature on absorption by two-phonon processes; rather, we are concerned primarily with frequencies sufficiently high that the number of phonons involved is the order of five or six. The two-phonon absorption spectra of crystals are rich in content, and greatly influenced by the details of the phonon spectrum of the crystal. We refer the reader elsewhere* for discussion of this topic and emphasize here the simple systematic features of the data that obtain when $\omega > 2\omega_M$.

After some introductory remarks, we turn to a discussion of the nature of the intrinsic multiphonon absorption in various classes of materials. At the extremely low absorption levels encountered in these studies, extrinsic processes can easily dominate the measured absorption coefficient. We conclude with a brief outline of the nature of the extrinsic processes.

The first modern experimental study of intrinsic lattice absorption by multiphonon processes is Rupprecht's (1964) remarkable set of measurements, which preceded the recent upsurge of activity in the area by nearly a decade and presents the principal notions that have dominated the recent discussions. Thus, we pause to summarize his work before we turn to a survey of more recent work.

Rupprecht examined the magnitude and frequency variation of the extinction coefficient $\kappa(\omega)$ in seven materials in the frequency range from 625 cm^{-1} to 2200 cm^{-1}. (The extinction constant $\kappa(\omega)$ is proportional to the wavelength times the absorption constant $\beta(\omega)$ of §1.) The materials examined were $BaTiO_3$, $SrTiO_3$, TiO_2, SiO_2, Al_2O_3, MgO, MgF_2 and CaF_2. The measurements were carried out at room temperature ($SrTiO_3$ was examined at nitrogen temperature also), with a method that examined attenuation of a beam transmitted through a film of the material under study.

*A number of papers on the topic of absorption by two-phonon processes appear in section C of "Lattice Dynamics", edited by R. F. Wallis (1965). In particular, see the paper by Bilz et al. on page 355 of this volume for data on the two- and three-phonon region in diamond structure semiconductors, and that by Karo et al. on page 387 for a discussion of NaCl structure ionic materials. One may find a general description of theoretical methods in anharmonic lattice dynamics and their application to computation of material properties in Cowley (1963).

In all cases examined, Rupprecht found strikingly simple frequency variation of the extinction coefficient. In addition to structure he interpreted as harmonics of the Reststrahl absorption, Rupprecht found a strong background (the dominant contribution to $\kappa(\omega)$) that was well fitted by the form

$$\kappa(\omega)=\kappa_0 \exp[-\omega/\omega_c], \tag{2.1}$$

where the prefactor κ_0 and characteristic frequency ω_c varied from material to material. In all cases, ω_c was found to lie between 85 cm^{-1} and 200 cm^{-1}, so ω_c clearly represents an average phonon frequency of some sort. In his paper, Rupprecht compares his data to the prediction of the simple form of the damped oscillator model, as we did in §1, and noted a similar dramatic failure of the damped oscillator model.

Rupprecht then proposed that the exponential variation of $\kappa(\omega)$ had its origin in multiphonon processes, where the incident photon is absorbed by breaking up into n phonons, presumably with wave vectors far out into the Brillouin zone. No detailed theory of this phenomenon was presented, however. The theory was not developed until Sparks and coworkers (Sparks 1971, 1972a, Sparks and Sham 1972a, b, 1973a) showed that the long high-frequency wing of the Reststrahl absorption band observed in alkali halides by several investigators could be explained by a first-principles calculation of multiphonon absorption. The long absorption wings were often described by a power law, $\beta \sim \omega^m$ for $m \gg 1$. Deutsch (1973) showed that his recent data fit an exponential better than a power law.

In the materials studied by Rupprecht, the absorption constant $\beta(\omega)$ was sufficiently large that it could be measured by direct examination of the attenuation of the incident beam. In many of the more recent studies, the absorption constant is so small that direct measurement of the attenuation is quite impossible, even though the path length through large samples may be many centimeters. As a consequence, most recent measurements require a more sophisticated method. In his recent and very extensive study of infrared absorption in the multiphonon region, Deutsch (1973) employed a dual-beam device with a sample of different thickness in each beam. An optical wedge is placed in one of the two beams, and a servomechanism adjusts the position of the wedge until the radiation transmitted in each beam is identical. This method has the virtue that the surface reflection losses need not be known to high accuracy for a reliable measurement of the absorption constant to be obtained. Deutsch was able to measure volume absorption constants β as small as 0.005 cm^{-1}. In a favorable case (Ge) where particularly long path lengths were available, values of β down to 0.002 cm^{-1} could be studied.

To improve further on the ability to measure the very small values of β encountered in very-high-purity materials in the 10 μm region of the spectrum, the method of laser calorimetry has been employed.* Here an incident laser beam is passed through the sample, and the temperature rise produced by absorption of the beam is found. By this method values of β in the range of 10^{-4} cm^{-1} or 10^{-5} cm^{-1} may be measured. Such values are quite inaccessible by usual spectroscopic methods, which are limited to 10^{-2} cm^{-1} typically. The major disadvantage of the laser calorimetry approach is that one can measure $\beta(\omega)$ only at those (discrete) frequencies at which laser radiation is available; the intensity of non-laser sources is simply too low to produce measureable heating. The requirement that lasers be employed makes detailed exploration of the frequency variation of $\beta(\omega)$ impossible.

With the above remarks in mind, we now turn to a description of experimental studies of particular materials.

2.2. Recent experimental studies of multiphonon absorption in different crystal classes

2.2.1. *Alkali halides*

Among the materials examined by Deutsch (1973) are the alkali halides LiF, KCl, NaCl and KBr. The data of Deutsch is remarkable for its accuracy; indeed the recent surge of theoretical activity in the problem of multiphonon absorption has been stimulated by the simple systematics that emerge from Deutsch's data.

Deutsch's data lies in the frequency region predominately above the two-phonon absorption bands of the material. In any case, the highest frequency at which the measurement can be performed is set by the limitation that the device can measure values of $\beta(\omega)$ no smaller than 0.005 cm^{-1}, unless particularly large crystals are available.

In fig. 2, we show Deutsch's data on the frequency variation and magnitude of $\beta(\omega)$ in NaCl. Some of the data (the crosses) have been taken from the AIP Handbook, a number of points (solid circles) are new measurements on high-purity material by Deutsch, and the encircled cross is a value measured by Horrigan and Rudko (1969) by the laser calorimetry method.

From a glance at fig. 2, it is evident that to high accuracy, $\beta(\omega)$ exhibits an exponential frequency variation of the form

$$\beta(\omega) = A \exp(-\omega/\omega_c). \tag{2.2}$$

*For a discussion of this method, see Hass et al. (1974).

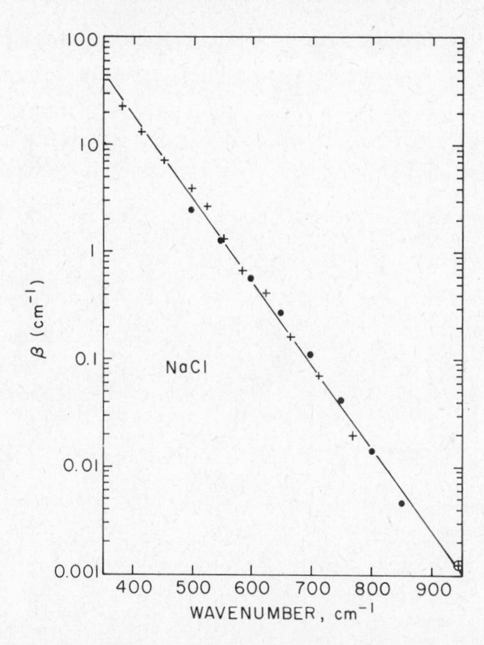

Fig. 2. The data from Sparks (1971, 1972a) on the magnitude and frequency variation of the volume absorption coefficient $\beta(\omega)$ in NaCl. The solid circles are new data reported by Deutsch, the crosses are taken from the AIP Handbook, and the encircled cross has been taken by Horrigan and Rudko (1969) by the laser calorimetry method.

Upon fitting the data to such an expression, Deutsch finds $\omega_c \sim 56\ \mathrm{cm}^{-1}$, a frequency about three times smaller than the Reststrahl frequency $\omega_{TO} \cong 166\ \mathrm{cm}^{-1}$.

Each of the four alkali halide crystals examined at room temperature by Deutsch showed the same type of exponential frequency variation for $\beta(\omega)$, although the value of ω_c and the exponential prefactor A in eq. (2.2) varied from case to case. We refer the reader to his paper for the quantitative information. (Recall that Rupprecht (1964) concluded that the *extinction* coefficient $\kappa(\omega)$ (imaginary part of the index of refraction) exhibited an exponential variation, while Deutsch assigns such a functional form to the *absorption* constant $\beta(\omega)$. We presume that Deutsch's data is the more accurate of the two, and Rupprecht's analysis may not be sensitive enough to distinguish between the frequency variation of $\kappa(\omega)$ and that of $\beta(\omega)$ when the data extends over a limited frequency range.)

In fig. 3, we show a figure constructed by Sparks* which for NaCl superimposes the data in Deutsch's paper with measurements reported

*See fig. 3 of Sparks (1972b). The previous tabulation of infrared absorption coefficients is being updated to include the vast amount of new experimental data. The results will be published by L. Case and M. Sparks.

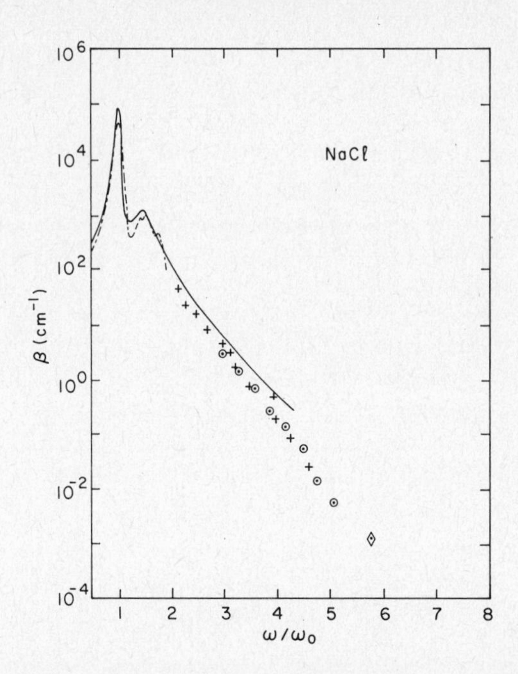

Fig. 3. The data from Hass et al. (1974) superimposed on that of Rupprecht (1964), for NaCl at room temperature. The data of Hass et al. (1974) covers the Reststrahl absorption line at $\omega_{TO} = 166\ \text{cm}^{-1}$ through the two-phonon region, while Deutsch's data extends into the multiphonon region.

earlier by Smart et al. (1965). The data of Smart et al. extends from low frequencies, through the strong Reststrahl absorption at the frequency of the transverse optical phonon, and through the two-phonon region. We see structure in the two-phonon region, but the frequency variation of $\beta(\omega)$ breaks out into the smooth exponential variation displayed in eq. (2.1) for $\omega > 2.5\ \omega_{TO}$.

The behavior discussed in detail for NaCl is typical of that obtained for all alkali halides examined to date at room temperature or above. One finds structure in the two-phonon region, but in the region $\omega > 3\ \omega_{TO}$, the frequency variation of $\beta(\omega)$ becomes smooth, and accurately fitted by the simple exponential form in eq. (2.1), with no hint of structure.

As the crystals are taken to low temperature, several alkali halides (KI, KBr, NaI and RbCl) that have been examined (Harrington et al. 1976) show clear evidence of structure in $\beta(\omega)$ in the three- and four-phonon regions. We shall discuss the origin of these structures in the next section, where they will prove of considerable theoretical interest.

We next turn our attention to the variation of $\beta(\omega)$ with temperature, with emphasis on the regime above room temperature. Barker (1972) has presented a study of the frequency and temperature variation of the absorption constant of KBr, NaCl and LiF. This study employed a spectroscopic method for measuring the absorption coefficient (rather than the more sensitive calorimetry methods mentioned earlier). The KCl data covers the three-phonon region, and extends into the four-phonon region.

Barker performs no analysis of his data. The measurements show that for room temperature up through the melting temperature, $\beta(\omega)$ is a smooth, featureless (monotonically decreasing) function of frequency. Also, at each frequency, $\beta(\omega)$ increases monotonically with temperature.

We wish to place emphasis on one important feature of Barker's results. His data covers the temperature regime from room temperature, up to and beyond the melting temperature. The frequency and temperature variation of $\beta(\omega)$ in the melt appear similar to that of the high-temperature solid; there is neither a discontinuity in $\beta(\omega)$ nor a change in its frequency variation upon melting.

These results, in combination with Deutsch's data, show that in the multiphonon region at room temperature or above, $\beta(\omega)$ is quite insensitive to the details of the phonon spectrum. The theorist has the hope that the gross, overall features of $\beta(\omega)$ in the multiphonon region can be accounted for through use of a rather schematic model of the solid. In §3, we shall see this is indeed the case. When specific structures do show up in $\beta(\omega)$, as they do in the low-temperature studies reported by Harrington et al. (1976) there is considerable interest in understanding its connection with the details of the phonon spectrum. We shall turn to this topic also in §3.

Barker's data on the temperature variation of the absorption coefficient extend only to 500 cm^{-1}. Harrington and Hass (1973) have reported a study of the temperature variation of $\beta(\omega)$ at the wavelength 10.6 μm of the dominant CO_2 laser line. They explore the temperature range from 300 K to 100 K, and the measurements were carried out for KCl, NaCl and NaF. The very weak absorption present at this frequency was measured through use of the laser calorimetry method discussed earlier in this section.

As we remarked in §1, the contribution to $\beta(\omega)$ from a process which involves n phonons should exhibit the temperature variation T^{n-1}, when the temperature is high compared to the Debye temperature. If we presume the mean number of phonons involved in the absorption process is $n \approx \omega/\omega_{TO}$, then for the three salts examined we expect (Harrington and Hass 1973) $n \cong 7$ (KCl), $n \cong 6$ (NaCl) and $n \cong 4$ (NaF). In all three cases examined, $\beta(\omega)$ exhibits a temperature variation much weaker than that expected from the simple T^{n-1} law. We shall see that such results follow from theories of multiphonon absorption when it is recognized that it is invalid to use a simple perturbation theoretic expansion of $\beta(\omega)$ in powers of the anharmonic coupling constant to calculate the absorption rate.

While the work of Harrington and Hass focuses on temperatures above room temperature, there have also been recent studies of the magnitude and frequency variation of $\beta(\omega)$ below room temperature, in the 10 μm frequency range. The first such study was reported by Pohl and Meier (1974), who examined the temperature variation of $\beta(\omega)$ for NaF from 4 K to 400 K, at the CO_2 laser frequencies 9.3 μm and 10.6 μm. They find $\beta(\omega)$ nearly independent of temperature below 90 K. Presumably then, below 90 K, the phonons are "frozen out" to the point where the Bose–Einstein thermal occupation factors influence $\beta(\omega)$ only weakly. Above 90 K, $\beta(\omega)$ rises with temperature, and the data of Pohl and Meier merge with the high-temperature data of Harrington and Hass very nicely in the range of temperatures where the two sets of data overlap.

Again through use of the laser calorimetry method, Rowe and Harrington (1976) have extended the measurement of $\beta(\omega)$ in both NaCl and KCl from room temperature to 100 K. The data is rather similar in a qualitative sense to that of Pohl and Meier (1974). In both KCl and NaCl, one sees clear evidence that $\beta(\omega)$ is approaching a temperature-independent value at the lowest temperatures explored. However, in contrast to the data of Pohl and Meier (1974), the low-temperature data of Rowe and Harrington does not appear to join smoothly onto the high-temperature data of Harrington and Hass, although superposition of the two sets gives a clear qualitative picture of the temperature variation* of $\beta(\omega)$. Thus, one should be cautious when fitting theories to the combined data of Harrington and Hass (1973) and Rowe and Harrington (1976) for NaCl and KCl.

The recent data on the temperature variation of $\beta(\omega)$ reported in Harrington and Hass (1973), Pohl and Meier (1974) and Rowe and Harrington (1976) is centered on frequencies equal or very close to the 10.6 μm CO_2 laser line. An intriguing study of the *frequency* variation of $\beta(\omega)$ at low temperatures in a number of alkali halides has been presented in the work mentioned earlier by Harrington et al. (1976). These authors explored the frequency variation of $\beta(\omega)$ at 50 K in several alkali halides through the region where the dominant contribution to $\beta(\omega)$ comes from two-phonon

*In figs. 8 and 10 of Rowe and Harrington (1976), their low-temperature data and the high-temperature data of Harrington and Hass are presented on the same graph, for NaCl and KCl, respectively. Extrapolation of the high-temperature data of Harrington and Hass (1973) down to low temperatures yields a value for $\beta(\omega)$ rather larger than that obtained from the measurements of Pohl and Meier (1974). Unfortunately, the temperature range explored in Rowe and Harrington (1976) does not overlap that of Harrington and Hass (1973), so a direct comparison of the two sets of data is difficult. One may contrast this situation by comparing the low-temperature data of Pohl and Meier on NaF with the high-temperature measurements of Harrington and Hass. As one sees from fig. 1 of Pohl and Meier (1974), the two sets of measurements are in excellent accord through the temperature regime where the two sets of measurements overlap.

processes (the lower range of frequencies explored in the work) through the region dominated by four-phonon processes (the higher ranges of frequencies explored). Five materials were examined in the study: KI, KBr, RbCl, NaI and NaCl. In four of the materials (KI, KBr, RbCl and NaI), striking and substantial departures from the exponential law of eq. (2.2) were found. Thus, while $\beta(\omega)$ has the smooth frequency variation displayed in eq. (2.1) for all alkali halides examined at elevated temperatures (room temperature and above), as the temperature of the material is lowered, clear structure appears that is obviously intimately related to the underlying phonon spectrum of the material.

It is noteworthy that no structure was found in NaCl. If one examines the phonon density of states for the four crystals which exhibit the structure, then there is either a distinct gap in frequency between the acoustical and optical phonon branches or a clearly defined minimum ("near gap") in the phonon density of states. No such feature appears in the phonon density of states in NaCl. Harrington et al. (1976) suggest that a "quasi-selection rule" is responsible for this structure which in all cases consists of excess absorption in a frequency region where the incident phonon may be absorbed by a process which involves three optical phonons. This quasi-selection rule was first mentioned by Eldridge (1972) and later emphasized by Duthler and Sparks (1973) in analyses of infrared absorption by two-phonon processes in alkali halides. A subsequent paper by Duthler (1976) examines its role in the multiphonon regime, and a detailed study of its influence on the three-phonon contribution to $\beta(\omega)$ appears in the very recent work of Eldridge and Staal (1977). We shall discuss it in §3.

We have discussed the experimental work on alkali halides in some detail because these results have provided the motivation for most of the relevant theoretical studies. To summarize briefly, the three principal experimental facts are (for the alkali halides)

(1) At room temperature and above, for frequencies beyond the region of two-phonon absorption, the frequency variation of the volume-absorption coefficient $\beta(\omega)$ is described very well by the simple empirical expression in eq. (2.2).

(2) At room temperature and above, $\beta(\omega)$ exhibits a temperature variation very much weaker than the T^{n-1} law of anharmonic perturbation theory. Here n is the number of phonons involved in the absorption process.

(3) At low temperatures (the order of 50K), clear structure in $\beta(\omega)$ develops in those materials with gaps or "near gaps" between the acoustical and optical phonon bands.

We next turn to other classes of materials, and the question of how universal the behavior outlined above is found to be.

2.2.2. *Alkaline earth fluorides*

In Deutsch's extensive study of room-temperature multiphonon absorption in insulators (Deutsch 1973), several of the alkaline earth fluorides (MgF_2, CaF_2, BaF_2 and SrF_2) were studied. In all of these cases, the volume absorption constant $\beta(\omega)$ has the exponential frequency dependence displayed in eq. (2.2). The materials with the lighter cations have Reststrahl frequencies considerably higher than found in a number of the alkali halides. Thus, the multiphonon tail can be followed out to rather high frequencies (2000 cm^{-1} for MgF_2), where the exponential law still obtains.

The temperature variation of the absorption constant in CaF_2, BaF_2 and SrF_2 has been measured by Lipson et al. (1976) for temperatures from room temperature to roughly 900 K. The results are qualitatively quite similar to those found in the alkali halides: the temperature variation of $\beta(\omega)$ is found to be substantially weaker than expected from the T^{n-1} law of elementary perturbation theory applied to the anharmonic lattice at high temperature.

The data cited above shows the behavior of $\beta(\omega)$ in the alkaline earth fluorides to be quite similar to the alkali halides. At this point, one may suspect the exponential law of eq. (2.2) is universally observed in all transparent solids in the multiphonon region, at room temperature and above. We now turn to examples which show this is not the case.

2.2.3. *Semiconductors*

Infrared absorption studies have been reported for a variety of cubic semiconductors for frequencies well beyond the two-phonon region. For example, Johnson (1965) has reported data on diamond, silicon and germanium for frequencies up to three times the maximum vibration frequency of the materials. Data on the zinc blende semiconductors GaAs and InAs were also reported by Johnson. Deutsch (1973) has studied CdTe, ZnSe and GaAs, and Turner and Reese (1962) studied AlSb in great detail.

For all of the materials described above, the frequency variation of $\beta(\omega)$ differs dramatically from the behavior typical of the alkali halides and the alkaline-earth fluorides. For these semiconductors, one finds sharp clearly defined structures that extend from the two-phonon region of the absorption spectrum out to the highest frequency studied. These structures appear superimposed on a monotonically decreasing background. In the alkali halides as the temperature is lowered, in some cases structure in $\beta(\omega)$ develops from the smooth and featureless room-temperature form, as we have seen. However, these features remain broad and gentle even at the lowest temperatures studied to date. In the semiconductors under discussion presently, the features are very sharp and well defined, even at room temperature.

Turner and Reese (1962) have presented a particularly complete analysis of data on the cubic material AlSb. Their data extends from the two-phonon region out to 9 μm, which fully includes the four-phonon region. At room temperature, a large number of fine structures can be resolved very clearly. In the three- and four-phonon region, which extends from 15 μm out to 9 μm, Turner and Reese identify twenty-eight distinct peaks and shoulders in their data.

In the two-phonon region, they find six clearly discernable structures, and note they can introduce four frequencies $\omega_{TO}, \omega_{LO}, \omega_{TA}, \omega_{LA}$ which may be combined in various fashions to fit the frequencies of the six features in the two-phonon region. Presumably these four frequencies correspond to large-wave-vector modes on the transverse-optical, longitudinal-optical, transverse-acoustic and longitudinal-acoustic branches of the phonon spectrum. After determining the four frequencies from the two-phonon data, Turner and Reese find that each of the twenty-eight features identified by them in the three- and four-phonon region can be fitted to suitable combinations of these four basic frequencies.

Quite clearly, even though there may be some ambiguity in certain of the specific assignments made by Turner and Reese, there is no question that the fine structure present throughout the three- and four-phonon regime requires explicit contact with the details of the phonon spectrum of the material for its interpretation. We have here (and in the other cubic semiconductors) a situation that differs qualitatively from that in the alkali halides. One sees no evidence of the simple exponential law of eq. (2.2) for the absorption constant in the multiphonon region. Instead rich and detailed structure persists out to the highest frequencies studied to date. The differences in the two types of absorption spectra are believed to result from the greater phonon line widths (i.e. stronger anharmonicity) present in the alkali halides.

2.2.4. *Other materials*

In Deutsch's (1973) extensive survey of infrared absorption in the multiphonon region, room-temperature data is presented for Al_2O_3. One finds a nearly smooth exponential frequency dependence for $\beta(\omega)$ we have discussed earlier in the subsection on the alkali halides. The data of Deutsch covers the frequency region from 1600 cm^{-1} to 3000 cm^{-1}, while the maximum vibration frequency in Al_2O_3 is near 750 cm^{-1}.

The frequency and temperature variation of $\beta(\omega)$ in Al_2O_3 has been reported by Billard et al. (1976). The data on the temperature variation of $\beta(\omega)$ extends from liquid nitrogen temperature to just over 2000 K, and for each of the temperatures studied, the data extends from 1000 cm^{-1} to 2700 cm^{-1}. This work represents the most complete study we have seen of the temperature variation of $\beta(\omega)$ in the true multiphonon region in that the temperature variation of $\beta(\omega)$ has been followed over a wide range of frequency for each temperature. This kind of study is much more satisfactory than laser calorimetry measurements at a single laser frequency. Of course, Billard and co-workers could carry out such a study for Al_2O_3 because the absorption constant is very much larger than in the alkali halides, and may be measured by direct methods. They plot the temperature variation of $\beta(\omega)$ for twelve frequencies between 1400 cm^{-1} and 2500 cm^{-1}. Below room temperature rather little temperature variation is found, while at high temperatures, $\beta(\omega)$ varies as $T^{\alpha(\omega)}$, where α is near unity at 1400 cm^{-1} and near 3 at 2500 cm^{-1}. This behavior is qualitatively consistent with the result expected from the simple argument presented in §1, where it was noted that perturbation theory leads to the temperature variation T^{n-1} for the n-phonon contribution to $\beta(\omega)$ at high temperatures.

So far, we have confined our attention to the behavior of $\beta(\omega)$ in crystalline materials. Maklad et al. (1974) have presented data on multiphonon absorption As_2S_3 glasses, and glasses made from As_2S_3–As_2Se_3 mixtures. Pronounced structure is observed in $\beta(\omega)$ throughout the portion of the multiphonon region explored in this study. The structure is very different from the sharp features found in semiconductors. In the glasses, one finds sequences of peaks in $\beta(\omega)$, each of which is quite broad. Maklad and co-workers interpret these structures in terms of multiphonon processes that originate in vibrations highly localized to particular chemical bond combinations present in the glasses. The behavior found for $\beta(\omega)$ in the glasses thus differs distinctly from that characteristic of either the alkali halides or the semiconductors.

2.3. Infrared absorption in the multiphonon region by extrinsic processes

All of the preceding discussion in the present section was concerned with the nature of the contribution to the volume absorption constant $\beta(\omega)$ from intrinsic processes in which the incident photon is absorbed by coupling to a substantial number n ($n \gtrsim 3$) of phonons associated with the host crystal. Great care must be exercised in many cases before one can justifiably conclude that a given measurement actually probes these intrinsic processes. The reason for this is that in the multiphonon region, the contribution to $\beta(\omega)$ from intrinsic processes is very small. The crystal is highly transparent, and this is the reason why one must resort to exotic experimental methods such as laser calorimetry to measure the absorption constant. When the intrinsic contribution to $\beta(\omega)$ is small, then the intrinsic absorption is easily masked by tiny residual absorptions induced by even a small amount of impurity or small degree of imperfection in a carefully prepared crystal.

We can appreciate the sensitivity of $\beta(\omega)$ to slight contamination by a simple numerical estimate. Consider an impurity with resonance frequency ω_0, and suppose it is present in concentration f. Suppose we have the bad luck to measure the absorption constant $\beta(\omega)$ right at the frequency ω_0. To obtain a rough estimate of the possible influence of the impurity absorption, we assume that the cross section and line width of the impurity are approximately the same as for the Reststrahl absorption, for which $\beta \cong 10^5\ \mathrm{cm}^{-1}$. Then an impurity concentration of 10^{-9} gives an impurity absorption $\beta_\mathrm{I} \cong 10^{-9} \times 10^5\ \mathrm{cm}^{-1} = 10^{-4}\ \mathrm{cm}^{-1}$, as strong as the intrinsic absorption of KCl at 10.6 μm and room temperature.

The above crude estimate shows that in highly transparent solids, even a tiny concentration as low as a few parts per billion of impurities which absorb in or near the region of interest can mask the intrinsic absorption processes. One requires material prepared with great care to realize absorption limited by intrinsic processes in the multiphonon region, in the highly transparent materials.

It is very difficult to identify absorption features from specific impurity centers in the highly transparent materials such as the alkali halides, even if one knows in a given case the absorption constant is impurity dominated. The impurity lines can be quite broad in ionic materials, owing to the strong coupling between the host phonons in the ionic material and the electric dipole active impurity absorption line. Also, as we have seen, the "dangerous" impurities may be present in only trace quantities, presumably as a residue from the material preparation procedure. Finally, absorption studies in highly transparent solids often employ the laser calorimetry method, as we have seen, so from such studies one has data on $\beta(\omega)$ only for one or at most a small number of laser frequencies. Duthler (1974) has presented a survey of specific impurities likely to "contaminate" measurements of $\beta(\omega)$, for the alkali halides potentially useful as optical elements in high-power laser devices. Flannery and Sparks (1977) have extended and updated Duthler's work. We refer the reader to these papers for a more complete discussion of this topic.

Another source of absorption that can partially override the intrinsic absorption is surface-induced absorption. The estimates presented above for impurity-induced absorption show that one or two atomic layers of impurities on the surfaces can mask the intrinsic absorption. A number of experimental studies show the presence of surface absorption quite clearly. In highly transparent solids, the absorption coefficient satisfies $\beta(\omega)L \ll 1$ for all laboratory-sized samples. Then in this limit, the amount of energy absorbed per unit time from a beam that passes through the sample should be proportional to the length, if only volume absorption were present. In practice, studies show the energy absorbed per unit time to consist of a contribution $\beta_s(\omega)$ independent of sample length, and a portion $L\beta(\omega)$ that scales linearly with length. A plot of the energy absorbed per unit time versus sample length is a straight line, but with non-zero intercept when extrapolated to $L=0$. We refer the reader to the paper by Rowe and Harrington (1976), where such data is displayed. The surfaces of the materials used in the optical studies are exposed to the atmosphere, and one may presume they contain a number of adsorbed impurities. We have seen that a very small number of impurities distributed through the bulk can mask the very weak intrinsic processes. The few atomic layers of contaminant surely present on the surface may play the same role, and it is not surprising to find it difficult to pin down the precise mechanism that controls surface absorption.

It has been pointed out by Sparks and Duthler (1973) that macroscopic inclusions can contribute substantially to absorption in the highly transparent regions. While such inclusions may be largely absent in the bulk of carefully prepared material, the surface of even cleaved material will contain damaged areas that may lead to enhanced surface absorption.

We have concluded the present section with a brief discussion of the extrinsic contributions to the absorption constant one encounters even in the best material so the reader may appreciate the care required to obtain data on intrinsic absorption in highly transparent materials. The excruciating care required to prepare the materials for these studies and the rather lengthy analyses required to remove spurious contributions to $\beta(\omega)$ are the reasons why only a limited amount of recent data is available on the most transparent materials (KCl, KBr) in the multiphonon regime.

3. *Theoretical studies of infrared absorption in the multiphonon region*

During the past five years, the theory of intrinsic multiphonon absorption has been pursued vigorously, to the point where the phenomenon may be understood quantitatively for particular crystals. Nearly all the theoretical papers are devoted to the alkali halides, because of their potential importance as window materials and the simple systematic features present in the data.

We begin with preliminary remarks of a general nature and then turn to a description of recent theoretical analyses. One must begin by constructing a Hamiltonian that describes the lattice dynamics of the crystal in question, including anharmonic terms. It is then necessary to couple the electromagnetic radiation to the lattice. The Hamiltonian thus consists of two distinct pieces, and we write

$$H = H_{\mathrm{L}} + H_{\mathrm{LR}}, \tag{3.1}$$

where H_{L} and H_{LR} describe the lattice vibrations (including anharmonic effects) and the coupling between the lattice and the electromagnetic radiation, respectively.

The complete form of H_{L} may be obtained if one knows (or can model) the potential energy of interaction $V(\boldsymbol{r}_{ij})$ between ion pairs in the crystal. The harmonic approximation yields the phonon spectrum of the crystal, and the anharmonic terms in the lattice Hamiltonian may be readily generated. The phonon spectra of many alkali halides are well known. For these materials, this procedure may be carried through with confidence to a high degree of sophistication, if required. The same is true of a number of other insulating crystals of simple structure.

It will be useful to comment on the form of H_{LR} in more detail. The interaction between the electromagnetic field and the lattice is of the form*

$$H_{\mathrm{LR}} = -\sum_{l\kappa} \boldsymbol{P}(l\kappa)\cdot\boldsymbol{E}(l\kappa), \tag{3.2}$$

where $\boldsymbol{E}(l\kappa)$ is the externally applied macroscopic electric field at site κ in unit cell $\boldsymbol{l}$, and $\boldsymbol{P}(l\kappa)$ is the electric-dipole moment of the ion at this site. The limit of long wavelengths is of interest here, and the electric field may be treated as spatially uniform, but oscillating in time with frequency ω.

The electric-dipole moment $\boldsymbol{P}(l\kappa)$ may be expanded in powers of the displacement $\boldsymbol{u}(l\kappa)$ of the ion from its equilibrium position. If we confine attention to the alkali halides for simplicity, this expansion has the form

$$\begin{aligned} P_\alpha(\boldsymbol{l}\kappa) = {} & e^*(\kappa)u_\alpha(\boldsymbol{l}\kappa) + \sum_{\beta\gamma}\sum_{\boldsymbol{l}'\kappa'} m^{(2)}_{\alpha\beta\gamma}(\boldsymbol{l}'-\boldsymbol{l};\kappa\kappa')u_\beta(\boldsymbol{l}'\kappa')u_\gamma(\boldsymbol{l}\kappa) \\ & + \sum_{\beta\gamma\delta}\sum_{\substack{\boldsymbol{l}'\boldsymbol{l}''\\ \kappa'\kappa''}} m^{(3)}_{\alpha\beta\gamma\delta}(\boldsymbol{l}'-\boldsymbol{l},\boldsymbol{l}''-\boldsymbol{l};\kappa\kappa'\kappa'')u_\beta(\boldsymbol{l}'\kappa')u_\gamma(\boldsymbol{l}''\kappa'')u_\delta(\boldsymbol{l}\kappa) + \dots, \end{aligned} \tag{3.3}$$

where the Greek subscripts refer to the Cartesian components of the subscripted quantities.

In eq. (3.3), $e^*(\kappa)$ is the Born or transverse effective charge, necessarily equal in magnitude but of opposite sign for the two ions in the unit cell.

*For a more detailed discussion, we refer the reader to Born and Huang (1954).

The parameter e^* in eq. (1.2) is the same Born effective charge, so the magnitude of this quantity is known from studies of the dielectric response (from the difference between the static dielectric constant, and the dielectric constant ε_∞ appropriate to frequencies well above the transverse optical phonon frequency). The non-linear terms in eq. (3.3) enter because the dipole moment is a non-linear function of displacement, with a dependence through the non-linear terms on its position relative to its neighbors.

If the expansion in eq. (3.3) could be terminated after the term linear in the displacement $\boldsymbol{u}(l\kappa)$, the multiphonon absorption rate could be calculated from knowledge only of e^*, and the lattice dynamics of the (anharmonic) lattice. In general, however, the non-linear terms in eq. (3.3) also contribute directly to the absorption processes in the multiphonon region.

We illustrate this in fig. 4, where three distinct contributions to the absorption constant in the three-phonon regime are displayed. If only the term linear in the displacement $\boldsymbol{u}(l\kappa)$ appears in eq. (3.3), then transverse electromagnetic radiation of long wavelength can excite only the transverse-optical phonon with wave vector near zero. Multiphonon absorption occurs because this mode can decay to a three-phonon final state through anharmonicity, as illustrated in fig. 4a. If, however, the non-linear terms in eq. (3.3) are present, the incident photon can couple directly to the same three-phonon final state that appears in fig. 4a. This is illustrated in fig. 4b. In the simultaneous presence of both anharmonicity and non-linear terms in the expansion of the electric-dipole moment, hybrid processes involving both interactions become possible. An example of such a hybrid process is presented in fig. 4c.

All three of the diagrams in fig. 4 couple the incident photon to the same final state, so the three processes interfere coherently. The possible role of the non-linear dipole moment and such interference effects were discussed first by Szigeti (1965), in his description of infrared absorption by two-phonon processes in the alkali halides.

From the preceding comments, we see that a complete calculation of infrared absorption by multiphonon processes requires knowledge not only of the lattice dynamics of the crystal, including anharmonic terms, but also of the non-linear variation of the electric-dipole moment with atomic displacement. Quite clearly, the non-linear terms in eq. (3.3) cannot be generated from the interatomic potentials used to set up the lattice Hamiltonian H_L, but require a picture of how the electronic charge distribution in the crystal responds to atomic displacements. This is poorly understood in quantitative terms at present, even for simple crystals. The role of the non-linear variation of the electric-dipole moment in multiphonon absorption remains an unsettled question, in our view.

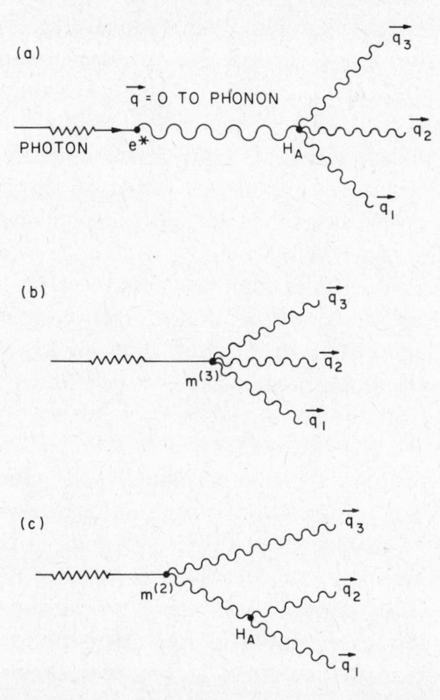

Fig. 4. Three distinct processes which contribute to the absorption coefficient in the three-phonon region. One has (a) a process where the incident photon couples to the long wavelength transverse-optical phonon, which then decays to a three-phonon state through the presence of anharmonicity (b) a process where the photon couples directly to the two-phonon state through non-linear variation of the electric-dipole moment on atomic displacement, and (c) a hybrid process.

With the above remarks as a framework, the subsequent discussion will be divided into three subsections. The first discusses theories which replace the crystal by a set of anharmonic Einstein oscillators, and the second explores theories based on a model phonon spectrum supplemented by anharmonicity. In the first two sections, we confine attention to studies which presume the dipole moment varies linearly with atomic displacement, and multiphonon absorption occurs only by virtue of anharmonicity in the lattice. The third subsection then explores the role of the non-linear terms in eq. (3.3), for both types of theory.

3.1. Multiphonon absorption by anharmonic oscillators with linear dipole moment

In §2, we saw that at room temperature and above, the frequency variation of $\beta(\omega)$ in the multiphonon region of frequencies is smooth and featureless for the alkali halides. The exponential frequency variation obtains in each case, even though the phonon spectra of the various crystals differ substantially. The behavior of $\beta(\omega)$ in the melt is very similar to that in the solid.

These observations suggest that one should be able to account well for the phenomena through use of models which treat the phonon spectrum of the solid in only a crude and schematic fashion. This has led a number of authors to mimic the real crystal by an array of anharmonic but uncoupled Einstein oscillators. The vibration frequency of the oscillator in the harmonic limit is chosen equal to an average phonon frequency of the material of interest, and the strength of its anharmonicity adjusted through an appropriate criterion. Such calculations in no sense constitute a fundamental theory of multiphonon absorption, since they avoid rather than address the reason for the insensitivity of $\beta(\omega)$ to structure in the phonon spectrum. However, with very few parameters and simple closed expressions, they can account remarkably well for the principal features of the data. Before we proceed to present results obtained in this manner, it may be worthwhile to comment on both the weaknesses and virtues of such models, since one finds papers that extend the calculations into regimes where even their qualitative validity comes into question.

These models are most suitable for the discussion of multiphonon absorption at elevated temperatures, where in the real crystal an abundant number of short wavelength optical and acoustical phonons are thermally excited. There is then rather little correlation between the vibrations of a given ion and its immediate neighbors, and it is intuitively reasonable to envision each unit cell to be an independently vibrating entity characterized by a small set of parameters. It seems rather questionable to extend the calculations to temperatures much below the Debye temperature, where the temperature variation of $\beta(\omega)$ is controlled by phonons with wavelength considerably longer than a lattice constant. A variety of factors, such as the wave vector and polarization dependence of anharmonic matrix elements begin to assert themselves here. We shall see an explicit illustration of such effects in the next subsection, where we discuss the origin of the structure found in $\beta(\omega)$ at low temperatures, in the alkali halides.

Thus, we regard the independent anharmonic oscillator models as a tool most useful for discussing the behavior of $\beta(\omega)$ at elevated temperature, where anharmonic effects may be substantial and more difficult to incorporate fully in more realistic models based on a perturbation theoretic treatment of anharmonicity. In the anharmonic oscillator models, while the phonon spectrum of the crystal is treated schematically, it is possible to avoid approximate treatments of anharmonicity, and treat it exactly. There seems little reason to treat anharmonic effects by approximate methods here, as some authors have, save to test approximation schemes developed for realistic crystal models against the exact solution of a simple case.

Maradudin and Mills (1973) and Mills and Maradudin (1973) have presented a detailed study of multiphonon absorption by anharmonic oscillators with linear electric dipole moment, with emphasis on the points raised in the preceding paragraphs. We summarize some of their principal results, and then turn to other calculations that have appeared in the literature, some of which amount to extensions of the basic scheme.

These authors argue that above the Debye temperature, where the specific heat assumes its classical temperature-independent value, the atomic motions in the solid may be treated by classical methods. Thus, in each unit cell, one imbeds a classical, but anharmonic oscillator characterized by the Hamiltonian

$$H = T + V(u), \tag{3.4}$$

where T is the kinetic energy, and $V(u)$ the potential energy, expressed as a function of the amplitude u of the vibrational motion. The fundamental task is to compute the infrared absorption spectrum of such an oscillator by classical methods.

This program can be carried out in closed form, for a variety of model potentials. A physically reasonable choice for which this may be done is the well-known Morse potential of molecular physics, where one has

$$V(u) = D(1 - \exp[-a(u - u_0)])^2. \tag{3.5}$$

The parameter u_0 is the equilibrium separation between the ions, so the shape of the potential near its minimum is controlled by the remaining two parameters, D and a. If the mass of the oscillator is fixed as the reduced mass of the unit cell, and u_0 at the interatomic separation, only two parameters (temperature independent) remain. For small $(u - u_0)$, $V(u) = a^2D(u - u_0)^2$, and by fixing the harmonic vibration frequency of the oscillator at that of the long wavelength transverse optical phonon, we are left with a single temperature-independent parameter with which to characterize the degree of anharmonicity present. At this point, the simplicity of the anharmonic oscillator model may be appreciated.

Maradudin and Mills find that for the Morse potential, the expression for the absorption constant is

$$\beta(\omega)=\frac{4\pi n_c e^{*2}}{Mc\sqrt{\varepsilon_\infty}}$$
$$\times\left\{\frac{4\pi}{\omega_0}\left(\frac{D}{k_B T}\right)^2\xi^2\sum_{n=n_m}^{\infty}\left(\frac{n-\xi}{n+\xi}\right)^n\frac{1}{n^3}\exp\left[-\frac{D}{k_B T}\left(1-\frac{\xi^2}{n^2}\right)\right]\right\}, \tag{3.6}$$

where in this expression, ω_0 is the vibration frequency of the oscillator in the harmonic limit, $\xi=\omega/\omega_0$, and n_m the first integer larger than ξ.

Quite clearly, the expression in eq. (3.6) does not have the simple exponential form of eq. (2.2). Nonetheless, a plot of $\beta(\omega)$ given by eq. (3.6) shows a behavior remarkably close to exponential, for ω in the range of $3\omega_0$ to $6\omega_0$. We illustrate this in fig. 5, at 900 K and for a value of D appropriate to NaCl ($k_B T/D=0.06$ at room temperature). One sees structure at low frequencies, and $\beta(\omega)$ settles into a smooth very nearly exponential behavior at the higher frequencies.

The reason for the near-exponential behavior of $\beta(\omega)$ may be appreciated by examining an unphysical, but nonetheless very simple limit of the expression of eq. (3.6). This is the limit of low temperatures, where the effect of anharmonicity is very small, and the absorption spectrum of the oscillator reduces to a line spectrum, with discreet absorptions at the overtones $n\omega_0$ of the fundamental frequency ω_0. In this limit, eq. (3.6) reduces to

$$\beta(\omega)=\frac{2\pi^2 n_c e^{*2}}{Mc\sqrt{\varepsilon_\infty}}\sum_{n=1}^{\infty}\gamma_n\delta(\omega-n\omega_0) \tag{3.7a}$$

with

$$\gamma_n=n!\left(\frac{k_B T}{4D}\right)^{n-1}. \tag{3.7b}$$

If the $n!$ were missing from eq. (3.7b), then eq. (3.7a) would describe a series of absorption peaks with integrated strength fit precisely by an envelope of exponential form. While the factor $n!$ increases strongly with n, its effect on the shape of the envelope is remarkably modest for small

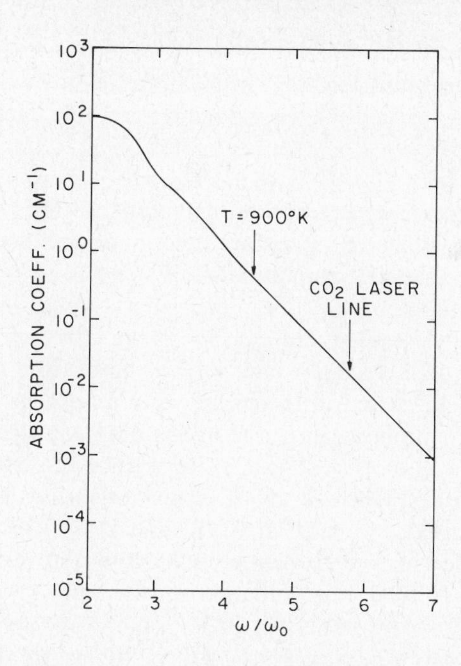

Fig. 5. Frequency dependence of the absorption coefficient for a Morse potential oscillator at $T=900\,K$, with the value of D chosen for NaCl. See Maradudin and Mills (1973).

values of $K_B T/D$, in the range $n=3$ to 6 or 7 of interest in the present problem. To see this, note

$$\log(\gamma_n) = -\log\left(\frac{4D}{k_B T}\right) - n\left\{\frac{1}{n}\log(n!) + \log\left(\frac{4D}{k_B T}\right)\right\}. \tag{3.8}$$

For NaCl at room temperature, $4D/k_B T \approx 60$, and the factor in curly brackets is rather insensitive to n, for n from 3 to 6. A plot of $\log\gamma_n$ gives very nearly a straight line in the region; this property persists in the full form of $\beta(\omega)$ given in eq. (3.6).

The above argument shows that something very near an exponential law will emerge over a *limited frequency range* for nearly any theory that associates a small coupling constant with each phonon absorbed. If g is such a coupling constant, the probability for absorbing n phonons will be $f(n)g^n$, where $f(n)$ may be highly model dependent. If g is small, and the ratio $\log f(n)/n$ slowly varying for n in the limited range of 3 to 6, the theory will yield a very nearly exponential variation of $\beta(\omega)$ in the range $3\omega_0$ to $6\omega_0$ where all the data of present interest lies, even though $\beta(\omega)$ may decrease by many decades as one moves through this range. This discussion suggests that one must proceed with caution, if one wishes to estimate the value of $\beta(\omega)$ at frequencies well above those where data exists through extrapolation by means of the exponential law in eq. (2.2).

In fig. 5, the value of D was chosen to reproduce the slope in the plot of $\log\beta(\omega)$ vs. ω at room temperature. At this point, all the basic parameters of the anharmonic oscillator model are determined, so both the temperature variation and absolute magnitude of $\beta(\omega)$ are fixed without the need of new parameters. From fig. 6 we present a comparison between the temperature variation predicted by eq. (3.6) at the CO_2 laser frequency 10.6 μm, and the data of Harrington and Hass in NaCl. The large deviations from the T^{n-1} law discussed in the earlier sections emerge nicely from the theory. In fig. 6, the prefactor in eq. (3.6) has been adjusted to fit the data at 700 K, but the value of D chosen to reproduce the room-temperature slope in the plot of $\log\beta(\omega)$ vs. ω yields an absolute value for $\beta(\omega)$ within a factor of 2 of the measured value at 900 K, and at the CO_2 laser line. This value of D may be checked further by estimating the thermal expansion coefficient of NaCl from the anharmonic oscillator model. This consistency check works remarkably well.

While the anharmonic oscillator model of multiphonon absorption is not a fundamental theory of the phenomenon, the model provides simple expressions for $\beta(\omega)$ with a small number of parameters. One finds the model works not only for NaCl, but for a variety of other alkali halides as well.

Other authors have applied anharmonic oscillator models to the analysis of data on multiphonon absorption, with success. For example, McGill et al. (1973) have presented a study of such a model by diagrammatic perturbation theory, combined with a numerical diagonalization of a fully quantum mechanical model Hamiltonian. They are led to a formula for $\beta(\omega)$ similar in structure to the limiting form in eq. (3.7a), but with a form for γ_n rather different in detail. By means of a suitably arranged fitting procedure, their model gives an account of the room temperature data of Deutsch's alkali halide data. In a subsequent paper, McGill and Winston (1974) account for the anomalous temperature variation of $\beta(\omega)$ (i.e. deviations from the T^{n-1} law) by arguing that the parameters which appear in their earlier formula, most particularly the frequency ω_0 of the oscillator in the harmonic limit, should be taken temperature dependent, as argued by Sparks and Sham (1973b). A description of the temperature variation of ω_0 is not contained in their model, but must be fed into the analysis from experimental data.

The anharmonic oscillator model of Maradudin and Mills has been extended by Rosenstock and co-workers to include a number of features they argue incorporate important features of the crystalline phonon

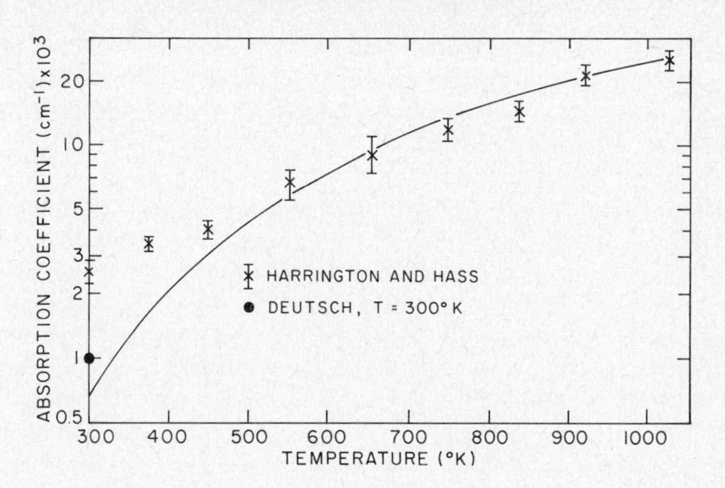

Fig. 6. A comparison for NaCl between the temperature variation for $\beta(\omega)$ given by the Morse potenial model at 10.6 μm, and the data of Harrington and Hass. See Maradudin and Mills (1973).

spectrum missing from the uncoupled oscillator model. Rosenstock (1974) begins by calculating the absorption of the Morse oscillator quantum mechanically, then averaging the resulting calculation over an array of Morse oscillators with harmonic vibration frequency distributed in a Debye-like fashion.

With Boyer et al. (1975), this procedure was extended to incorporate features of the real phonon spectrum. This was done by repeated convolution of the phonon density of states, to produce the density of states for two-phonon, three-phonon ... excitations. The absorption coefficient is calculated by weighting the n-phonon density of states with the matrix element of a Morse potential oscillator from the ground level, to the nth excited states. This provides an average oscillator strength for all possible n-phonon transitions, with the absorption coefficient formed by summing over all processes allowed by energy conservation.

One interesting feature of the calculation by Boyer et al. (1975) is the prediction of clear structure in $\beta(\omega)$ in those materials with a gap or pronounced minimum in the one-phonon density of states between the acoustical and optical branches. Such gaps or minima occur in materials such as KI, where there is a substantial difference in mass between the two constituents.

In §2, we saw that the measurements of Harrington et al. (1976) show clear structure in precisely these materials. However, the nature of the structures found in these studies differs substantially from that expected from the calculations reported by Boyer et al. (1975). The origin of the structure resides not in features of the one-phonon density of states which persist in high-order convolutions, but rather in a systematic polarization dependence of the matrix element that couples the photon to the multiphonon final state. We discuss this in the next subsection.

Namjoshi and Mitra have reported a series of calculations of the frequency (Namjoshi and Mitra 1974a) and temperature variation (Namjoshi and Mitra 1974b) of $\beta(\omega)$ that differ in detail, but are similar in spirit to those of Rosenstock and Boyer. One may regard both sets of theories as "hybrid theories", which feed information about the phonon spectrum of the real crystal into a phenomenological prescription for calculation of $\beta(\omega)$. While Namjoshi and Mitra make use of the properties of an isolated anharmonic oscillator in their work, they construct the n-phonon density of states at frequency ω by a convolution procedure, then introduce an average oscillator strength for coupling to n-phonon states from phenomenological arguments.

3.2. First-principles approaches to the theory of multiphonon absorption, with linear dipole moment

We next turn to theories of infrared absorption in the multiphonon region which proceed from first principles, along the lines delineated in the opening remarks of the present section. That is, one begins by constructing a microscopic description of the lattice dynamics of the crystal in question, with anharmonicity included. Upon supposing further that the electric-dipole moment varies linearly with atomic displacement (that the non-linear terms in eq. (3.3) may be ignored), $\beta(\omega)$ is calculated through use of many-body perturbation theory, or some other approximate method.

Two groups have presented theories of multiphonon absorption from this first principles point of view. In a series of papers, Sparks and his collaborators Sham and Duthler approach the problem through use of diagrammatic perturbation theory. They account very nicely for the frequency and temperature variation of $\beta(\omega)$ found in the alkali halides and also the structure found at low temperatures (Harrington et al. 1976). The explanation of the structure in $\beta(\omega)$ is a particularly notable achievement since its origin lies in the dependence of the anharmonic matrix elements on the phonon branch index, as shown by Duthler. As a result, a proper explanation of this structure cannot emerge from any of the phenomenological models discussed in the preceding subsection.

Bendow and his collaborators proceed very differently. They do not employ perturbation theory to treat the anharmonicity, but instead through an approximation scheme obtain a closed expression for $\beta(\omega)$ in terms of the correlation function $\langle u_\alpha(\boldsymbol{l}\kappa,t)u_\beta(\boldsymbol{l}'\kappa',0)\rangle$ which may be calculated once the phonon spectrum is known. The expression for $\beta(\omega)$ produced by this method is not easy to evaluate for a realistic phonon spectrum. As a consequence, quantitative calculations have been reported only for model phonon spectra of schematic form. Also, as in many approximate theories of many-body problems based on a decoupling procedure, it is not clear what the criterion is for validity of the scheme. In the correlation function approach of Bendow et al., some subsets of diagrams of the many-body perturbation theory are summed to infinite order, but in any finite order one finds diagrams omitted that are the same order of magnitude as those retained.

We discuss first the series of papers by Sparks, Sham and Duthler, and then turn to more detailed remarks on the work by Bendow and co-workers.

The application of the method of many-body perturbation theory to the problem of multiphonon absorption at first seems a formidable task, since one may write down a very large number of diagrams which couple the incident photon to a given multiphonon final state. The success of the procedure rests first on a systematic classification of the diagrams that isolates only a small subset as important, and secondly upon introduction of a scheme that enables the final complex formulae to be evaluated numerically.

In an early paper, Sparks and Sham (1973a) address both of these questions. We begin with a summary of their analysis.

One begins by writing $\beta(\omega)$ in the form

$$\beta(\omega)=\sum_{n=1}^{\infty}\beta_n(\omega), \tag{3.9}$$

where $\beta_n(\omega)$ is the contribution to $\beta(\omega)$ by n-phonon processes. Necessarily $\beta_n(\omega)$ vanishes for $\omega > n\omega_{\mathrm{M}}$, where ω_{M} is the maximum phonon frequency in the crystal.

The basic structure one must analyze is illustrated in fig. 7a. When the electric-dipole moment varies linearly with atomic displacement, the photon couples only to the transverse optical phonon with wave vector (near zero) equal to that of the photon. This transverse-optical phonon, referred to as the fundamental phonon below, subsequently decays to an n-phonon final state through the action of anharmonicity. The calculation is quite equivalent to a microscopic calculation of the damping factor γ of eq. (1.4), with γ endowed with the frequency dependence that emerges from a proper microscopic description of this parameter.

The fundamental expansion parameter of the many-body perturbation theory, following Van Hove (1953), is $\varepsilon=(\Delta r^2)^{1/2}/a_0$, with Δr^2 the mean-square displacement of ions from their equilibrium position and a_0 the

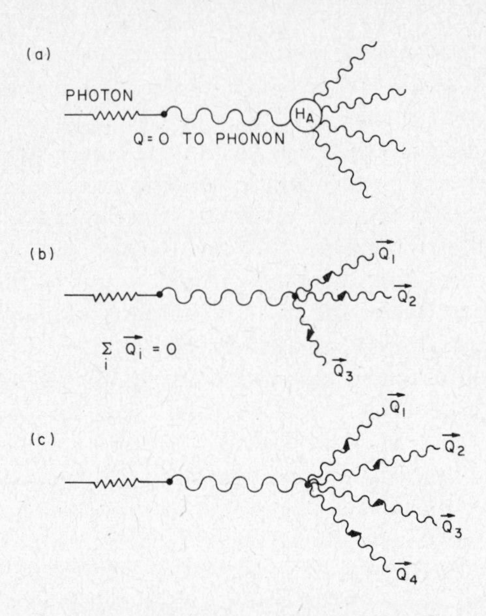

Fig. 7. A summary of the diagrams encountered in the many-body perturbation theory. One has (a) the fundamental absorption process when the electric-dipole moment varies linearly with atomic displacement, and in (b) and (c) two diagrams that contribute to $\beta(\omega)$ in the frequency regime from $2\omega_M$ to $3\omega_M$, where ω_M is the maximum vibration frequency of the crystal.

lattice constant. The diagrams of many-body perturbation theory may be compared by counting powers of ε contained implicitly in each matrix element, then noting that $\varepsilon \ll 1$. For the alkali halides, $\varepsilon \approx 0.05$ at room temperature, so one is well into the small ε limit.

We show an example of this procedure in figs. 7b and 7c, where we give two diagrams that contribute to $\beta(\omega)$ in the frequency regime below $3\omega_M$. The matrix element of the anharmonic portion H_A of the lattice Hamiltonian in fig. 7c is smaller than that in fig. 7b by one power of ε. Phase space considerations suggest the contribution to $\beta_3(\omega)$ from the diagram in fig. 7b is a maximum at frequencies near, but somewhat below $3\omega_M$*. In this frequency regime, the contribution from fig. 7c, though finite, is negligibly small, since the controlling matrix element is smaller by the factor of ε. Through extension of this argument, one concludes that attention may be confined to only those diagrams in $\beta_n(\omega)$ where the fundamental phonon decays by *emitting n* phonons. Sparks and Sham refer to this process as the

*Also, the anharmonic matrix element vanishes in the limit of zero wave vector, for any phonon on an acoustical branch. In addition to phase space considerations, this further diminishes the role of low frequency acoustical modes.

n-phonon summation process, since the energy carried by the n final-state phonons is the sum of the energy of each phonon which participates.

In fig. 7b, the fundamental phonon couples directly to the three final-state phonons, through terms in H_A quartic in the atomic displacement $U(l\kappa)$. The relevant matrix element $V_3(Q_{TO}, Q_1, \dots, Q_3)$ is the bare vertex for the anharmonic decay process, in the language of many-body perturbation theory. There are in addition vertex corrections the same order in ε as the bare vertex itself. An example is given in fig. 8. Thus, the matrix element for decay of the transverse-optical phonon to the three final-state phonons is written $\Lambda_3 V_3(Q_{TO}, \dots, Q_3)$ with vertex corrections incorporated in Λ_3. In the discussion of Sparks and Sham, Λ_3 is presumed a constant independent of wave vector and branch index of the participating phonons, although it is a function of these quantities in general.

With the above remarks in hand, the contribution to $\beta(\omega)$ from n-phonon summation processes assumes the form given in eq. (1.3) with γ replaced by $\gamma_n(\omega)$, with

$$\gamma_n(\omega) = \frac{2\pi}{\hbar^2\omega} n!(1+n_\omega)^n \sum_{Q_1 \dots Q_n} \Lambda_n^2 |V_n(0, \dots, Q_n)|^2 \tilde{n}_n \times \Delta\left(\sum_{j=1}^{n} Q_j\right) \delta\left(\omega - \sum_j \omega_{Q_j}\right). \tag{3.10}$$

Here $n!$ is a combinatorial factor, $n_\omega = [\exp(\hbar\omega/k_B T) - 1]^{-1}$ is the Bose–Einstein function, $\Delta(Q)$ vanishes unless Q equals a reciprocal lattice vector and is unity then, and $\tilde{n}_n = \prod_{j=1}^{n}(n_j+1)/(n_\omega+1)$ with n_j the thermal occupation of the mode Q_j. The wave vector Q_{TO} of the fundamental optical phonon has been set to zero on the right-hand side of eq. (3.10).

The next step is to learn how to evaluate expressions such as eq. (3.10) for the case where the number of final state phonons is large. Before we proceed, note that evaluation of the sums is trivial for an Einstein spectrum, for which the results reduce to those obtained later by other investigators.

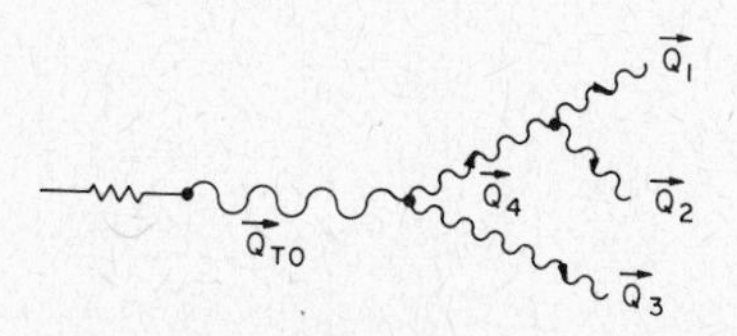

Fig. 8. A diagram which contributes to $\beta_3(\omega)$ which is the same order in ε as the diagram in fig. 7b.

With Λ_n approximated by a number independent of the wave vector and branch indices, evaluation of eq. (3.10) is assisted greatly by two approximations:

(i) The dominant contribution to H_A comes from the repulsive, short-ranged part of the ion–ion interaction. As a consequence, it turns out that $V_n(Q_{TO}, Q_1, \ldots, Q_n)$ may be written as a sum of separable terms:

$$V_n(Q_{TO}, Q_1, \ldots, Q_n) = \sum_{\gamma=1}^{6} \prod_{Q_j} v_n^{\gamma}(Q_j). \tag{3.11}$$

The index γ refers to each of the six nearest neighbors, but the calculation may be arranged so the contribution from only two of the neighbors enters.

(ii) The restriction of wave-vector conservation is ignored by replacing the factor $\Delta(\Sigma_j Q_j)$ by simply N^{-1}, where N is the number of unit cells in the crystal.

Assumption (i) has been used in a number of past theoretical studies of anharmonic effects in crystals, and is justified by the straightforward arguments presented by Sparks and Sham, and earlier authors. While assumption (ii) is clearly valid when the number n of final-state phonons is very large, it is not obvious at the outset what value of n is required for its validity. One is assisted here by a calculation carried out many years ago by Kane (1967), who examined a very different physical phenomenon, but was led to evaluate a formula with algebraic structure similar to that found for the three-phonon contributions to $\beta(\omega)$. Kane evaluates his expression numerically with wave-vector conservation fully included, and a second time with it ignored. The results of the two calculations agree very well. By drawing analogy with Kane's work, one can argue that assumption (ii) should introduce little error, even in the three-phonon region.

With the two assumptions above, the evaluation of $\gamma_n(\omega)$ reduces to the evaluation of a single n-fold convolution of the form

$$\sum_n (\omega) = \int \mathrm{d}\omega_{Q_1} \ldots \mathrm{d}\omega_{Q_n} f(\omega_{Q_1}) \ldots f(\omega_{Q_n}) \delta\left(\omega - \sum_j \omega_{Q_j}\right), \tag{3.12}$$

where $f(\omega_{Q_1})$ is proportional to the phonon density of states, and the coefficient $v_n^{\gamma}(Q_1)$ of eq. (3.11).

At first glance, the expression in eq. (3.12) bears a close resemblance to those which form the basis of phenomenological attempts to incorporate features of the real phonon spectrum into the calculation of $\beta(\omega)$. We discussed two such theories (Namjoshi and Mitra 1974a, b, Rosenstock 1974, Boyer et al. 1975) in the preceding subsection. The fundamental difference is that the form of $f(\omega_Q)$ in eq. (3.12) emerges from a systematic microscopic analysis, and is also much richer in content than the forms postulated in the phenomenological theories. For example, the phonon eigenvectors appear in the expression for $f(\omega_Q)$, so the theory incorporates polarization effects explicitly.

While integrals such as that in eq. (3.12) can be evaluated directly, even for rather large values of n, one last step produces an analytic expression that represents $\Sigma_n(\omega)$, for $n \geqslant 3$. The central limit theorem asserts that integrals such as that in eq. (3.12) approach a Gaussian at large n. The theorem also prescribes the width of the Gaussian, the frequency about which it is centered and the prefactor. Through use of this theorem, for $\beta_n(\omega)$ Sparks and Sham find

$$\beta_n(\omega) = \Lambda_n^2 b_n(\omega, T) \exp\left[-(\omega - n\bar{\omega})^2 / 2n\Delta^2 \right], \tag{3.13}$$

where explicit expressions are obtained for $b_n(\omega, T)$, $\bar{\omega}$ and Δ. The frequency $\bar{\omega}$ and Δ, and the parameters in $b_n(\omega, T)$ may all be related to well-defined averages over the phonon spectrum. Estimates for $\bar{\omega}$ and Δ show $\bar{\omega}$ to be comparable to, but somewhat less than, the maximum frequency ω_M, while $\Delta \sim \omega_M/5$. The prefactor $b_n(\omega, T)$ is proportional to ε^{2n-2}, and is a smooth function of ω.

Upon superimposing Gaussians for $\beta_n(\omega)$ to form the absorption coefficient $\beta(\omega)$, as in eq. (3.9), Sparks and Sham find a smooth, nearly exponential behavior for $\beta(\omega)$. In a subsequent publication (Sparks and Sham 1974), they note that their method can produce an *analytic* formula for $\beta(\omega)$ with the exponential frequency variation of eq. (2.2). This occurs only for a special dependence of the combination $\Lambda_n^2 b_n(\omega, T)$ on n. The required dependence is in fact quite close to that which emerges from the microscopic calculation presented in their first paper.

In this approach, deviations from the T^{n-1} law for the temperature variation of $\beta(\omega)$ follow when the temperature variation of the phonon frequencies, lattice constant and other physical parameters are fed into the formulae (Sparks and Sham 1973b). The temperature variation of $\beta(\omega)$ produced by introducing these effects into the theory is in good accord with the data of Harrington and Hass (1973).

The papers of Sparks and Sham thus provide a procedure for deriving useable and rather simple formulae for $\beta(\omega)$, with the factors that enter related to the fundamental anharmonic coupling constants of lattice dynamics, and prescribed averages over the phonon spectrum. One quantitative uncertainty is the magnitude of the vertex corrections described by Λ_n. Sparks and Sham find for NaCl the values $\Lambda_4^2 = 1.93$, $\Lambda_5^2 = 3.40$ and $\Lambda_6^2 = 7.38$. The vertex corrections thus influence the magnitude and frequency variation of $\beta(\omega)$ in an important way, when n is in the range of 4 or higher. In a recent paper, Duthler (1976) has recalculated Λ_4 and Λ_5 for NaI by a different method, to find the very much smaller values $\Lambda_4^2=1.08$ and $\Lambda_5^2=1.166$, to suggest the vertex corrections are quite negligibly small. Recall that the corrections to the basic vertex contribute to Λ_n-1, so the vertex corrections to Λ_5 are estimated by Duthler to be a full order of magnitude smaller than the estimate of Sparks and Sham.

Since the microscopic theory of Sparks and Sham incorporates the dependence of the anharmonic matrix elements on phonon wave vector and branch index, one can begin to assess which phonon combinations contribute most importantly to $\beta(\omega)$.

Such questions were addressed initially by Eldridge (1972) and again by Duthler and Sparks (1973). These authors explore $\beta(\omega)$ in the region where two-phonon processes make the dominant contribution to $\beta(\omega)$, to find systematic features referred to as "quasi-selection rules". From group theoretic arguments one may deduce a rigorous selection rule for decay of the fundamental phonon by two-phonon processes. This selection rule states that for rock salt structure materials, the anharmonic matrix element which couples the fundamental mode to two phonons on the same *branch* must vanish identically. Thus, decay of the fundamental mode to two longitudinal-optical modes is strictly forbidden, but decay to one transverse-optical mode and one longitudinal-optical mode is allowed. Duthler and Sparks find that, for the model used by them, the matrix elements which control decay to *any* two optical modes, and for decay to *any* two acoustical modes are much smaller than those that control decay to one optical and one acoustical mode. The term "quasi-selection rule" is introduced to describe these systematic features, since the matrix elements that control the "quasi-forbidden" processes are not zero, but simply small. The quasi-selection rule is obeyed particularly well in crystals with a gap between the acoustical and optical branches. Duthler and Sparks interpret structure in the two-phonon absorption bands of NaI by invoking the quasi-selection rule.

Duthler (1976) has extended this study into the multiphonon region to explain the structures in $\beta(\omega)$ observed by Harrington et al. (1976) in alkali halides below room temperature. The data shows a clear feature in $\beta(\omega)$ in the three-phonon region. This bump shows that there is excess absorption in the three-phonon region, over what is expected from the exponential law in eq. (2.2). Upon extending the quasi-selection rule into the multiphonon region, Duthler finds the contribution to $\beta(\omega)$ from processes which involve an *odd* number of optical modes is substantial, while the contribution from those that involve an *even* number of optical modes is small. This accounts nicely for the appearance of the structure in $\beta(\omega)$ at low temperature, where anharmonic broadening of the participating phonons becomes small. In the theory, the quasi-selection rule is expected to be most important in those alkali halides where there is a gap or near-gap between the acoustical- and optical-phonon branches. This happens when the mass difference between the two atomic constituents is large. As we have seen, it is in such alkali halides that the structure is observed in the data.

The papers of Sham, Sparks and Duthler show how one may approach and carry through calculations of $\beta(\omega)$ with realistic models of the phonon spectrum and crystalline anharmonicity as the basic input. It is pleasing to see structures related to the details of the matrix elements appear in $\beta(\omega)$; their explanation is a major success of this approach.

Bendow and collaborators approach the calculation of $\beta(\omega)$ from a different point of view (Bendow 1973, Bendow et al. 1973). They proceed by an approximation procedure which avoids a perturbation expansion in the parameter ε of anharmonic perturbation theory. They begin by writing the potential energy V between the vibrating ions in the form

$$V=\frac{1}{2}\sum_{ij} v(\boldsymbol{R}_i-\boldsymbol{R}_j)=\frac{1}{2}\sum_{\boldsymbol{q}}\sum_{ij} v(\boldsymbol{q})\exp\left[\mathrm{i}\boldsymbol{q}\cdot(\boldsymbol{R}_i-\boldsymbol{R}_j)\right], \tag{3.14}$$

where $\boldsymbol{R}_i$ and $\boldsymbol{R}_j$ are the instantaneous positions of the ions. Upon writing

$$\boldsymbol{R}_i=\boldsymbol{R}_i^{(0)}+\boldsymbol{u}_i \tag{3.15}$$

where $\boldsymbol{R}_i^{(0)}$ is the equilibrium position of ion i, and $\boldsymbol{u}_i$ its displacement from its equilibrium site, we have

$$V=\frac{1}{2}\sum_{\boldsymbol{q}}\sum_{ij} v(\boldsymbol{q})\exp\left[\mathrm{i}\boldsymbol{q}\cdot\left(\boldsymbol{R}_i^{(0)}-\boldsymbol{R}_j^{(0)}\right)\right]\exp\left[\mathrm{i}\boldsymbol{q}\cdot(\boldsymbol{u}_i-\boldsymbol{u}_j)\right]. \tag{3.16}$$

Bendow et al. (1973) express $\beta(\omega)$ directly in terms of V, rather than expanding the right-hand side of eq. (3.16) in powers of $\boldsymbol{u}_i-\boldsymbol{u}_j$. They find $\beta(\omega)$ to be proportional to a quantity $\mathcal{T}_0(\omega)$, where

$$\mathcal{T}_0(\omega)=\int_{-\infty}^{+\infty}\mathrm{d}t\,\mathrm{e}^{-\mathrm{i}\omega t}\,\mathcal{T}_0(t) \tag{3.17}$$

and

$$\begin{aligned}\mathcal{T}_0(t)=\sum_{\boldsymbol{q}\boldsymbol{q}'}\sum_{ijkl}&\lambda(\boldsymbol{q}\boldsymbol{q}')v(\boldsymbol{q})v(\boldsymbol{q}')\exp\left[\mathrm{i}\boldsymbol{q}\cdot\left(\boldsymbol{R}_i^{(0)}-\boldsymbol{R}_j^{(0)}\right)\right]\\&\times\exp\left[\mathrm{i}\boldsymbol{q}'\cdot\left(\boldsymbol{R}_k^{(0)}-\boldsymbol{R}_l^{(0)}\right)\right]\langle\exp(\mathrm{i}\boldsymbol{q}\cdot[\boldsymbol{u}_i(t)-\boldsymbol{u}_j(t)])\\&\times\exp(\mathrm{i}\boldsymbol{q}'\cdot[\boldsymbol{u}_k(0)-\boldsymbol{u}_l(0)])\rangle.\end{aligned} \tag{3.18}$$

In eq. (3.18) $\lambda(\boldsymbol{qq}')$ may be expressed in terms of the eigenfrequencies and eigenvectors of the phonons of wave vector $\boldsymbol{q}$ and $\boldsymbol{q}'$. The angular brackets denote a statistical average over the enclosed operators. This average is taken with respect to the density matrix $\rho = Z^{-1}\exp(-\beta H)$, with H the full vibrational Hamiltonian including anharmonicity, and $Z = \mathrm{Tr}(\rho)$ is the partition function. The expression Bendow et al. (1973) form for $\beta(\omega)$ from $\mathcal{T}_0(\omega)$ is then rigorous save for the neglect of the non-linear variation of the electric dipole moment with displacement.

The statistical average in eq. (3.18) is difficult to carry out, particularly if one wishes to avoid a perturbation expansion in anharmonic terms. Bendow et al. proceed by replacing the density matrix of the anharmonic crystal by that of the crystal in the harmonic approximation, $\rho_0 = Z_0^{-1}\exp(-\beta H_0)$. Then the statistical average is readily carried out through use of the Baker–Hausdorff theorem (Hausdorff 1906) to give

$$\langle \exp(\mathrm{i}\boldsymbol{A}\cdot\boldsymbol{u}_1)\exp(-\mathrm{i}\boldsymbol{B}\cdot\boldsymbol{u}_2)\rangle_0 = \exp\Big[-\tfrac{1}{2}\boldsymbol{A}\cdot\langle \boldsymbol{u}_1\boldsymbol{u}_1\rangle_0\cdot\boldsymbol{A} - \tfrac{1}{2}\boldsymbol{B}\cdot\langle \boldsymbol{u}_2\boldsymbol{u}_2\rangle\cdot\boldsymbol{B} + \boldsymbol{A}\cdot\langle \boldsymbol{u}_1\boldsymbol{u}_2\rangle_0\cdot\boldsymbol{B}\Big], \qquad (3.19)$$

where the average over ρ_0 is denoted by appending a subscript zero to the angular brackets. The final step is to note that the correlation function $\langle \boldsymbol{u}_i(t)\boldsymbol{u}_j(0)\rangle$ may be expressed in terms of the eigenvector $\boldsymbol{e}(\boldsymbol{q}\lambda)$, frequency $\omega(\boldsymbol{q}\lambda)$ and Bose–Einstein function $n(\boldsymbol{q}\lambda)$ of the mode in branch λ, with wave vector $\boldsymbol{q}$:

$$\langle \boldsymbol{u}_i(t)\boldsymbol{u}_j(0)\rangle = \frac{1}{2N(M_iM_j)^{1/2}}\sum_{\boldsymbol{q}\lambda}\hat{e}(\boldsymbol{q}\lambda)\hat{e}(\boldsymbol{q}\lambda) \times\Big[(1+n(\boldsymbol{q}\lambda))\exp\big[\mathrm{i}\boldsymbol{q}\cdot(\boldsymbol{R}_i-\boldsymbol{R}_j)-\mathrm{i}\omega(\boldsymbol{q}\lambda)t\big] + n(\boldsymbol{q}\lambda)\exp\big[-\mathrm{i}\boldsymbol{q}\cdot(\boldsymbol{R}_i-\boldsymbol{R}_j)-\mathrm{i}\omega(\boldsymbol{q}\lambda)t\big]\Big] \qquad (3.20)$$

The final expression for $\beta(\omega)$ obtained by this method is complex in appearance, and we refer the reader to the paper by Bendow et al. (1973) for its full form. The remarks here illustrate its general mathematical structure, which is strongly reminiscent of the scattering cross section for scattering of thermal neutrons from harmonic crystals.*

It is rather difficult to carry through a calculation of $\beta(\omega)$ with the above prescription, unfortunately. From eq. (3.17) and eq. (3.18), one may appreciate this from appearance of the double sum or wave vector followed by a multiple sum over lattice sites. From eq. (3.19), one sees the

*See ch. 19 of Kittel (1963).

correlation function in eq. (3.20) appears in the exponent, so the Fourier transform in eq. (3.17) is non-trivial. Nonetheless, through introducing a sequence of approximations, Bendow and co-workers (Bendow 1973, Bendow et al. 1973) have evaluated the expression for some simple and schematic phonon spectra. The assumptions replace the form factors $v(\boldsymbol{q})$ in eq. (3.18) by a Gaussian form, replace the eigenvectors $\boldsymbol{e}(\boldsymbol{q}j)$ everywhere in eq. (3.20) by $\boldsymbol{q}=0$ optical-mode eigenvectors, and then a choice of model phonon spectra allows analytic evaluation of portions of the expression. The most detailed study of the form for $\beta(\omega)$ produced by this model is found in Bendow's (1973) paper. Among the many curves presented, one notes a number with the smooth, nearly exponential behavior for $\beta(\omega)$ present in the alkali halides, and many with deep, pronounced structure out to the seven-phonon region. Unfortunately, there is no direct comparison of the absolute magnitude and frequency variation of $\beta(\omega)$ with data, with input parameters chosen from independent considerations.

A study of $\beta(\omega)$ by the method of Bendow and co-workers could delineate the regime of validity of the uncoupled anharmonic oscillator models. This is because the correlation function $\langle \boldsymbol{u}_i(t)\boldsymbol{u}_j(0)\rangle$ appears directly in the final expression. In principle, through a complete study of $\beta(\omega)$ with use of realistic eigenvectors and phonon spectrum, one can ask how importantly correlations in the vibrational motion of ions in different unit cells influence $\beta(\omega)$. Our earlier remarks suggest on physical grounds that at high temperatures, correlations within the unit cell (i.e. only over distances comparable to a lattice constant) are important, with correlations over longer distances important below the Debye temperature. It would be interesting to explore this issue by calculating $\beta(\omega)$ for a given phonon spectrum first by including correlations between nearest neighbors, then nearest neighbors plus next-nearest neighbors, etc.

3.3. The effect of the non-linear variation of electric-dipole moment on multiphonon absorption

All of the theories discussed in the two preceding subsections presume that the electric-dipole moment of an ion in a given unit cell varies linearly with atomic displacement. We now turn to a description of the role of the non-linear terms in the expansion of eq. (3.3).

The role of the non-linear variation of the electric-dipole moment was first explored by Szigetti (1965) in his description of absorption by two-phonon processes in the alkali halides. More recent papers discussed below extend his calculation into the multiphonon region.

Szigetti finds that in the presence of the non-linear electric moment, there is quantum mechanical interference between the contribution to $\beta(\omega)$

from lattice anharmonicity and that from the non-linear variation of the electric moment. We illustrate this in fig. 9. In the presence of only anharmonicity, as discussed in the preceding subsection, the photon can couple to the lattice only by exciting the fundamental phonon, which subsequently decays to the two-phonon final state by action of the crystalline anharmonicity. We illustrate this in fig. 9a. In the presence of the non-linear electric moment, the photon may couple directly to the *same* two-phonon final state, as illustrated in fig. 9b. In quantum mechanics, in such a circumstance one must add the matrix elements for each process together before squaring to form the transition rate. Thus, there are interference effects between the two processes which, as we shall see below, can have a dramatic effect on the frequency variation of $\beta(\omega)$.

In Szigetti's notation modified slightly, in the presence of both anharmonicity and the non-linear electric moment, the absorption constant $\beta(\omega)$ in the two-phonon region can be written

$$\beta_2(\omega)=\frac{\pi^2\hbar}{V}\sum_{ij}\frac{1}{\omega_i\omega_j}\left[\frac{\alpha_0 b_{0ij}}{\omega_0^2-\omega^2}-\beta_{ij}\right]^2$$
$$\times\left[\left(1+\bar{n}_i+\bar{n}_j\right)\delta(\omega-\omega_i-\omega_j)+\left(\bar{n}_i-\bar{n}_j\right)\delta(\omega-\omega_j+\omega_i)\right], \quad (3.21)$$

where V is the crystal volume, i and j refer to both the wave vector and branch index of the participating phonons, and ω_0 is the fundamental phonon frequency. Both sum and difference processes are involved in eq.

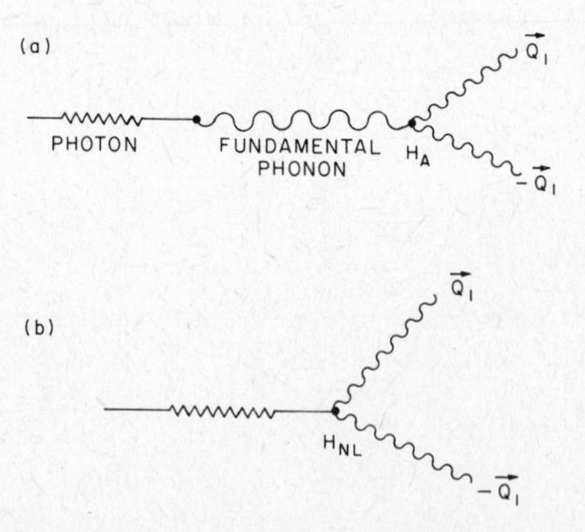

Fig. 9. The processes examined by Szigetti in his theory of absorption by two-phonon processes in the alkali halides. In (a), we have the process responsible for two-phonon absorption in the absence of non-linear variation of the electric moment. In (b), the process induced by the non-linear electric moment is given. The two contributions to $\beta(\omega)$ interfere coherently.

(3.21). Finally, α_0 is the coupling constant (proportional to the transverse effective charge e^*) that couples the photon to the fundamental phonon, b_{0ij} the anharmonic matrix element that controls decay of the fundamental phonon to the two-phonon final state and β_{ij} a measure of the quadratic terms in the expansion of the electric-dipole moments in powers of the atomic displacement.

The most striking feature of eq. (3.21) is the frequency variation of the effective matrix element. Since the interference between the two processes in fig. 9 depends strongly on frequency, if the fundamental coupling constants have the proper sign, $\beta(\omega)$ may *vanish* as a consequence. Thus, inclusion of the non-linear electric moment in the calculation can not only alter the magnitude of $\beta(\omega)$, but affect its frequency dependence profoundly.

In his paper, Szigetti estimates the magnitude and sign of the coupling constants in eq. (3.21) to conclude the interference is destructive, with $\beta(\omega)$ driven to zero near the higher end of the two-phonon region, or possibly a bit beyond. There is no evidence of such a dramatic feature in the data on $\beta(\omega)$. It is possible in principle that the two terms interfere in a constructive rather than a destructive manner for photon frequencies above that of the fundamental phonon. If this were so, then Szigetti's considerations show that in the two-phonon regime, the magnitude of $\beta(\omega)$ may be influenced importantly by the presence of the non-linear electric moment, but there will be no dramatic effect on its frequency dependence.

There have been two calculations which extend Szigetti's theory of two-phonon absorption into the multiphonon region of frequencies. Mills and Maradudin (1974) extended their description of multiphonon absorption by a classical Morse potential oscillator to the case where the electric-dipole moment of the oscillator is a non-linear function of displacement. The form used for the spatial variation of the electric-dipole moment allows a closed form expression to be obtained for $\beta(\omega)$. Sparks (1974) has explored the question for a one-dimensional lattice dynamical model, with particular attention to comparison of the predicted form for $\beta(\omega)$ with the alkali halide data.

Bendow et al. (1974) also study the effect of the non-linear electric moment on $\beta(\omega)$. While these authors begin with a general expression for $\beta(\omega)$ similar to that which formed the basis of their earlier treatment of absorption induced by anharmonicity, the approximations introduced by them (either retention of only quadratic terms in the cumulant expansion, or use of the density matrix for the harmonic lattice) fail to incorporate the interference effect in their theory.

Since certain simple and systematic features of the role of the non-linear electric moment in multiphonon absorption emerge in the analysis of the classical Morse potential oscillator, we begin with a description of the results of the analysis of Mills and Maradudin, then turn to Sparks' description of the lattice dynamical model.

The interaction of the anharmonic oscillator with the external electric field has the form $-P(u)E(t)$, where $P(u)$ is the electric-dipole moment of the oscillator with displacement u and $E(t)$ the electric field of the infrared radiation. The fundamental quantity that enters the analysis is the effective charge $e^*_T(u) = \mathrm{d}p/\mathrm{d}u$. Mills and Maradudin presume the effective charge of the oscillator has the spatial dependence

$$e^*_T(u) = e^*_T\left[1 + R\left\{1 - \exp(-sa[u-u_0])\right\}\right], \tag{3.22}$$

where e^*_T is the transverse effective charge for small amplitude motions, and a is the range parameter that appears in eq. (3.5). Finally R and s are dimensionless parameters that characterize the magnitude and range of the non-linear contribution to the effective charge. For the form of $e^*_T(u)$ in eq. (3.22), a closed form for $\beta(\omega)$ obtains in the classical limit.

If $(u-u_0)$ is small, the exponential in eq. (3.22) may be expanded to give

$$e^*_T(u) = e^*_T[1 + Rsa(u-u_0) + \dots]. \tag{3.23}$$

Thus, when $R>0$, as the constituents of the molecule are separated, $e^*_T(u)$ initially increases *above* the value e^*_T appropriate to the limit of small displacements; the molecule becomes more ionic as it is stretched. The converse is true for $R<0$. On the basis of a simple argument, Mills and Maradudin estimate that $R = +0.35$ for NaCl. While this number is only a crude estimate, on physical grounds one expects to find s near unity, and the magnitude of R somewhere between 0.1 and 1.0.

In fig. 10a we reproduce the numerical calculations of Mills and Maradudin for $\beta(\omega)$ at 900 K, for the classical Morse potential oscillator with parameters characteristic of NaCl, and $R = +0.35$. The curve labeled $s=0$ reproduces the form for $\beta(\omega)$ with linear electric-dipole moment (fig. 5 of the present article). Dramatic interference dips in $\beta(\omega)$ very similar to those expected from extrapolation of Szigetti's results in eq. (3.21) to higher frequency are the most prominent feature in the curves. Quite clearly, nothing even qualitatively similar to these dips appears in any of the data reported at this time.

For $s=1$ and various values of R, the form predicted for $\beta(\omega)$ is displayed in fig. 10b. For *negative* values of R, the interference is constructive and no dip appears. For R positive, the dip is moved to frequencies beyond those explored experimentally when R is the order of 0.1 or less. Note that for the smallest non-zero values of R explored ($R=0.1$), where no dip appears for $\omega<\omega_0$, in the multiphonon regime there is a very

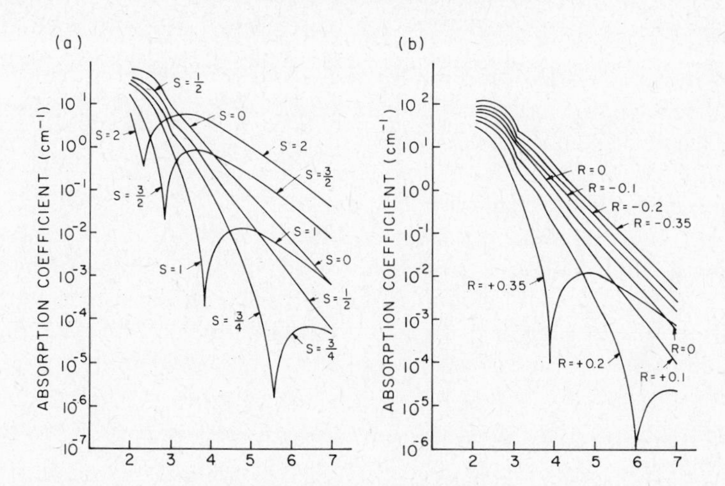

Fig. 10. The effect of the non-linear moment on the absorption constant $\beta(\omega)$ of the classical Morse potential oscillator for (a) $R=+0.35$ and various values of s, and (b) $s=1$ and various values of R.

substantial effect of the non-linear moment on the magnitude of $\beta(\omega)$. In practice, one does not expect to find R as small as 0.1, and at $\omega=6\omega_0$ the numerical value of $\beta(\omega)$ for $R=+0.1$ is an order of magnitude smaller than that for $R=-0.1$. Thus, if we set the question of the interference dip aside for the moment, this calculation of Mills and Maradudin suggests that the presence of even a modest non-linear variation of the electric-dipole moment will influence the magnitude (but not the temperature variation) of $\beta(\omega)$ in the multiphonon region of frequencies.

Sparks' study of the effect on the non-linear electric moment on multiphonon absorption begins by examining the form of the matrix element for coupling to an n-phonon final state, for a one-dimensional diatomic line. The positive ions are treated as rigid point charges, with a shell model description used for the negative ions. Incorporation of anharmonicity into the "springs" which couple the shell model constituents leads to an electric moment that varies in a non-linear fashion with atomic displacement. After a study of the structure of the effective matrix element, which is remarkably similar in form to that found by Szigetti (eq. (3.21)), Sparks computes $\beta(\omega)$ by grafting the matrix element onto the earlier results of Sparks and Sham.

Sparks' conclusions are consistent with Szigetti's earlier study of the two-phonon region, and with those of Mills and Maradudin. The principal results are:

(i) Inclusion of the non-linear moment leaves the temperature dependent of $\beta(\omega)$ unaffected.

(ii) The interference effect explored by Szigetti persists into the multiphonon region. Sparks' shell model allows the coupling constants to be related back to the microscopic parameters in the model. Sparks find destructive interference between the two contributions to $\beta(\omega)$ for his basic model (as did Szigetti), with a pronounced minimum in the region where data presently available shows $\beta(\omega)$ is smooth.

(iii) As one would expect from extrapolation of eq. (3.21) to higher frequencies, the non-linear electric moment gives the dominant contribution to $\beta(\omega)$ in the multiphonon regime. This is also consistent with the analysis of the anharmonic oscillator model by Mills and Maradudin. Inclusion of the non-linear electric moment in the calculation of $\beta(\omega)$ can increase its value by perhaps two orders of magnitude over the value appropriate to the absorption rate induced by anharmonicity alone, by the time $\omega \cong 6\omega_0$.

As remarked earlier, no evidence of the interference induced minimum is found in data presently available. There are two reasons why the dip may be absent. One is that the models may err in predicting destructive interference. A smooth frequency variation of $\beta(\omega)$ obtains if the interference is constructive. Sparks discusses modifications of his basic model which can produce constructive rather than destructive interference, for example. Mills and Maradudin propose that if the ionic character of the bond decreases with increasing interatomic character, constructive interference results. A second proposal, advanced by Sparks, is that the interference dip may occur in the two-phonon region where considerable structure in $\beta(\omega)$ is present. If the zero in $\beta(\omega)$ is reduced to a minimum by lifetime effects, the result may be undramatic and overlooked in the analysis of the data.

At the time of this writing, the influence of the non-linear electric moment on $\beta(\omega)$ in the multiphonon region is poorly understood. The absence of a pronounced interference dip in the data, and the rather good agreement between calculated and measured values of $\beta(\omega)$ both suggest that in the alkali halides, absorption induced by the non-linear electric moment is a small fraction of the total, for frequencies explored experimentally to date. One is made uncomfortable by the model calculations (Szigetti 1965, Mills and Maradudin 1974, Sparks 1974) which suggest its inclusion should have an appreciable effect.

It would be highly desirable to devise a means to study the non-linear electric moment directly, to measure the basic coupling constants that enter $\beta(\omega)$.

A method for obtaining such information has been proposed by Hellwarth and Mangir (1974) through use of the temperature variation of a sum rule on the imaginary part of the dielectric constant. This sum rule, and a number of its properties, was first discussed is a paper by Wallis and Maradudin (1962).

If we consider a cubic crystal, the sum rule takes the following form. Let P_x be the x component of the electric-dipole moment of the unit cell produced by displacement of the ions, and let $u_\alpha(\kappa)$ be the αth Cartesian component of the displacement of the atom at side κ in a particular unit cell, with M_κ its mass. Then if $\varepsilon_2(\omega)$ is the imaginary part of the dielectric constant within the lattice (multiphonon) absorption bands we have

$$\int_0^\infty \mathrm{d}\omega\,\omega\varepsilon_2(\omega) = \frac{2\pi^2}{V_\mathrm{c}} \sum_\kappa \sum_\alpha \frac{1}{M_\kappa} \left\langle \left(\frac{\delta P_x}{\delta u_\alpha(\kappa)} \right)^2 \right\rangle. \tag{3.24}$$

The angular brackets denote a statistical average over the density matrix of the anharmonic crystal. The integral in eq. (3.24) extends over all frequencies where lattice absorption dominates $\varepsilon_2(\omega)$, but does not include electronic absorption. If the damped harmonic oscillator form of the dielectric constant used to obtain eq. (1.2) is inserted into the right-hand side of eq. (3.24), one finds

$$\int_0^\infty \mathrm{d}\omega\,\omega\varepsilon_2(\omega) = \frac{4\pi^2 e^{*2}}{MV_\mathrm{c}}, \tag{3.25}$$

with M the reduced mass of the unit cell.

Hellwarth and Mangir argue that if the full form of the electric moment in eq. (3.3) is inserted into eq. (3.24), explicitly temperature-dependent corrections to the right-hand side of eq. (3.25) appear. These explicitly temperature-dependent corrections have their origin in the non-linear terms in eq. (3.3). Thus, if one can measure the temperature variation of the integral on the left of eq. (3.24) and that of the effective charge e^*, one may obtain direct information about the non-linear terms in the expansion of the electric-dipole moment in powers of the atomic displacement.

Unfortunately, there has been no systematic analysis of the right-hand side of eq. (3.24) for a full lattice dynamical model, save for brief remarks in the original paper by Wallis and Maradudin (1962). In their paper, Hellwarth and Mangir evaluate the right-hand side for an anharmonic oscillator, then use data on LiF to infer the strength of the non-linear electric moment.

One may express two reservations about the use of the sum rule. The first is a practical concern. In the alkali halides, the dominant contribution to the integral comes from the strong Reststrahl band, where the non-linear electric moment plays at most a minor role. In the multiphonon region, $\varepsilon_2(\omega)$ is orders of magnitude smaller than within the Reststrahl band, so the frequency region of greatest concern makes a small contribution to the integral. An expansion of the right-hand side of the sum rule in powers of the temperature at high temperatures shows the first temperature correction (Hellwarth and Mangir 1974, Wallis and Maradudin 1962) in alkali halides is controlled by the quadratic terms in eq. (3.3). It is not clear if these terms are the dominant contribution to the electric-moment-induced absorption in the multiphonon region, or if the principal correction comes from higher-order terms not present in the leading temperature correction to the sum rule. While the proposal by Hellwarth and Mangir to use the sum rule in eq. (3.24) is most intriguing, our view is that further theoretical analysis of the sum rule is required before it may be applied with confidence at more than a semi-quantitative level.

We conclude this chapter on this note. The recent burst of experimental and theoretical activity on lattice absorption in the multiphonon region has led to a clear understanding of the physics of the phenomenon. Simple models give a rough handle on the principal features of the data, and in parallel with this we have seen successful theories emerge founded on first principles lattice dynamical calculations. The one issue not fully resolved is the role of the absorption induced by the non-linear variation of electric-dipole moment with atomic displacement. The theories which ignore this part of the absorption appear in good accord with experiment, and the data provides no hint of the interference dip expected from the theoretical models. Yet the theoretical estimates suggest that in the multiphonon region, the absorption induced by the non-linear electric moment should be at least comparable to, and most likely substantially larger than, that induced by anharmonicity alone.

Acknowledgement

At an early stage of the writing, one of us (D. L. M.) found conversations and correspondence with Dr. Marvin Hass most useful.

References

Barker, A. J. (1972), J. Phys. C **5**, 2276.

Barker, A. S. (1968), Phys. Rev. **165**, 917.

Barker, A. S. and J. J. Hopfield (1966), Phys. Rev. **145**, 391.

Bendow, B. (1973), Phys. Rev. B **8**, 5821.

Bendow, B. (1978), Multiphonon Infrared Absorption in the Highly Transparent Regime of Solids: A Review, *Solid state physics*, ed. by F. Seitz and D. Turnbull (Academic Press, New York), Vol. 33, p. 249.

Bendow, B., S. C. Ying and S. P. Yukon (1973), Phys. Rev. B **8**, 1679.

Bendow, B., S. P. Yukon and S. C. Ying (1974), Phys. Rev. B **10**, 2286.

Billard, D., F. Gervais and B. Piriou (1976), Phys. Stat. Sol. (b) **75**, 117.

Born, M. and K. Huang (1954), *Dynamical theory of crystal lattices* (Oxford University Press, London).

Boyer, L. L., J. A. Harrington, M. Hass and H. B. Rosenstock (1975), Phys. Rev. B **11**, 1665.
Cowley, R. A. (1963), Adv. Phys. **12**, 421.
Deutsch, T. F. (1973), J. Phys. Chem. Solids **34**, 2091.
Dow, J. and D. Redfield (1972), Phys. Rev. B **5**, 594.
Duthler, C. J. (1974), J. Appl. Phys. **45**, 2668.
Duthler, C. J. (1976), Phys. Rev. B **14**, 4606.
Duthler, C. J. and M. Sparks (1973), Phys. Rev. B **9**, 830.
Eldridge, J. E. (1972), Phys. Rev. B **6**, 1510.
Eldridge, J. E. and P. R. Staal (1977), Phys. Rev. B **16**, 3834.
Flannery, M. and M. Sparks (1977), Extrinsic Absorption in Infrared Laser-Window Materials, presented at the *Ninth damage symposium on optical materials for high power lasers*, Boulder, Colorado, Oct. 4–6, 1977.
Harrington, J. A. and M. Hass (1973), Phys. Rev. Lett. **31**, 710.
Harrington, J. A., C. J. Duthler, F. W. Patten and M. Hass (1976), Solid State Commun. **18**, 1043.
Hass, M., J. W. Davisson, P. H. Klein and L. L. Boyer (1974), J. Appl. Phys. **45**, 3959.
Hausdorff, (1906), Ber. Verhandl. Sachs. Akad. Wiss. Leipzig Math.-Phys. Kl, **58**, 19.
Hellwarth, R. and M. Mangir (1974), Phys. Rev. B **10**, 1635.
Horrigan, F. and R. Rudko (1969), Materials for High Power CO_2 Lasers, Final Technical Report under Contract No. DAAH01-69-C-0038, Raytheon Research Division, Waltham, Mass. (Internal Number S-1170 (1969)).
Johnson, F. A. (1965), Progress in Semiconductors **9**, 181.
Kane, E. O. (1967), Phys. Rev. **159**, 624.
Kittel, C. (1963), *Quantum theory of solids* (Wiley, New York).
Kittel, C. (1971), *Introduction to solid state physics*, fourth ed. (Wiley, New York).
Lipson, H. G., B. Bendow, N. Massa and S. S. Mitra (1976), Phys. Rev. B **13**, 2614.
Maklad, M. S., R. K. Mohr, R. E. Howard, P. B. Macedo and C. T. Moynihan (1974), Solid State Commun. **15**, 855.
Maradudin, A. A. and D. L. Mills (1973), Phys. Rev. Lett. **31**, 718.
McGill, T. and H. V. Winston (1974), Solid State Commun. **13**, 1459.
McGill, T. C., R. W. Hellwarth, M. Mangir and H. V. Winston (1973), J. Phys. Chem. Solids **34**, 2105.
Mills, D. L. and A. A. Maradudin (1973), Phys. Rev. B **8**, 1617.
Mills, D. L. and A. A. Maradudin (1974), Phys. Rev. B **10**, 1713.
Namjoshi, K. V. and S. S. Mitra (1974a), Phys. Rev. B **9**, 815.
Namjoshi, K. and S. S. Mitra (1974b), Solid State Commun. **15**, 317.
Pohl, D. W. and Peter F. Meier (1974), Phys. Rev. Lett. **32**, 58.
Rockni, M. et al. (1972), Solid State Commun. **10**, 103.
Rosenstock, H. B. (1974), Phys. Rev. B **9**, 1973.
Rowe, J. M. and J. A. Harrington (1976), Phys. Rev. B **14**, 5442.
Rupprecht, G. (1964), Phys. Rev. Lett. **12**, 580.
Smart, C., G. R. Wilkinson, A. M. Karo and J. R. Hardy (1965) in *Lattice dynamics*, ed. by R. F. Wallis (Pergamon Press, Oxford), p. 387.
Sparks, M. (1971), Introduction to the High-Power Infrared Window Material Problem, *AFCRL conference on high-power infrared laser window materials*.
Sparks, M. (1972a), Theoretical Studies of High-Power Infrared Window Materials, Xonics Quarterly Technical Progress Report No. 1, Contract DAHC15-72-0219, March 1972.
Sparks, M. (1972b), Theoretical Studies of High Power Infrared Window Materials, Final Technical Report for ARPA Contract No. DAHC 15-72-C-0129, Dec. 1972.
Sparks, M. (1974), Phys. Rev. B **10**, 2581.
Sparks, M. and C. J. Duthler (1973), J. Appl. Phys. **44**, 3038.

Sparks, M. and L. J. Sham (1972a), Theory of Multiphonon Infrared Absorption, *AFCRL conference on high-power ir laser window materials*, Hyannis, Mass., Oct. 30–Nov. 1, 1972.
Sparks, M. and L. J. Sham (1972b), Solid State Commun. **11**, 1451.
Sparks, M. and L. J. Sham (1973a), Phys. Rev. B **8**, 3037.
Sparks, M. and L. J. Sham (1973b), Phys. Rev. Lett. **31**, 714.
Sparks, M. and L. J. Sham (1974), Phys. Rev. B **9**, 827.
Szigetti, B. (1965), in *Lattice dynamics*, ed. by R. F. Wallis (Pergamon Press, Oxford), p. 405.
Turner, W. J. and W. E. Reese (1962), Phys. Rev. **127**, 126.
Ushioda, S. and J. D. McMullen (1972), Solid State Commun. **11**, 299.
Van Hove, L. (1953), Phys. Rev. **89**, 1189.
Wallis, R. F. (1965), *Lattice dynamics* (Pergamon Press, New York).
Wallis, R. F. and A. A. Maradudin (1962), *Report of the international conference on the physics of semiconductors* (Exeter, 1962), p. 490.

Author Index

Subject Index